S0-CBI-039
WITHDRAWN
CALTECH LIBRARY SERVICES

# Estimation of Natural Groundwater Recharge

edited by

I. Simmers

Institute of Earth Sciences,
Free University, Amsterdam,
The Netherlands

D. Reidel Publishing Company

Dordrecht / Boston / Lancaster / Tokyo

Published in cooperation with NATO Scientific Affairs Division

Proceedings of the NATO Advanced Research Workshop on
Estimation of Natural Recharge of Groundwater
(with special reference to Arid- and Semi-Arid Regions)
Antalya (Side), Turkey
8-15 March 1987

Library of Congress Cataloging in Publication Data

NATO Advanced Research Workshop on Estimation of Natural Recharge of Groundwater (1987: Antalya, Turkey)

Estimation of natural groundwater recharge / edited by I. Simmers.

p. cm — (NATO ASI series. Series C, Mathematical and physical sciences ; vol. 222)

"Proceedings of the NATO Advanced Research Workshop on Estimation of Natural Recharge of Groundwater . . . Antalya (Side), Turkey, 8–15 March, 1987"—T.p. verso.

"Published in cooperation with NATO Scientific Affairs Division."

Includes index.

ISBN 90–277–2632–9

1. Groundwater flow—Congresses. I. Simmers, I. (Ian), 1937– . II. North Atlantic Treaty Organization. Scientific Affairs Division. III. Title. IV. Series: NATO ASI series. Series C, Mathematical and physical sciences; no. 222.

GB1197.7.N37 1987
551.49—dc 19 87–26854

---

Published by D. Reidel Publishing Company
P.O. Box 17, 3300 AA Dordrecht, Holland

Sold and distributed in the U.S.A. and Canada
by Kluwer Academic Publishers,
101 Philip Drive, Assinippi Park, Norwell, MA 02061, U.S.A.

In all other countries, sold and distributed
by Kluwer Academic Publishers Group,
P.O. Box 322, 3300 AH Dordrecht, Holland

D. Reidel Publishing Company is a member of the Kluwer Academic Publishers Group

---



Printed in The Netherlands

## ALL ASI SERIES BOOKS TO BE PUBLISHED AS A RESULT OF ACTIVITIES OF THE SPECIAL PROGRAMME ON GLOBAL TRANSPORT MECHANISMS IN THE GEO-SCIENCES

This book contains the proceedings of a NATO Advanced Research Workshop held within the programme of activities of the NATO Special Programme on Global Transport Mechanisms in the Geo-Sciences running from 1983 to 1988 as part of the activities of the NATO Science Committee.

Other books previously published as a result of the activities of the Special Programme are as follows:

BUAT-MENARD, P. (ed.) - The Role of Air-Sea Exchange in Geochemical Cycling (C185) 1986

CAZENAVE, A. (Ed.) - Earth Rotation: Solved and Unsolved Problems (C187) 1986

WILLEBRAND, J. and ANDERSON, D.L.T. (Eds.) - Large-Scale Transport Processes in Oceans and Atmosphere (C190) 1986

NICOLIS, C. and NICOLIS, G (Eds.) - Irreversible Phenomena and Dynamical Systems Analysis in Geosciences (C192) 1986

PARSONS, I. (Ed.) - Origins of Igneous Layering (C196) 1987

LOPER, E. (Ed.) - Structure and Dynamics of Partially Solidified Systems (E125) 1987

VAUGHAN, R.A. (Ed.) - Remote Sensing Applications in Meteorology and Climatology (C201) 1987

BERGER, W.H. and LABEYRIE, L.D. (Eds.) - Abrupt Climatic Change - Evidence and Implications (C216) 1987

VISCONTI, G. and GARCIA, R. (Eds.) - Transport Processes in the Middle Atmosphere (C213) - 1987

SIMMERS, I. (Ed.) - Estimation of Natural Recharge of Groundwater (C222) 1987

HELGESON, H.C. (Ed.) - Chemical Transport in Metasomatic Processes (C218) 1987

Table of Contents

***Groundwater recharge estimation (Part 2) : numerical modelling techniques***



***Applications and case studies***

## Preface

In view of the rapidly expanding urban, industrial and agricultural water requirements in many areas and the normally associated critical unreliability of surface water supplies in arid and semi-arid zones, groundwater exploration and use is of fundamental importance for logical economic development. Two interrelated facets should be evident in all such groundwater projects :

(a) definition of groundwater recharge mechanisms and characteristics for identified geological formations, in order to determine whether exploitation in the long-term involves 'mining' of an essentially 'fossil' resource or withdrawal from a dynamic supply. A solution to this aspect is essential for development of a resource management policy;

(b) determination of recharge variability in time and space to thus enable determination of total aquifer input and to quantify such practical aspects as 'minimum risk' waste disposal locations and artificial recharge potential via (e.g.) devegetation or engineering works.

However, current international developments relating to natural recharge indicate the following 'problems' ;

- *no single comprehensive estimation technique can yet be identified from the spectrum of methods available; all are reported to give suspect results.*

Although recharge mechanisms are reasonably well known, deficiencies are evident in quantifying the various elements - in this respect the use of tracer techniques and remote sensing offer interesting potential. Differences in sources and processes of recharge in humid climates compared with arid/semi-arid areas mean that applicability of available estimation techniques will be different. Under arid conditions recharge is intermittent and concentrated in small areas. These conditions thus influence the choice of estimation method; viz, methods in which areal parameters are used are more suited to humid climates, whereas point methods are preferable for more detailed studies of arid area recharge. The need to proceed from a well defined conceptualisation of different recharge processes is thus essential, as is the need to use more than one technique for result verification.

- *most reported studies relate to ad hoc site specific solutions and available data input quantity and quality remains extremely variable.*

In many instances, especially in developing countries, there are inadequate data to obtain reasonable recharge model calibration. However, in most cases, if a flexible approach to project design or management strategy is adopted, the necessary data can be collected within the operation of normal groundwater production. This implies that by adopting a series of basic master hydrogeological conceptual models, the development of quantitative computerised models is achieved in a stepwise manner allowing resource development without loss of time or money.

Issues relating to space/time variability and result regionalisation create a series of as yet unresolved problems, clearly indicating a focus for further investigation. For example, over some quite large areas recharge appears to show little lateral variability, while in other, apparently similar areas it can range over at least an order of magnitude. Furthermore, the various methods of recharge estimation are applicable over different spatial and temporal scales and the results can thus often not be compared. A descriptive insight into recharge spatial variability is available by way of standard hydrogeological mapping techniques, but these do not produce the quantitative information ultimately required for resource management.

Using these current 'problem' issues as a framework, the present Workshop was held to critically address such specific topics as recharge determination methodology, estimation with inadequate data and regionalisation of point observations. Attended by 42 participants from 17 countries, the scientific programme was divided into sessions on 'recharge concepts', 'recharge techniques' (2 sessions), 'applications and case studies' and a 'comparative analysis with humid zone recharge'. To stimulate discussion, answers were also sought to the following specific questions posed in the opening session :

- is sufficient known about the mechanisms of recharge ?
- under what limiting conditions (climatic, physiographic, data availability) is each basic recharge estimation method of value and why ?
- how can recent developments in remote sensing best supplement traditional recharge estimation techniques ?
- what are the present deficiencies in information between recharge theory and application, and how best can these be bridged for the 'practioner'?

The following volume represents the formal proceedings of this meeting, and contains delivered keynote addresses and a selection of the offered papers. Each relates to one or more of the above stated 'problems' and more specific questions, the whole providing a timely commentary on the 'state-of-the-art' with respect to groundwater recharge estimation for (semi-)arid areas. In the interests of rapid publication editorial ammendments to the individual contributions have been kept to a minimum.

Finally and on behalf of the organising group, I would like to express my appreciation of NATO Science Committee support for our Antalya workshop - without this support we would have been quite unable to mount either the programme offered or the proceedings now presented.

Prof.Dr. Ian Simmers
Director, NATO-ARW on
'estimation of natural
recharge of groundwater'

12 August, 1987.

# GROUNDWATER RECHARGE CONCEPTS

# GROUNDWATER RECHARGE CONCEPTS

J. Balek
Stavební geologie n.p.
Gorkého nám. 7
113 09 Prague 1
Czechoslovakia

ABSTRACT. Groundwater recharge is usually considered a process of water movement downward through the saturated zone under the forces of gravity or in a direction determined by the hydraulic conditions. In a broader sense the process of recharge is more complicated. Perhaps one of the most important factors is the time delay between the period in which the meteoric water enters the saturated zone and the time period in which it is manifested as an effectively exploitable groundwater source.

## 1. PROCESSES INVOLVED

Natural recharge of groundwater may occur by precipitation, by rivers and canals and by lakes. In practically all these sources all water is of a meteoric origin. The so called juvenile water of volcanic, magmatic and cosmic origin contributes to groundwater recharge only exceptionally. Generally, as a groundwater recharge is considered that amount of water which contributes to the temporary or permanent increase of groundwater resources, the complexity of the process has to be considered from the time interval at which the water transformed by the interception enters the soil profile. When water is ponded on the soil surface, the entry of water into the soil is very high; however, it drops after a short period of time to a steady state called the final infiltrability. In some works we find the infiltration process defined as the result of the permeability of the entire soil profile and the final infiltrability closely related to the saturated soil conductivity. Indeed, the soil becomes a very decisive factor in the groundwater recharge process.

The soil crust is often comprised of two parts : a very thin non-porous layer and a thicker layer of inwashed fine particles. Thus the crust and its physical properties may become the first decisive factor in the soil infiltration. The soil texture and structure play another significant role within the soil profile; the thickness of the unsaturated zone is also very important. In some regions this is determined by the transpirational process, in others by the depth of the groundwater level and saturated zone; here it is regulated by the groundwater level rise. Evapotranspiration is another important factor in groundwater recharge. Bare soil, for example, may soak up the first portion



*I. Simmers (ed.), Estimation of Natural Groundwater Recharge, 3–9.*

of heavy rain, then swell and become waterproof. Water infiltration in physically very similar soil, covered by more or less thick vegetation, is delayed by the process of interception, and percolation is regulated through the water uptake of plants. Thus seasonal changes of infiltration and groundwater recharge are pronounced in agricultural fields, forests, etc.

Soil moisture plays another important role. The rate of infiltration is higher when soil is dry; later on, after the capillary pores have been filled, it drops. Furthermore, the infiltration conditions change significantly when soil is frozen. The water content plays a particularly significant role in frozen soils; some soils become more permeable when dry and frozen.

While infiltration is analysed as a surface phenomenon, the quantity of water that passes a given point in the soil profile in a given time is sometimes called percolation. This process occurs when the soil moisture process has reached field capacity. Usually this becomes possible when the infiltration rate is in excess of the transpiration rate. Sometimes the percolation is also called the transmission rate.

Todd (1967) gave to the term "percolating rate" a legal implication by considering that some water laws of the past recognised percolating waters as "... those moving slowly through the ground but not a definite underground stream", in contrary to water courses "... occurring either on the surface or underground".

Another relevant term is the seepage which is reserved for the movement of water through the porous media to the ground surface or surface water bodies, or vice versa.

Storm seepage is recognised as that part of rainfall which infiltrates into the soil surface and moves laterally through the soil horizons directly to the streams as ephemeral, shallow or perched groundwater above the main groundwater level. This process is also considered by some authors as identical with subsurface flow.

## 2. THE TIME FACTOR IN GROUNDWATER RECHARGE

In principal, the following types of groundwater recharge can be recognised: short-term recharge, seasonal recharge, perennial recharge and historical recharge.

### 2.1. Short-term Recharge

This occurs occasionally after a heavy rainfall, mainly in regions without marked wet and dry seasons. Such a type of recharge was described by Sophocleous and Perry (1985) who measured the recharge process in Kansas, U.S.A., on unconsolidated deposits without a surface drainage pattern. They concluded that antecedent moisture conditions, thickness of the aquifer and the nature of the unsaturated zone were major factors affecting recharge. They observed a significant variation in recharge ranging from less than 2.5 mm to 154 mm.

Sharma and Hughes (1985) studied the groundwater recharge in Western Australia; they concluded that about 50 % of recharge occurs through so called preferred pathways by-passing the soil profile. Being unable to quantify the considerable variability in the recharge and preferred pathways, they proposed conducting further research into this problem.

## 2.2. Seasonal Recharge

This occurs regularly, e.g. at the beginning of the snowmelt period in temperate regions or during the wet period in wet and dry regions. Occasionally when an unfavourable soil moisture development has taken place, the recharge may be poor or may not occur at all.

For a determination of such a type of recharge Kitching et al. (1977) took the lysimetric approach. In the Bunter Sandstone in England, he found 136 mm of the 600 mm rainfall to be groundwater recharge. Morel and Wright (1978) published a ten year record of groundwater recharge for the Chalk in West Suffolk with results which coincided with those mentioned above. However, in Bunter sandstone Bath et al. (1978) analysed the age of water and concluded that a large proportion of water was recharged already during late Pleistocene and early Holocene in a period when rainfall was delivered from continental air masses.

In Czech sandstone (Balek, 1987) it was found by a lysimetric approach that of the 600 - 800 mm of annual rainfall less than 10 % contributed to groundwater through a loess soil, while on afforested rocks with a shallow soil profile it was more than 50 %. However, by a comparative measurement of tritium and C 14 it was found that only part of the baseflow is formed by water of recent origin, while the other part was at least 6,000 years old. This indicates that the validity of water balance equations needs to be re-examined, because even the recent water is derived from precipitation which has been fallen six years ago.

In temperate regions such as Czechoslovakia, groundwater recharge usually occurs during the winter and spring period; however, some minor recharges can be traced throughout a year. The total duration of the recharge period is about 80 - 100 days. During such a period the groundwater level does not have to rise steadily. At the end of the period when the groundwter outflow is in excess of the recharge the groundwater level becomes stable or is falling.

## 2.3. Perennial Recharge

A continuous perennial recharge may occur in parts of the humid tropics where there exists nearly permanent downward flow of water. Under natural conditions it is rather exceptional. Nevertheless, many hydraulic models are based on the assumption of a constant recharge into the aquifer. Kitching considering the above fact stated : "...input data to groundwater models consists of the hydrogeological framework of the aquifer including boundary conditions, abstractions, outflows and recharge; however, in many cases the estimation of recharge constitutes one of the largest errors introduced into the model". Morel and Wright (1978) found in England that even there recharge may continue for eight months; however, during exceptionally dry periods it may cease for a period of 18 months.

A continuous recharge is typical for man-made canals (Smiles, Knight, 1979) and to some lakes in temperate regions (Winter, 1976).

Artificially, a similar effect can be achieved through an artificial recharge; however, this process is beyond the scope of this paper.

## 2.4. Historical Recharge

Such a type of recharge occurred a long time ago and contributed to the formation of the present groundwater resources. It is closely related with what is known as groundwater residence time, defined as the time that has elapsed between the time interval in which a given volume of water was recharged and the interval at which it has reached the groundwater table (Campana and Simpson, 1984). Sometimes the time interval at the end of which a given volume of recharged water has been transformed into the baseflow is decisive. Perhaps the longest residence time of water, equal to 350,000 years, was found by Airey et al. (1978) in the Great Artesian Basin of Australia. The authors concluded that a steady piston flow approximation adequately described groundwater transport of meteoric water and that the contribution from leakage was negligible. Campana and Simpson (1984) used for the same purpose so called discrete state compartment models applied in conjunction with environmental isotopes.

# 3. SPATIAL VARIABILITY OF THE RECHARGE

Soil and aquifer properties including recharge processes vary from place to place both laterally and vertically. For regionalisation purposes the boundaries are plotted rather abruptly and the characteristics are determined from average or single measurements within the boundaries. Such variability can be traced in microscale and macroscale. Very little is known about the gradual changes existing in subregions and regions. Actually, they can be regarded as properties of the soil/aquifer system.

In a vertical approach Besbes and de Marsily (1984) attempted to define and quantify the difference between infiltration and recharge in an aquifer. They concluded that an average infiltration and average recharge are identical over a long time period and the distinction accounts only for the time delay and smoothing that percolation through the unsaturated zone imposes in transforming infiltration into recharge. It should be added that there is a great difference between the actual infiltration and potential infiltration as indicated by the infiltrometers.

Another role in the recharge process is played by the changing slope of the surface and slope of the boundary between soil layers and weathered rock. This factor has been considered significant in the theory of infiltration barriers (Andersen, Christiansen, 1986). According to that theory when the boundaries between two layers are sufficiently steep, the infiltrating water is drained through the upper fine textured layer laterally. It has been found that already a slope steeper than 1 % can be sufficient to form the capillary barrier, provided the top layer has a high hydraulic conductivity and sufficient capillary rise.

Vertical spatial variability of the profile is also important when recharge areas and preferred pathways are traced. More intensive groundwater recharge can be expected where the soil profile is rather shallow or non-existent and when denuded rocks contain many natural depressions, fissures and fracturs. This indicates that the validity of analytical methods based on Darcy's law needs to be re-examined at many locations.

Traditionally in hydrological projects, infiltration/recharge analysis has been based on a comparison of rainfall and baseflow measurements. The result is a watershed infiltration non-comparable with the soil infiltration because in the former an interception impact is incorporated and thus results of hydrological and soil physics methods often deviate.

Also an analysis based on the rainfall total as an indicator of the recharge, sometimes introduced in hydraulic models, should be replaced by a joint analysis of all involved factors. For such a purpose so called localised correlation combined with map overlay techniques appear to be feasible.

In statistical approaches two known features can be utilised:

- the absence of repeating pattern (the larger the area or the more intensive the sampling, the m ore complex is the variation);
- the point-to-point variation in a sample, reflecting the real variation in the soil-aquifer system.

Thus the so called regionalised variable theory can be applied, based on the assumption that values of a closely spaced aquifer/soil property are likely to be similar, whereas those at a distance are not. From several statistics methods the procedure known as kriging, in which the semivariogram is used for spatial interpretation, should be remembered.

Also the concept based on the comparison of the recharge process and optical/physical properties of the surface, which utilises remote sensing techniques, appears to be feasible.

## 4. ARID AND SEMI-ARID REGIONS

All types of groundwater recharge may occur in these regions; however, far fewer experiments have become available. Dincer (1974), for example, proved in Saudi Arabia that about 20 mm of rainfall had annually infiltrated the sand dunes at least to a depth of 7 metres; however, there was no indication whether the recharge had reached the deeper aquifers.

In those areas the natural recharge by streams is more frequent. The sources are usually intermittent or ephemeral streams when a temporary river water level is above the groundwater level - if any. The regime of perennial streams during dry periods is formed by the baseflow and the contribution to the groundwater sources - a reverse process - occurs usually during flood periods.

A seasonal recharge of groundwater from exogeneous sources through rivers was described by Fritz et al. (1981) in Pampa de Tamarugal, Chile, where water resources originate in the upper altitudes of the Andes and water for recharge comes from surface infiltration of flood water through creek channels.

Water from ephemeral streams of the wadi type forms locally important sources in many parts of North Africa. A distributed parameter wadi-aquifer model was developed by Illangasekare et al. (1984).

Other important sources of groundwater recharge are formed in arid areas where ephemeral rivers terminate in depressions known in Africa as sebkhats or chotts. Here the topography of the region plays a dominant role and thus through a diversion of the river channels the recharge can be at least partly regulated. However, when the infiltration from intermittent lakes in the depressions is delayed, considerable water loss is caused by the evaporation from temporary lake levels.

Locally important sources of water infiltrating under the river sandy bottom are found at the rims of wet and dry regions. In Zambia, for instance, locally utilisable water is dug from those sandy bottoms and a simple underground dam or system of dams can contribute to the stabilisation of recharged water.

Perhaps the most significant sources of groundwater in arid regions originate from the recharge of historical origin. In Sahara and elsewhere it is commonly agreed that a significant groundwater recharge occurred there earlier than 20,000 BP and between 10,000 - 2,000 BP, while the recharge concepts for the period between 20,000 and 10,000 BP are contradictory.

Also the concept of the vertical recharge of historically and/or regionally exogeneous water may significantly influence the estimate. De Vries (1984) concluded for the Kalahari desert that an active recharge can take place only in the areas where thickness of sand cover does not exceed six metres, otherwise seasonal retention and subsequent evapotranspiration would consume the penetrating water. Under such circumstances the current depletion goes to the account ofgroundwater storage formed during a major wet period 4,000 years ago.

## 5. CONCLUSIONS

The present state of knowledge in the field of groundwater recharge does not allow any particular preferred concept for future research. However, results which have been achieved elsewhere are at least worthwhile for selecting particularly important topics. One of them seems to be analysis of the disproportion between the concept of regionally smoothed properties of the soil/aquifer system on one hand and possible occurrence of preferred pathways on the other. These conclusions can lead to a qualitatively new point of view on the vertical flow in unsaturated and saturated zones. Also the combined effect of vertical and horizontal recharge time of residence and its variability needs to be analysed in relation to the validity of water balance equations. A joint impact of all parameters influencing the recharge processes should be analysed by statistical and deterministic methods. In particular the role of changing vegetational pattern in the formation of groundwater recharge should be examined.

As far as arid and semi-arid regions are concerned, more experiments and measurements of groundwater recharge have to be performed under ecological conditions of aridity. Such an approach may result in more definite conclusions on the regional validity of various concepts developed under temperate conditions. Groundwater recharge processes should be analysed in the regions with sparse localised water supply with the aim to develop effective methods of recharge regulation.

## 6. REFERENCES


Andersen L.J., Christiansen J.C., 1986. The Capillary Barrier. Geol. Survey of Denmark, Int. Rep. No. 20 - 1986, 26 p.

Airey P., Calf. G.E., Campbell B.L., Hartley P.E., Roman D., 1978. Aspects of the Isotope Hydrology of the Great Artesian Basin, Australia. Isotope Hydrology, IAEA Vienna, 205 - 219.

Balek J., 1986. "Groundwater Resources Assessment by Remote Sensing". Int. Workshop on Natural Recharge of Groundwater. Antalya.

Bath A.H., Edmunds W.M., Andrews S.N., 1978. "Paleoclimatology Deduced from the Hydrochemistry of Triassic Sandstone Aquifers, U.K.". Isotope Hydrology, IAEA Vienna 545 - 566.

Besbes M., de Marsily C.E., 1984. "From Infiltration to Recharge : Use of a Parametric Transfer Function". Journal of Hydrology, 74, 271 - 293.

Campana M.E., Simpson E.S., 1984."Groundwater Residence Time and Recharge Rates Using a Discrete State Compartment Model and C14 Data". Journal of Hydrology, 72, 171 - 185.

De Vries J.J., 1984. "Holocene Depletion and Active Recharge of the Kalahari Groundwaters - a Review and Indicative Model". Journal of Hydrology, 70, 221 - 232.

Fritz D., Suzuki Ol, Silva C., 1981. "Isotope Hydrology of Groundwaters in the Pampa del Tamarugal, Chile". Journal of Hydrology, 53, 161 - 184.

Illangasekâre T.H., Morel-Seytoux H.J., Koval E.J., 1984. Design of a Physically-Based Distributed Parameter Model for Arid Zone Surface Groundwater Management. Journal of Hydrology Vol. 74, 213 - 232.

Kitching R., Sheaver T.R., Shedlock S.L., 1977. "Recharge to Bunter Sandstone Determined from Lysimeters. Journal of Hydrology, 74, 93 - 109.

Sharma M.L., Hughes M.W., 1985. "Groundwater Recharge Estimation Using Chloride, Deuterium and Oxygen - 18 Profiles in the Deep Coastal Sands of Western Australia". Journal of Hydrology, 81, 93 - 109.

Smiles D.E., Knight J.H., 1979. "The Transient Water Beneath a Leaking Canal". Journal of Hydrology, 44, 149 - 162.

Morel E.H., Wright C.E., 1978. "Methods of Estimating Natural Groundwater Recharge". Reading, U.K., Central Water Plan. Unit, Techn. Note No. 25, 40 p.

Sophocleous M., Perry C.A., 1985. "Experimental Studies in Natural Groundwater Recharge Dynamics". Journal of Hydrology 81, 297 - 332.

Todd D.K., 1967. Groundwater Hydrolgy. John Wiley and Sons, New York, 336 p.

Winter T.C., 1976. Numerical Simulation Analysis of the Interation of Lakes and Groundwater. USGS Prof. P. 1001, 95 p.

# AN UNEXPECTED FACTOR AFFECTING RECHARGE FROM EPHEMERAL RIVER FLOWS IN SWA/NAMIBIA

S Crerar, R G Fry, PM Slater, G van Langenhove and D Wheeler
Department of Water Affairs
Private Bag 13193
Windhoek
9000
SWA/Namibia

ABSTRACT. The processes controlling recharge to alluvial groundwater have presented complex and even intractable hurdles to accurate estimates of this parameter. Casual observations in ephemeral rivers in SWA/Namibia have indicated that variable recharge quantities result from similar flood events. For this reason the Department of Water Affairs initiated a research project to identify and investigate relevant parameters. To achieve these objectives three carefully chosen natural channel sections were identified for instrumentation, a simulated channel was constructed under laboratory conditions and the basis of a mathematical model laid down. This paper comments on the field and laboratory instrumentation applied and the results derived from the first fully instrumented natural river channel for which two seasons of flood flow and resultant recharge data are available. These results are compared with laboratory trials for a range of flow regimes. Whilst this work is at an early stage it has already become clear that silt carried by flood waters can effectively seal the alluvial surface even during the flood event at unexpectedly high flow velocities. Thus the other processes controlling recharge in the unsaturated zone may become relatively unimportant for much of the duration of any given flood event.

## 1. INTRODUCTION

Due to the prevalence of semi-arid conditions in SWA/Namibia 67% of water supplied is derived from sub-surface sources. Of this 18.6% is generated by abstraction from unconsolidated alluvial aquifers. Superficially this percentage appears small. However communities numbering many thousands of people are totally reliant upon such sources. Typically their reliability is based on adequate occurrence of recharge from ephemeral river flow. Basic supply scheme monitoring procedures relating to flood volumes and resultant groundwater level rises highlighted anomalous situations. No direct correlation was seen between the availability of surface water and resultant recharge. For this reason the Department of Water Affairs initiated a research



*I. Simmers (ed.), Estimation of Natural Groundwater Recharge, 11–28.*

project to establish which of the many parameters known to affect recharge represents the major control, the ultimate goal being accurate evaluation of alluvial aquifer reliability and possibly the enhancement of recharge.

The problem was tackled via three main thrusts. Firstly the establishment of field experimental sites, secondly construction of a physical hydraulic model and thirdly via computer simulation. An initial review of published data indicated that much emphasis and therefore work has been devoted to an understanding of unsaturated flow processes. Thus this project aligned itself to satisfactory measurement of the parameters controlling unsaturated flow. However it was clear from the outset that inhibition of recharge due to the presence of an antecedent silt layer might prove to be a joker in the deck. This paper sets out to establish that from field and laboratory observation silt layers are not merely an initial condition to be overcome in the early stages of flooding. Silt may be deposited during a flood event inhibiting recharge at relatively high flow velocities. Under such conditions the characteristic of the unsaturated zone does not represent the major factor controlling recharge in alluvial beds of ephemeral streams.

## 2. EXPERIMENTAL SITE LOCATION AND LAYOUT

At the outset experimental site criteria were set up as follows:

(i) The alluvial channel should have well defined boundaries.
(ii) Unconsolidated alluvial thickness should be at least 5 metres.
(iii) Clay content in the alluvium should be low.
(iv) Geological structures within the containing country rock should be minimal to avoid extraneous alluvial groundwater loss or gain.
(v) The channel should offer high potential for full width flow.
(vi) A location suitable for establishment of an accurate surface flow gauging station should exist.
(vii) There should be no significant surface flow tributaries within the experimental section.

Figure 1 illustrates the location of the first of three experimental sites identified which is at Gross Barmen on the Swakop River. This site displays all the required attributes and it is from here that the field data presented in this paper is drawn. Figure 2 shows the form and distribution of instrumentation to be discussed. The chosen site is 1,8 kilometres long and typically 60 metres in width. It is fed by a catchment area of 5930 kilometres$^2$ from which a high annual expectation of flood flow has been identified. It should be noted that this is in a semi-arid regime, and that the river is usually dry for upwards of 340 days a year.

Prior to instrumentation the site was subjected to the following surveys in order to adequately identify the geometry of the alluvial body.

(i) Detailed resistivity depth probing at 10m spacing on lateral traverses 10m apart.
(ii) Auger drilling: one per traverse as a control on geophysics.

(iii) Detailed topographic survey of all geophysical traverse lines identifying XY co-ordinates and elevation above mean sea level.

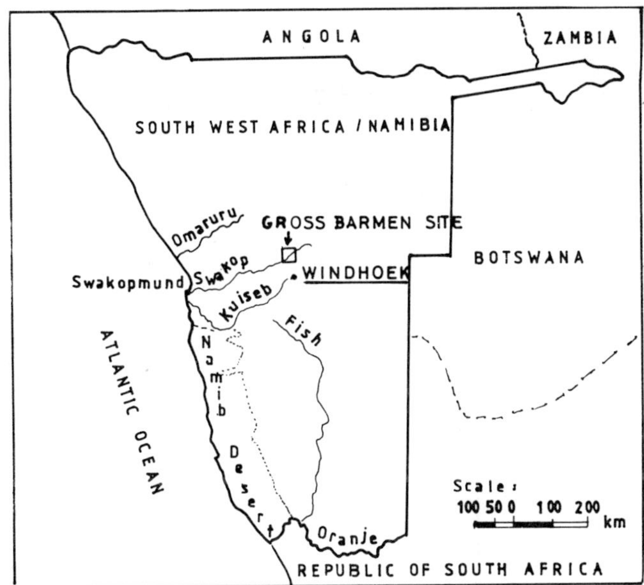


Figure 1. Map of Namibia showing Location of Experimental Site

Detailed cross-section and longitudinal geometry were digitised via an H.P. 9816S computer enabling establishment of a graphical relationship between average groundwater level and saturated alluvial aquifer volume.

## 2.1 Field Instrumentation

2.1.1 Flood Gauge. This structure comprises a master shaft to bed level obtaining full flood record and a slave shaft with 20 centimetre bed clearance providing back-up record during the important flood peak segment of the record. Both shafts are equipped with autographic recorders. (See Figure 2 for location.)

2.1.2 Observation Boreholes. Thirty-eight 50mm diameter observation boreholes were drilled to facilitate detailed identification of the form of the water table, and subsequently measured with at least a monthly frequency. Four 150mm boreholes were drilled, two equipped with continuous wire wound screen and two as fully penetrating observation boreholes, to facilitate test pumping. Establishment of the saturated hydraulic characteristics indicated that a specific yield of 3% was generally applicable. Alluvial samples retrieved at 1

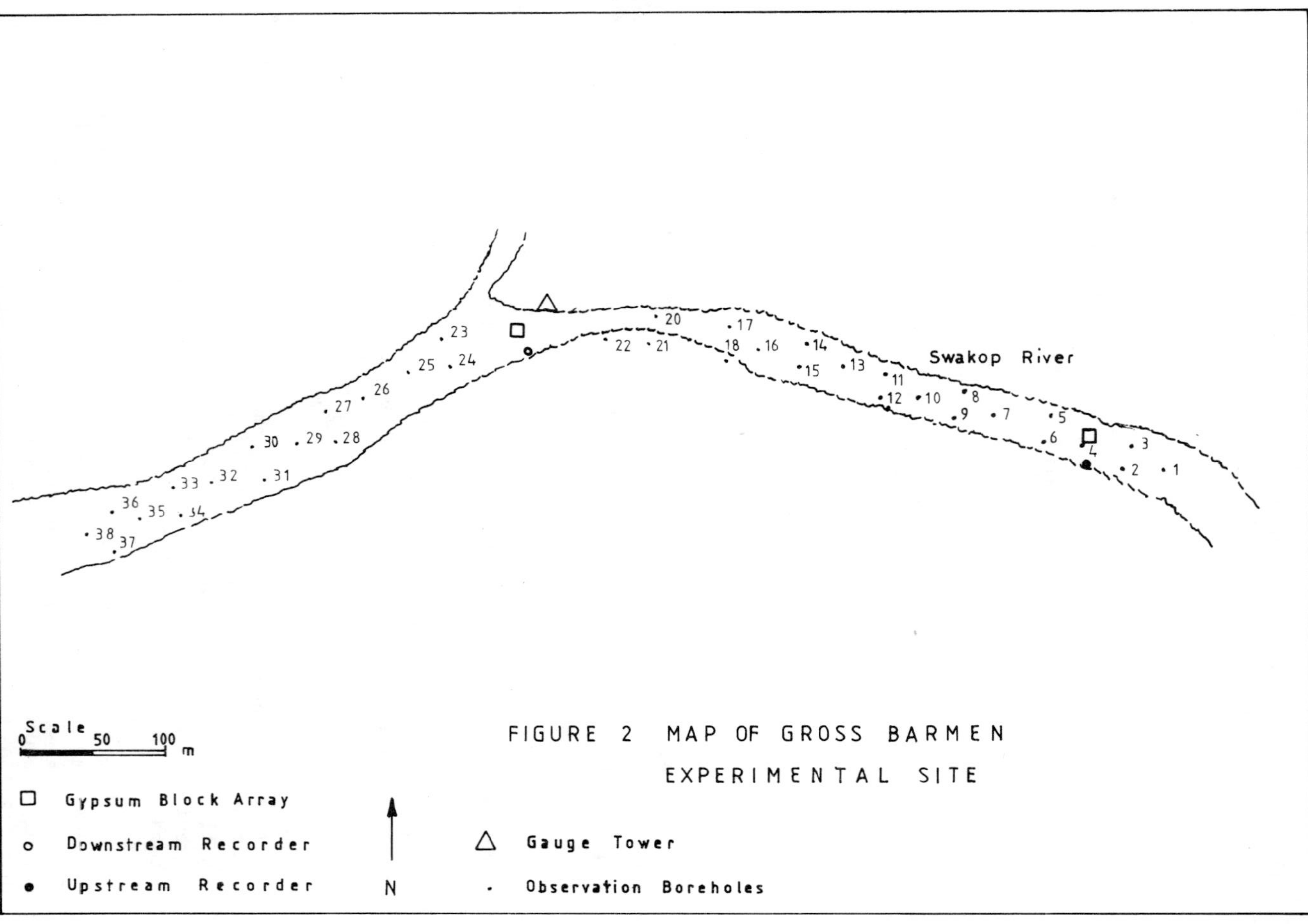


FIGURE 2 MAP OF GROSS BARMEN EXPERIMENTAL SITE

metre intervals were subjected to sieve analysis and a calculation of theoretical permeability compared well with test pumping results. The tested boreholes were equipped with pressure transducer operated autographic recorders providing a 97% continuous record at the locations shown on figure 2. All installations were included in the topographic survey.

2.1.3 Ancillary Equipment. Clearly as a function of the entire project, unsaturated zone parameters are of considerable importance. For this reason neutron probe access tubes were installed to a depth of 3 metres associated with specifically designed gypsum block arrays. Details of these installations and their results will be presented in later papers.

2.2 Field Results

Data gathered over two flood seasons were analysed and are presented in TABLE I.

TABLE I: Recharge Analyses at Gross Barmen

| Date | Flood Volume ($m^3$) | Average Flood Velocity (m/sec) | Recharge ($m^3$) | % Recharge per Kilometre |
|---|---|---|---|---|
| 25.1.85* | 27 600 | 0.05 | 3 400 | 6.9 |
| 27.1.85 | 290 000 | 0.31 | 5 700 | 1.1 |
| 28.1.85 | 30 000+ | - | 400 *f* | 0.6 |
| 16.11.85* | 47 000 | 0.08 | 1 800° | 2.1 |
| 29.11.85 | 312 000 | 0.19 | 5 100 | 0.9 |
| 15.12.85 | 5 000+ | - | 0 | 0 |
| 18.1.86 | 135 000 | 0.18 | 1 200 | 0.5 |
| 20.1.86 | 144 000 | 0.22 | 1 200 | 0.5 |
| 27.1.86 | 300 000 | 1.0 | - *f* | - |

* First flood of season
\+ Flood peak not recorded by gauge
*f* Aquifer fully recharged during flood
° Estimated ± 10% due to flood occurrence during recorder maintenance

These results demonstrate considerable variation in recharge not directly correlateable with flood volume. However it is noticeable that for the first flood in each season a much higher percentage recharge occurs.

It has proved possible to analyse recharge rate related to time into flood where a full continuous record was obtained. These results are presented in Figures 3 and 4. Determination of the initial recharge rate from the time lapse between flood peak arrival and

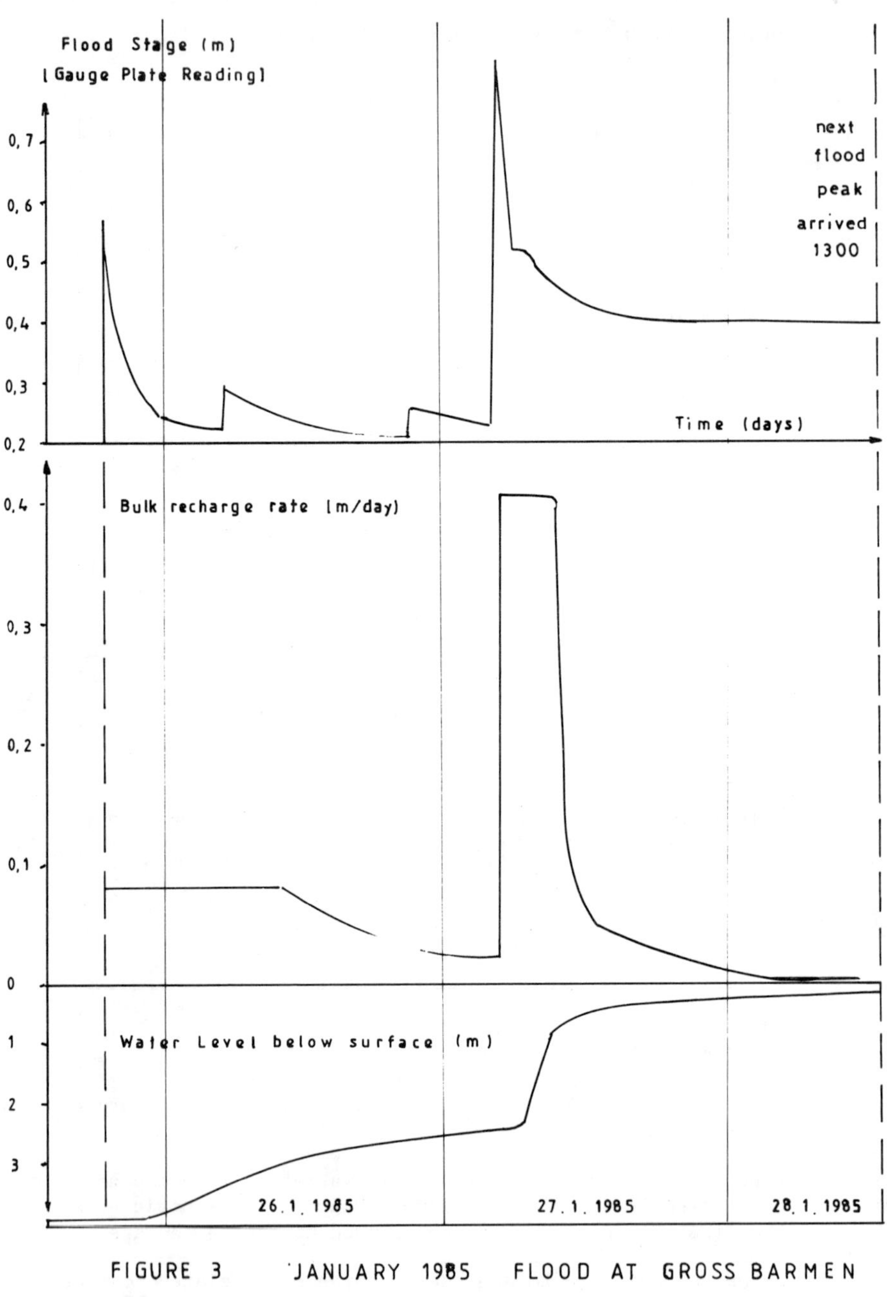


FIGURE 3 JANUARY 1985 FLOOD AT GROSS BARMEN

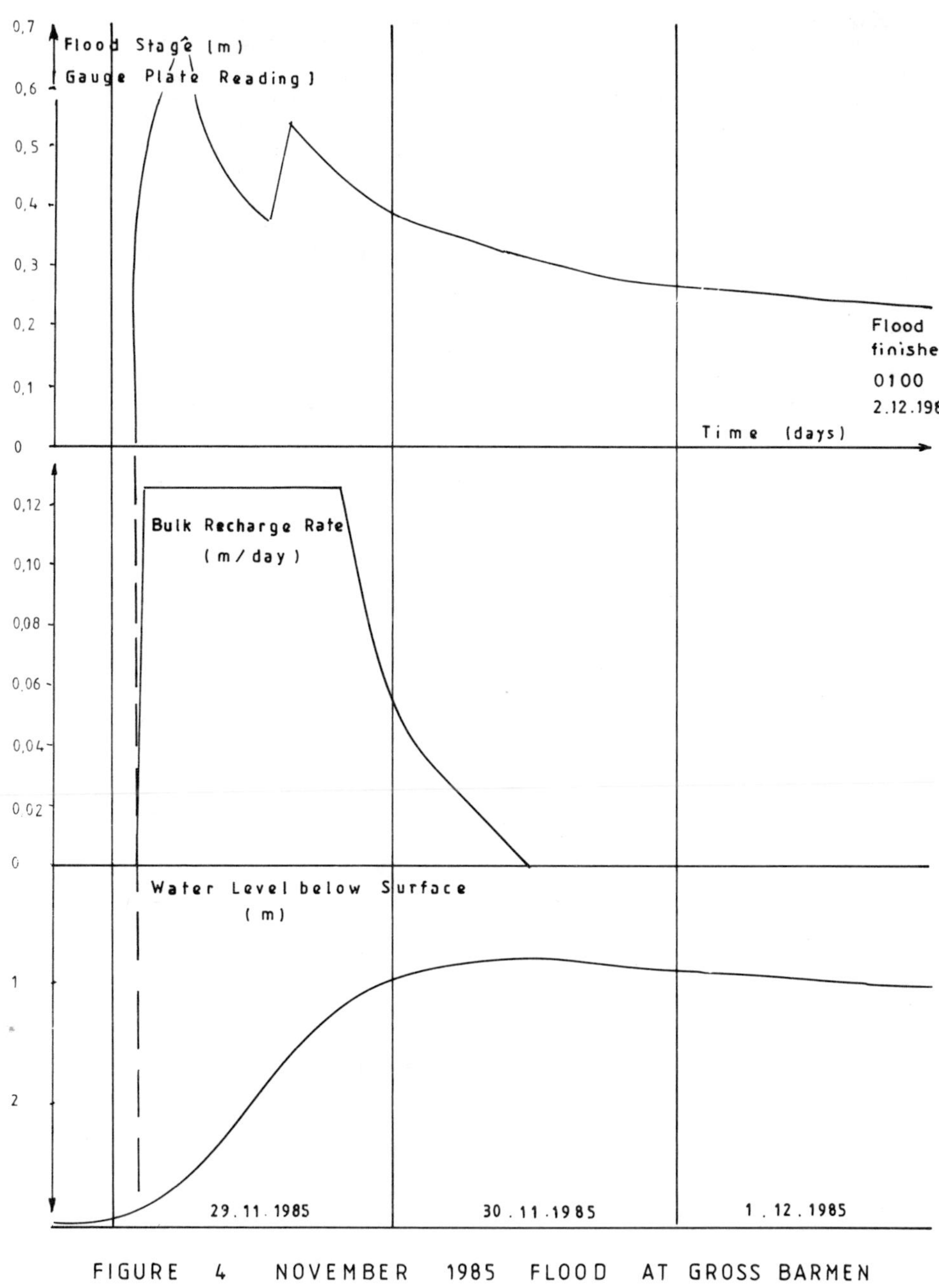


FIGURE 4 NOVEMBER 1985 FLOOD AT GROSS BARMEN

groundwater table reaction gave higher than expected values, ranging from 28 m/day on the 27/1/85 to 67 m/day for the flood of 29/11/85. This apparent recharge rate taking place along preferred pathways declined within an hour to the bulk recharge rate depicted. Bulk rates are determined from division of the recharge volume for a particular period by the average surface area of river flow.

All three floods result in initially steady rates of bulk recharge lasting at least 4 hours. This timescale negates the hypothesis mentioned in the introduction that antecedent silt conditions are important in inhibiting recharge. The rate of this steady bulk recharge is related to peak flood flow velocity as shown in Figure 5. Subsequently, when the flood stage has declined to 30-40 centimetres below the peak, a marked decline in recharge rate (in one case to zero as shown in Figure 4) occurs although the aquifer is not fully saturated and the flood is still in progress. Clearly recharge has been anomalously inhibited subsequent to achievement of steady state infiltation capacity.

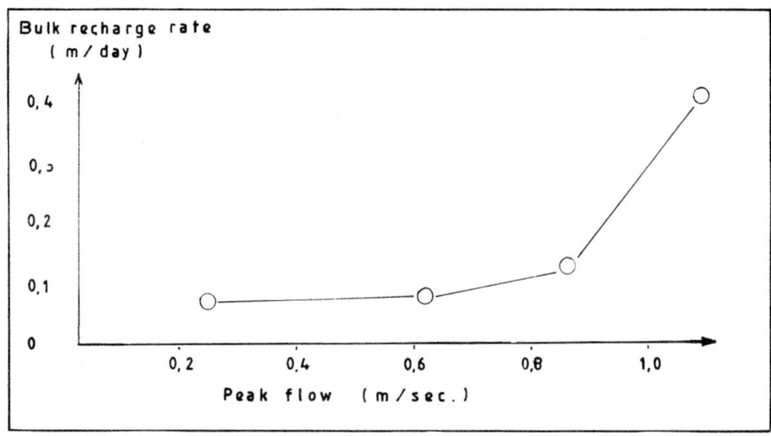


Figure 5: Recharge Rates in Different Floods

Although the nature of the sand/water interface cannot be adequately observed during a flood, water sampling from a flood on 27/1/86 shows that a significant silt load is carried (see Figure 6). It was felt that this suspended silt load could be important in controlling recharge. Hence a series of laboratory experiments to investigate the effects of varying silt loads and flood flow velocities was established in a simulated river section using alluvium obtained from the Gross Barmen experimental site.

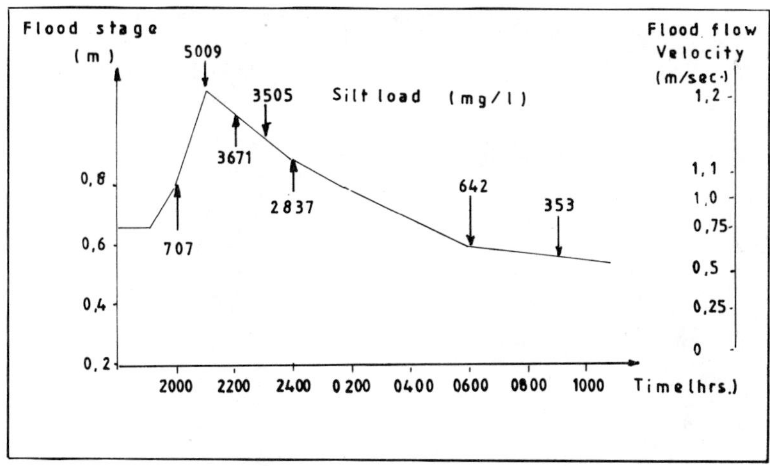


Figure 6: Silt Load of Flood Event 27/1/86 - 28/1/86

## 3. LABORATORY INVESTIGATIONS

### 3.1 Experimental set-up

3.1.1 Test rig. The test rig (see Figure 7) essentially consists of a 0,20m wide, 1m deep and 6m long channel made out of steel. It is typically half-filled with the alluvium, over which simulated flood flows may pass.

Over a 1m length the walls are made of perspex to allow visual observations, and over half of this a 0,45m deep test box is added to provide a realistic depth to which the water can infiltrate. The bottom of the channel and box are perforated with 5mm diameter holes to allow vertical throughflow.

The water was circulated to and from a small sump using a centrifugal pump. Water velocities and depths in the channel could be adjusted by one or a combination of the following actions:-

(i) adjusting the opening of the valve on the supply line.
(ii) adjusting the height of the gate.
(iii) changing the bed slope, either by lifting the whole structure, or by adding sand.

3.1.2 Measurement of variables. The standard laboratory equipment used was:

(i) point gauges for the determination of water depths and water pressures (indirectly via connecting pipe and measuring cylinder).
(ii) surveying equipment for the determination of bed levels before and after test runs to monitor deposition and erosion patterns.
(iii) Ott 50mm diameter propellors for the measurement of water velocities.

Figure 7: Test Rig for Laboratory Investigations

Because of the time lag involved in the indirect measurement of water pressures using measuring cylinders and point gauges, and because of the impracticalities of taking simultaneous readings at several points, a multi-channel data acquisition system and ten pressure transducers were installed for all but the preliminary exploratory runs. Seven of the transducers were installed in a vertical line down the side of the perspex channel and box. The other two were positioned as detailed in TABLE II.

TABLE II

| Transducer number | | Vertical Position relative to original bed levels (m) | |
|---|---|---|---|
| 0 | | + 0,030 | |
| 1 | Middle | - 0,025 | |
| 2 | of | - 0,175 | |
| 3 | box | - 0,325 | |
| 4 | | - 0,475 | |
| 5 | | - 0,625 | |
| 6 | | - 0,875 | |
| 7 | 0,5m upstream | - 0,025 | (*) |
| 8 | | - 0,500 | |
| 9 | Not linked | | (+) |

(*) "7" is to check possible problems with "1"

(+) "9" is intended to check the stability of the electronics of the system.

3.1.3 Practical Methodology. In order to obtain reproducible and consistent results the following points with respect to methodology were taken into account:

(i) Before every test run the bed profile was surveyed and if necessary sand was removed or added in order to provide the correct amount of material for the required slope. The top layer (2-3 cm) of silty sand left by the previous run was then removed (necessary to prevent hysteresis) and added again upstream at the beginning of the run (for continuity).

(ii) During each run silt samples (4 at half hour intervals) were taken in the turbulent well-mixed outflow.

(iii) Observations of leakage from the base of the channel (through the drilled holes) were made, and outflow through the 5 holes in the bottom of the test box was measured using a calibrated orifice.

(iv) The duration of each test run was 2 hours and the bed was allowed to dry out partially for a period of at least 2 days between runs.

## 3.2 Results and Observations

3.2.1 The presence of a "silt layer" and its effect. At lower velocities (TABLE III summarises the velocities and bottom slopes investigated under different suspended silt loads) the formation of a thin continuous silt layer could be visually observed through the perspex wall, and its effect on the infiltration could be monitored directly (from the throughflow) and indirectly (from the pressure drop measured by the sub-surface transducers).

However, at higher velocities these visual observations indicated that no such continuous silt layer was formed on top of the bed, since now a continuous movement of sand grains could be seen.

TABLE III Flow rates and velocities investigated

| Target bed slope (percent) | Velocity (m/s) | Flow Rate ($m^3/s$) |
|---|---|---|
| 1 | 0,60 | 0,008 |
| 1 | 0,75 | 0,011 |
| 1,5 | 1,00 | 0,016 |
| 2 | 1,20 | 0,020 |

Nevertheless, the magnitude of throughflow and pressure drop indicated that for certain conditions infiltration was still being inhibited. Two possible explanations are:

(i) The top layer of sand is slowly clogged by silt in the infiltrating water.

(ii) A very thin continuous silt layer is formed under the moving sand.

3.2.2 Behaviour of transducers nearest to the surface. A good example of the rather unexpected reaction of the pressure transducer under the surface can be seen in Figure 8.

(i) Initially, as would be expected. the pressure rises upon the arrival of the infiltrating water.

(ii) Then very suddenly a sharp drop in pressure occurs, in this case even to negative values.

(iii) There is a gradual recovery, but not to the original values.

This behavior is easily explained by the formation of an inpermeable silt layer on top of the bed. During the first few minutes water is infiltrating quickly, saturating the sand at a relatively high pressure; then the silt layer forms, cutting off further supply from above. The water deeper in the sand still moves further downwards and through the bottom, thereby effectively "sucking" the higher zones into showing negative values of pressure. Only later is there a slow recovery to what should become a small positive pressure.

3.2.3 Behaviour of other pressure transducers. Figure 10 shows a comparison of all transducer readings during a test on 16/9/86.

Although there are some differences at the very beginning of the test due to transition to full saturation, the subsequent drop in pressure is very consistent overall. The transducers higher up in the channel (Nos 1, 2, 3 and 7) react en masse showing that the whole body of water is confined between the bed and the bottom. As there is no noticeable gradient, there is no flow (confirming the sealing effect of the silt layer). The lowest of the transducers (no. 6) also drops but not as much as the higher ones, and remains stable at that level. After 55 minutes a negative pressure has been established and transducers 1 - 6 show an upward gradient. This induces flow which causes the general recovery of pressure towards the uniform small negative pressures seen after 115 minutes.

It appears from this and other evidence that once a continuous silt layer is formed. it effectively reduces infiltration to nil or to a negligible value compared to that which would occur in the case of free infiltration in "clean" sand.

3.2.4 Runs with permanent silt layers. An example of the quick formation of a silt layer has already been shown in Figure 8. In many cases a continuous silt layer only forms much later, and at a much slower pace, as shown in Figure 9. However, the final impact of such a layer is as significant as in the previous case: the pressure drop through this layer exceeds the pressure loss in the sand by an order of magnitude.

3.2.5 Runs with intermittent silt layers. In some cases a clear trend towards the formation of a silt layer was abruptly reversed, giving a sudden rise in subsurface pressure followed by a rise in outflow (see Figure 11). The most feasible explanation for this is that the silt layer is disrupted locally and that the whole system is suddenly exposed to higher pressure.

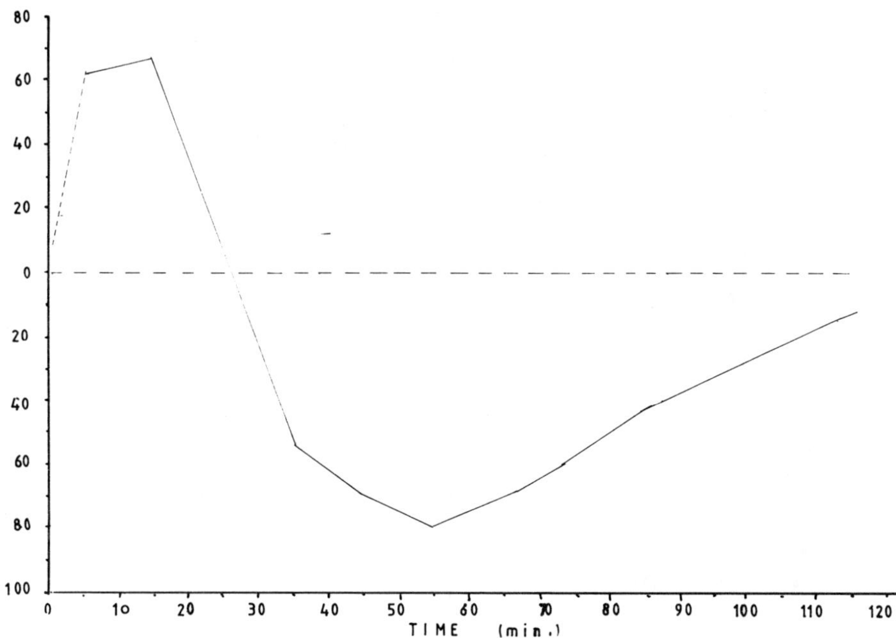


FIGURE 8 SUBSURFACE PRESSURE DURING RUN 18/9/86

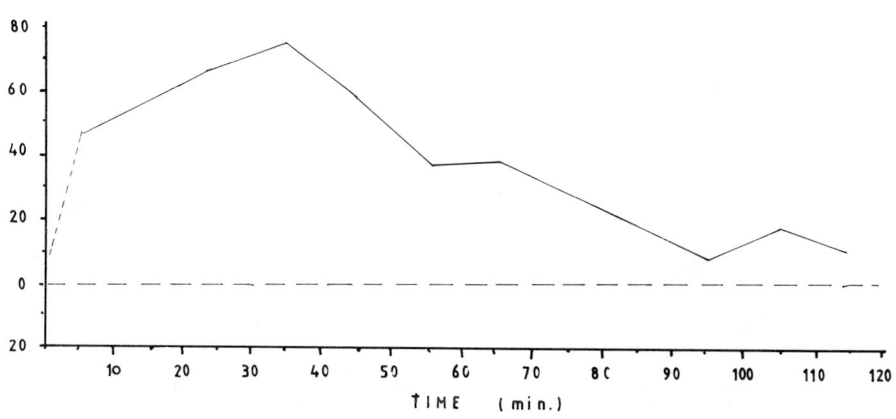


FIGURE 9 SUBSURFACE PRESSURE DURING RUN 11/6/86

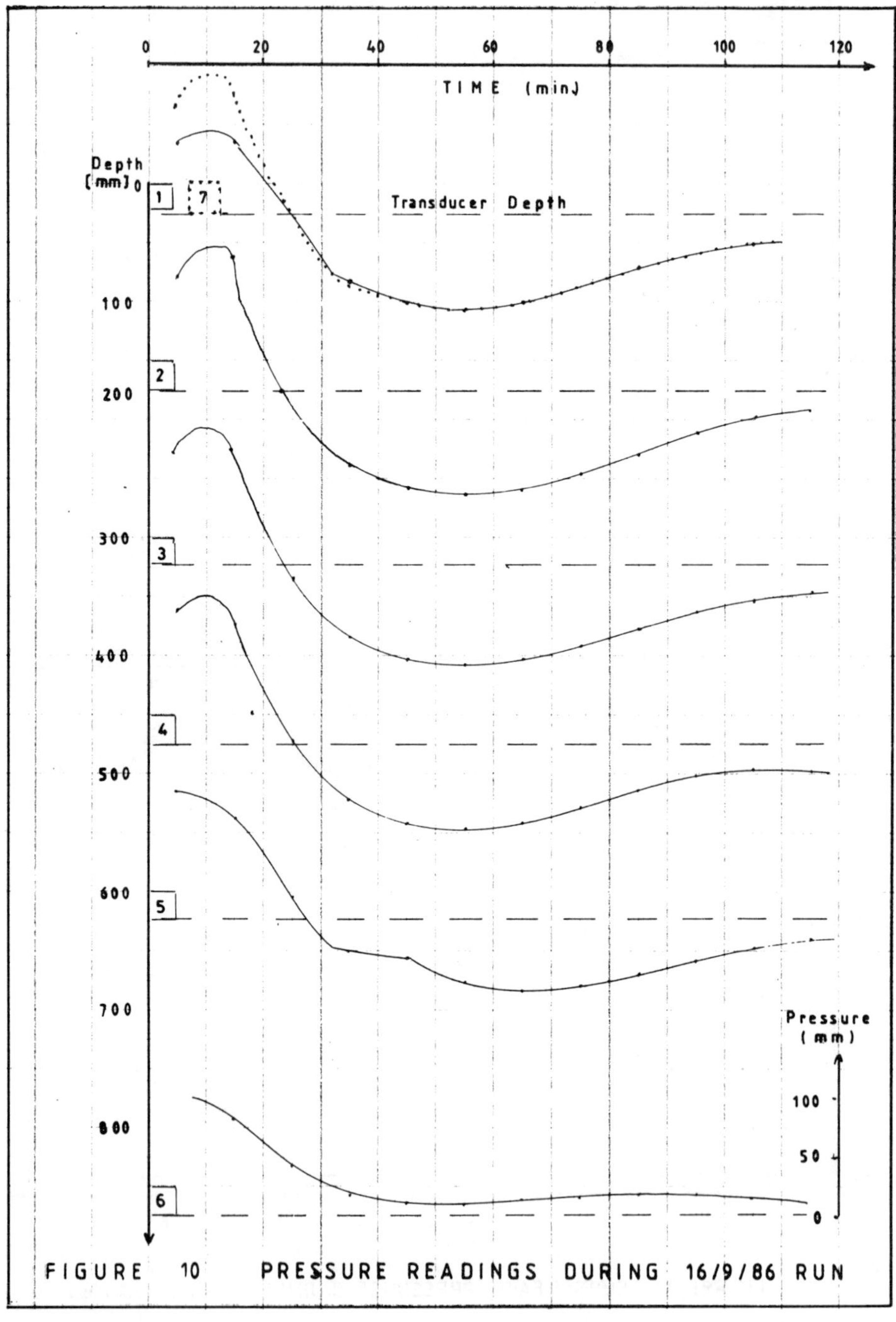


FIGURE 10 PRESSURE READINGS DURING 16/9/86 RUN

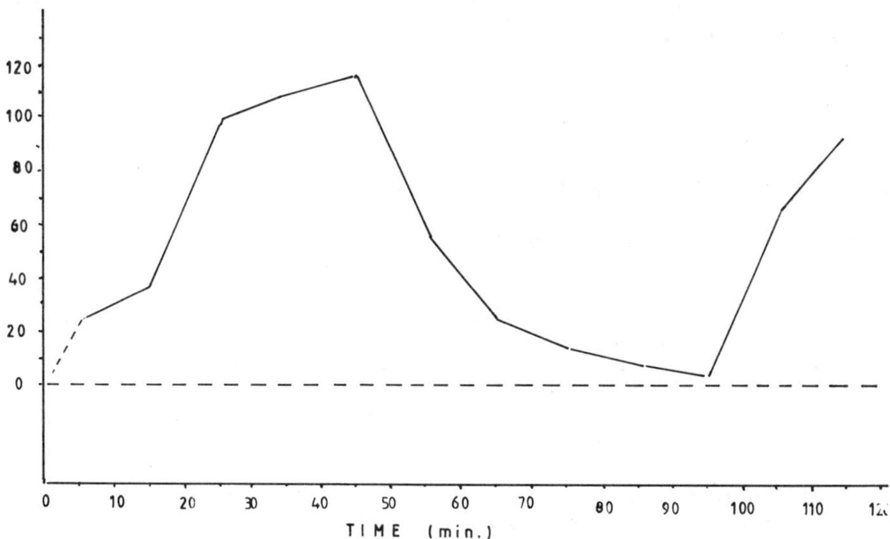


FIGURE 11 SUBSURFACE PRESSURE DURING RUN 9/10/86

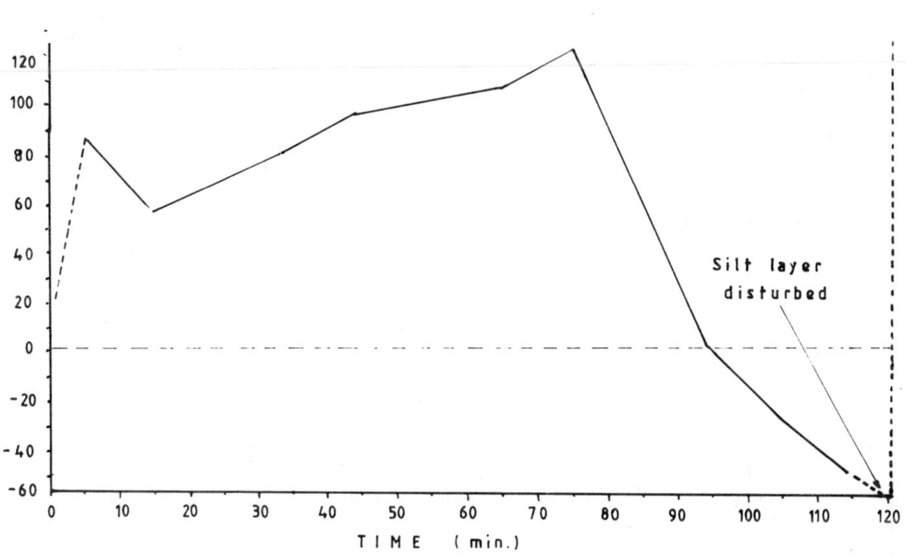


FIGURE 12 SUBSURFACE PRESSURE DURING RUN 21/11/86

3.2.6 Runs with interference. On a few occasions when a silt layer was suspected to be present at the end of a test period, the top layer in the test zone was artifically disturbed and the test continued. As a result the pressure rose immediately and the outflow from the bottom of the channel downstream of the test box, which had ceased, started again (see Figure 12).

This clearly demonstrates that the infiltration is inhibited by the top layer.

### 3.3 Summary of Laboratory Results and Discussion.

An overview of results is presented in TABLE IV.

(i) Silt carried by flood flows clearly inhibits recharge.

(ii) Contrary to expectations the magnitude of the silt load does not appear to represent the major factor controlling the extent of recharge inhibition.

(iii) Runs done under almost identical conditions result in very different patterns of events. However in all cases blocking occurs at some stage and remains in evidence for most or all of the remainder of the run.

(iv) In general the higher the silt load the more quickly the blocking occurs, and the higher the observed temporary underpressure. However, at lower silt loads, the blocking mechanism is not prevented but only delayed.

(v) The intermittent breaking up of the blocking mechanism appears to occur rather randomly. Probably the test period of 2 hours is not long enough to conclude that this would never happen in the cases where it was not observed.

(vi) This breaking up appears to be favoured by higher velocities and can be induced very easily by local disturbances, for which the model is not adequately representative due to its inherent limitations as far as width and length are concerned.

The ultimate conclusion is that, for the observed range of flow velocities and for any reasonable silt load - clogging of the top layer will occur and reduce infiltration to a negligible value. This clogged layer will break up intermittently under conditions which are difficult to quantify, at this stage - after which free infiltration is temporarily restored.

## 4. COMPARISON OF FIELD AND LABORATORY RESULTS

Using silt loads of the order shown in Figure 6 (typical of actual floods analysed) the laboratory experiment has shown that precipitation of silt is effective in inhibiting recharge.

During the floods of 27/1/85 and 29/11/85 shown in Figures 3 and 4 respectively the recharge rate suddenly dropped when flood flow velocity declined below 0.36 m/sec. The explanation for this is that above a certain velocity silt layer break-up occurs. Conversely therefore below this threshold velocity the silt layer becomes increasingly continuous. Further laboratory experimentation at lower velocities is now envisaged. Although the remaining flood illustrated

**TABLE IV: OVERVIEW OF LABORATORY RESULTS**

| FLOW DEPTH (mm) | FLOW VELOCITY (m/s) | SILT LOAD (mg/ℓ) AVERAGE | PRESSURE IN "1" (mm) | | FINAL OUTFLOW ($\times 10^{-3}$ L/S | COMMENTS | INTERPRETATION |
|---|---|---|---|---|---|---|---|
| | | | MINIMUM | FINAL | | | |
| 65 - 75 | 0.60 | 651 | 49 | 56 | 10.05 | pump giving problems | intermittent breaking up of silt |
| | | 1001 | 22 | 22 | 6.95 | | slowlyforming silt layer |
| | | 1180 | 54 | 119 | 9.90 | temporary over pressure | unclear |
| | | 1867 | -15 | - 8 | 8.10 | | quickly forming silt layer |
| 70 - 80 | 0.75 | 384 | 1 | 1 | 2.00(1) | temporary over pressure | slowly forming silt layer |
| | | 477 | 9 | 10 | 2.00(1) | | slowly forming silt layer |
| | | 417 | 36 | 43 | 8.65 | temporary over pressure | slowly forming silt layer |
| | | 551 | -32 | -32 | 7.35 | temporary over pressure | slowly forming silt layer |
| | | 731 | -51 | -51 | 7.10 | temporary over pressure | quickly forming silt layer |
| | | 791 | 39 | 40 | 8.00 | | intermittent breaking up of silt |
| | | 851 | -51 | -51 | 7.00 | | quickly forming silt layer |
| | | 3546 | -80 | -16 | 6.50 | | very quickly forming silt layer |
| 75 - 85 | 1.00 | 2144 | -51 | -51 | 7.35 | | slowly forming silt layer |
| | | 2244 | 96 | 99 | 8.90 | overpressure during most of test | unclear |
| | | 2143 | 8 | 8 | 2.00(1) | | slowly forming silt layer |
| | | 2189 | 50 | 50 | 9.15 | temporary overpressure | intermittent breaking up of silt |
| | | 2019 | 3 | 93 | 8.15 | temporary overpressure | intermittent breaking up of silt |
| | | 4281 | 0 | 12 | 4.40 | | quickly forming silt layer |
| | | 4925 | 9 | 9 | 4.55 | | slowly forming silt layer |
| 80 - 90 | 1.20 | | 18 | 28 | 9.90 | | slowly forming silt layer |
| | | 1195 | -53 | 143 | 9.60 | temporary overpressure | intermittent breaking up of silt |
| | | 2060 | 3 | 3 | 7.15 | temporary overpressure | intermittent breaking up of silt |
| | | 3206 | 3 | 59 | 3.14 | | intermittent breaking up of silt |
| | | 3827 | 42 | 42 | 5.70 | | slowly forming silt layer |
| | | 3380 | 21 | 56 | 5.30 | | intermittent breaking up of silt |
| | | 3595 | - 7 | 91 | 2.00(1) | temporary overpressure | intermittent breaking up of silt |
| | | 7034 | 18 | 18 | 5.50 | temporary overpressure | intermittent breaking up of silt |
| | | 7685 | 35 | 52 | 3.40 | temporary overpressure | intermittent breaking up of silt |
| | | 8997 | - 3 | 58 | 4.25 | | intermittent breaking up of silt |

in Figure 3 (26/1/85) demonstrated velocities above this threshold only for the first hour, recharge continued at a steady slow rate for fifteen hours. It is believed that this flood carried very little silt, as has been observed at the experimental site for other small first floods of a season.

## 5. SUMMARY

This work shows that the formation during flood events of an impermeable silt layer is an extremely important factor in the determination of recharge to unconsolidated alluvium underlying ephemeral rivers. It was found surprising that this has not received detailed treatment in previous literature, with the exception of a brief reference to silt layer inhibition of sediment transport in a permanent stream in Alaska (Harrison and Clayton, 1970). It is planned to continue project work in the following areas:-

(i) Instrumentation of two further field sites exhibiting different flood silt load characteristics and the potential for inhibition of recharge due to entrapped air.

(ii) Continuation of laboratory experiments aimed at establishment of threshold velocities and silt load conditions facilitating recharge.

(iii) Finalization of a computer model of infiltration through the unsaturated zone based upon the work of Morel Seytoux (1984), and incorporating the effects of a silt layer.

## REFERENCES

Harrison and Clayton, (1970)
'Effects of Groundwater Seepage on Fluvial Processes'
Bulletin of the Geological Society of America, Vol **81**, p 1217-1226

Morel Seytoux, (1984)
'A Two Phase Numerical Model for the Prediction of Infiltration'
Publication of Colorado Water Resources Research Institute.

## ACKNOWLEDGEMENTS

The authors wish to acknowledge the Department of Water Affairs for providing funds for this research project and encouraging publication of this paper, and the Project Steering Committee for its valuable advice.

# ON THE CONTINUITY OF AQUIFER SYSTEMS ON THE CRYSTALLINE BASEMENT OF BURKINA FASO

J.J. van der Sommen 1)
W. Geirnaert 1)2)

1) Iwaco bv P.O. Box 183, Rotterdam, The Netherlands
2) Institute of Earth Sciences, Amsterdam Free University, P.O. Box 7161, Amsterdam, The Netherlands.

ABSTRACT. Two types of aquifer are present on the West African Shield: those in the hardrock, connected with fault and fissure systems, and the weathered mantle aquifer. This hydraulically connected aquifer system is generally regarded as highly discontinuous with individual compartments in which isolated groundwater circulation occurs. Extensive drilling for rural and urban water supply in Burkina Faso, Niger and Ghana has, however, brought evidence for the presence of a continuous piezometric surface over large areas. Hydrochemical and isotope data and the cyclic fluctuation of groundwater levels provide evidence for a regional recharge-discharge system. The importance of this regional system increases from north to south in tandem with higher precipitation. The hypothesis of a substantial lateral flow in the fault and fissure system was applied in the study for the location of new well fields for the capital of Burkina Faso, Ouagadougou.

## INTRODUCTION

Extensive drilling for rural water supply programs on the West African Shield has produced large amounts of data on weathered mantle and fractured bedrock aquifers. Generally these data are only used for statistical studies to optimalize the procedures of siting wells and not to obtain a better insight into the hydrogeological conditions of these aquifers. A better insight is, however, urgently needed, not only to predict the long term behavior of the growing number of wells but also to meet the increasing demands for new wells in both quality and quantity (large yields for urban water supply, inadmissibility of groundwater mining, reduced distance to the consumer).

Up to the present time, scientifically oriented hydrogeological investigations have focussed mainly on small scale river basins in which individual fractures are being studied and pumping tests carried out. Geophysical studies are also performed to improve siting techniques and to find means of predicting the yield of boreholes (Grisseman and Ludwig 1986). Environmental isotope and hydrochemical studies are the only



*I. Simmers (ed.), Estimation of Natural Groundwater Recharge, 29–45.*

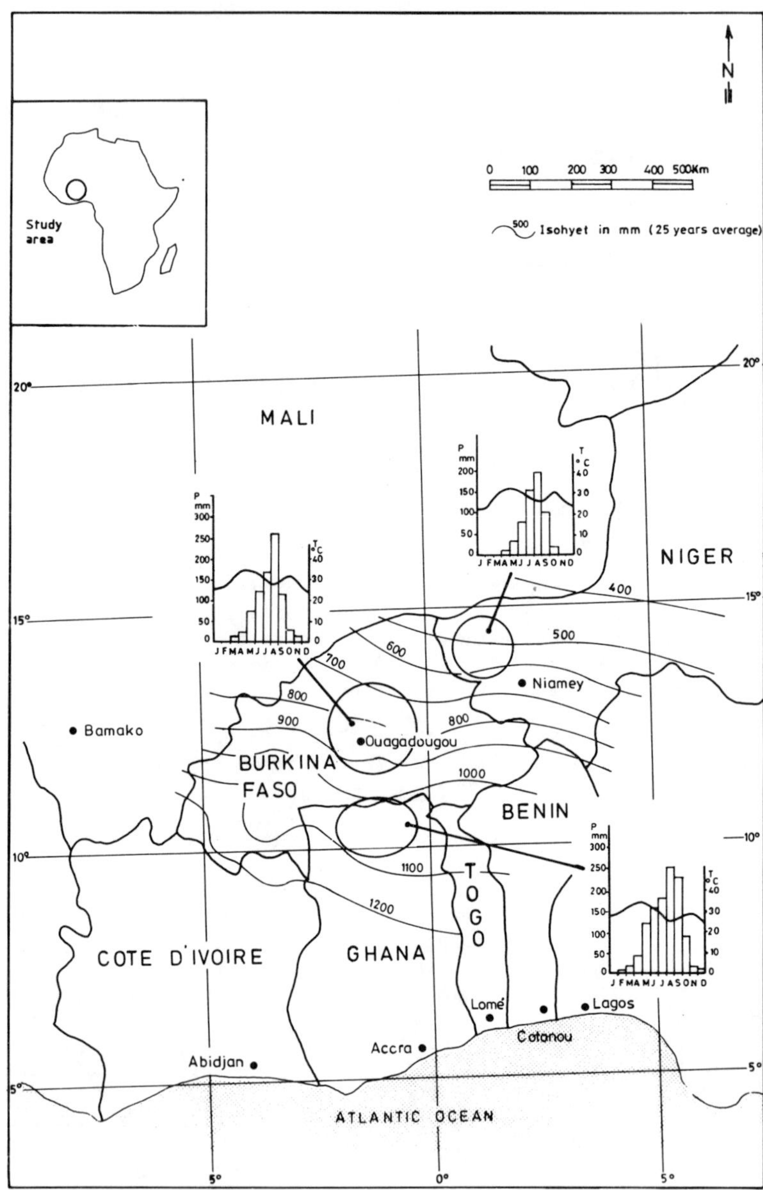


Figure 1 Location map and climatological data

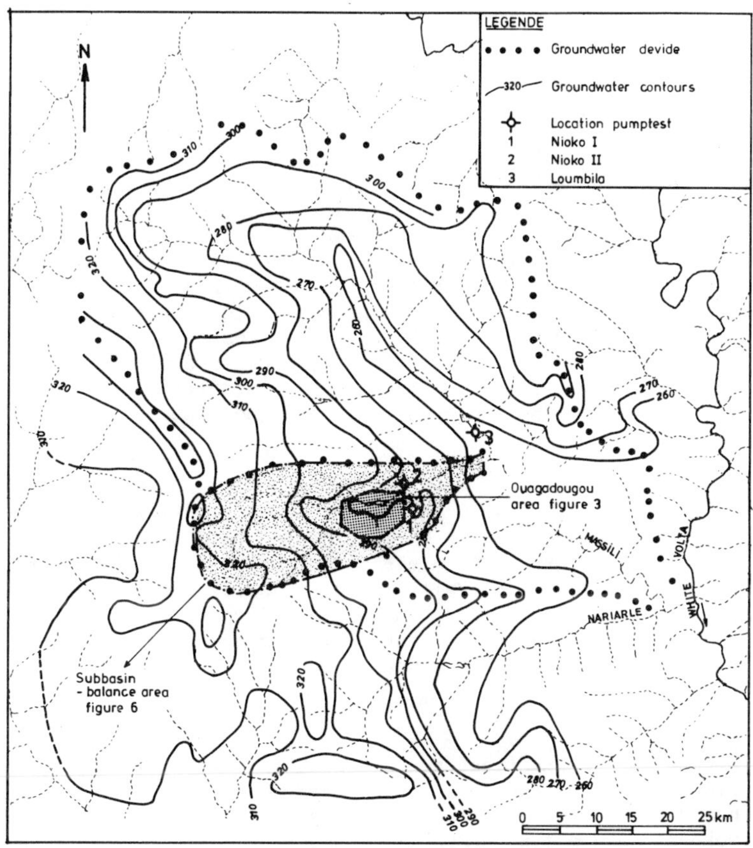


Figure 2 Groundwater level contour map of the Massili basin

really regional studies in the area to date. The larger interest in small scale studies compared to regional studies stems not only from lack of data but also from the generally held opinion that hard rock aquifers are highly discontinuous (Diluca and Muller 1985) and do not permit a classical hydrogeological approach.

A study of the groundwater potential of the immediate surroundings of the capital of Burkina Faso, Ouagadougou, (Iwaco 1986) in which 50 new boreholes were made, together with 850 boreholes from recent and ongoing rural water supply programs (Iwaco 1987) has shown that the groundwater surface forms a continuous pattern comparable to unconsolidated aquifers.

This apparent continuity of weathered mantle and fractured bedrock aquifers on the crystalline basement will be discussed with respect to recharge.

## HYDROGEOLOGICAL FRAMEWORK

### Geology

Burkina Faso and large parts of the neighboring countries of Niger and Ghana are underlain by Precambrian igneous and metamorphic rocks forming part of the West African Shield. They consist primarily of granites, schists and basic rocks (amphibolite) . Intrusions of pegmatite, quartz and aplite are frequent as well as dolerites. The whole of the crystalline basement has been subjected to faulting and fracturing by Panafrican tectonic movements. Although extensive outcrops occur, large areas are covered by a more or less continuous weathered layer with a maximum thickness of over 100m but generally less than 30m. The degree of weathering is dependent on lithology and intensity of fracturing.

Recent surface deposits are of minor importance and consist mainly of alluvial deposits confined to stream beds, lateritic layers and sand-dunes, restricted to northern Burkina Faso and Niger.

### Climate

Rainfall is between 500 mm/y in the north and 1100 mm/y in the south (fig. 1). The average over a 50 year period in Ouagadougou is 850 mm/y. The area is characterized by a marked dry and wet season; 75% of the total annual precipitation falls in July, August and September. Potential evaporation exceeds 2000 mm/y. Actual evapotranspiration will be higher in the south where vegetation density is higher and also because the groundwater surface is located nearer to the surface than in the north.

### Aquifer system definition

Aquifers on the basement can be conceptualized as a two component system. One component of the aquifer, the weathered mantle, consists of a more or less homogeneous cover ranging from clay to clayey sand with a generally high porosity and a low permeability and thus having a predominant reservoir function. The second component is made up of fractured bedrock having a low porosity and high permeability and is thus strongly conductive. A hydraulic contact between the two components is present. In general, dug wells in the weathered mantle and boreholes in the bedrock, show equal groundwater levels.

The presence of groundwater in exploitable quantities is determined separately from climatological factors and soil infiltration characteristics, by the nature of the bedrock, the amount of weathering and the intensity of fracturing. In granitic rocks the weathered mantle will be more uniform and less clayey than in schists and will also frequently show a transition zone of decomposed rock between the clayey weathered zone and the unweathered bedrock. This transition zone is generally well developed in fractures.

In general, well yields are low, around 2.5 $m^3$/hr, although in rural water supply programs a yield of 0.7 $m^3$/hr is required for installation of a handpump. In 10% of wells, yields exceed 5 $m^3$/hr. High borehole yields are often related to the presence of quartz or pegmatite dikes and veins caused by a more intense fracturing than in the neighboring rock and a deeper weathering.

## GROUNDWATER LEVEL CONTOUR MAPS

As part of a study concerning the water supply of Ouagadougou a groundwater survey was carried out including a photo-hydrogeological study, exploration geophysics, test-hole drilling and long duration pump tests. In an initial phase a well inventory was carried out including more than 850 wells. The area of investigation (fig. 2 ) covered the basins of the Massili and Nariarle, both sub-basins of the White Volta river. The total surface area is about 4000 $km^2$ and the altitude lies between 275 and 350m above sea level. The Massili basin consist entirely of migmatites and granites and is covered with a weathered layer, on average, 20m thick. Only in the downstream part are outcrops present.

A regional groundwater level contour map was made of this area on the basis of topographic maps with a scale of 1:50.000 and 1: 200.000. Dry season water levels were taken of wells capturing the hardrock aquifer. Groundwater levels were found between 5 and 30m below surface. The map on figure 2, however, gives only a general picture as the accuracy of topographic height determination was only +/- 5m. Notwithstanding these large errors the map reveals a consistent picture of contours following the topography. Hydraulic gradients are small and directed towards the White Volta river. The drainage axis is a major NW-SE directed fault zone in the Massili river valley.

A detailed groundwater-level contour map (fig. 3) was made of the Ouagadougou area, with a surface area of 175 $km^2$. Topographic heights were obtained from 1:25.000 and 1:10.000 scale maps, the latter with a contour interval of 2m, and from altimeter measurements. An accuracy of +/- 2 m was thus obtained, suggesting a regional hydraulic connection within the aquifer system. Groundwater flow is directed towards the artificial lake in the northern part of the city and to the secondary stream channels which are determined by major fractures (fig. 4). The contour map shows that apparently no significant infiltration from the lake takes place. This was confirmed by isotope samples taken from wells near the lake which gave only slightly enriched oxygen-18 values. This lack of infiltration from the lake might be due to clogging of the preferent recharge openings.

The hypothesis of the existence of a regional flow system will be further discussed by way of a rough groundwater balance of a sub-basin of the Massili river.

## GROUNDWATER BALANCE

A simplified model of the groundwater balance for the sub-basin of the Massili basin of 325 $km^2$ in which Ouagadougou is located, is presented in figure 5. The equation is given by:

$$AR = EX + ET + DW + DH + dS \qquad (1)$$

in which:

AR = actual recharge
EX = groundwater extraction
ET = evapotranspiration
DW = lateral groundwater flow in the weathered zone
DH = lateral groundwater flow in the hardrock
dS = change in groundwater storage

Estimated numerical values of the above parameters are given in figure 6.

Actual Recharge

Actual recharge will be defined for the present study as the volume of water reaching the saturated zone.

Environmental isotope studies in Burkina Faso (Geirnaert et al 1984) and Niger (Ousmane et al 1983) show that in most groundwater samples tritium is present, giving evidence that recharge is actually taking place. Low oxygen-18 and deuterium values in these areas demonstrate that direct recharge took place during periods of intensive rainstorms as no

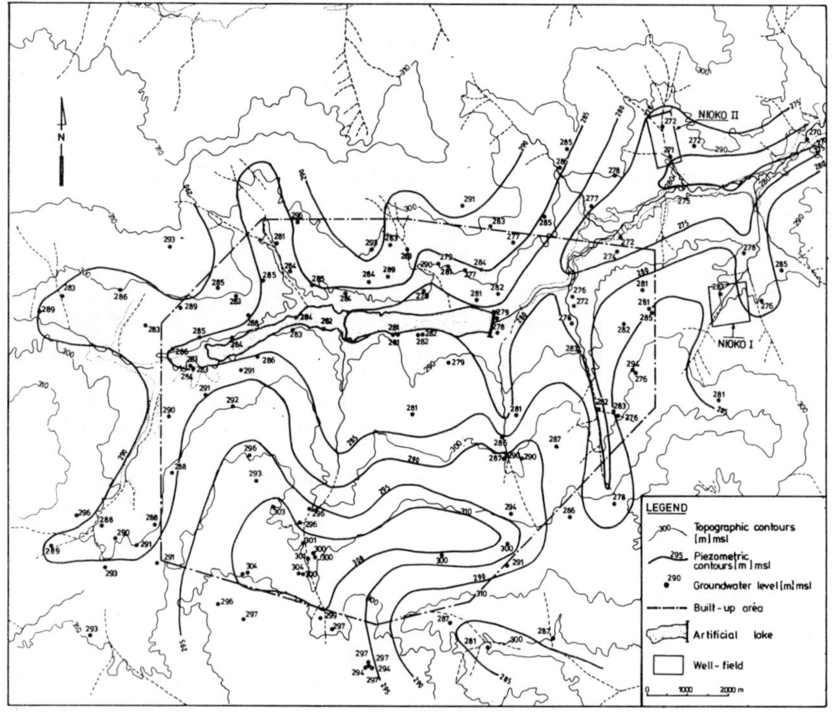


Figure 3 Groundwater level contour map of the Ougadougou area

evaporation is evident. Additional evidence was obtained from monitoring the groundwater chemistry during the rainy season which showed that the groundwater rise was accompanied by dilution. (van der Sommen and Geirnaert, in prep.)

In addition, annual fluctuations of the water table give evidence of natural recharge. In northern Ghana records of over 10 years are available for 30 observation wells (Wardrop 1980). Average fluctuations are strongly influenced by lower precipitation in the past decennium. Up to the rainy season of 1984 a continuous lowering of water level took place; due to increased precipitation the level has recovered on average 1m since then and is now only 1.3 m lower than the average end of rainy season levels in the nineteen-seventies.

Estimates of natural recharge based on water balance calculations, water table fluctuations and isotope analysis all show values in the range of 2 to 16% of total annual precipitation, ie. 17 to 136 mm/yr. In Ghana a value of 3.5% corresponding with 35 mm/y was derived from the study of the long term behavior of urban well fields (Wardrop 1980). For the present discussion we will take 2 to 3 % as a conservative estimate

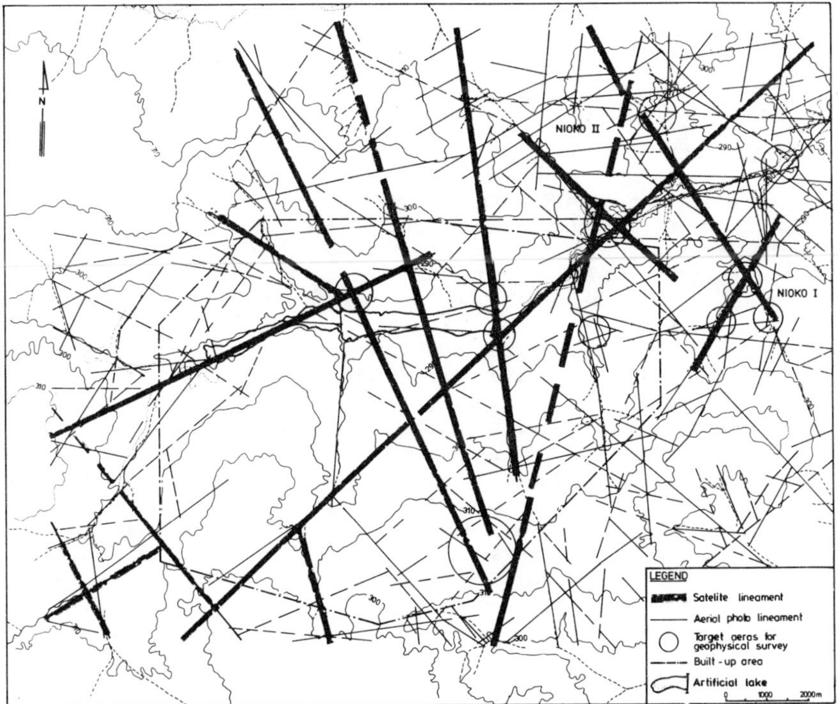


Figure 4 Lineaments and targets for groundwater exploration

for recharge in the Massili basin, corresponding to a water depth of 17 to 25 mm/yr.

Groundwater extraction

Groundwater extraction by boreholes and dug wells within the sub-basin is estimated at 0.7 to 1.2 x $10^6$ $m^3$/yr, which corresponds to a water depth of 2 to 4 mm/yr.

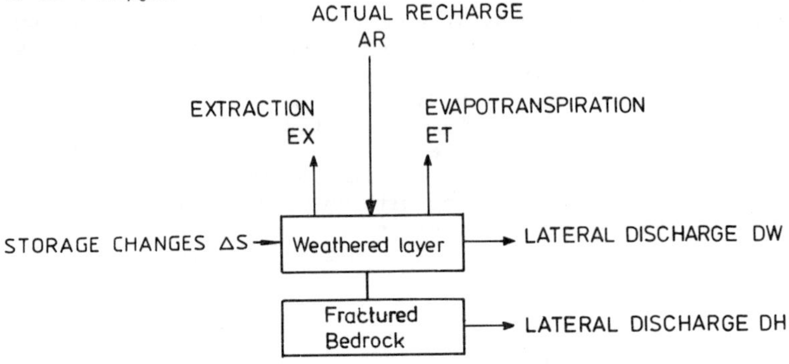


$$AR = EX + ET + DW + DH + \Delta S$$

Figure 5 Simplified recharge-discharge model

Evapotranspiration

Few data on evapotranspiration from groundwater in Burkina Faso are available. Chloride concentration in groundwater can however, give some idea of the importance of evapotranspiration. Chloride input in the groundwater system originates from rainfall as no lithological Cl sources are present. In rainfall and throughfall a maximum of 1 mg/l Cl is given by Roose et al (1981). If evapotranspiration is an important discharge mechanism than much higher values would be expected in groundwater. In Central Burkina Faso 72 chloride determinations were available (Iwaco 1986) ranging from 0.5 to 7.8 mg/l. Concentrations above 5 mg/l could be related to contamination as nitrate values were also high. The average value for uncontaminated samples is 2.3 mg/l (Table I).
In southwestern Niger low chloride values were also encountered with an average of 3.0 mg/l. In Ghana's northern region higher values were reported (Wardrop 1980). The average of 29 representative samples from more than 1000 chemical analyses was 9.3 mg/l with a range of 2 to 25 mg/l. It must, however, be noted that nitrate contents were also higher than in Burkina Faso so that some contamination cannot be excluded. As groundwater levels are higher than in Burkina Faso (average 6m below ground level) and vegetation is denser, evapotranspiration is also expected to be higher.

Chloride concentrations in groundwater in Burkina Faso show that evapotranspiration is not likely to be an important discharge process. Its importance seems to increase from north to south with higher groundwater tables and denser vegetation.

Lateral groundwater flow in the weathered mantle

Groundwater flow in the weathered mantle can be estimated by taking an average hydraulic gradient of 1 : 600 (fig. 3) and a transmissivity value of 20 $m^2$/day (maximum value from pumping test data). Drillers logs and geo-electrical measurements gave a more or less continuous alteration zone with an average saturated thickness of 10m. The outflow of the basin can thus be estimated over a section of 10 km at 0.12 x $10^6$ $m^3$/yr, corresponding to a water depth of 0.4 mm/yr. The lateral flow through the weathered mantle is therefore relatively unimportant.

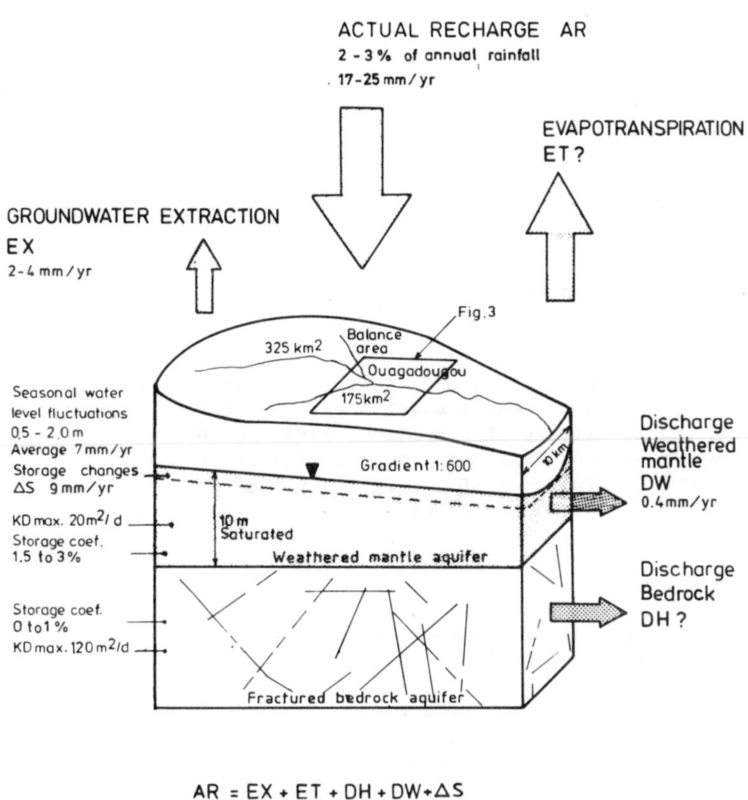


Figure 6 Water balance of a sub-basin of the Massili river

Lateral groundwater flow in the fractured bedrock

Groundwater flow in the fractured bedrock is difficult to quantify. Aquifer geometry, transmissivity and storage coefficients obtained from individual pumptests cannot be applied on a regional scale as boundary effects are almost invariably present.

For a given rock mass an equivalent porous medium permeability can be calculated (Boehmer and Boonstra 1986). A porous medium permeability of 21 m/day is equivalent to 3 fractures of $10^{-3}$ m width or 3000 fractures of $10^{-4}$ m width. In drillers logs the occurrence of open fractures of various cm width is often described, showing that overall permeability of the fractured rock mass can be considerable.

Change in groundwater storage

In general, in water balance calculations it is assumed that on a yearly basis groundwater storage changes are negligible. However it appears that during the last decennium the groundwater table in the Sahelian countries has continuously declined. Two observation wells situated in the Ouagadougou area clearly show a trend of declining groundwater levels since 1978. From fig. 7 the average yearly fall of the groundwater level is estimated at 0.45 m/yr. This represents a water depth of 9 mm/yr as groundwater extraction from dug wells and boreholes in this area is estimated at 2 to 4 mm /yr. the possibility that some of the groundwater level decline is caused by groundwater extraction or by a worsening of soil infiltration capacity characteristics due to urbanization cannot be excluded. The largest share of groundwater level decline must however, be attributed to the low precipitation totals during the years of measurement. In this respect it is interesting to note that during the poor rainy season in 1984 a small rise took place and during the 1985 and 1986 rainy seasons with a normal total amount of precipitation a large rise was recorded. The seasonal rise in groundwater level is on the average 0.8 m. With an estimated storage coefficient of 0.02 this represents a water depth of 16mm.

## DISCUSSION

Using the magnitude of the groundwater balance terms as presented above, a solution of equation (1) reads as follows:

$$ET + DH = AR - EX - DW + dS$$

in which:

AR = 17 to 25 mm/yr
EX = 2 to 4 mm/yr
DW = 0.4 mm/yr
dS = 9 mm/yr

Total amount of discharge by evapotranspiration and lateral groundwater flow in the fractures of the bedrock for the Ouagadougou area is

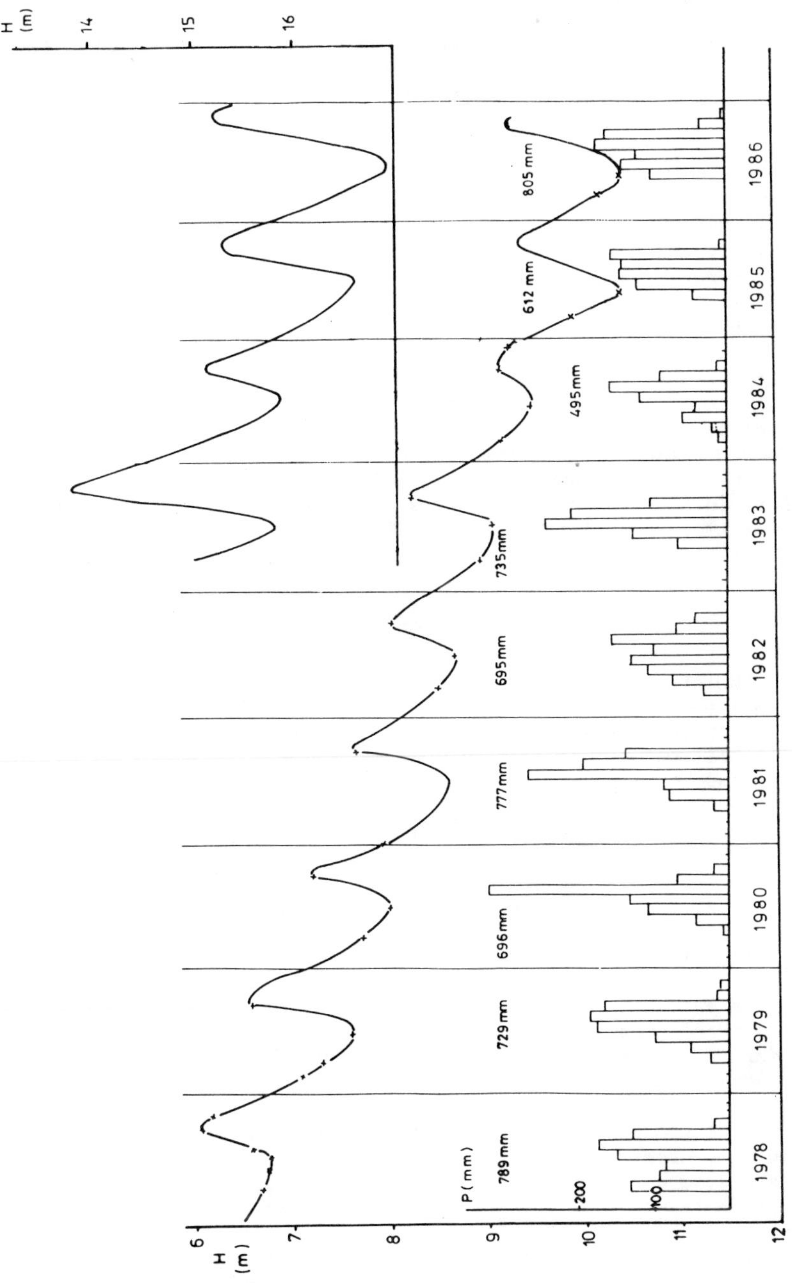


Figure 7 Hydrographs of the Ouagadougou area

approximately 25 mm/yr. As stated before, chloride data indicates low evapotranspiration rates of groundwater. We assume therefore that lateral flow through the fractures is an important discharge mechanism.

## CONTINUITY OF AQUIFER SYSTEM AND REGIONAL TREND

In table I a comparison is made between northern Ghana, the above discussed central region of Burkina Faso and southwestern Niger. The depth to water-table decreases to the south from 13 to 14 m in Burkina Faso and Niger to about 6 m in Ghana. The equal values in Burkina Faso and Niger are influenced by the kind of wells used for the comparison; in Niger rural water supply wells were used, located mainly in topographically low areas (small river valleys) and in Burkina Faso urban water supply wells, which were distributed more randomly. In the north the thickness of the saturated alteration zone is highly variable and often reduces to zero.

As a consequence of the higher water table in Ghana more water is available for evapotranspiration, resulting in a higher chloride concentration in the groundwater.

The increase in saturated thickness of the weathered mantle and the more intense weathering process, widening fractures and increasing overall permeability, will give towards the south more continuity of the aquifer system.

## SELECTION OF NEW WELL FIELDS FOR OUAGADOUGOU

In the study for the location of new well fields for Ouagadougou (Iwaco 1986) the hypothesis of the presence of a substantial lateral flow through the fractured bedrock was applied. Targets for geophysical exploration were selected in depressions in the piezometric contour map, corresponding to the valley bottoms of minor rivers. The geophysical survey was followed by an extensive test drilling program which resulted in the installation of two well fields: Nioko I with 4 wells of 15 to 46 $m^3$/hr and a total expected yield of 1500 $m^3$/day and Nioko II with 5 wells of 18 to 95 $m^3$/hr and also an expected total yield of 1500 $m^3$/day. In both well fields various observation wells were installed in both weathered mantle and bedrock and long duration pumptests carried out (table II). Both tests showed a radial flow pattern with a relatively fast contribution from the weathered zone (fig.8 and 9). The head difference between the observation wells in the weathered mantle and fractured bedrock indicates a vertical flow from the mantle into the bedrock. The importance of this vertical flow is described by Rushton (1986).

According to the groundwater level contour map good prospects for lateral groundwater flow are also present in the Massili valley proper. In this area, however, extensive outcrops are present precluding a thick saturated weathered mantle. Finding large yielding wells is thus more difficult. Geophysical prospecting and test drilling were also carried out in this area. One borehole was found with a good yield of 18 $m^3$/hr and a long duration pumptest was carried out (table II). After 96 hours of pumping the drawdown suddenly increased from 15.5 m to 39m. The level of

Table I Regional comparison of hydrogeological data

| | South | Central | North |
|---|---|---|---|
| | Ghana (1) | Burkina Faso (2) | Niger (3,4) |
| Number of wells: | 1581 | 238 | 137 |
| Static water level (m): | 6 | 14 | 13 |
| Weathered layer thickness (m): | >26 * | 30 | 19 |
| Saturated thickness weathered layer: | >19 * | 13 | 5 |
| Chloride concentration mg/l: | 9.3 (29) | 2.3 (72) | 3.0 (18) |

(29) number of chemical analyses
* wells did not reach bedrock
1) Wardrop (1980)
2) Iwaco (1986)
3) Iwaco (1983)
4) Ousmane ea. (1983)

Table II Data from pumping tests in the Ouagadougou area

| Well field: | Nioko I | Nioko II | Loumbila |
|---|---|---|---|
| Code number pumping well: | 30 | 41 | 17 |
| Duration pump test (hr): | 150 | 75 | 150 |
| Discharge ($m^3$/hr): | 18 | 20 | 15 |
| Transmissivity ($m^2$/day): | 60-70 | 40-75 | 10 |
| Draw down at end of pump test (m): | 9.6 | 8 | >39 |
| Storage coefficient | | | |
| - hardrock: | $4.10^{-4}$ | $3.10^{-4}$ | |
| - weathered mantle: | $2.10^{-2}$ | | |

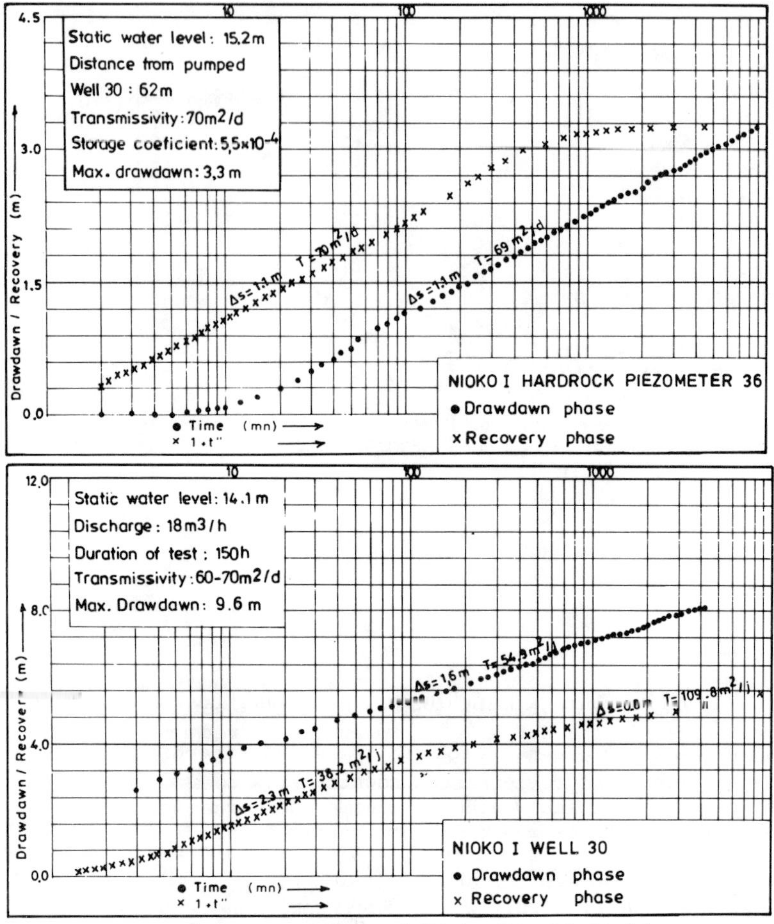


Figure 8 Pumping test of Nioko I well field

the sudden drop corresponds with the main water entry found during drilling. From piezometer data it is clear that no radial flow pattern is present here and that boundary effects occur (fig. 10).

CONCLUSIONS

Two components of one aquifer system can be distinguished in Burkina Faso: the alteration zone and the fractured bedrock aquifer. Where a saturated weathered mantle is present over large areas the aquifer may be considered as continuous. It was found that even in regions with a thin or absent saturated weathered mantle, continuity in the aquifer system must be present.

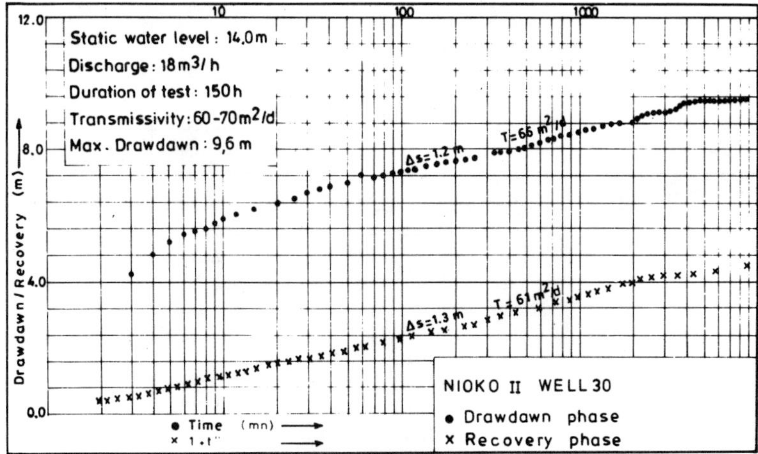


Figure 9 Pumping test of Nioko II well field

Fluctuations of the groundwater table and piezometric contour maps have shown that a regional recharge-discharge system is present. A rough groundwater balance in the central part of the country and chloride concentrations in the groundwater provide evidence that the main discharge system is lateral groundwater flow. This flow is concentrated in the fractured bedrock as permeabilities of the weathered mantle are very low. The volume of lateral flow is of the same order of magnitude as the actual recharge. In the north, with lower actual recharge and only a thin or absent saturated weathered mantle, continuity of the aquifer system is less evident but regional groundwater flow does occur.

In the south, with a thick weathered mantle and higher groundwater tables, weathering processes are more intense resulting in a higher degree of continuity.

The study for the future water supply of Ouagadougou has shown that for high yield urban water supply wells, both components of the aquifer system should be well developed, ie. a thick saturated weathered mantle and fractured bedrock with good lateral continuity. Exploration for urban wells should therefore be focussed on first delineating areas with a thick saturated mantle and subsequently locating within this area the broken zones.

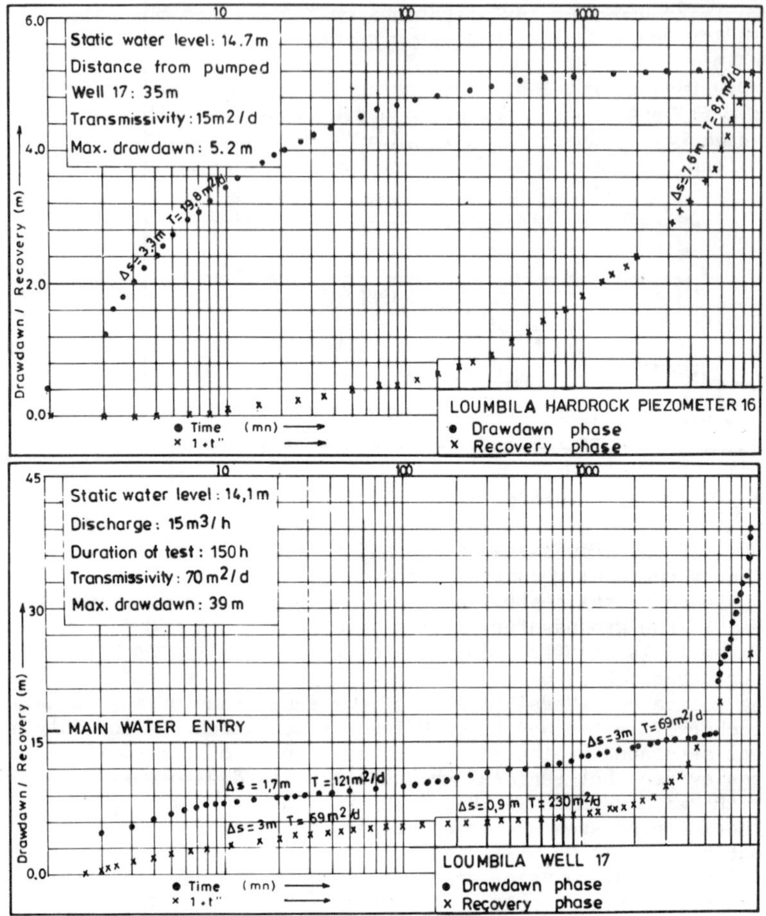


Figure 10 Pumping test of Loumbila well

LITERATURE

Boehmer, W.K.and Boonstra, J. 1986. Flow to wells in intrusive dikes. PhD thesis, Amsterdam Free University.

Boukari, M., 1984. `Contribution a l'étude hydrogéologique des régions du socle de l'Afrique Intertropicale: l'exèmple de Dassa-Zoume (Bénin méridional) ' Jln African Earth Sciences. Vol. 6 no. 3.

Diluca, C. and Muller, W. 1985. Evaluation hydrogeologique des projets d'hydraulique en terrains cristallins du bouclier Ouest Africain. CIEH série hydrogéologie, Ouagadougou, Burkina Faso.

Geirnaert, W., Groen, M., van der Sommen, J.J. 1984. `Isotope studies as a final stage in groundwater investigations on the African Shield' Challenges in African hydrology and water resources. IAH publ. 144.

Grissemann, C., Ludwig, R., 1986. `Recherche des critères et d'une méthodologie d'implantation de forages d'hydraulique sémi-urbaine en zone cristalline du bouclier Ouest-Africain' Bundesanstalt für Geowissenschaften und Rohstoffe, Hannover.

Iwaco 1983. `Projet hydraulique villageoise dans la région du Liptako de la Republique du Niger'.

Iwaco 1986.`Approvisionnement en eau potable de la ville de Ouagadougou, étude hydrogeologique'

Iwaco 1987. `Hydrogeological report on the 600-wells program' in prep.

Leusink, A., Tiano, B., 1985. `Observations du niveau de la nappe des eaux souterraines et de sa composition chimique et isotopique en zone de socle cristallin au Burkina Faso.' Bull. de Liaison du CIEH. no. 62.

Ousmane, B., 1978. `Contribution a l'étude hydrogéologique des régions du socle du Sahel' these 3eme cycle, Universite du Montpellier.

Ousmane, B., Fontes, J.Ch., Aranyossy, J.F.,Joseph, A., 1983. `Hydrologie isotopique et hydrochimie des aquifères discontinus de la bande sahélienne et de l'Air (Niger)'. Isotope hydrology IAEA Vienna.

Roose, E., 1981. `Dynamique actuelle de sols ferrallitiques et ferrugineux tropicaux d'Afrique Occidentale'. Doc. Orstom no. 130.

Rushton, K.R., 1986.` Vertical flow in heavily exploited hard rock and alluvial aquifers'. Groundwater, Vol 24 no. 6.

Wardrop, W.L. and associates Ltd., 1980. `Ghana Upper Region water supply project'.

# GROUNDWATER RECHARGE ESTIMATION
## (Part 1)
## PHYSICAL/CHEMICAL METHODS

# A REVIEW OF SOME OF THE PHYSICAL, CHEMICAL AND ISOTOPIC TECHNIQUES AVAILABLE FOR ESTIMATING GROUNDWATER RECHARGE

G.B. Allison
CSIRO, Division of Soils
Private Bag 2, P.O.
GLEN OSMOND, S.A. 5064

## 1. INTRODUCTION

Groundwater recharge means different things to different people. For example, to an agronomist, water which moves beneath the root zone of crops represents a loss in yield and so should be minimised. Those who are interested in water resources, take the opposite view. A few of the reasons for studying natural groundwater recharge are: to determine the safe yield of a groundwater system; to assess the extent of development of secondary salinisation following land clearing; and, for those interested in storage of waste materials, to identify areas of very low groundwater recharge. Only natural recharge, either local or localized will be considered here. Local (or diffuse) recharge is defined as that reaching the water table by percolation of precipitation in excess of evapotranspiration, through the unsaturated zone. Localized recharge occurs following runoff and subsequent ponded infiltration through low-lying areas, streams or lakes.

Although artificial recharge and that induced by irrigation practices will not be discussed here, many of the techniques mentioned are applicable to evaluation of these.

This paper is concerned with chemical, isotopic and physical methods for estimating groundwater recharge in semi-arid areas (where annual precipitation is less than about 700 mm). In such areas, the native vegetation has often developed root systems such that almost all of the precipitation is consumed by evapotranspiration. As a result, the local recharge flux is very low, but constant in time. Following clearing of native vegetation and establishment of comparatively shallow-rooted pastures and crops, local recharge increases, and, because of the variability of rainfall in many semi-arid areas, it will also be highly variable in time. In these cases, physical methods, which rely on direct measurements of hydrological parameters, are problematic because, in the first case, the fluxes are low and changes in these parameters will be small and difficult to detect. In the second case, the variability is such that measurements must be made over several years to obtain an estimate of mean values. Thus, for estimation of local recharge in semi-arid areas, chemical and isotopic methods show more promise than physical methods and this review is biassed towards that prejudice. Physical methods assume more



*I. Simmers (ed.), Estimation of Natural Groundwater Recharge, 49–72.*

importance in wetter areas and for estimation of artificial recharge where the fluxes are higher and the inputs more constant in time.

Many of the groundwater systems of interest contain water of very great residence time, for example, aquifers beneath the deserts of northern Africa (Burdon, 1977; Lloyd and Farag, 1978) and the Great Artesian Basin in Australia (Airey *et al.*, 1983). In the former studies, it was suggested that the present hydraulic head distribution is a relic of recharge in the Pleistocene and in the latter chlorine-36 dating has suggested groundwater residence times of the order of $10^6$ years. Thus, if we are to be able to manage these and similar large groundwater systems it is necessary to obtain information about palaeo-recharge.

In computing the water balance of an aquifer, estimates of discharge may lead to indirect estimates of groundwater recharge. Like recharge, such discharge may be localised, as spring or river flow, or diffuse and occur over large areas where the water table is at or near the surface.

Valuable qualitative information about sources of groundwater and mechanisms of recharge can be obtained by comparing the water chemistry or isotopic composition of possible water sources. Although such studies will be referred to, the qualitative aspects of recharge estimation will be stressed in this paper.

Recent reviews of a range of techniques for, and problems associated with, estimation of groundwater recharge have been given by Peck (1979), Lloyd (1981) and Allison (1981). In addition, Keith *et al.* (1982) have prepared a bibliography of recharge studies in semi arid areas. In their review on the use of environmental isotopes for modelling in hydrology, Dincer and Davis (1984) outline some of the techniques available for recharge estimation using these isotopes.

Below I cover, non-exhaustively, some aspects of both local and localized recharge. The important areas of palaeo-recharge and groundwater discharge are not discussed.

## 2. LOCAL RECHARGE

To estimate local recharge some techniques rely on measurements made in the unsaturated zone and for others, in the saturated zone. In this section, I will concentrate on the unsaturated zone as techniques based on the saturated zone may also be applicable to localized recharge and these are discussed in Section 4. One of the major problems in studying the unsaturated zone in semi-arid areas is that measurements need to be made beneath the zone where uptake of water by roots is significant. For annual crops and pastures, this is not a great problem, as rooting depths usually will be less than 2 m and loss of water from depths greater than this will be negligible. However, native vegetation may have living roots to great depth. We have found healthy roots 10 mm in diameter 17 m beneath the surface in a region where the mean annual rainfall is ~300 mm $a^{-1}$.

## 2.1 Physical Methods for Estimation of Local Recharge

In principle, one of the simplest methods which has been used for estimating local recharge, R, is the use of empirical expressions of the type

$$R = k_1 (P - k_2), \tag{1}$$

where P is precipitation and $k_1$ and $k_2$ are constants for a particular area. Such expressions have been used with varying degrees of success, and are probably most useful for making "first-guess" estimates of recharge where annual recharge is fairly high, say greater than 50 mm $a^{-1}$.

Methods which rely on estimates of soil physical parameters fall into 3 main classes, being

(i) soil water balance

(ii) zero-flux plane method

(iii) estimation of water fluxes beneath the root zone using unsaturated hydraulic conductivity and the gradient in water potential.

An extensive literature is available on these techniques and it will be touched on only briefly here.

(i) *The soil water balance* can be represented, in the absence of significant surface runoff, by:

$$R = P - E_a + \Delta S \tag{2}$$

where R is recharge, P is precipitation, $E_a$ is actual evapotranspiration and $\Delta S$ is the change in soil water storage, which usually may be ignored if calculations are made on an annual basis. A range of techniques for estimation of $E_a$ based on Penman-type equations (Howard and Lloyd, 1979; Rushton and Ward, 1979) and other methods (e.g. Alley, 1984) have been used. The data requirement of these methods is large and Howard and Lloyd (1979) suggest that large errors can occur unless the accounting period is less than 10 days. Several of the models used require a root constant which is related to the soil moisture deficit established in the soil and many workers have increased their estimates of recharge by reducing this parameter. This often leads to prediction of hydrological phenomena which are not observed in the field (Rushton and Ward, 1979). The water balance technique has been used extensively in temperate areas. However, it is unlikely to be successful in semi-arid areas because of long periods of less than potential evaporation, when errors in $E_a$ are greatest and P and $E_a$ are nearly equal.

(ii) *The zero-flux plane (ZFP) method* relies on location of a plane of zero hydraulic gradient in the soil profile. Recharge over a time interval is obtained by summation of the changes in water content below the plane. The position of the ZFP is usually located by installation of tensiometers. Unfortunately the method breaks down in periods of high infiltration when the hydraulic gradient becomes positive downwards throughout the profile. In periods when this occurs either a soil water balance, or the techniques described in (iii) below, must be used (Wellings, 1984; Sophocleous and Perry, 1985).

Use of this technique can give good estimates of recharge for periods of the year when the ZFP exists as has been shown by studies in the UK (e.g. Wellings, 1984). However, it suffers from the same problems as (i) or (iii) if annual values of recharge are required, as either of these techniques must be used when the ZFP disappears.

(iii) Two detailed and recent studies have reported attempts to use *unsaturated zone hydraulic conductivity*, K(θ) or K(ψ), and *hydraulic potential data*, ψ(θ), to solve Darcy's equation in the unsaturated zone, and hence estimate soil water flux over periods in excess of 12 months (Sophocleous and Perry, 1985; Stephens and Knowlton, 1986). The latter paper is particularly appropriate for this workshop as it reports measurements at a semi-arid site where the precipitation is ~180 mm $a^{-1}$.

If the water flux is calculated at such a depth in the profile that no further abstraction by roots occurs, then the flux will be equal to groundwater recharge. Both of the studies referred to above were carried out in unconsolidated sands, which should be ideal for making soil physical measurements. However, both studies report that their recharge estimates should only be regarded as rough approximations. Particular problems noted were that the K(θ) or K(ψ) relationships were difficult and time consuming to determine, both in the field and in the laboratory, with difficulty and uncertainty increasing with soil dryness. Slight differences in measured water content translate into large differences in unsaturated hydraulic conductivity. Hysteresis in the K-ψ relationship is also a problem and Stephens and Knowlton use K-ψ values obtained during drainage and they note that the recharge they calculate during the wetting cycle would be an over-estimate of the actual value.

The work of Stephens and Knowlton highlights the problem of using these techniques for estimation of groundwater recharge, especially where the fluxes are low. They found that the annual recharge flux was either 7 or 37 mm $a^{-1}$ depending on how they computed their mean hydraulic conductivity.

Lysimetry has long been used successfully to make direct estimates of groundwater recharge in temperate areas. However lysimeters are expensive and permanent instruments, which, unless they extend to the water table, will modify the soil moisture regime in the soil because drainage can occur only when a water-table develops at the base of the lysimeter. This last factor, however, is unlikely to be a problem if the lysimeter is relatively deep and the vegetation shallow-rooted as, for example, that described by Kitching and Shearer (1982). Alternatively, a suction can be applied at the base of the lysimeter in order to simulate more closely the natural environment.

All of the techniques discussed above present difficulties in application in arid and semi-arid areas, principally because recharge fluxes are likely to be low and highly variable from year to year, necessitating long periods of measurement. Lysimetry has the potential to overcome the problems of low flux, and if deep enough, variations from year to year, but the problems of often great lateral variability in local recharge, expense and modification of the soil moisture regime remain unsolved. However, all of these techniques can yield valuable information about recharge mechanisms. For example, Wellings (1984) tensiometer data show that matrix flow dominated above the Chalk aquifer in England over his period measurement, by showing that the matric potential of soil water was almost always more negative than that required to permit fissure flow. Many lysimeter studies have given information about the mechanisms of solute and water transport (for example Stichler *et al.*, 1984; Fontes, 1983).

## 2.2 Natural Tracers for Estimation of Local Recharge

Natural tracers have both advantages and disadvantages over the methods described in Section 2.1. On the positive side they represent a spatially uniform (at least to a first approximation) input to the soil water/groundwater system. In many cases, a history of fallout in precipitation is known, and since the aquifer and the overlying unsaturated zone usually store sufficient water to represent many years of recharge, an historical record of recharge may be derived. The principal disadvantage of isotopes is that they may offer only an indirect measure of recharge (cf iii above), and the mechanism of infiltration may affect interpretation of results.

The natural tracers most commonly used in recharge studies are $^{3}H$, $^{14}C$, $^{15}N$, $^{18}O$, $^{2}H$, $^{13}C$ and chloride. Of these, the first two are radioactive, with half lives of ~12.3 and 5700 years respectively. Their concentrations in input to the hydrologic cycle have been modified greatly by nuclear testing. Chlorine-36, also radioactive, naturally occurring and produced by nuclear testing has been used in only a few studies, but it may become of wider interest as more analytical facilities become available. Its half life of $\sim 3 \times 10^{5}$ years makes it attractive for studying groundwater systems with low recharge rates. Input concentrations of the other isotopes mentioned above have also changed in time, but over a much longer time scale, due to changes in temperature and rainfall patterns. As far as I am aware, little is known of the temporal changes in the fallout of chloride.

Of the tracers mentioned above, $^{3}H$, $^{2}H$ and $^{18}O$ most accurately simulate the movement of water because they form part of the water molecule. In most soils $Cl^{-}$ and $NO_3^{-}$ move as the water does, but in some heavier textured soils anion exclusion may be a problem (Gvirtzman *et al.*, 1986), and the tracer will appear to move more rapidly than the parcel of water being traced.

During percolation of water through the unsaturated zone, the water in the large pores will move more rapidly than that in the small pores. However, because the distance between large and small pores is usually small, diffusion will cause the concentration of a tracer to become equal in all pores (Zimmermann *et al.*, 1967). As a consequence of this, all molecules of a perfect tracer will move downward with the same apparent velocity. In this way piston flow results, and the most recent water input will overly that already in the system.

Such flow models can be used with success in describing the flow in the unsaturated zone. For example, the diffusional redistribution of tritium away from fissure or fracture patterns which give rise to the secondary porosity in the English Chalk, can be used to explain the differences between the hydraulic response of the aquifer and the observed tritium profiles in the unsaturated zone (Foster, 1975). In other materials with bimodal porosity, derivatives of this type of model are being used widely to explain the distribution of solutes in the unsaturated zone (Grisak and Pickens, 1980).

Although piston flow is often able to explain the behaviour of tracers in the field, there is an increasing number of examples which suggest that water movement in the unsaturated zone occurs along preferred pathways, especially when deep-rooted native vegetation is present. For example, Fig.1 shows depth profiles of tritum in water of the unsaturated zone at two sites in Australia. At each site, tritium profiles were measured beneath native forest and at an adjoining location where the native forest had been cleared for pasture many years earlier. In each case, independent evidence suggested that the recharge rate beneath the pasture was considerably higher than that beneath the native vegetation. However, at both sites shown in Fig.1, measureable tritium occurs to a greater depth, and the profiles contain more tritium beneath the native vegetation. Of particular interest are the profiles shown in Fig.1a, where recharge beneath the native vegetation was found to be ~0.1 mm $a^{-1}$ and measurable tritium was found to ~12 m. Beneath the nearby pasture recharge was much higher (~3 mm $a^{-1}$), but no tritium was found beneath 3 m (Allison and Hughes, 1983). These data are strongly indicative of preferred flow of water along living roots and make interpretation of tritium profiles (see next section) very difficult.

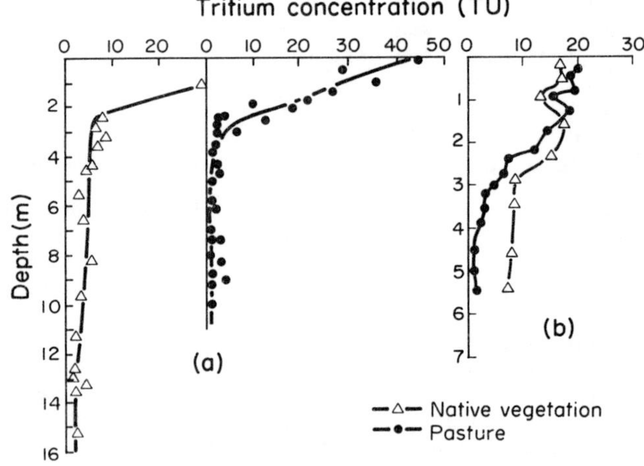


Fig. 1 Depth profiles of tritium concentration in soil water at two sites in Australia. Mean annual recharge beneath pasture: (a) 3 mm $a^{-1}$, (b) 60 mm $a^{-1}$. (a) is from Allison and Hughes (1983).

Another example of preferred flow paths, principally along channels of living roots is given by Johnston *et al.* (1983), who studied the movement of dye tracer beneath a native forest in Western Australia.

2.2.1 *Tritium.* Many studies on the estimation of recharge using natural tritium in the unsaturated zone have been given in the literature (Smith *et al.*, 1970; Andersen and Sevel, 1974; Allison and Hughes, 1974; Bredenkamp *et al.*, 1974; Foster and Smith-Carington, 1980). Almost all of the studies reported have made use of the fact that the peak in tritium in precipitation due to nuclear fallout has been preserved in the unsaturated zone. Fig.2 gives a comparison of the depth profiles of tritium obtained in both the northern and southern hemispheres together with the tritium fallout in rainfall at the two sites. Estimates of mean annual recharge have been made at both sites even though the tritium fallout has been much lower in the southern hemisphere.

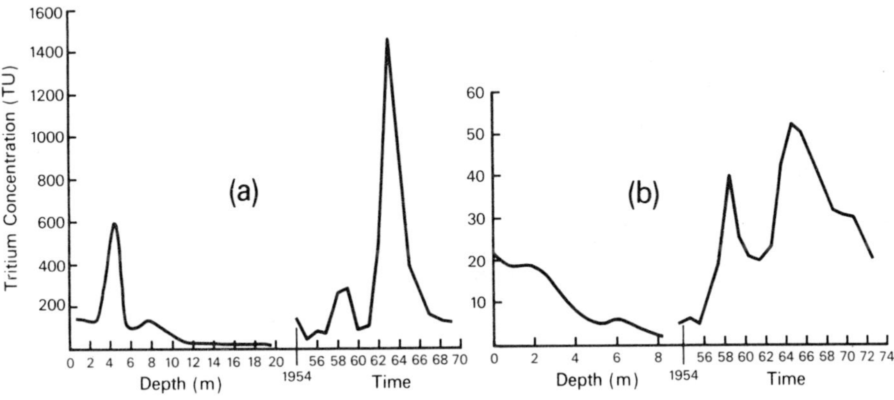


Fig. 2 A comparison of depth profiles of tritium in soil water with tritium fallout in rainfall for (a) the northern hemisphere (after Smith *et al.*, 1970) and (b) the southern hemisphere (after Allison and Hughes, 1974).

As Smith *et al.* (1970) showed, besides enabling estimates of recharge to be made, tritium profiles can also be used to give information about mechanisms of flow in the unsaturated zone. They showed that about 15% of the recharge water moved rapidly down the profile, bypassing the layered inputs which accounted for the remainder of the recharge flux.

Three techniques have been used for estimating recharge rate from tritium profiles in the unsaturated zone:

(i) from the position of the fallout peak. This necessitates that the recharging water moves by piston flow, as it is assumed that the water in the profile above the fallout peak represents recharge since the time of the fallout peak. Any bypass or fissure flow will result in the recharge rate being under-estimated. Because much of the water in the root zone will be used by the vegetation and will not become recharge, the amount of water in the profile should be estimated at the time of year when the soil moisture deficit is at its maximum.

(ii) from the shape of the tritium peak in the soil. In principle, this should be a more reliable method than that described above, as information about flow mechanisms can also be obtained. However it is necessary to develop a model which either describes the tritium concentration of water lost from the soil surface (Smith *et al.*, 1970) or estimates of the tritium input function to the unsaturated zone at the bottom of the root zone (Allison and Hughes, 1974). In order to obtain estimates of mean annual recharge $\bar{R}$, a weighting function $\omega_i$ which takes into account year-to-year variation of recharge $R_i$ $(=\omega_i\bar{R})$ is needed. Comparison of calculated and observed tritium profiles are made for a range of values of $\bar{R}$ to obtain the best fit. Smith *et al.* (1970) obtained values of $\omega_i$ by estimating evapotranspiration; however, in many semi-arid areas this is not possible, and other weighting functions must be developed. Expressions based on eq. (1) or fluctuations in groundwater level have been used. A comparison of several possible weighting schemes has been given by Allison and Hughes (1978).

(iii) from the total amount of tritium, $T_t$, stored in the profile. This is given by

$$T_t = \int_0^\infty T(z)\theta(z)dz, \qquad (3)$$

where $T(z)$ is the tritium concentration of water in the unsaturated zone at distance z beneath the surface, and $\theta(z)$ the volumetric water content. In practice, this integration should be carried out to a depth such that water representing pre-1950 recharge is intersected. Mean annual recharge can then be estimated using

$$T_t = \bar{R} \sum_{i=1}^{\infty} \omega_i T_{pi} \exp(-t\lambda), \qquad (4)$$

where $T_{pi}$ is the tritium concentration of recharge water i years before the present and $\lambda$ is the decay constant for tritium.

The technique described in (iii) is probably the most useful for routine use, as the estimate of $\bar{R}$ is independent of the distribution of tritium in the profile and thus non-ideal percolation of water can be coped with. As well as this, few samples need to be analysed, as the detailed depth - tritium concentration profiles needed by method (ii), are not required.

To estimate recharge from tritium profiles, the tritium concentration of effective input to the soil water system needs to be known. A promising method, discussed recently by Grabczak *et al.* (1985), is to make use of seasonal variations of the $^{18}O$ and $^{2}H$ content of precipitation to estimate which precipitation is responsible for recharge and hence obtain a tritium input function.

In the methods i-iii above, it has been assumed that the unsaturated zone has been deep enough, or recharge small enough, such that 20-30 years recharge can be held above the water-table. For cases where this is not so, and recent recharge reaches the water table, Atakan *et al.* (1974) have demonstrated how recharge estimates can be obtained by detailed sampling beneath the water-table and taking lateral flow into account.

Recent work by Gvirtzman and Margaritz (1986) has demonstrated the usefulness of tritium profiles for estimating recharge beneath an irrigated area. Their tritium profiles show remarkable depth discrimination and enable annual recharge cycles to be seen. Such discrimination indicates a very low effective diffusion coefficient and hence a low value for the tortuosity (0.05-0.11). Another interesting outcome of this study is the very high percentage (~50%) of "immobile water" which is not readily exchanged with the "mobile" water. This leads to tritium profiles which are quite different from those predicted by piston flow.

The techniques described here are not valid in areas where the root zone is deep and flow regimes like those suggested by Fig.1 occur. Also, it has been suggested that in arid areas, where soils are sandy and have a high air-filled porosity, it is possible for water vapour of high tritium concentration to diffuse into soil water or groundwater systems. Estimates of recharge made using tritium in such areas therefore would be anomalously high. Careful consideration should be given to this possibility when interpreting data such as those given by Dincer *et al.* (1974).

2.2.2 *Chloride*. Input of chloride occurs at the soil surface both in rainfall and as dry fallout. Samples which have been collected from most stations include both wet and dry fallout, although the dry deposition may be modified by the collection vessel. The chloride may be of either atmospheric or terrestrial origin but, for the hydrologic studies described here, the net accession is required. Thus in areas which are uniform laterally it is possible that only chloride of oceanic origin should be considered. Hutton (1976) showed that chloride of atmosphic origin dominated up to 300 km from the coast in southern Australia. Blackburn and McLeod (1983) suggested that about 50% of the chloride deposition at distances of the order of 700 km from the coast was of oceanic origin. Thus the location of an area of interest must be taken into account when using chloride to estimate recharge.

Most plant species do not take up significant quantities of chloride from soil water, thus chloride is concentrated by evapotranspiration in the root zone. In the simplest case of piston flow of water in the unsaturated zone, the chloride concentration in soil water should increase through the root zone to a constant value (Gardner, 1967). Provided the water table is deep or has the same chloride concentration as the soil water, a depth-concentration profile such as shown by curve (a) in Fig.3 should result.

Under steady state conditions, following Eriksson and Khunakasem (1969), the flux of chloride at the surface is equal to the flux of chloride beneath the root zone, thus

$$\overline{PC}_p = \overline{RC}_s, \tag{5}$$

where $C_p$ and $C_s$ are the chloride concentrations in precipitation and soil water respectively. Since $\bar{R}$ is required, and because there will be strong correlation between R and $C_s$, the mean flux beneath the root zone can be written as:

$$\overline{RC_s} = \bar{R}\bar{C}_s + \overline{\Delta R \Delta C_s} \qquad (6)$$

where Δ, refers to the deviation from the mean. Combining eq.5 and 6 gives

$$\bar{R} = (\overline{PC_p} - \overline{\Delta R \Delta C_s})/\bar{C}_s. \qquad (7)$$

Allison and Hughes (1978) found that $\overline{\Delta R \Delta C_s}$ was 10-15% of $\overline{RC_s}$ in an area where rainfall is ~700 mm $a^{-1}$ and recharge is 50-100 mm $a^{-1}$. However, in more arid areas where recharge is expected to be lower and more variable, this term may be considerably larger.

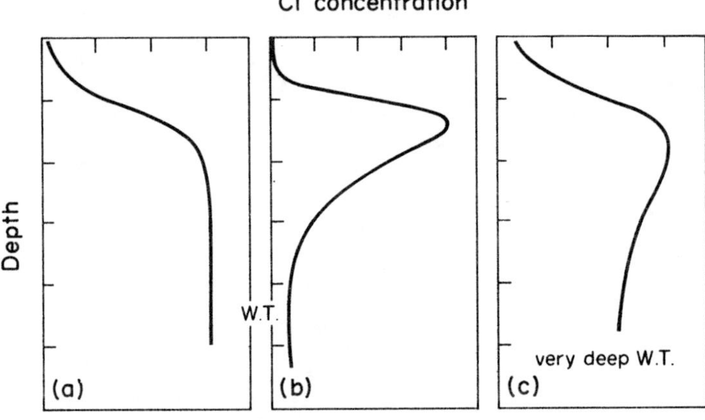


Fig. 3 Schematic depth profiles of the chloride concentration of soil water
(a) Piston flow with abstraction of water by roots
(b) Abstraction of water by roots, but with either preferred flow of water to beneath the root zone, or diffusive loss of chloride to the water table
(c) A profile which may reflect the recharge history of the site.

This chloride mass balance technique in the unsaturated zone has been used successfully to evaluate recharge in a range of environments (Edmunds and Walton 1980; Allison and Hughes, 1978; Sharma and Hughes, 1985). However, many of the depth profiles of chloride concentration in soil water show a more complex shape than that shown as curve (a) in Fig.3. Some idealized examples of these more complex shapes are given for comparison in Fig.3(b and c). Examples of a range of field profiles is given by Dimmock *et al.* (1974).

Peck *et al.* (1981) and Johnston (1983) developed several models in an attempt to interpret "bulge-type" chloride profiles (see Fig.3(b) for example) in terms of recharge rate. The equation which forms the basis of their models is

$$\theta \frac{\partial C}{\partial t} = \partial(D_s \frac{\partial C}{\partial z}) / \partial z - q_w \partial C/\partial z + S - CW/\rho \qquad (8)$$

where $\theta$ is volumetric water content
$C$ is chloride concentration in soil water
$t$ is time
$D_s$ is the effective diffusion coefficient of chloride in soil water
$q_w$ is the volume flux density of soil water
$z$ is depth
$S$ is solute source-strength $[ML^{-3}T^{-1}]$
$W$ is water source-strength $[ML^{-3}T^{-1}]$
$\rho$ is density of the soil solution

When chloride concentrations in the profile were low (~2000 g $m^{-3}$ at the peak), the diffusive contribution to the solute flux was considerably less than that due to convection. In this case $W<0$ above the peak in chloride and $W>0$ below it. This was interpreted as being due to extraction of water by roots to 5-7 m depth, and beneath this, flow in preferred pathways which bypass the soil matrix.

When chloride concentrations in the soil solution are higher (>8000 g $m^{-3}$) and recharge rates are correspondingly lower, the diffusive transport of chloride is comparable with that by convection. Whereas, in the previous case, it is not necessary to have accurate information about $D_s$, in this case its value is critical to interpretation of the profile in terms of water flux. The effective diffusion coefficient of chloride in soil water can be written as

$$D_s = \theta \tau D \qquad (9)$$

Where $\tau$ is tortuosity and D is the diffusivity of chloride (usually as NaCl) in free water. In the unsaturated zone $\tau$ varies over a wide range and this highlights the difficulty in interpreting "bulge-type" profiles in semi-arid areas where recharge fluxes may be low. The literature on the relationship between $D_s$ and water content has been reviewed recently by C.D. Johnston (unpublished thesis, 1987).

Further work on interpretation of the "bulge" type profiles in terms of water flux, has been reported by Watson (1982).

As mentioned earlier, perennial native vegetation usually develops much deeper root systems than the annual vegetation which often replaces it following agricultural development. Allison and Hughes (1983) found chloride profiles of the type shown as (c) in Fig.3, beneath native eucalyptus vegetation where maximum chloride concentrations in the soil solution were ~14000 g $m^{-3}$. Very similar profiles were found at several locations beneath the native vegetation. Chloride profiles, taken beneath land cleared for agriculture about 80 years previously, again showed very similar profiles, but with the very high concentrations displaced deeper in the profile. This provides good evidence of piston flow and these workers were able to use this downward displacement to estimate that recharge had increased from <0.1 mm $a^{-1}$ beneath native vegetation to ~3 mm $a^{-1}$ following clearing. A possible explanation of the shape of the chloride profile shown as Fig.3c is that it represents a record of the change in recharge with time (Allison *et al.*, 1985).

2.2.3 *Oxygen-18 and Deuterium.* These stable isotopes have proved particularly valuable for determining the origin of groundwater, and consequently some hundreds of studies have been reported on this topic. Use has been made of the variation of the isotopic composition of rainfall with elevation (Arnason, 1977). Variations of isotopic composition with rainfall intensity and the changes which occur following evaporation have led to a determination of the possible sources of groundwater in arid areas (Gat, 1984; Gat and Naor, 1979 and Issar *et al.*, 1983).

Relatively few studies using stable isotopes have been carried out in the unsaturated zone. Thoma *et al.* (1979) found that the seasonal variations of the $^2H$ composition of rainfall were preserved in a sand dune in France. A knowledge of water content would then enable estimates of recharge to be made. Bath *et al.* (1982) and Saxena and Dressie (1984) found that in some of their profiles, $^{18}O$ and/or $^2H$ in soil water showed cyclical variations in depth which corresponded to seasonal variations in rainfall. Knowledge of water content and an assumption of piston flow enabled estimates of local recharge to be made.

The three studies referred to above were all carried out in temperate areas where recharge is ~200 mm $a^{-1}$ or more. In drier climates, strongly positive values of either $\delta_2$ or $\delta_{18}$ may occur near the surface due to evaporation through the soil surface (Münnich *et al.*, 1980), leading to the possibility of identifying an annual marker in the soil water. However depth profiles of $\delta_{18}$ and $\delta_2$ in soil water, shown in Fig.4, where the mean annual recharge is estimated from chloride profiles to be ~50 mm $a^{-1}$ and ~3 mm $a^{-1}$, do not show cyclic variations in either of these isotopes. The most likely reason for the absence of such peaks in the soil water profile is that if piston flow occurs, the low recharge

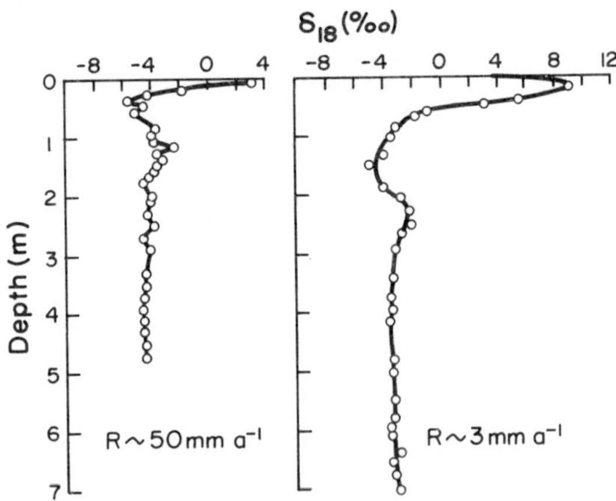


Fig. 4 Depth profiles of $^{18}O$ δ-values ($\delta_{18}$) in soil water at sites in southern Australia where the mean annual recharge is 50 mm $a^{-1}$ and 3 mm $a^{-1}$.

will ensure that the peaks and troughs in δ-values will be separated by only short distances in the soil profile and diffusional redistribution in both liquid and vapour phases will occur. Allison *et al.* (1984) developed a technique based on the displacement of the $\delta_{18}$ and $\delta_2$ values of soil water from the local meteoric line, which may enable estimates of local recharge to be made under conditions of low recharge. They showed that the displacement of $\delta_2$ ($\Delta\delta_2$) from the meteoric line should be linearly related to $R^{-\frac{1}{2}}$ and found this relationship to hold for the four sites they studied. The nature of the relationship between $\Delta\delta_2$ and R leads to more precise estimates as recharge becomes lower, but the technique requires further testing.

2.2.4 *Other 'Natural' Tracers*. Measurement of the bomb-pulse of $^{36}Cl$ has been carried out at one site in a semi-arid region (Phillips *et al.*, 1984). They found that the $^{36}Cl$ fallout peak was very closely reflected in the $^{36}Cl$ peak in soil water. However, because the recharge flux was so low, the peak in $^{36}Cl$ was still within the root zone at about 1 m depth, and its position could not be used to estimate the recharge rate. A chloride mass balance at their site gave a recharge flux at a depth of 5 m beneath of surface of ~0.02 mm $a^{-1}$. A cumulative chloride-depth profile suggests that the bottom of the effective root zone at their site was ~1.5 m and that it would take greater than 125 years for the bomb peak in $^{36}Cl$ to reach this depth. Thus the bomb peak of $^{36}Cl$, like tritium is not an ideal tracer for areas of low recharge. It will probably be best suited for regions where the local recharge is expected to be higher than about 50 mm $a^{-1}$, but this will depend upon the water holding capacity of the soil.

Studies are underway in southern Australia to assess the usefulness of decay of prebomb $^{36}Cl$ in deep unsaturated zones where recharge fluxes are <0.1 mm $a^{-1}$ and it is expected that >100,000 years recharge are stored in these unsaturated zones.

Another involuntary tracer which may be used to give information on rate of water movement in the unsaturated zone is nitrate, which has come into increasing agricultural use in many areas since about 1950. Since some of this added nitrate is leached below the root zone, the change from higher to lower concentrations of nitrate in soil water at depth is an indication of the position in the profile of recharge originating at the time of increased use of nitrate. A knowledge of the amount of water stored in the profile should then enable an estimate of $\bar{R}$, since the use of nitrate started, to be made.

## 2.3 Applied Tracers

In contrast with the tracers discussed in the previous section, which are input each year to the unsaturated zone in precipitation, use of an applied tracer necessitates a one-off application, followed by sampling of the tracer pulse in time. Ideally, the tracer should be applied beneath the root zone and a sufficient time allowed to elapse between injection and sampling to allow the depth interval traversed by the tracer peak to be measured accurately. These conditions make the technique valuable in temperate regions where the rooting depth is shallow and recharge rates high. In semi-arid regions where roots are usually deeper and recharge fluxes lower, this technique will be less useful and will not be discussed in detail here. Tracers which have been used with success are $^{18}O$, $^{2}H$, $^{3}H$ and bromide (Blume *et al.*, 1967; Zimmermann *et al.*, 1967; Saxena and Dressie, 1984; Athavale *et al.*, 1980 and Sharma *et al.*, 1985). Sharma *et al.* (1985) applied bromide at the soil surface and their data gives very strong evidence of non-piston flow through the root zone of native vegetation in a sandy soil.

## 3. VARIABILITY OF LOCAL RECHARGE

It is now becoming realised that there can be considerable variation in rates of local recharge over the scale of a few metres in many soil types (Johnston, 1987; Sharma and Hughes, 1985). Fig.5 shows

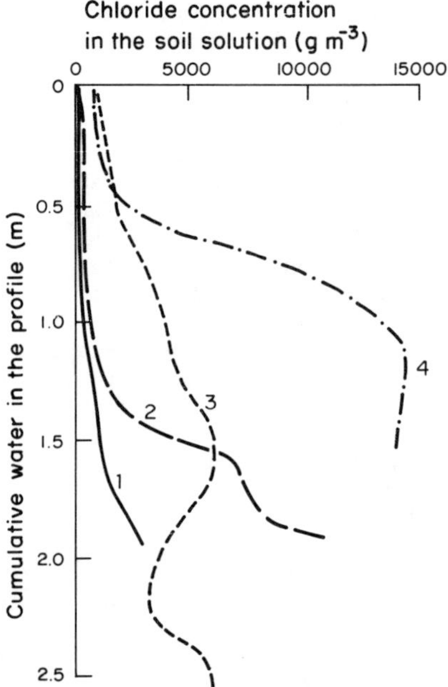


Fig. 5 The relationship between the chloride concentration of soil water and cumulative water stored in the profile in a field, cleared about 50 years ago, under a pasture-crop rotation. All sites are within a circle of radius 100 m. Rainfall is ~300 mm a$^{-1}$.

four profiles of chloride concentration in soil water from what appears to be a uniform field, where the surface soils were sands and sandy loams (Allison and Hughes, unpub. data). All sites were within a circle of radius 100 m and it is believed that the large variations in chloride concentration reflect correspondingly large changes in recharge rate. The native vegetation was removed from the field about 50 years ago and information from many neighbouring uncleared sites suggests that chloride profiles were uniform beneath the native eucalyptus vegetation. At other sites which appear similar, chloride profiles in both cleared and uncleared sites are equally uniform (Allison and Hughes, 1983).

The reason for this variability is unclear. It may be due to changes in the water holding capacity of the soil or alternatively the infiltration characteristics of the surface may be such that runoff followed by localised recharge may occur. Whatever the reason, its identification, quantification, and development of statistical and probably soil and geomorphological techniques for estimating recharge over a sizable area remain seminal problems in recharge evaluation.

This small scale variability of local recharge is not always a problem. Athavale *et al.* (1980) were able to make what appears to be a reliable estimate of local recharge for an area of ~1600 km$^2$ in India from estimates at 28 sites. Allison and Hughes (1978) found that natural tritium profiles and hence estimates of local recharge varied from one soil type to another, but did not vary greatly within a soil type. They were able to use tritium profiles together with a soil map to estimate total local recharge to an area of similar size to the Indian work. Foster and Smith-Carington (1980) found that tritium profiles up to 10 km apart in Norfolk showed "remarkable lateral uniformity".

At present it appears impossible to predict those areas where spatial variability will be high. Therefore, there seems to be no alternative but to make estimates of local recharge at many sites within an area of interest to assess the degree of spatial variability.

Frequency-domain electromagnetic (EM) or transient electromagnetic (TEM) methods have been used with success to assess soil salinity (Williams and Baker, 1982; Buselli *et al.*, 1986). However, because these techniques respond primarily to soil conductivity, the best correlation between soil conductivity, as measured by the instrument, and some field characteristic, is expected to be a function of both water content and salt concentration. Unfortunately, as shown in section 2, the size of the recharge flux is proportional to the inverse of chloride (or salt) concentration in soil water, $C_s$, only. In fact, recent work by Hughes *et al.* (1987) found, by regression, that $C_s\theta^2$ for the depth interval 0-6 m explained 84% of the variance of the measured soil conductivity for the EM instrument they used. There was no correlation between $C_s$ and soil conductivity. Nevertheless it appears that such techniques may be useful for qualitatively identifying areas of higher recharge (low conductivity).

## 4. LOCALIZED RECHARGE

As aridity increases, local recharge is likely to become less important and localised recharge more important in terms of total recharge to an aquifer. Localized recharge may occur from rivers, ephemeral streams or wadis; by runoff to closed depressions which may or may not be karstic; lakes and swamps; and alluvial fans on floodplains. Some of the variability referred to in the previous section may be due to localized recharge. There is some unpublished evidence that lateral subsurface flow in arid dune fields may occur, leading to the possibility of enhanced infiltration in the dune valleys (Allison, Aranyossy and Fontes, unpub. observations, 1985 and D.B. Stephens, priv. comm., 1985).

There have been many qualitative studies of localized recharge, using both chemical and isotopic signatures in groundwater. In many of these studies, the most likely of a range of possible sources of recharge have been identified. As before, I will concentrate here on the quantitative aspects of such studies.

An assessment of the importance of localized recharge may be obtained by investigating the variability in groundwater of a chemical or isotopic parameter, whose concentration is determined by recharge rate or mechanism. For example, Eriksson (1976) studied the frequency distribution of chloride concentration in groundwater samples. He showed that if local recharge provided the major source of water to the aquifer, the chloride concentration of the water samples should be log-normally distributed. Eriksson found this to hold for the aquifer in India which he studied. Leaney and Allison (1986) treated their groundwater data as suggested by Eriksson and found, for most of the aquifer under study, that the chloride concentrations fitted a log-normal distribution. The recharge rate calculated from these groundwater chloride data agreed well with estimates of recharge made using data from the unsaturated zone, suggesting that local recharge was indeed the dominant mechanism of recharge. However the log-normal distribution of chloride data became highly skewed if data was included from an area where numerous features responsible for localized recharge had been identified (Allison *et al.*, 1985).

In a relatively homogeneous unconfined aquifer, which has recharge distributed over its surface (either local or localised), a depth profile of groundwater taken from beneath the water table will contain chemical and isotopic signals reflecting recharge upgradient, with the deepest samples giving information about recharge furthest from the bore. In most real aquifers where stratification, fracturing or dual porosity etc may be important, a simple profile is unlikely to result. However a mixed sample from a bore may contain information about recharge over a considerable area. The lateral separation between sites of localized recharge will be important, and if a bore or bores can intersect streamlines originating from many sites of localized recharge, then it may be possible to get aquifer-scale information from groundwater samples. In many situations a combination of local and localised recharge will occur. If however, the recharge features are far apart it may be necessary to evaluate recharge from each separately. These two possibilities are considered below.

### 4.1 Widely Distributed Localized Recharge

This may occur in conjunction with local recharge in karstic areas, dunefields and other areas consisting of many closed basins or alluvial fans. It may be identified by investigating the distribution functions of various tracers as discussed above or by undertaking studies on the surface or in the unsaturated zone (Allison *et al.*, 1985, Gunn, 1983).

If the localized recharge is widely distributed over the aquifer, a range of models for interpretation of environmental tracers in groundwater has been developed. Zuber (1984,1986) and Maloszewski and Zuber (1982) have reviewed many of the models which have been developed. One of the simplest, and probably one of the most widely used with tritium and carbon-14 data, is the model which assumes complete mixing within the aquifer (Hubert, 1970). Although this model is often physically unreasonable, useful results may be obtained if the samples are taken from fully penetrating bores. A derivative of this model is the multibox model which has been used to model both dispersion and recharge which varies over the surface of the aquifer (Przewlocki and Yurtsever, 1974; Campana and Simpson, 1984). With these and all other models concerned with interpretation of environmental isotope data, additional information about the aquifer is needed to enable an estimate of recharge to be made. For the above models this is the volume of water stored in the aquifer.

Chloride has been used successfully to estimate the water balance, and hence recharge, for whole catchments (Peck and Hurle, 1973; Claassen *et al.*, 1986). Probably the greatest difficulty with this technique is the assumption that no chloride is being derived by weathering within the catchment. Change in land use (and hence recharge rate) can dramatically affect assumptions about hydrologic steady state (Peck and Hurle, 1973).

## 4.2 Localized Recharge from Single Features

This may occur through the base of perennial streams or wadis. Allemmoz and Olive (1980) used a neutron moisture meter to show that a considerable increase in water content up to 5 m beneath a wadi bed occurred following a flood event. Groundwater samples, taken from near the wadis, were analysed for tritium and the tritium concentrations interpreted in terms of a completely mixed aquifer. This assumption is obviously not valid for water samples taken some distance from a wadi bed; however, the fully mixed model may be used to give first estimates of the ratio of recharge to aquifer volume.

A more reliable technique involves collection of groundwater samples at a range of depths beneath the water table and distances from the recharge feature. Analysis of the samples for either tritium or carbon-14 could then be used to calculate the rate at which groundwater is moving away from the recharge feature. The problem of correction of the carbon-14 data for the various exchange and or disolution processes has been reviewed on several occasions and will not be discussed here. Verhagen (1984) studied the isotope geochemistry of groundwater from lines of bores either side of a river which was thought to be a potential source of recharge to an underlying aquifer. He found no systematic trend of carbon-14 activity of groundwater with distance from the river. This indicated that modern-day seepage from the river was unlikely to be a significant contributor to groundwater recharge.

In a study to assess the importance of infiltration from the Chimbo River in Ecuador to an underlying aquifer, Payne and Schroeder (1979) measured the $^{18}O$ and deuterium composition of rainfall, groundwater and river water. They found that the frequency distribution of δ-values of groundwaters was highly skewed and, because the catchment of the Chimbo River was at a higher elevation than the area studied, they were able to identify infiltrated river water and water locally recharged by rainfall. After making some simplifying assumptions about groundwater flow they were able to show that about one third of the groundwater recharge was derived from the river. This method is likely to be applicable in many arid areas as river water often has its source a considerable distance from the aquifer under study.

In systems of high recharge rate and where there is a seasonality in the isotopic composition of the river, it may be possible to use the technique developed by Stichler *et al.* (1986). By using the (varying) $\delta_{18}$ signal of the Danube River as the input function and the time course of $\delta_{18}$ in bores on an island in the river as the response function, they were able to show that between 80 and 95% of the groundwater beneath the island originated from the river. From the lag between the isotopic composition of the river and that in boreholes, they were able to calculate mean transit times of groundwater. If an estimate of water stored had been available this would have enabled the recharge from the river to be calculated.

## 5. CONCLUSIONS

In this review, only some of the techniques available for estimating groundwater recharge have been discussed. Sophisticated analytical techniques have been developed to measure a range of environmental tracers, not mentioned here, ranging from radio-isotopes of the noble gases to trace quantities of fluorocarbons. Concentrations of many of these have been interpreted to make estimates of groundwater recharge.

The discussion has concentrated on recharge to unconfined aquifers in areas of reasonably low rainfall ($<700$ mm $a^{-1}$). It is likely that, the greater the aridity, the smaller and more variable the recharge flux will become. This suggests that chemical and isotopic methods are likely to be more successful than the physical methods which rely on a direct measurement of a water flux. An additional advantage of tracers is that their diffusivity is much less variable with change in water content or soil type than the diffusivity of soil water. Thus tracer behaviour represents a much more robust indicator of water movement in a porous medium than does the solution of the equations of water flow, especially in the unsaturated zone.

In estimation of recharge a tracer like tritium relies on estimation of the amount of the tracer beneath the soil surface, thus the precision of the estimate of recharge will increase with recharge rate. In contrast, for a tracer like chloride, the concentration of which is inversely proportional to recharge rate, the precision of the estimate will increase with decrease in recharge rate. Tritium fallout in precipitation is relatively well known globally from the IAEA network, but the same is not true for net chloride accession, and lack of chloride data in precipitation can be a severe limitation to its usefulness. Even though the concentration of tritium in precipitation has fallen to almost pre-bomb levels, and the fallout peak in the hydrologic system has been reduced greatly in magnitude by the combined effects of decay and dispersion, new low-background counters have ensured its continuing usefulness.

The estimation of recharge rate from investigation of profiles of tracer concentration (e.g. chloride) under some conditions, relies critically on a knowledge of the tracer diffusivity and hence on tortuosity. Estimation of this parameter remains a critical problem.

Probably the most difficult and important problem to be overcome in estimation of recharge is the prediction and assessment of its variability. Over some quite large areas it appears to show little lateral variability, while in other, apparently similar areas it can range over at least an order of magnitude. Use of rapid and non destructive techniques such as electromagnetic methods may offer some hope here.

## 6. REFERENCES

Airey, P.L., Bentley, H., Calf, G.E., Davis, S.N., Elmore, D., Gove, H., Habermahl, M.A., Phillips, F., Smith, J. and Torgersen, T. (1983). Isotope hydrology of the Great Artesian Basin, Australia. *Proc. Int. Conf. on Groundwater and Man.* Vol.1, Aust. Water Resour. Council Conf. Ser. No.8, Aust. Govt. Pub. Service, Canberra, pp.1-11.

Allemmoz, M. and Olive, Ph. (1980). Recharge of groundwater in arid areas: case of the Djeffara Plain in Tripolitania, Libyan Arab Jamahiriya. *Arid Zone Hydrology: Investigations with Isotope Techniques.* Proc. Advisory Group Meeting, Vienna, 1979, IAEA, pp.181-191.

Alley, W.M. (1984). On the treatment of evapotranspiration, soil moisture accounting, and aquifer recharge in monthly water balance models. *Water Resour. Res.*, 20, 1137-1149.

Allison, G.B. (1981). The use of natural isotopes for measurement of groundwater recharge. *Proc. Groundwater Recharge Conf.*, Townsville 1980, Aust. Water Resour. Council, Conf. Ser. No.3, Aust. Govt. Pub. Serv., Canberra, pp.203-214.

Allison, G.B., Barnes, C.J., Hughes, M.W. and Leaney, F.W.J. (1984). Effect of climate and vegetation on oxygen-18 and deuterium profiles in soils. *Isotope Hydrology 1983.* Proc. Symp. Vienna, IAEA pp.105-123.

Allison, G.B. and Hughes, M.W. (1974). Environmental tritium in the unsaturated zone: estimation of recharge to an unconfined aquifer. *Isotope Techniques in Groundwater Hydrology* IAEA, Vienna, pp.57-72.

Allison, G.B. and Hughes, M.W. (1978). The use of environmental chloride and tritium to estimate total local recharge to an unconfined aquifer. *Aust. J. Soil Res.*, 16, 181-195.

Allison, G.B. and Hughes, M.W. (1983). The use of natural tracers as indicators of soil water movement in a temperate semi-arid region. *J. Hydrol.*, 60, 157-173.

Allison, G.B., Stone, W.J. and Hughes, M.W. (1985). Recharge in karst and dune elements of a semi-arid landscape as indicated by natural isotopes and chloride. *J. Hydrol.*, 76, 1-25.

Andersen, L.J. and Sevel, T. (1974). Six years environmental tritium profiles in the unsaturated and saturated zones, Grønhoj, Denmark. *Isotope Techniques in Groundwater Hydrology*, IAEA, Vienna, pp.3-20.

Arnason, B. (1977). Hot groundwater systems in Iceland traced by deuterium. Nordic Hydrol., 8, 92-102.

Atakan, Y., Roether, W., Münnich, K-O and Matthess, G. (1974). The Sandhaussen shallow groundwater experiment. *Isotope Techniques in Groundwater Hydrology*, IAEA, Vienna, pp.21-43.

Athavale, R.N., Murti, C.S. and Chand, R. (1980). Estimation of recharge to the phreatic aquifers of the Lower Maner Basin, India, by using the tritium infection method. *J. Hydrol.*, 45, 185-202.

Bath, A.H., Darling, W.G. and Brunsden, A.P. (1982). The stable isotopic composition of infiltration moisture in the unsaturated zone of English Chalk. *Stable Isotopes* (H.L. Schmidt *et al.* Eds). Elsevier, Amsterdam, pp.161-166.

Blackburn, G. and McLeod, S. (1983). Salinity of atmospheric precipitation. in the Murray-Darling Drainage Division, Australia. *Aust. J. Soil Res.*, 21, 411-434.

Blume, H.P., Zimmerman, U. and Münnich, K-O. (1967). Tritium tagging of soil moisture: the water balance of forest soils. *Isotope and Radiation Techniques in Soil Physics and Irrigation Studies.* Proc. Symp. Istanbul, IAEA, Vienna pp.315.

Bredenkamp, D.B., Schutte, J.M. and DuToit, G.J. (1974). Recharge of a dolomitic aquifer as determined from tritium profiles. *Isotope Techniques in Groundwater Hydrology*, IAEA, Vienna, pp.73-94.

Burdon, D.J. (1977). Flow of fossil groundwater. *Q. J. Eng. Geol.* 10, 97-124.

Buselli, G., Barber, C. and Williamson, D.R. (1986). The mapping of groundwater contamination and soil salinity by electromagnetic methods. *Hydrology and Water Resources Symp.*, Brisbane, Inst. Eng., Aust., Pub. No. 86/13, pp.317-322.

Campana, M.E. and Simpson, E.S. (1984). Groundwater residence times and recharge rates using a discrete-state compartment model and $^{14}C$ data. *J. Hydrol.*, 72, 171-185.

Claassen, H.C., Reddy, M.M. and Halm, D.R. (1986). Use of the chloride ion in determining hydrologic basin water budgets - a 3 year case study in the San-Juan Mountains, Colorado, USA., *J. Hydrol.*, 85, 49-71.

Dimmock, G.M., Bettenay, E. and Mulcahy, M.J. (1974). Salt content of lateritic profiles in the Darling Range, Western Australia. *Aust. J. Soil Res.*, 12, 63-69.

Dincer, T., Al-Mugrin, A. and Zimmermann, U. (1974). Study of the infiltration and recharge through sand dunes in arid zones with special reference to stable isotopes and thermonuclear tritium. *J. Hydrol.*, 23, 79-109.

Dincer, T. and Davis, C.H. (1984). Application of environmental isotope tracers to modeling in hydrology. *J. Hydrol.* 68, 95-113.

Edmunds, W.M. and Walton, N.R.G. (1980). A geochemical and isotopic approach to recharge evaluation in semi-arid zones - past and present. *Application of Isotope Techniques in Arid Zone Hydrology.* Proc. Advisory Group Meeting, Vienna, 1978, IAEA, pp.47-68.

Eriksson, E. (1976). The distribution of salinity in groundwaters of the Delhi region and recharge rates of groundwater. *Interpretation of Environmental Isotope and Hydrochemical Data in Groundwater Hydrology*. Proc. Advisory Meeting, Vienna 1975, IAEA, pp.171-177.

Eriksson, E. and Khunakasem, V. (1969). Chloride concentrations in groundwater, recharge rate and rate of deposition of chloride in the Israel coastal plain. *J. Hydrol.*, 7, 178-197.

Fontes, J.Ch., (1983). Examples of isotope studies of the unsaturated zone. *Rep. Inst. Geol. Sci.*, 82/6, 60-70.

Foster, S.S.D. (1975). The Chalk groundwater tritium anomoly - a possible explanation. *J. Hydrol.*, 25, 159-165.

Foster, S.S.D. and Smith-Carington, A. (1980). The interpretation of tritium in the Chalk unsaturated zone. *J. Hydrol.*, 46, 343-364.

Gardner, W.R. (1967). Water uptake and salt distribution patterns in saline soils. *Isotope and Radiation Techniques in Soil Physics and Irrigation Studies*. Proc. Symp. Istanbul, IAEA, Vienna. pp.335-340.

Gat, J.R. (1984). Role of the zone of aeration in the recharge and establishment of the chemical character of desert groundwaters. *Recent Investigations of the Zone of Aeration*. Proc. Symp. Munich (Eds Udluft, P. *et al.*) Tech. Univ. Munich, 2, 487-498.

Gat, J.R. and Naor, H. (1979). The relationship between salinity and the recharge/discharge mechanism in arid low lands. *The Hydrology of Areas of Low Precipitation*. Proc. Canberra Symp. IAHS Pub. No.128, pp.307-312.

Grabczak, J., Maloszewski, P., Rozanki, K. and Zuber, A. (1985). Estimation of the tritium input function with the aid of stable isotopes. Unpub. manuscript.

Grisak, G.E. and Pickens, J.F. (1980). Solute transport through fractured media. 1. The effect of matrix diffusion. *Water Resour. Res.*, 16, 719-730.

Gunn, J. (1983). Point recharge of limestone aquifers - a model from New Zealand Karst. *J. Hydrol.*, 61, 19-29.

Gvirtzman, H. and Margaritz, M. (1986). Investigation of water movement in the unsaturated zone under an irrigated area using environmental tritium. *Water Resour. Res.*, 22, 635-642.

Gvirtzman, H., Ronen, D. and Margaritz, M. (1986). Anion exclusion during transport through the unsaturated zone. *J. Hydrol.*, 87, 267-283.

Howard, K.W.F. and Lloyd, J.W. (1979). The sensitivity of parameters in the Penman evaporation equations and direct recharge balance. *J. Hydrol.*, 41, 329-344.

Hubert, P., Marcé, A., Olive, P. and Siwertz, E. (1970). Etude par le tritium de la dynamique des eaux souterraines. *C.R. Acad. Sci.* 270D, 908-911.

Hughes, M.W., Allison, G.B. and Cook, P.G. (1987). The calibration and use of electromagnetic induction meters in recharge studies. To be submitted *Aust. J. Earth Sci.* (draft avail.).

Hutton, J.T. (1976). Chloride in rainwater in relation to distance from the ocean. *Search*, 7, 207-208.

Issar, A., Nativ, R., Karnieli, A. and Gat, J.R. (1984). Isotopic evidence of the origin of groundwater in arid zones. *Isotope Hydrology 1983*, Proc. Symp. Vienna, IAEA, pp.85-104.

Johnston, C.D. (1983). Estimation of groundwater recharge from the distribution of chloride in deeply weathered profiles from south-west Western Australia. *Internat. Conf. on Groundwater and Man*, Sydney. pp.143-152.

Johnston, C.D. (1987). Preferred water flow and localised recharge in a variable regolith. *J. Hydrol.*, Submitted.

Johnston, C.D., Hurle, D.H., Hudson, D.R. and Height, M.I. (1983). Water movement through preferred paths in lateritic profiles of the Darling Plateau, Western Australia. *Groundwater Research Tech. Paper* No.1, CSIRO, Aust., 34 pp.

Keith, S.J., Paylore, P., DaCook, K.J. and Wilson, L.G. (1982). Bibliography on groundwater recharge in arid and semi-arid areas. *Arizona Water Resources Research Centre*, Tuscon, 158 p.

Kitching, R. and Shearer, T.R. (1982). Construction and operation of a large undisturbed lysimeter to measure recharge to the Chalk aquifer, England. *J. Hydrol.*, 58, 267-277.

Leaney, F.W. and Allison, G.B. (1986). Carbon-14 and stable isotope data for an area in the Murray Basin: its use in estimating recharge. *J. Hydrol.*, 88, 129-145.

Lloyd, J.W. (1981). A review of various problems in the estimation of groundwater recharge. *Proc. Groundwater Recharge Conf.*, Townsville 1980, Aust. Water Resour. Council, Conf. Ser. No.3, Aust. Govt. Pub. Serv., Canberra, pp.1-25.

Lloyd, J.W. and Farag, M.H. (1978). Fossil groundwater gradients in arid regional sedimentary basins. *Groundwater* 16, 388-393.

Maloszewski, P. and Zuber, A. (1982). Determining the turnover time of groundwater systems with the aid of environmental tracers. 1. Models and their applicability. *J. Hydrol.*, 57, 207-231.

Münnich, K-O, Sonntag, C., Christmann, D. and Thoma, G. (1980). Isotope fractionation due to evaporation from sand dunes. *Mitt. D.D.R. Zentralinst. Isotop.-Strahlenforsch*, 29, 319-332.

Payne, B.R. and Schroeter, P. (1979). Importance of infiltration from the Chimbo River in Ecuador to Groundwater. *Isotope Hydrology 1978*. Proc. Symp. Neuherberg, IAEA, Vol.1, pp.145-156.

Peck, A.J. (1979). Groundwater recharge and loss. *The Hydrology of Areas of Low Precipitation*, Proc. Canberra Symp. Dec. 1979. IASH Pub. No.128 pp.361-370.

Peck, A.J. and Hurle, D.H. (1973). Chloride balance of some farmed and forested catchments in southwestern Australia. *Water Resour. Res.*, 9, 648-657.

Peck, A.J., Johnston, C.D. and Williamson, D.R. (1981). Analyses of solute distributions in deeply weathered soils. *Agric. Water. Manag.* 4, 83-102.

Phillips, F.M., Trotman, K.N., Bentley, H.W., Davis, S.N. and Elmore, D. (1984). Chlorine-36 from atmospheric nuclear weapons testing as a hydrologic tracer in the zone of aeration in arid climates. *Recent Investigations in the Zone of Aeration*, Proc. Symp. Munich (Eds Udluft *et al.*), Tech. Univ. Munich, 1, 47-56.

Przewlocki, K. and Yurtsever, Y. (1974). Some conceptual models and digital simulation approach in the use of tracers in hydrological systems. *Isotope Techniques in Groundwater Hydrology*, Proc. Symp., Vienna, IAEA, pp.425-448.

Rushton, K.R. and Ward, C. (1979). The estimation of groundwater recharge. *J. Hydrol.*, 41, 345-361.

Saxena, R.K. and Dressie, Z. (1984). Estimation of groundwater recharge and moisture movement in sandy formations by tracing natural oxygen-18 and injected tritium profiles in the unsaturated zone. *Isotope Hydrology 1983*, Proc. Symp. Vienna, IAEA, pp.139-150.

Sharma, M.H., Cresswell, I.D. and Watson, J.D. (1985). Estimates of natural groundwater recharge from the depth distribution of an applied tracer. *Proc. 21st Internat. Assoc. Hydraulic Res.*, Melbourne, pp.65-70.

Sharma, M.L. and Hughes, M.W. (1985). Groundwater recharge estimation using chloride, deuterium and oxygen-18 profiles in the deep coastal sands of Western Australia. *J. Hydrol.* 81, 93-109.

Smith, D.B., Wearn, P.L., Richards, H.J. and Rowe, P.C. (1970). Water movement in the unsaturated zone of high and low permeability strata by measuring natural tritium. *Isotope Hydrology*. IAEA, Vienna pp.73-87.

Sophocleous, M. and Perry, C.A. (1985). Experimental studies in natural groundwater recharge dynamics: the analysis of observed recharge events. *J. Hydrol.*, 81, 297-332.

Stephens, D.B. and Knowlton, R. (1986). Soil water movement and recharge through sand at a semi-arid site in New Mexico. *Water Resour. Res.*, 22, 881-889.

Stichler, W., Moser, H. and Schroeder, M. (1984). Measurements of seepage velocity in a sand lysimeter by means of $^{18}O$ content. *Recent Investigations in the Zone of Aeration*. Proc. Symp. Munich (Eds Udluft, P. *et al.*). Tech. Univ. Munich, 1, 191-204.

Stichler, W., Maloszewski, P. and Moser, H. (1986). Modelling of river water infiltration using oxygen-18 data. *J. Hydrol.*, 83, 355-365.

Thoma, G., Esser, N., Sonntag, C., Weiss, W., Rudolph, J. and Leveque, P. (1979). New technique of in-situ soil-moisture sampling for environmental isotope analysis applied at Pilat sand dune near Bordeaux. *Isotope Hydrology 1978*. Proc. Symp. Neuherberg, IAEA, Vol.2, pp.753-766.

Verhagen, B.T. (1984). Environmental isotope study of a groundwater supply project in the Kalahari of Gordonia. *Isotope Hydrology 1983*. Proc. Symp. Vienna, IAEA. pp.415-432.

Watson, J.D. (1982). Analysis of salinity profiles in lateritic soils of southwestern Australia. *Aust. J. Soil Res.*, 20, 37-49.

Wellings, S.R. (1984). Recharge of the Chalk aquifer at a site in Hampshire, England, 1. Water balance and unsaturated flow. *J. Hydrol.*, 69, 259-273.

Williams, B.G. and Baker, G.C. (1982). An electromagnetic induction technique for reconnaissance surveys of soil salinity hazards. *Aust. J. Soil Res.*, 20, 107-118.

Zimmermann, U., Münnich, K-O and Roether, W. (1967). Downward movement of soil moisture traced by means of hydrogen isotopes. *Geophys. Monograph* No.11 (Ed. G.E. Stout) Amer. Geophys. Union, pp.28-35.

Zuber, A. (1984). Review of existing mathematical models for interpretation of tracer data in hydrology. *Mathematical Models for the Interpretation of Tracer Data in Groundwater Hydrology.* Proc. Advisory Group Meeting, Vienna, IAEA.

Zuber, A. (1986). Mathematical models for the interpretation of environmental radio-isotopes in groundwater systems. *Handbook of Environmental Geochemistry, Vol.2, The Terrestrial Environment, B* (Fritz, P. and Fontes, J-Ch. eds), Elsevier, pp.1-59.

# EVAPORATION IN ARID AND SEMI-ARID REGIONS

H.A.R. de Bruin
Agricultural University of Wageningen,
Department of Physics and Meteorology,
Duivendaal 2,
6701 AP Wageningen,
The Netherlands.

## ABSTRACT

A survey is given of the meteorological aspects of evaporation in arid and semi-arid regions.
The main features of existing routine methods are discussed.
It is pointed out that most models for evaporation are developed for temperate crops. The main differences between temperate and (semi-)arid regions are described.
Special attention will be paid to the use of remote sensing techniques, for the determination of evaporation.

## 1 INTRODUCTION

It is the objective of this paper to give a general survey of the meteorological aspects of evaporation, E, in arid and semi-arid regions. Since a significant part of precipitation evaporates back into the atmosphere, E is very important for the determination of the available groundwater in these areas.

Most of past research on evaporation has concentrated on temperate climates, and only a few studies are devoted to evaporation in (semi-)arid zones. In particular, there are very few direct observations of E in dry regions.
The differences between evaporation in temperate and (semi-)arid climates are discussed in section 2.

The next section deals with the theory of evaporation. Firstly, the Penman-Monteith equation is discussed, which describes succesfully most aspects of E of temperate crops. Furthermore, it is explained that the Penman-Monteith equation (in the following referred to as P-M-equation) cannot be used in a predictive way. This is due to the fact that this equation describes E in terms of *dependent* variables, i.e. E is interrelated with the saturation deficit that appears in the right-hand side of the P-M-equation. This effect can be accounted for by considering the interaction between the surface fluxes of sensible heat and water vapour on one hand and the tem-



*I. Simmers (ed.), Estimation of Natural Groundwater Recharge, 73–88.*

perature and humidity of the overlying planetary boundary layer (PBL) on the other. In this way it can be understood why in the temperate zones E is determined primarily by radiation: it is almost independent of windspeed and it is much less dependent on the surface resistance than predicted by plant-physiologists.

An important feature of (semi-)arid regions is that they are seldom covered completely by vegetation. As a result the P-M-equation cannot be applied, without further ado, in these area. For instance, now the evaporation from the soil must be taken into account. Moreover, quantities or features such as net radiation, surface roughness, atmospheric stability, the energy balance at the surface and the role of advection need a different description in (semi-)arid areas. These aspects will be discussed briefly in section 4, in which also attention is paid to direct measurements of E under dry conditions.

It is felt by many researchers that the only way to determine evaporation operationally on a regional scale is to use remotely sensed data. This idea, which is based on intuition rather than physical principles, has been elaborated in the past decade. In section 5 some remote-sensing techniques for the estimation of E are discussed and a personal view is given to this approach. Also the determination of precipitation by means of remote-sensing is considered briefly.

## 2 DIFFERENCES BETWEEN TEMPERATE AND (SEMI-)ARID REGIONS

It will be clear that most differences between these two regions arise from differences in precipitation, P. To quantify this, P has to be compared with the water equivalent of the energy, A, available for evaporation and sensible heat flux. In many cases A equals the net radiation minus soil heat flux density. The temperate climates are characterized by the fact that $P>A/\lambda$ (for the symbols and units see Appendix I), whereas in the (semi-)arid regions, generally, P is smaller than $A/\lambda$. For reasons explained later, here P is not compared with the "potential evaporation", more commonly used by hydrologists and in agrometeorology. It can be said that in the temperate climates the limiting factor for evaporation is the available energy, whereas E is limited by the available water in (semi-) arid regions.

Vegetation adapts to its environment and, consequently, temperate crops cover, usually, the ground completely, whereas arid crops do not. This has great impact on the description of evaporation. For example, the evaporation from the soil is - generally - negligibly small in the temperate regions, but can be dominant in arid zones.

Also, the spatial and temporal distribution of precipitation is different for the two regions. In the temperate zones precipitation is distributed reasonably uniform both in time and space. On the contrary the distribution of rainfall in (semi-)arid regions is very irregular. There precipitation primarily occurs in the form of convective storms, which are confined to the wet season. In this respect, it is better to use the term *periodic-arid* instead of *semi-arid* because these convective storms can produce a much

higher rainfall intensity than those in temperate climates.
The amount of precipitation that reaches the deeper ground layers is highly dependent on the spatial and temporal distribution of rainfall during the wet season. Then, over short time intervals precipitation can easily exceed the water equivalent of the available energy and it is possible that a part of the rainfall will reach the deeper soil layers.
Due to the non-uniform spatial rainfall distribution, the ground will be wetted very irregularly after a rainfall event. For that reason it is to be expected that in (semi-)arid regions *advection* influences evaporation. Relatively dry and warm air will flow from the drier regions over the wetter ones. Then, over the latter sensible heat will be extracted from the air by which additional energy is available for evaporation. Generally, advection is less important in the temperate zones, where precipitation is distributed much more uniformly*.

## 3 THEORY

Using a similar approach to Penman (1948), i.e. combining the bulkaerodynamic equations for the vertical transfer of sensible heat and water vapour and the surface energy-balance equation, Monteith (1965) derived a formula for transpiration and evaporation from a vegetated surface that nowadays is known as the Penman-Monteith equation. For the theoretical background the reader is referred to Monteith (1981) and Brutsaert (1982).
The P-M-equation can be written (McNaughton and Jarvis, 1983):

$$\lambda E = \frac{s}{s+\gamma} A + \frac{\rho c_p (D_i - D_{eq})}{(s+\gamma) r_a + \gamma r_c} \qquad (1)$$

where the *equilibrium saturation deficit*, $D_{eq}$, is given by

$$D_{eq} = \frac{s}{s+\gamma} \frac{\gamma r_c}{\rho c_p} A \qquad (2)$$

For the symbols see Appendix I.

Experiences show that Eq. (1) succesfully describes the transpiration and interception loss from different kinds of vegetation such as tall, rough forests, arable crops, heathland and grassland.

In Monteith's concept the vegetation layer is described in a very simple way, i.e. the canopy is treated as if it were one 'big leaf' to which a canopy resistance $r_c$ is assigned. This 'big leaf' has the same albedo and surface roughness as the actual crop. Moreover, it is assumed that within the 'stomata' of the 'big leaf' the water vapour pressure equals the saturated value at mean surface temperature $T_s$. Since the temperature gradient within the canopy can be very pronounced, $T_s$ is difficult to determine. Because in the Penman-Monteith approach $T_s$ is eliminated, its

* Due to land use advection is also often important in temperate agricultural areas, but then it occurs on a smaller scale.

exact value is not too important. However, if the P-M-equation is used in combination with a remotely sensed value of $T_s$ the problem of the precise definition of $T_s$ becomes crucial (see later).

The crude description of the vegetation layer in Monteith's model reflects also on the determination and definition of the aerodynamic resistance $r_a$. It is beyond the scope of this paper to discuss this problem here in detail, but it is noted that $r_a$ depends, besides on wind speed at reference level, also on the roughness lengths for heat, water vapour and momentum, the zero-displacement height(s) and the stability of the overlying air. Several of the surface parameters are ill-defined and difficult to estimate. The canopy resistance, $r_c$, is a complex function of incoming solar radiation, saturation deficit and other parameters such as soil moisture. The relationship between $r_c$ and these environmental quantities varies from species to species. It is not possible to measure $r_c$ directly; usually, it is obtained through the P-M-equation by measuring all other quantities, inclusive E. Note that than $r_a$ must be estimated also. Often this is done in a crude way, so that several values for $r_c$ published in literature are biased due to errors made in $r_a$.

Something has to be said about the 'old'-Penman equation used extensively by hydrologists for the estimation of regional evaporation. As shown by e.g. Thom and Oliver (1977) the success of this approach is rather fortuitous, since the canopy resistance is ignored and a wrong expression for $r_a$ is used. For short crops errors are cancelling out. However, the 'old'-Penman equation is not able to describe transpiration or interception from rough, tall vegetations properly.

Micro-meteorological observations over temperate, most arable, crops reveal that their transpiration rate depends strongly on net radiation. Morover, it appears that the second term in the right of the P-M-equation is typically about one-fourth the size of the first term, leading to

$$\lambda E \approx \alpha \frac{s}{s+\gamma} A, \qquad (3)$$

where $\alpha$ is a coefficient of value of about 1.2 - 1.3. Eq. (3) was first proposed by Priestley and Taylor (1972). For large 'saturated' land and water surfaces Eq. (3) describes the actual evaporation loss surprisingly well. For a review of literature devoted to this subject see Brutsaert (1982, p. 219).

What is the reason that the Priestley-Taylor approach works so well? An explanation for this can be found in the fact that E is interrelated with the saturation deficit $D_i$. At the surface sensible heat and water vapour are brought into the atmosphere. As a result the temperature and humidity of the planetary boundary layer (PBL) will increase. Consequently, $D_i$ of the PBL is affected. In turn, this will influence E so that E and $D_i$ are interrelated.

An important consequence of this feature is that the P-M-equation does not express E in terms of independent quantities. It is a *descriptive* equation, and it cannot be used in a *predictive* way. For instance, it is not able to predict under dry conditions what E will be if the water supply is plentiful, because for that a value of $D_i$ is needed under these 'potential' conditions. This implies that all estimates of the so-called 'potential evaporation' based on the P-M-equation or related formula (as that by Penman), using the actual $D_i$ observed under dry conditions, must be considered unrealistic.

The interaction between E and $D_i$ can be accounted for by coupling the P-Me-quation to a PBL-model. This has been done by Perrier (1980) and McNaugthon and Jarvis (1983). These authors assumed that the PBL is impermeable at its top. They showed that then $D_i$ approaches $D_{eq}$ (see Eq. (2)), and the Priestley-Taylor parameter $\alpha$ goes to 1. In reality, the PBL is not impermeable at the top. The PBL height rises during daytime as long as the PBL is heated at the surface. As a result, there is entrainment of relatively warm and dry air at the top of the PBL from the stable layer aloft. This process is described by the model of Tennekes (1973). De Bruin (1983) coupled a simplified version of Tennekes' model to the P-M-equation and obtained in this way a model for regional evaporation in which the interrelation between E and $D_i$ is accounted for. In Fig. 1 one result of De Bruin's model is shown, i.e. the calculated diurnal variation of the Priestley-Taylor parameter $\alpha$ is drawn for different values of the canopy resistance $r_c$. It is seen from these graphs that a 3-fold change of $r_c$ from e.g. 30 to 90 $sm^{-1}$ causes a change in $\alpha$ and thus in E of only 20%. This is due to the interaction between the surface fluxes of sensible heat and water vapour with the saturation deficit $D_i$ of the overlying PBL. Most temperate arable crops for which soil moisture seldom is a limiting factor have a $r_c$ in the range

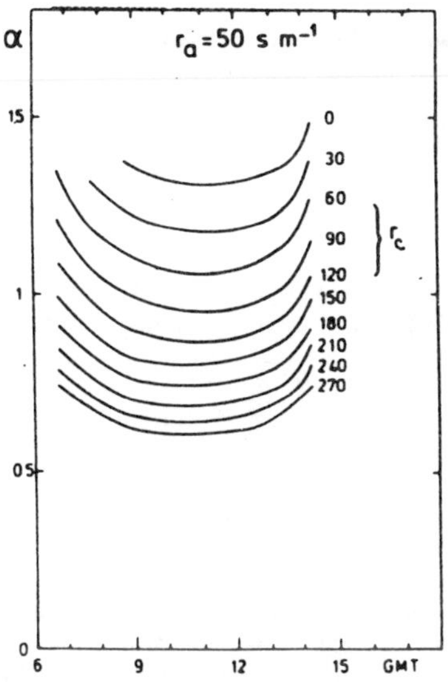


Fig.1 Daytime variation of the computed Priestley-Taylor parameter $\alpha$. (After De Bruin, 1983)

of 30-90 $sm^{-1}$. So E depends in first instance on the so-called equilibrium evaporation rate, $E_{eq} = s/(s+\gamma)A$. Note that $s/(s+\gamma)$ is a function of temperature only. Some PBL-features are described too simple by De Bruin's model; however, a more complete approach, published recently by McNaughton and Spriggs (1986) gives similar results to those shown in Fig. 1 (see also Jarvis and McNaughton, 1986).

Much less is known about natural vegetation in the temperate climates, which generally grow on poor soils that are not suitable for agriculture. Wallace et al. (1983) did work on heather in Yorkshire (U.K.) (For simplicity heather is considered here natural).
They found relatively high values for $r_c$ of about 100-150 $sm^{-1}$, so that the transpiration rate of heather is considerably less than that of e.g. grass. This is confirmed by recent measurements in the Netherlands (Bosveld, personal communication). The interception loss from heather, however, is greater than that for grass due to its greater aerodynamic roughness. This behaviour is very similar to that of coniferous forests (see e.g. McNaughton and Jarvis, 1983).

## 4 EVAPORATION IN (SEMI-)ARID REGIONS

### 4.1 General

In the previous section the most recent ideas about evaporation have been reviewed briefly. However, since these are based mainly on work done in the temperate zones, it is questionable whether the obtained results can be used in arid or (semi-)arid regions.

The main differences between the two climatological regions are given in section 2. In this section these difference will be discussed in more detail. Because only a very few direct observations are available from evaporation in (semi-)arid regions many of our considerations will be speculative.

### 4.2 Sparse vegetation

An important difference between temperate and dry land crops is their ground coverage. In (semi-)arid regions crops hardly ever cover the ground completely, certainly not at the end of the dry season. Then the 'big leaf' approach, that resulted in the P-M-equation, does not apply any longer. Recently, a group at the Institute of Hydrology (Wallingford, U.K.) did direct measurements of evaporation from a sparse dryland crop, millet, in Niger (Wallace et al., 1986). They used the eddy-correlation technique for the determination of total evaporation, and porometers and lysimeters to evaluate transpiration and soil evaporation separately. Some preliminary results are shown in Figs. 2-4.
In Fig. 2 the total daily actual evaporation as measured with the eddy-correlation method is compared with two estimates of 'potential' evaporation using a pan and the 'old'-Penman formula respectively. Also the rainfall is shown. It is seen clearly that actual and 'potential' evaporation rates are poorly related. Just after rainfall their correlation is negative. Then E increases, but the "potential" rates decreases due to a decrease of Di. It

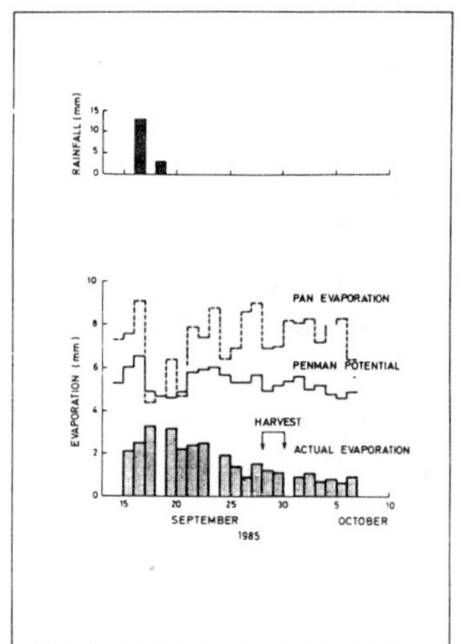


Fig.2 A comparison of the observed total evaporation with the 'old'-Penman potential and pan evaporation. (After Wallace et al., 1986)

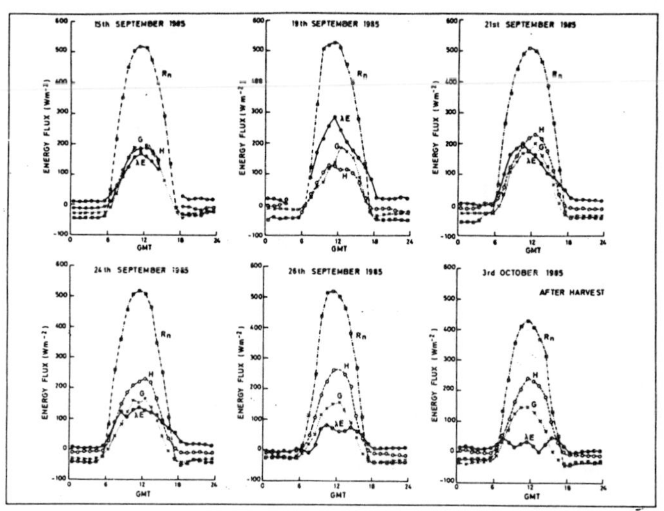


Fig.3 The components of the energy balance of a millet crop for six days; $R_n$ is net radiation, G, H, $\lambda E$ are the soil heat, sensible and latent heat flux density respectively. (After Wallace et al., 1986)

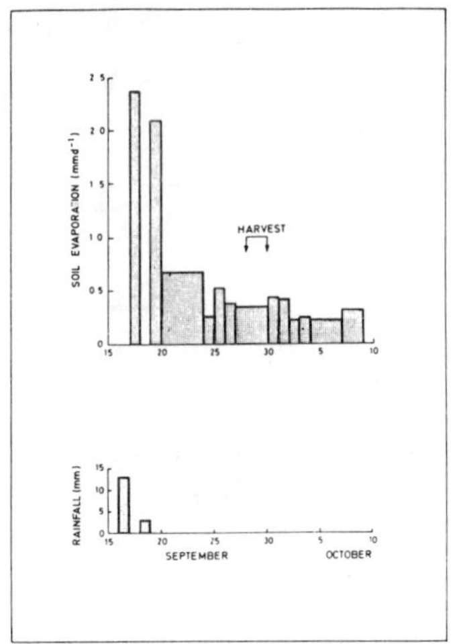


Fig.4 The observed soil evaporation and rainfall. (After Wallace et al., 1986)

illustrates clearly the interrelation between E and $D_i$. Fig. 3 shows the diurnal variation of the four components of the energy-balance at the Earth's surface for six selected days. This gives an impression how $\lambda E$ compares to net radiation ($R_n$), sensible heat and soil heat flux (H and G respectively). In Fig. 4 the observed daily/soil evaporation is depicted for the same period as that for Fig. 2.

It is too early to draw final conclusions from these measurements, but some features can be mentioned.

a. Soil evaporation is important, especially after rainfall.
b. Soil heat flux is similar in size to e.g. $\lambda E$ and it cannot be neglected as often is done in temperate regions.
c. The total evaporation is small compared to the 'potential' rate. This is due to the low leaf area index of the crop.
d. The use of the concept of 'potential' evaporation determined with a pan or the 'old'-Penman formula is questionable.

Some of these points are discussed in more detail in the next sub-sections.

4.2.1 Soil evaporation and soil heat flux

The physical processes related to soil evaporation and soil heat flux appear to be very complicated and are not yet fully understood. The vertical transport of water in the soil takes place both in the liquid and the vapour phase. It is related to gravity and gradients of pressure, temperature, soil moisture and soil concentration and also to the soil heat flux. It is beyond the scope of this paper to discuss these complicated processes.

Distinction has to be made between cases with and without a groundwater table (e.g. Brutsaert, 1982).

Experiments show that for the case where no water table is present, the cumulative evaporation from a bare soil is proportional to the square root of the time t after a rainfall event at t=o (Brutsaert, 1982; Wallace et al, 1986). Whether this rule can be used for practical calculations is still unclear.

Menenti (1984) studied the evaporation in desert areas called *playas*, where groundwater reaches the surface. He found evaporation rates of 2-4 mm day$^{-1}$ from a playa in Lybia and estimated the total evaporation loss to be 220 mm per year. Moreover, Menenti presented a modification of the Penman-Monteith equation, introducing a resistance for water vapour transport for the soil layer between the surface and the water table.

An important point is that for dry bare soils the energy-balance equation for the Earth's surface does not contain the latent heat flux density, $\lambda E$, because the phase change from liquid water to vapour does not take place at the surface. Monteith (1981) avoided this problem by assuming that the top soil layers are isothermal. This approach has been used by Shuttleworth and Wallace (1985) in their model for evaporation from sparse crops.

The vertical transport of heat in the soil is, besides normal thermal conductivity, also related to the vertical movement of liquid water and water vapour. In practical applications these effects are accounted for by using an apparent thermal conductivity.
A recent description of the different aspects of heat, water and vapour movements in bare soils is given by Ten Berge (1986). He presented also a coupled PBL-surface model and considered the application of remote sensing for e.g. the estimation of evaporation.

Concluding this subsection, it is remarked that at the moment there do not exist simple methods, requiring routine input data only, to estimate evaporation and heat flux from bare soils, whereas these quantities are extremely important in (semi-)arid zones. Therefore, future research must, in the opinion of the present author, be devoted to these features.

### 4.2.2 Miscellaneous micrometeorological aspects

In this sub-section we will point out very briefly some micrometeorological features of evaporation from sparse dryland crops, which are different from vegetation covering the ground completely.

i. Roughness length and zero-plane displacement

Experience shows that the aerodynamic roughness length of a sparse crop is greater than that of a crop of the same height and with the same architecture, covering the ground completely (Businger, 1975). Moreover, it must be expected that the roughness lengths for momentum, heat and watervapour

(Brutsaert, 1982), which often are assumed to be equal using P-M-equation for temperate crops (Thom and Oliver, 1977) are significantly different for sparse vegetation.

This applies also for the zero-plane displacement (see e.g. Hicks and Everett, 1979). All these aspects reflect on the determination of the aerodynamic resistance $r_a$ that appears in the P-M-equation.

ii. Stability effects

In dry regions the sensible heat flux at the surface is, usually, relatively large. For that reason stability effects on $r_a$ become very important and these should be taken into account. Since the stability shows a pronounced diurnal variation this complicates significantly practical calculations. For more information see e.g. Brutsaert (1982) and Menenti (1986).

iii. Net radiation

Although the evaporation from dryland crops is less dependent on the available energy, net radiation is still an important quantity. Often it is not measured directly and must be estimated from standard weather data. Most of the empirical relations developed for this purpose (see Brutsaert, 1982) apply, however, primarily to temperate climates. It is to be expected that deviations will occur when using these on sparse dryland crops. For example, dust will influence the incoming shortwave radiation (Brinkman and McGregor, 1983), whereas the often low humidity of the air will affect the incoming longwave radiation. Moreover, albedo can show a large annual variation and the outgoing longwave radiation will often show a great diurnal variation due to variations of the surface temperature.

## 4.3 Advection

It is argued in section 2 that due to the irregular spatial and temporal rainfall distribution, advection can be important for evaporation in (semi-)arid regions. With respect to this feature, there is a lack of direct measurements also. To the knowledge of the present author the most recent measurements of evaporation under advective conditions using adequate equipment, were done in Australia by Lang et al (1983). They observed a downward sensible heatflux of more than 100 $Wm^{-2}$ over an irrigated rice field, surrounded by extensive dry areas, at 300 m from the leading edge. Consequently, $\lambda E$ is signifcantly larger than the net radiation minus the soil heat flux density, showing how important advection can be. Also, they found a significant difference between the turbulent exchange coefficients for water vapour and sensible heat, under these conditions. In the P-M-equation these are assumed to be the same. As a result the turbulent transfer of sensible heat and water vapour can not be described any longer with simple first-order closure models. Recently, Kroon (1984) used the second-order closure model by Rao et al (1974) to investigate the influence of advection on evaporation. He found significant effects.

It still is not clear over which distance from the leading edge advection is important. In literature figures are mentioned varying from 100 m to

several kilometers.
In the opinion of the present author in future work the interaction with the PBL has to be taken into account. If this appears to be significant, advection will be important over several kilometers, since the PBL-height is typical 1-2 km.

## 4.4 Direct measurements

The use of direct measuring techniques for evaporation, which are developed and tested over temperate crops, meets difficulties under dry conditions too. Here we confine ourselves to two 'meteorological' methods that are generally used.

a) Bowen ratio method

Since in this method it is assumed that the turbulent exchange coefficients for heat and water vapour are equal, the use of this method is questionable in advective conditions (see above). Moreover, under very dry conditions the vertical profile of humidity must be determined very accurately, while the soil heat flux must be measured precisely as well.

b) Eddy-correlation method

As pointed out by Webb et al. (1980) the eddy-correlation observations of E must be corrected for contribution of the sensible heat flux to the mean vertical wind speed, under dry conditions.

# 5 THE USE OF REMOTE SENSING

For many practical applications in hydrology, e.g. for groundwater problems, evaporation has to be known on a regional scale. This is difficult to achieve with the existing observation techniques, because they provide only point measurements. Therefore, in the last decade much attention has been paid to the use of remote sensing for the determination of evaporation. This research has been stimulated by the launch of several operational satellites monitoring the Earth's surface with time interval of half an hour (geostationary satellite like METEOSAT) to 5-15 days (polar satellites, with high resolution imagery like LANDSAT and SPOT).
The general principle of remote sensing is that we want to evaluate the state of one or several physical quantities at the Earth's surface with the aid of some electromagnetic property (e.g. colour, polarization etc.). This is possible only if this state is a unique function of this property. However, in many cases, such a unique function does not exist.

For the determination of soil moisture of bare soils use can be made of the fact that at microwave wavelength, between 1 and 50 cm, the dielectric constant of water is an order of magnitude larger than that of dry soil. By using either passive (radiometry) or active (RADAR) microwave techniques, in principle, soil moisture can be observed (e.g. Schmugge et al, 1978). Methods have been developed to determine soil moisture from the diurnal variation of the remotely sensed surface temperature (the thermal inertia method). Recently, Ten Berge (1986) showed that this approach is very inaccurate.

The electromagnetic property of the Earth's surface, which is used in most remote sensing techniques to determine evaporation, is the emitted radiation by the surface at wavelengths in the 'atmospheric window', i.e. between about 8 and 12 $\mu$m. From this the surface temperature $T_s$, can be derived, which in turn is related to evaporation. It must be realised that several problems arise using this approach:

a. For the determination of $T_s$ one has to know the emissivity of the surface and atmospheric corrections must be made. For land surfaces the accuracy of $T_s$ that can be achieved in practice is about 3K (Ten Berge, 1986).

b. Since large vertical temperature gradients can exist within the canopy of vegetation, it is difficult to define exactly $T_s$ itself or to verify the remotely sensed value of $T_s$ with independent methods.

c. In most techniques it is assumed that the different components of the energy balance are governed by a single surface temperature. In reality this is not true. For example, the temperature that effectively determines the emitted longwave radiation is not necessarily equal to that which governs the sensible heat flux.

d. Evaporation is no unique function of $T_s$.

e. Often there is a scale problem. The remotely observed $T_s$ concerns in many case areas which are so large that they are not homogeneous. The physical meaning of this observed $T_s$ is unclear. This effect complicates also comparisons with ground truth measurements which are taken at one point.

From the above it will be clear that the methods for the determination of evaporation using remotely sensed $T_s$ are (semi-)empirical in nature. Therefore, they are applicable only for the conditions under which they are developed and tested.

In most studies the difference between $T_s$ and air temperature, $T_a$, at standard height (about 2 m) is used to determine the sensible heat flux density, H. Then E is derived from the energy balance equation. For more details see e.g. Dugdale (1983).
The problem with this approach is that ($T_s$-$T_a$) is a dependent variable: if H changes ($T_s$-$T_a$) will change. For that reason, the air temperature at 2 m above the surface will vary horizontally over non-uniform terrain and it cannot be taken equal to the $T_a$ measured at a nearby weather station as usually is done. Moreover, ($T_s$-$T_a$) is very sensitive to measuring errors. We recall that the error in $T_s$ is about 3K. In arid regions this will be worse, due to the atmospheric corrections, which there can be very large. Dugdale (1983) reported corrections of more than 20K for METEOSAT data.

Models based on ($T_s$-$T_a$) estimates have been developed by Soer (1980), Price (1982), England et al (1983) and others.

The most simple approach is that by Jackson et al. (1977), who found an empirical relation between mean daily evaporation and the midday difference between the surface temperature and the air temperature at reference level. This method is refined by Seguin and Itier (1983) who showed that the ratio of daily mean and instantaneous net radiation as well as the aerodynamic resistance $r_a$ play a role in this relationship. Recently, Nieuwenhuis et al. (1985) proposed a modification of the approach by Jackson et al. These authors assumed that the difference between the actual and potential evapotranspiration is proportional to the difference in surface temperature of the field of interest and that of a nearby terrain that is transpiring at a potential rate. The proportional constant appears to be strongly dependent on the surface roughness length. This result was obtained by using the TEGRA model developed by Soer (1980), who combined the surface energybalance equation to a model for liquid water flux in the soil-plant system. He assumes that transpiration equals the total evaporation, so he neglects the soil evaporation.
The advantage of the aproach of Nieuwenhuis et al. (1985) is that it does not try to determine E in an absolute sense, which is very difficult for reasons mentioned above, but in a relative way. Only differences in evaporation are determined. Menenti (1984) estimated evaporation from desert areas, using remotely sensed data, also relative to a reference surface.

To avoid the problem that air temperature is a dependent variable and cannot be taken from a nearby weather station, Klaassen and Van den Berg (1985) and Ten Berge (1986) developed coupled PBL-surface models. For practical calculations these models may be too complicated; however, they can serve to verify the assumption made in the more simple approaches (Ten Berge, 1985).

In the opinion of the present author, the remote sensing techniques to estimate E are still too inaccurate to be applicable in (semi-)arid regions. This opinion is shared by e.g. Dugdale and Milford (1986).

Because in these areas precipitation determines evaporation to a great extent, it is worthwhile to mention here also remote sensing techniques to estimate rainfall.
Most succesful are the methods for convective storms. For these there exists a relation between the temperature structure of the top of the cloud (which can be determined with IR imagery) and the rainfall rate; see e.g. Scofield (1984) and Dugdale et al. (1986).
It appears that in this way reasonable rainfall estimates can be obtained with an accuracy similar or even better than that achieved with an existing raingauge network. Due to the variability of rainfall the accuracy of the latter is low, especially for short time intervals (Dugdale, 1983).

Finally, it is noted that, probably, remote sensing techniques allow the determination of the geohydrological properties of the Earth's surface. This can be very useful for groundwater problems.

## 6 CONCLUDING REMARKS

In this paper a brief review is given of different aspects of evaporation in (semi-)arid regions. It must be concluded that most techniques developed so far for the determination of this quantity, cannot be applied in these regions without further ado due to the special conditions. This implies that much work has to be done in future in order to improve our knowledge about evaporation under dry conditions.

## LITERATURE


Brinkman, A.W. and J. McGregor, 1983: Solar radiation in dense Saharan aerosol in Northern Nigeria, Quart. *J.R. Met. Soc., 109*, 831-847.

Brutsaert, W.H., 1982; *Evaporation into the atmosphere*. Reidel Publ. Comp., pp. 299.

Businger, J.A., 1975: Aerodynamics of vegetated surfaces. *In Heat and mass Transfer in the Biosphere*, D.A. de Vries and N.H. Afgan (Eds.), 139-167.

De Bruin, H.A.R., 1983: A model for the Priestley-Taylor $\alpha$. *J. Climate and Appl. Meteor, 22*, 572-578.

Dugdale, D. 1983, *Mesure de l'humidité du sol et de l'évaporation*. In: Application de la télédétection à l'agro-météorologie opérationelle dans les pays semi-arides. Séminaire Agrhymet, ESA, 51-55.

Dugdale,G, 1983. *Estimation des précipitation*. In: proc. Applications de la télédétection à l'agro-météorologie opérationelle dans les pays semi-arides, ESA publ, 55-71.

Dugdale, G., 1986: *The calibration and interpretation of Meteosat based estimates of Sahelian Rainfall*. In: Proc. 6th Meteosat Users' Meeting, ESA to be published.

Dugdale, G. and J.R. Milford, 1986: *Applications of Meteosat data in agriculture and hydrology*. Proc. 6th Meteosat Users' Meeting, Amsterdam, 25-27 november 1986. To be published by ESA.

England, T.E., R. Gombeer, E. Hechinger, R.W. Herscky, A. Rosema and L. Stroosnijder, 1983: The group Agromet Montoring Project (GAMP): Application of Meteosat data for rainfall, evaporation, soil-moisture and plant-growth monitoring in Afrika. *ESA Journal 7*, 169-188.

Hicks, B.B. and R.G. Everett, 1979: Comments on 'Turbulent exchange coefficients for sensible heat and water vapour under advective conditions' *J..Appl. Meteor., 18*, 381-382.

Jackson, R.D., R.J. Reginata and S.B. Idso, 1977: Wheat canopy temperature: a practical tool for evaluating water requirements. *Water Resour. Res., 13*, 651.

Jarvis, P.G. and K.G. McNaughton, 1986: Stomatal Control of transpiration: Scaling up from leaf to region. *Adv. in Ecological Res., 15*, 1-49. Academic Press (London).

Kroon, L.J.M., 1984. Profile derived fluxes above inhomogeneous terrain: a numerical approach. *PH-D thesis Agricultural Univ. Wageningen*.

McNaughton, K.G. and P. Jarvis 1983. Predicting effects of vegetation changes on transpiration and evaporation. In: *Water Deficit and Plant Growth*, Vol. VII. Academic Press.

McNaughton, K.G. and T.W. Spriggs, 1986: A mixed layer model for regional evaporation. *Boundary-layer Meteor.*

Monteith, J.L., 1965: Evaportion and environment. *Symp Sor. Exp. Biol., 19*, 205-234.

Monteith, J.L. 1981: Evaporation and surface temperature, *Quart. J.R. Met. Soc., 107*, 1-27.

Nieuwenhuis, G.J.A., E.H. Smidt and H.A.M. Thunissen, 1985: Estimation of regional evapotranspiration of arable crops from thermal infrared images. *Int. J. Remote sensing, 6*, 1319-1334.

Penman, H.L., 1948: Natural evaporation from open water, bare soil and grass, *Proc. R. Soc. London. Ser. A 193*, 120-146.

Perrier, A., 1980: Étude micro-climatique des rélation entre les propriété's de surface et les caractéristique de l'air: Application aux échanges régionaux. *Météorologie et Environnement*, EVRY (France) Octobre.

Priestley, C.H.B. and R.J. Taylor, 1972. On the assessment of surface heat flux and evaporation using large-scale parameters. *Mon. Weather Rev., 100*, 81-92.

Price, J.C., 1982. On the use of satellite data to infer surface fluxes at meteorological scales. *J. Appl. Meteor., 21*, 1111-1122.

Rao, K.S., J.C. Wyngaard and O.R. Coté, 1974: Local advection of momentum heat and moisture in micrometeorology. *J. of Atm. Sci.*, 29, 304-310.

Schmugge, Th., F.T. Ulaby and E.G. Njoku, 1978: *Microwave observations of soil moisture: review and prognosis*. Rep. Goddard Space Flight Center, Greenbelt, Maryland.

Scofield, R.A., 1984: satellite-based estimates of heavy precipitation. SPEI Vol 481, *Recent advances in civil space remote sensing*.

Seguin, B. and B. Itier, 1983: Using midday surface temperature to estimate daily evaporation from satellite thermal IR data. *Int. J. of Remote Sensing, 4*, 371.

Shuttleworth, W.J. and J.S. Wallace, 1985. Evaporation from sparse crops-an energy combination theory. *Quart. J.R. Met. Soc., 111*, 839-855.

Soer, G.J.R., 1980. Estimation of Regional Evapotranspirtion and soil moisture conditions using remotely sensed crop surface temperatures. *Rem. Sens. of Envir., 9*, 27-45.

Tennekes, H., 1973: A model for the dynamics of the inversion above a convective boundary layer. *J. Atmos. Sci., 30*, 558-567.

Thom, A.S. and H.R. Oliver, 1977: On Penman's equation for estimating regional evaporation. *Quart. J.R. Met. Soc., 103*, 345.

Wallace, J.S., J.H.C. Gash, D.D. McNiel and M.V.K. Sivakumar, 1986. Measurements and prediction of actual evaporation from sparse dryland crops. *Scientific Report Institute of Hydrology*.

Wallace, J.S., J.M. Roberts and A.M. Roberts, 1982: Evaporation from heather moorland in North Yokshire, England. *Proc. Symp. Hydrolog. Res. Basins, Bern*, 397-405.

Webb, E.K., G.I. Pearman and R. Leuning, 1980: Correction of flux measurements for density effects due to heat and water vapour transfer. *Quart, J.R.*, Met. Soc., *106*, 85-100.

Ten Berge, H.F.M., 1986. Heat and water transfer at the bare soil surface. Ph-D thesis, University of Wageningen.

APPENDIX I

| Symbol | Definition | Units |
|---|---|---|
| $c_p$ | specific heat of air at constant pressure | $J\ kg^{-1}\ K^{-1}$ |
| $r_a$ | aerodynamic resistance for transport of heat and water vapour of the air layer between the surface and screen height | $s\ m^{-1}$ |
| $r_c$ | canopy or surface resistance | $s\ m^{-1}$ |
| s | slope of the saturation specific humidity versus temperature curve at screen temperature T | $K^{-1}$ |
| A | available energy = net radiation minus (soil heat flux density + heat storage term for the vegetation layer) | $W\ m^{-2}$ |
| $D_i$ | saturation (specific humidity) deficit = saturation spec. hum. minus actual spec. hum. | |
| $D_{eq}$ | equilibrium saturation deficit see Eq. (2) | |
| E | evaporation (rate) | $kg\ m^{-2}\ s^{-1}$ (or $mm\ day^{-1}$) |
| P | precipitation | $kg\ m^{-2}\ s^{-1}$ (or $mm\ day^{-1}$) |
| $T_a$ | air temperature at screen height | K or °C |
| $T_s$ | surface temperature | K or °C |
| $\gamma$ | psychrometric constant ($c_p/\lambda$) | $K^{-1}$ |
| $\lambda$ | latent heat of vaporization of water | $J\ kg^{-1}$ |
| $\rho$ | density of air | $kg\ m^{-3}$ |

# SATELLITE REMOTE SENSING AND ENERGY BALANCE MODELLING FOR WATER BALANCE ASSESSMENT IN (SEMI-) ARID REGIONS

A.A. van de Griend
Institute of Earth Sciences, Free University
P.O. Box 7161
1007 MC Amsterdam

and R.J. Gurney
NASA-Goddard Space Flight Center
Hydrological Sciences Branch, Code 624
Lab. for Terrestrial Physics,
Greenbelt, MD 20771 624
U.S.A.

ABSTRACT. The terms of the water balance in semi-arid regions may be monitored using different types of remotely sensed information from satellites. Such an integrated approach focusses on the possibilities of monitoring the soil moisture status and evapotranspiration over time from the combination of (a) thermal infrared, (b) visible and near infrared and (c) passive microwave remote sensing.

Application of such an approach is expected to give a better insight into the spatial and temporal variability of soil moisture content and evapotranspiration, and therefore may contribute to a better understanding of large scale recharge phenomena in semi-arid regions. Actual recharge depends on subsurface hydrogeological conditions for percolation to underlying aquifers, and may vary between localized concentrated percolation on the one hand and more regional diffuse percolation on the other. Thus a subsurface scaling problem comes to the fore which is beyond the remote sensing approach described in this paper which focusses on the average surface conditions at pixel scale, as a possible boundary condition for recharge.

The surface energy balance may be modelled using thermal infrared surface temperature observations and large scale near surface meteorological information, which allows the evapotranspiration rate to be estimated and the soil moisture status to be inferred together with stress conditions of the vegetation. This requires a remotely sensed estimate of the vegetation cover and biomass which may be derived from visible and NIR signatures. Separately, the soil moisture status of the top soil may be derived from passive microwave signatures.

This paper illustrates the approach of using all these data together and describes some aspects of an energy and mass balance model to be used in the data analysis for scarcely vegetated areas, such as in semi-arid regions.



*I. Simmers (ed.), Estimation of Natural Groundwater Recharge, 89–116.*

## 1. INTRODUCTION

One of the major problems in water resources management in semi-arid regions is the lack of fundamental knowledge of the spatially and temporally variable hydrological status of the surface (such as soil moisture content), and related terms of the water balance (such as evapotranspiration and natural recharge).

Although many local scale field studies have been performed (see e.g. De Vries, 1984) they do not recognise the problems of the translation of point information to regional scale information by appropriate scaling and regionalisation.

With remote sensing from space now becoming regularly available, monitoring of the physical hydrological status of the surface becomes possible in several ways, using different types of remotely sensed signatures, each leading to a different type of information. However, an integrated approach, using the different types of remotely sensed signatures together has undoubtedly advantages over separate applications of each type of data. Such an integrated approach will be applied to the semi-arid regions of Botswana and focusses on the following items:

1. Estimation of evapotranspiration and soil moisture content by surface energy balance modelling, using thermal infrared signatures and additional estimates of the vegetation characteristics such as leaf area index (LAI), the latter being extracted from (3).

2. Estimation of soil moisture content in soils, mainly covered with low natural savanna vegetations, intermixed with agricultural fields, using passive microwave signatures.

3. Estimation of the characteristics of the vegetation (such as biomass and LAI) using visible and near-infrared signatures.

These research items are mutually supplementary in the sense that (3) gives crucial information to perform (1) and (2), whereas (1) gives additional information on the physiological condition of the vegetation.

Although the basic theory of the physical processes leading to remotely sensible signatures is fairly well understood, application of the above items is not straight-forward. Both energy balance and microwave models have been tested and calibrated predominantly using point or local ground truth observation, assuming spatially homogeneous conditions. Although such models have been applied to pixel scale signatures (e.g. Owe et al., 1986 : Raffy & Becker, 1986) little is known about the influence of spatial (within pixel) variability of surface characteristics on model performance and its consequences for the inverse problem (see e.g. Raffy & Becker, 1985).

Therefore, application of remotely sensed signatures to infer pixel scale physical hydrological information, requires fundamental solutions for a series of basic problems, which can be summarized as follows :

(a) development (c.q. calibration) of hydrometeorology and/or hydrology driven surface energy balance models, c.q. microwave emission models for pixel scale parameterization and subsequent calibration using remotely sensed signatures;

(b) development of a methodology for "model oriented" regionalization of point c.q. local scale ground truth for pixel scale calibration;

(c) development of a methodology for discrimination of physiographically significant uniform landscape units, (1) for subsequent selection of control areas for collection of ground truth and (2) as a basis for regional scale application of remote sensing models.

The above items form the backbone of a proposed cooperative research program of the Free University[1], NASA/GSFC[2] (U.S.A.), the Agricultural University[3] and the Meteorological Service of Botswana which aims at operationalization of the integrated use of satellite remote sensing for adequate monitoring of the hydrological status of the surface.
In the following sections the ways of using the different remotely sensed satellite signatures will be explained, together with their potentialities and restrictions in view of the preferable and the actually available spatial and temporal resolutions of the signatures.

## 2. PASSIVE MICROWAVE REMOTE SENSING

### 2.1 Introduction

Microwave remote sensing can be used to estimate the soil moisture content of the upper-most surface layer of the earth. The strong dependence of the dielectric properties of the earth's surface layer on moisture content - and its influence on the emissivity - forms the basis of the technology. Therefore, measurement of the emission of microwave energy in this region of the electromagnetic spectrum (about 0.75-80 cm wavelength, see Fig. 1) is especially qualified to estimate the moisture status of the earth's surface.

A very important advantage of this spectral region is the almost complete independance of atmospheric conditions, such as rain or clouds, which makes microwave remote sensing an "all weather" technology.

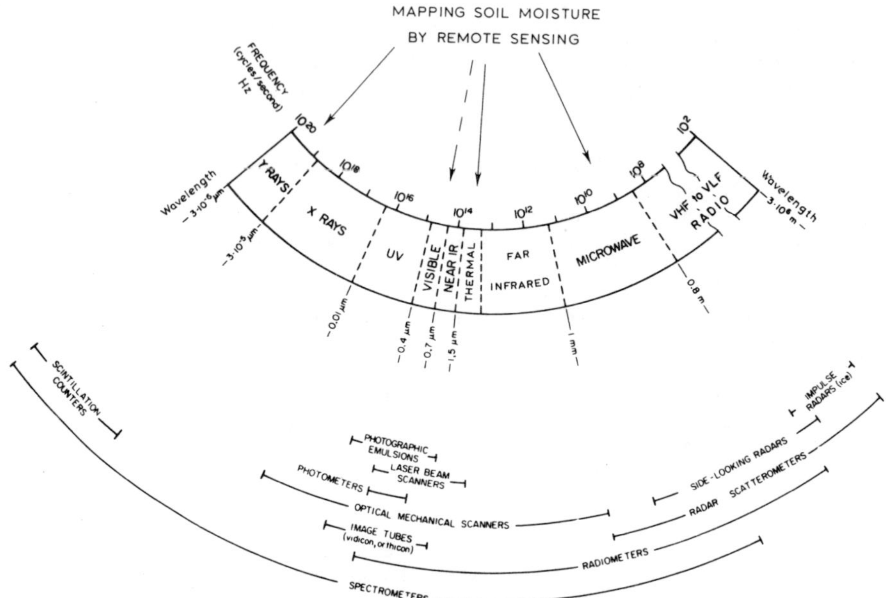


Fig. 1 The electromagnetic spectrum and relevant regions for mapping soil moisture by remote sensing
(From : Van de Griend & Engman, 1985).

## 2.2 Theory and applications

Microwave remote sensing is based on the measurement of emitted energy from the surface. At microwave wavelengths this intensity is proportional to the product of the temperature ($T_s$) and the emissivity ($\varepsilon_s$) of the surface (Rayleigh-Jeans approximation) and is commonly expressed as the brightness temperature ($T_B$), where

$$T_B = \varepsilon_s \; T_s \tag{1}$$

As a result of reflection and transmission, the actually measured value of $T_B$ by a radiometer at some height (h) above the surface equals (Schmugge et al., 1986) :

$$T_B = \tau(r \; T_{sky} + \varepsilon_s T_s) + T_{atm} \tag{2}$$

where r is the reflectivity of the surface and $\tau$ is the atmospheric transmission. The first term is the reflected sky brightness temperature, the second term is the emission from the surface ($1\text{-}r = \varepsilon_s$, the emissivity) whereas the third term is the contribution from the atmosphere (atmosphere brightness temperature). The different components are represented schematically in Fig. 2. The influence of atmospheric effects, however, is generally small.

The emissivity $\varepsilon_s$ is strongly dependent on the dielectric properties which, in turn, are a function of soil moisture content. However, in addition to moisture content, microwave emission is dependent on a series of other surface characteristics such as vegetation cover, surface roughness etc. as indicated in Fig. 3. These influences have been well documented in past research activities and have been summarised by Schmugge et al. (1986).

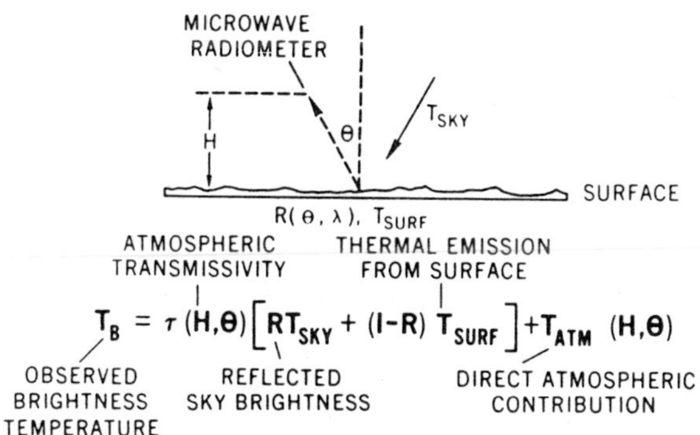


Fig. 2 Schematic diagram of the sources of microwave radiation measured by a radiometer (From : Schmugge, 1985).

## Skin depth

A very important characteristic of microwave radiation is that the energy is emitted not only from the surface, but from the top layer with a certain "skin-depth". This skin-depth is primarily a function of wavelength and soil moisture content (Fig. 4). Since the energy-emittance depth depends on wavelength, and thus on frequency, each frequency has its own depth range of penetration.

Radiometers flown on aircraft have demonstrated the potential for soil moisture determination (see e.g. Burke and Schmugge, 1982; Jackson et al., 1984). Figure 5 illustrates some of the results obtained from aircraft flights.

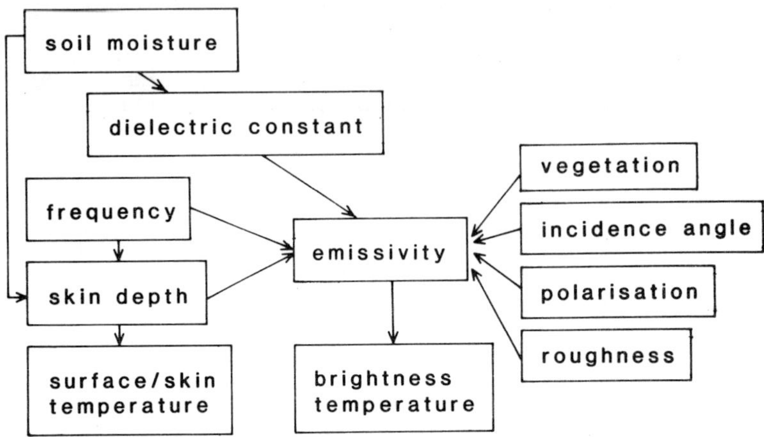


Fig. 3 Schematic representation of the various factors influencing the relationship between soil moisture content and $T_B$.

Surface roughness and vegetation

From the series of target characteristics given in Fig. 3, there are two which can have a significant effect on the measurement of soil moisture, i.e. surface roughness and vegetation. Roughness reduces the reflectivity, and therefore increases the emissivity of the surface (see e.g. Choudhury et al., 1979). Roughness, however reduces the sensitivity of emissivity to soil moisture variation and thus reduces the range of $T_B$ from wet to dry soils. Vegetation tends to absorb some of the emitted energy. The magnitude, however, depends on wavelength and water content of the vegetation. At the wavelengths used most frequently experimentally, 21 cm (L-band), 6.3 cm (C-band) and 2.8 cm (X-band), only the L-band penetrates low vegetation (say < 1.5 m) significantly and therefore its emission is directly a function of soil moisture. The shorter wavelengths are scattered and absorbed and the data must be corrected for vegetation scattering (Burke & Schmugge, 1982).

Satellite applications

Wang (1985) studied the sensitivity of satellite derived estimates of $T_B$ at different wavelengths to variations in moisture conditions for different vegetation types. Because actually measured soil moisture data are seldomly available for large scale calibration

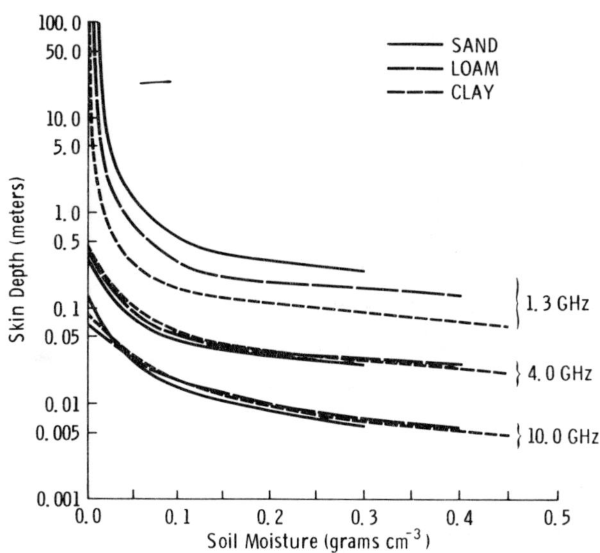


Fig. 4 Skin depth in a soil with variable water content for different frequencies (From : Cihlar & Ulaby, 1975).

(such as with satellite remote sensing) he used an API (Antecedent Precipitation Index) as a measure for the soil moisture status. In this study he compared Skylab 1.4 GHz ( $\lambda$ = 21 cm), and two channels of the Nimbus-7 Scanning Multichannel Microwave Radiometer (SMMR), i.e. 6.6 GHz ( $\lambda$ = 4.5 cm) and 10.7 GHz ( $\lambda$ = 2.8 cm), for two regions, i.e. range land with crops and forest. The 1.4 GHz gave the highest sensitivity and even seems to penetrate forest canopies (see Fig. 6). Further, it was shown that, notwithstanding the theoretical low skin-depth and the partial scattering due to the vegetation cover, even the 6.6 GHz showed a significant correlation between $T_B$ and API for range land vegetation (see Fig. 6). Therefore, an independent means of estimating the biomass of vegetation cover is expected to improve the soil moisture determination significantly. This, however, requires physically based models describing the radiation transfer through vegetation layers in the microwave regions, such as those described by Ulaby et al. (1981, 1982).

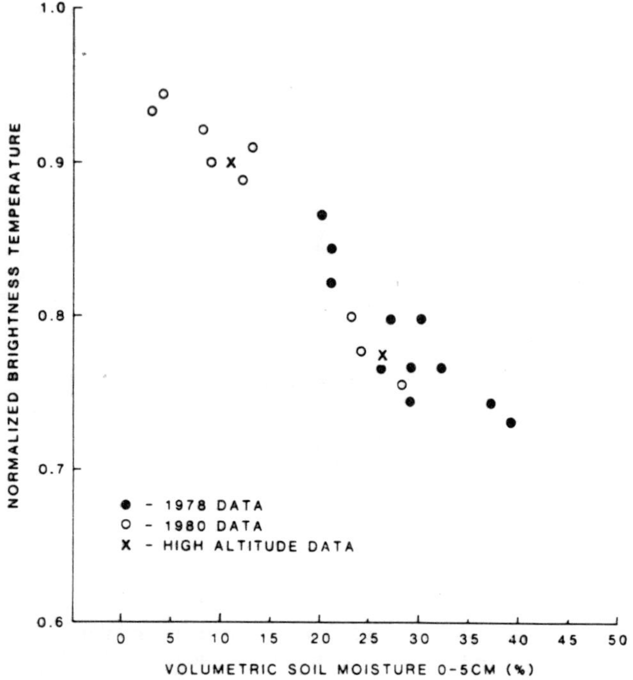


Fig. 5 Passive microwave data for Oklahoma watersheds, L-band, 0° look angle and H-polarization (From Jackson et al., 1984).

Limitations

A major limitation of microwave remote sensing may be the restricted spatial resolution from any significant altitude. This is because the thermal emission is of very low intensity and its measurement therefore requires very sensitive radiometers.

In general, the angular resolution of the radiometer is a function of the size of the antenna. This angular resolution is approximately $\lambda/D$, where D is the size of the antenna and forms a major factor in evaluating the potential use from space. For example, a 1.4 GHz ($\lambda$ = 21 cm) radiometer with a 10 m antenna mounted on a satellite in a 500 km orbit would have a spatial resolution of 10 km. Potentialities of such a system are obviously in the field of large scale monitoring of soil moisture conditions and their spatial variations. Although microwave remote sensing will not lead to profile soil moisture estimates as accurately as achievable at a point by conventional methods, the main advantage of the method is the possibility of measuring large scale soil moisture conditions on a repetitive basis.

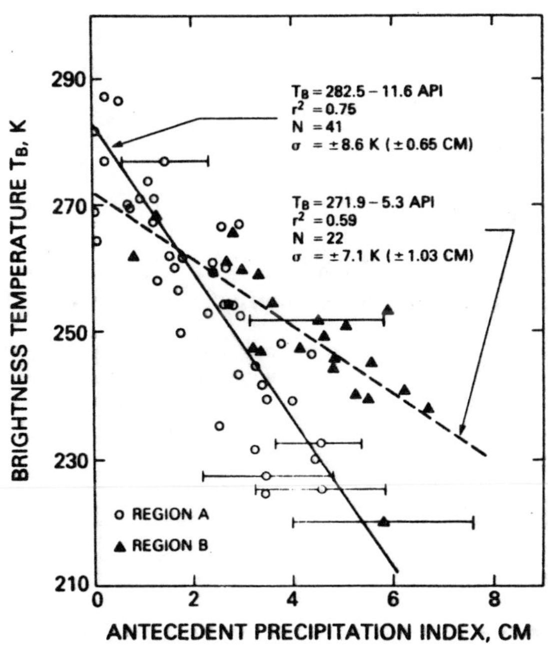


Fig. 6[a] $T_B$ measured by Sky Lab. 1.4 GHz radiometer plotted versus API. (Region A : range land; Region B : forest). From : Wang, 1985.

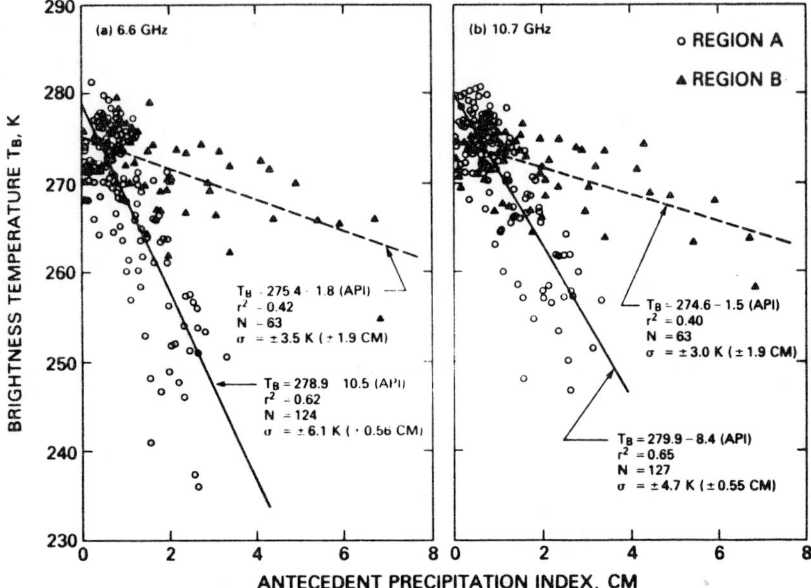


Fig. 6[b] $T_B$ measured by SMMR plotted versus API for (a) 6.6. GHz and (b) 10.7 GHz. (Region A : range land; Region B : forest). From : Wang, 1985.

## 3. THERMAL INFRARED REMOTE SENSING

### 3.1 Introduction

Thermal infrared remote sensing is based on the measurement of emitted energy from the surface in the atmospheric window of wavelength range 10.5-12.5 μm (see Fig. 1). It can be used to determine the surface temperature of landforms or the vegetation canopy. Surface temperature variations result from the balance of radiant, latent, sensible and soil heat fluxes. Heat fluxes are determined by spectral characteristics of the surface, meteorological conditions and thermal properties of the ground. Therefore, application of thermal infrared remote sensing to infer information on the physical status of the surface (such as soil moisture content) and on the magnitude of the fluxes (e.g. latent heat of evapo(transpi)ration) comes down to solving the energy balance (see e.g. Van de Griend et al., 1985) :

$$G = H + LE + R_{net} \tag{3}$$

where G, H and LE are the ground heat, sensible heat and latent heat fluxes, and $R_{net}$ is the net radiation.

The ground heat flux (G) is determined by thermal properties of the ground, and may be modeled using the coupled moisture and heat diffusion equations by Philip and De Vries (1957) whereas the soil moisture flux is dictated by soil-water potential, soil physical characteristics and plant-resistances to match the atmospheric demand (Camillo et al., 1983).

The latent and sensible heat fluxes may be modeled using standard models such as :

$$LE = - \frac{\rho c_p}{\gamma} \frac{e_s - e_a}{r_s + r_a} \tag{4}$$

$$H = - \rho c_p \frac{(T_s - T_a)}{r_a} \tag{5}$$

where $\rho$ is the density of the air, $c_p$ the specific heat, $\gamma$ the psychrometric constant, $e_s$ the surface vapor pressure, $e_a$ the air vapor pressure, $r_s$ the surface resistance, and $r_a$ the aerodynamic resistance. Corrections for unstable conditions may be included using Businger-Dyer's resistance formulation based on Monin-Obukhov's stability criterion (see e.g. Businger et al., 1971; Dyer, 1974; Brutsaert, 1982).

The net radiation is usually divided into long-wave and short-wave (solar) components, where both components can either be measured or modeled using standard models (see e.g. Eagleson, 1970).

For evaporation (bare soil or bare soil fraction) the resistance $r_s$ is a surface diffusion resistance and can be modeled as a function of soil water potential (Camillo & Gurney, 1986). For transpiration the resistance $r_s$ is the bulk stomatal resistance and can be modeled as a function of plant water potential (see e.g. Van de Griend et al., 1985) and as a function of plant physiological characteristics and available short wave radiation within the canopy (see e.g. Sellers, 1986).

### 3.2 The inverse problem

The components of the energy-balance equation can be defined as functions of known weather, soil and plant conditions, with the only unknown being the (apparent) surface temperature, i.e. the remotely sensed composite temperature of the surface. In the inverse problem, therefore, solution of the energy-balance by matching the modeled surface temperature to the remotely sensed surface temperature leads to estimates of the individual terms, of which LE (latent heat of evapotranspiration) and G (ground heat flux) are of direct interest for hydrological applications. This approach has been used in different grades of sophistication both for bare soils (Rosema, 1975; Price, 1977, 1980, 1982; Camillo et al., 1983; Monenti, 1984) and vegetated surfaces (Soer, 1980; Gurney & Camillo, 1984; Van de Griend et al., 1985).

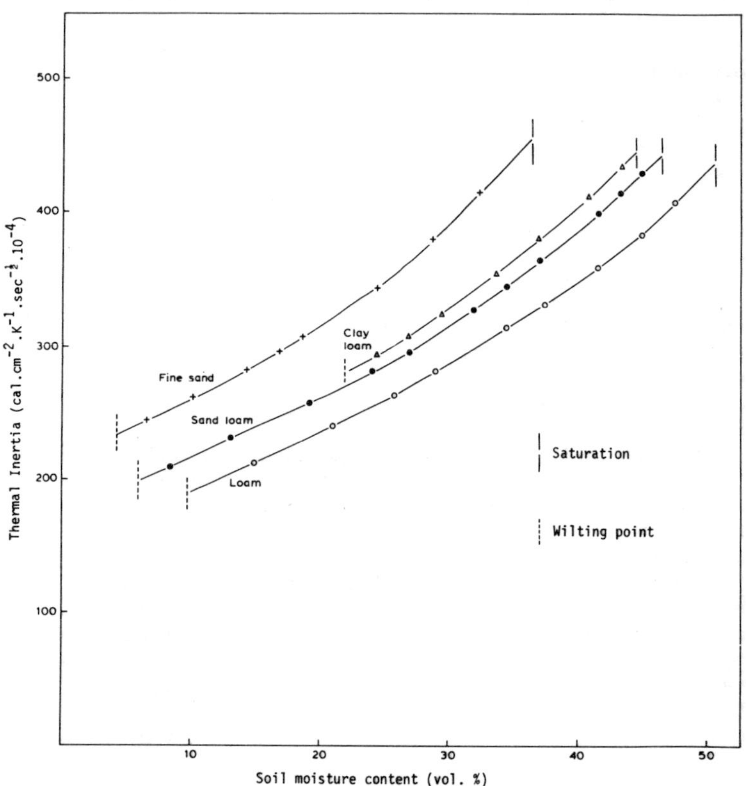


Fig. 7 Thermal inertia (TI) versus volumetric soil moisture content for four different standard soil types (From : Van de Griend et al., 1985).

Thermal properties of the upper soil layers are usually expressed as "thermal inertia", TI = $\sqrt{\rho c \lambda}$, which depends on thermal conductivity ($\lambda$), specific heat (c) and density ($\rho$) of the subsurface. It represents the ability to absorp and transmit heat, and is a function and of both soil type and moisture content (Fig. 7). The relation between TI and soil moisture as a function of soil type can be derived using the De Vries models (1975) for thermal conductivity $\lambda_s$ and volumetric heat capacity ($C_s$), where $C_s = \rho$ c.

From the energy balance (3) in which :

$G = f(T_s, TI)$
$H = f(T_s, r_a)$
$LE = f(T_s, r_a, r_s)$
$R_{net} = f(T_s)$

it follows directly that for a fixed meteorological condition the daily course of the energy-balance is determined by the daily courses of $T_s$, LE and the value of TI, which is invariable during equilibrium conditions. Therefore, if daily maximum and minimum surface temperatures would characterise the daily course sufficiently well, there should be a near-unique relationship between the diurnal surface temperature on the one hand and LE and TI on the other, or

$$T_s(\text{diurnal}) = f(E, TI).$$

Such a relationship is shown in Fig. 8. From sensitivity studies (Van de Griend et al., 1985, 1986) it was found that one day-time and one night-time surface temperature observation are sufficient to estimate both E and TI.

Application of thermal-infrared remote sensing to derive the terms of the energy-balance over large areas, using signatures gathered by satellites, however, is not straight forward. A series of problems, both in the field of theoretical modelling as well as in the field of practical application, should be recognised. Some of these aspects will be briefly discussed below.

### 3.2.1 Problems of advection and inhomogeneous areas for the inverse problems

A major problem in applying surface energy balance models for large inhomogeneous areas is the phenomenon of advection. Both heat and latent heat fluxes are functions of temperature and vapor pressure gradients respectively between the surface and at some height in the atmosphere above. In homogeneous areas the atmospheric vapor pressure is influenced by evapotranspiration and the atmospheric boundary layer tends to reach a quasi-equilibrium situation of heat and moisture exchange within the soil-vegetation-atmosphere continuum.

In inhomogeneous areas, such as with spatially variable soil moisture or vegetation conditions, advection of air from one area into another implies automatically spatially variable conditions of heat and moisture fluxes, especially along the boundaries of homogeneous areas.

Sophisticated surface energy balance models use observed meteorological parameters to derive LE, H and G from surface temperature observations, and have been tested mainly for homogeneous areas and with local scale ground truth. Application of such models to larger areas obviously produces a problem of

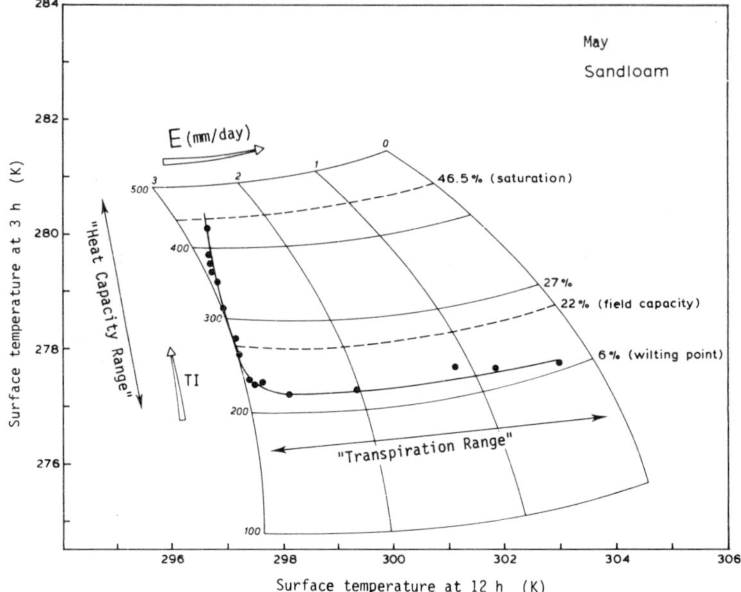


Fig. 8 Maximum and minimum daily surface temperature as a function of thermal inertia (TI) and transpiration (E). The curve drawn through the points represents the compiled relationship for the soil type Sand-loam.

scaling. With satellite remote sensing the scale of the problems is determined by the spatial resolution of the remotely sensed signatures usually expressed in terms of pixel-dimensions. Therefore future research should be focussed on a series of basic problems related to (1) the within pixel variability (internal variability) of surface physical and micro-meteorological conditions, (2) their influence on pixel scale model performance and (3) development of a methodology for regionalisation of local scale ground truth for pixel scale calibration and application.

Solutions for these problems therefore form a prerequisite for solving the inverse problems.

### 3.2.2 Aerodynamic resistance for sensible and latent heat fluxes

The roughness length $z_o$ is an extremely dominant parameter in the aerodynamic resistance formulation for sensible and latent heat :

$$r_a = \frac{[\ln(\frac{z-d}{z_{o_m}})-P_1]\cdot[\ln(\frac{z-d}{z_{o_h}})-P_2]}{k^2\, u\, (\frac{k_v}{k_m})} \tag{8}$$

where z is the height above the surface, d is the displacement length, k is Von Kármán's constant, u is the wind speed at height z, $P_1$ and $P_2$ are Monin-Obukhov stability corrections, $k_v$ and $k_m$ are diffusivities for vapor and momentum respectively, and $z_{o_m}$ and $z_{o_h}$ are roughness lengths for momentum and heat.

In areas with horizontal variability of surface characteristics, such as with isolated plants or groups of plants, the roughness lengths for sensible heat ($z_{o_h}$) and latent heat ($z_{o_v}$) are not by definition the same. The inequality $z_{o_h} \neq z_{o_v}$ is explained from the different sources for the bulk of heat and latent heat fluxes. The plant leaves form the principle source for the latent heat fluxes whereas the bulk sensible heat originates from the open bare soil spots. The same probably counts for the displacement height d.

The main problem from a remote sensing point of view, however, is the fact that even for homogeneous areas $z_o$ cannot be determined in the field other than by matching observed heat and latent heat flux profiles and wind profiles to modeled profiles. For inhomogeneous areas this becomes practically impossible since $z_o$ varies from one place to another. Therefore, probably the best way to estimate $z_o$ for pixel-scale applications is to use $z_o$ as an optimization parameter in the energy balance model by matching the observed diurnal course of the (apparent) surface temperature to the remotely sensed surface temperatures such as done in Van de Griend & Camillo (1986) and Camillo & Gurney (1986). Those studies were done for low grass vegetation and bare soil respectively, where the sources for sensible and latent heat are approximately at the same height.
Under circumstances where $z_{o_h} \neq z_{o_v}$, such as in inhomogeneous sparsely vegetated regions this problem might be approached analogously in combination with empirical evidence of the ratio $z_{o_h}/z_{o_v}$ for different vegetation associations or biotopes, preferably on the basis of field experiments in representative areas.

### 3.2.3 Sensible and latent heat fluxes from bare soils

For bare soils, the sensible and latent heat fluxes also have different sources, except where the soil surface is wet. Sensible heat flux originates at the surface, whereas latent heat fluxes occur in the soil profile itself due to a phase change from liquid to vapor and vice versa. Vapor transport into the atmosphere then originates from the soil-air spaces through the top layer of the profile.

Phase changes are dominated by temperature changes and temperature gradients, at least up to a PF = 4.2 (i.e. wilting point potential) since vapor pressure in the soil-air spaces is saturated in this range. At suctions < PF = 4.2 vapor pressure is a function of both suction and temperature (see e.g. Eagleson, 1970).

Latent heat exchanges in the profile can be modeled using the earlier mentioned Philip & De Vries (1957) equations for heat and moisture diffusion. Menenti (1984) derived a modified approach to model the heat and vapor fluxes based on the concept of the evaporating front. In both models the heat and moisture fluxes in the profile are described in terms of temperature and moisture gradients, and related phase change processes, whereas the transportation of vapor from the surface into the air occurs directly from the vapor phase in the topsoil into the air above, according to

$$E_s = - \frac{\rho c_p}{L\gamma} \frac{e_s - e_a}{r_a + r_s} \tag{9}$$

Here $r_s$ is a surface diffusion resistance, which was modeled by Camillo & Gurney (1986) as a function of the soil moisture content in the top 5 cm of the soil.

### 3.2.4 The accuracy of surface temperature measurement by satellites

Remote measurement of the surface temperature is based on measurement of the emitted longwave radiation which is a function of the radiative surface temperature according to Stefan-Bolzman :

$$\uparrow R_{Long} = \varepsilon_s \, \sigma \, T_s^4 \tag{10}$$

where $\varepsilon_s$ is the emissivity and $\sigma$ is Stefan-Bolzman's constant ($\sigma = 5.669 \; 10^{-8} \; w/m^2/k^4$)

Although sensors have been developed to measure the black-body radiation (i.e. for $\varepsilon_s = 1$) with an accuracy equivalent to 0.1 k, problems arise from atmospheric influences such as from atmospheric moisture and dust. Absorption by atmospheric vapor is lowest in the window from 10.5 - 12.5 μm, but may be significant under certain conditions. Both empirical and physical corrections have been developed for atmospheric corrections (see e.g. Clough et al., 1980).

Menenti (1984) discussed the uncertainties of the correction for atmospheric water vapor using a radiative transfer model. Such corrections are based on temperature, atmospheric pressure and vapor pressure profiles of the atmosphere which can be determined using radio-soundings. Menenti (1984) concluded that due to a series of uncertainties (spatial variability of vertical profiles and optical depths, the use of "near-by" radio soundings, uncertainty about absorption coefficients of atmospheric vapor etc.), the accuracy of the correction is of the same order of magnitude as the correction itself and therefore the atmospheric correction procedures, using radio soundings, have no operational value.

More recent developments are the use of multi-frequency atmospheric sounders on board earth observation satellites to derive corresponding atmospheric profiles for correction of surface signatures. The use of such information would improve the correction procedures significantly. In this respect the EOS (Earth Observation System) should be mentioned which has been designed especially for energy-balance and hydrological applications (NASA, 1986) and includes a special sounder for atmospheric corrections for the various wavelengths. It is still in the planning-phase but indicates near future developments in the field of energy-balance modeling from space.

As long as atmospheric correction techniques have no practical value, direct comparison of satellite measured signatures and ground truth seems a necessary control procedure. This however is not straight-forward and extremely difficult in inhomogeneous areas. In such environments pixel representative sampling would require low-flying aeroplanes to do observations such as done by Menenti (1984).

Despite the theoretical uncertainties, and scale problems with respect to the collection of ground truth, the correspondence between satellite derived surface temperature and "ground truth", collected by handheld radiometer and low flying aircraft, as reported by Menenti (1984) is extremely good (Fig. 9). The differences are small, and certainly within the error of ground truth resulting from within pixel variability of the surface temperature. It is generally expected that the surface temperature can be estimated under favourable conditions with an accuracy of 1-1.5 k.

The surface-emissivity $\varepsilon_s$

One of the uncertainties in estimating the surface temperature from space is the emissivity of the surface $\varepsilon_s$, in eq. 10. The emissivity of most natural surfaces varies between 0.92 and 0.98 (Smith, 1983). Although tables have been published for average values for certain surface types the exact value is difficult to establish. The resulting error in the computed surface temperature, however, is relatively small. The total outgoing longwave radiation follows from :

$$\uparrow R_{Long} = \varepsilon_s \ \sigma \ T_s^4 + (1-\varepsilon_s) \ \varepsilon_a \ \sigma \ T_a^4 \qquad (11)$$

where $\varepsilon_s$, $\varepsilon_a$ and $T_s$, $T_a$ are the emissivities and temperatures of the surface and the air respectively. In this equation the second term is the emitted longwave radiation and the third term is the

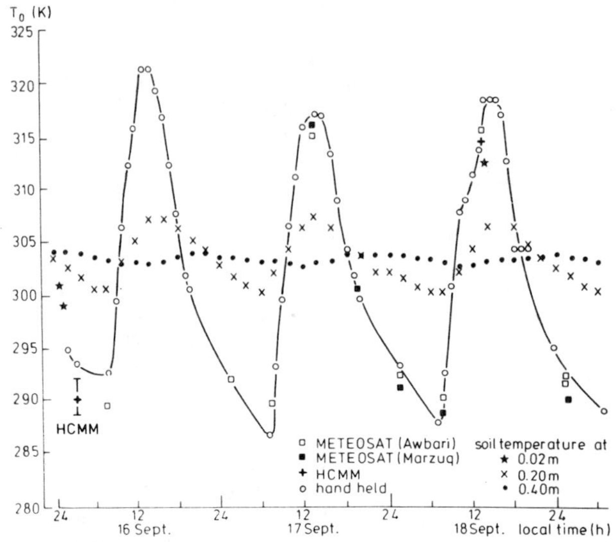


Fig. 9 Surface radiation temperature $T_o$ (K) as measured by METEOSTAT and HCMM radiometers, and on the ground by hand-held radiometers (From : Menenti, 1984).

reflected longwave sky radiation. It follows directly from this equation that an error in $\varepsilon_s$ is compensated in large part by the reflectivity r, where $r = 1 - \varepsilon_s$. The atmospheric $e_a$ can be determined empirically as a function of atmospheric vapor pressure and air temperature (see e.g. Hatfield et al., 1983) and varies between 0.80 for clear sky and 0.95 for complete overcast conditions (see e.g. Eagleson, 1970).

## 3.3 The (apparent) surface temperature in sparsely vegetated areas

### 3.3.1 Introduction

For low vegetation with complete coverage, the vegetation may be represented as a simple uni-layer, under the assumption that the aerodynamics of momentum, heat and latent heat exchange within the canopy is of minor influence on the apparent surface temperature sensed from above (see e.g. Soer, 1980; Nieuwenhuis, 1981). For higher vegetation, however, especially with incomplete coverage, light penetration and light scattering (Ross, 1975) as well as the aerodynamics within the canopy determine the vertical temperature distribution within the vegetation and the soil underneath.

This temperature distribution determines the apparent surface temperature, which is a composite temperature of the surface fractions exposed to the sensor, with the sensor looking down either at or off nadir.

Estimation of the surface energy balance term (eq. 3) from observed apparent surface temperatures, therefore requires a surface energy balance model with a canopy representation which simulates the vertical temperature distribution of the foliage elements.

### 3.3.2 Modelling the (apparent) surface temperature

In order to simulate the apparent surface temperature of (sparsely) vegetation surfaces a multi-layer canopy-model has been developed (Van de Griend, 1987). This model simulates the role of the vegetation in the exchange of heat and moisture between the soil and the atmosphere and thus allows simulation of the influence of the vegetation layer on the sensitivity of the apparent surface temperature to differences in thermal inertia of the soil. It is therefore a helpful tool for the inverse problem, i.e. estimation of heat and latent heat fluxes and thermal inertia from observed apparent surface temperatures.

The model is based (1) on the simultaneous solution of the energy balance equation $R_{net} = H + LE$, defined for each individual horizontal layer of the canopy, (2) on light penetration and scattering theory within canopies, and (3) on a description of the aerodynamic resistances for heat and latent heat exchange within and between the individual layers of the canopy. The vegetation canopy forms an integral part of the soil-vegetation-atmosphere continuum.

### 3.3.3 Light penetration and scattering in the vegetation canopy

Light penetration and scattering in the vegetation layer is especially relevant in terms of available energy for heat and latent heat exchange. The depth of penetration is in large part a function of plant structure (leaf distribution, leaf orientation and leaf inclination) and cumulative leaf area index (L).

The influence of plant structures for radiative transfer has been parameterised by Ross (1975) in terms of so-called light penetration functions :

$$\tau_s (L, \beta) = \exp(- L \frac{G(\beta)}{\sin \beta}) \qquad (12)$$

$$\tau_d (L) = 2 \int_{o}^{\pi/2} \exp \left[-L \frac{G(\Theta)}{\cos \Theta}\right] \cos \Theta \sin \Theta \, d\Theta \tag{13}$$

where $\tau_s$ and $\tau_d$ are the fractional penetrations at depth L for direct solar and diffuse radiation respectively, $G(\beta)$ is the G-function, giving the projection of unit foliage area in the direction of the sun ($\beta$ being the solar angle) or any other direction $\Theta$.

The G-function is a characteristic of a certain vegetation type and is independent of L. Therefore eq. 12 describes the light penetration of direct solar radiation at some depth L as a function of solar angle and vegetation type, and eq. 13 for diffuse sky radiation.

The light penetration functions of Ross (1975) were used by Dickinson (1983) to formulate the two-stream approximation method of radiative transfer in vegetative canopies. This method describes the upward and downward diffuse radiative fluxes as a function of L, normalised by the incident flux. Sellers (1985) used Dickinson's formulation and derived a set of analytical solutions for the upward and downward diffuse fluxes considering situations where the soil surface may play a part in the scattering of direct beam radiation (i.e. scarse canopy).

Combination of Sellers equation for diffuse radiation and equations 12 and 13 have been used to compute $R_{net}$ for the individual layers in the multi-layer canopy model.

### 3.3.4 Leaf Area Index (LAI) and cumulative leaf area index (L)

An important aspect of vegetation characteristics is the vertical distribution of foliage area density a(z) with dimension $m^2/m^3$. The cumulative leaf area index L is defined as

$$L(z) = \int_{z}^{h} a(z) dz \tag{14}$$

where h and z are respectively the height of the canopy and a height within the canopy above the soil surface.

The leaf area index is then defined as

$$LAI = \int_{o}^{h} a(z) dz \tag{15}$$

As argued above cumulative leaf area index L and plant structure play an important role for the available energy within the canopy. Plant structure determines the light penetration as a function of L. Although the determination of plant structure is difficult at this moment by means of remote sensing, the use of oblique looking sensors in specific wave lengths may lead to such possibilities. However, since the vegetation structure is characteristic for certain species, described by the G-function, it probably does not need to be determined by remote sensing as long as the vegetation type in the area of application is known.

In the multi-layer canopy model, L is used to define the thickness of the individual layers, which together have a thickness of LAI. Therefore LAI should preferably be determined from remote sensing.

## 4. VISIBLE and NIR Remote Sensing

### 4.1 Vegetation and Leaf Area Index (LAI)

Determination of vegetation characteristics from remote sensing signatures is primarily based on measurement of reflectance characteristics in the visible and near infrared portions of the spection (Fig. 1). In the visible spectral region (0.58-0.68 μm) chlorophyll is responsible for considerable absorption of incoming radiation whereas in the near infrared spectral region (0.73-1.1 μm) mesophyl leaf structure leads to considerable reflectance. This contrast in both channels has been used extensively for studying different land surfaces and vegetation characteristics. In order to represent this contrast in one index several ratio transforms have been proposed, of which the normalised difference vegetation index (NDVI) has been shown to be highly correlated with vegetation parameters such as green-leaf biomass and green-leaf area (Curran, 1980; Holben et al. 1980, Tucker et al. 1981, 1985, Justice et al. 1985). It is defined as :

$$NDVI = (Ch2 - Ch1)/(Ch2 + Ch1) \tag{16}$$

where Ch1 represents data for the visible channel (0.58-0.68 μm) and Ch2 represents data from the near-infrared channel (0.73-1.1 μm).

Radiative transfer models have been used to simulate reflection characteristics of vegetation canopies as a function of canopy-structure, physiological parameters and LAI (Sellers, 1985; Asrar, 1986). Those models focus on the photo-synthetically active radiation (PAR) in the range of 0.4-0.7 μm and on the above-mentioned visible and NIR channels. Asrar et al. (1986) used such models to devise simple empirical models and compared measured and estimated LAI-values for prairie-grass canopies.

The correlation between spectral reflectance and vegetation indices decreases with increasing vegetation biomass and height of the vegetation stand due to the fact that penetration of direct radiation decreases exponentially with depth (eq. 12). The correlation therefore is higher for relatively low vegetation types such as natural grass vegetations of savanna and prairies (Asrar et al., 1986).

### 4.2 Albedo

Albedo, or the reflectivity in the visible and NIR region, plays an important role in the energy balance of the surface. Although average values for specific surface types are known from the literature it should preferably be estimated remotely.

Menenti (1984) compared albedo measurements in a desert area with albedo-values determined from Landsat and HCMM (Heat Capacity Mapping Mission) data and found the results to be very close. This result is encouraging for (scarsely) vegetated surfaces such as in semi-arid regions, where pixel-scale ground-truth on albedo is difficult to gather due to the spatial variability and is probably only possible using aircraft.

### 4.3 Discrimination of physiographically uniform landscape units for large scale application of remote sensing

Application of physically based remote sensing techniques to larger (multiple pixel) areas automatically introduces the problem of between-pixel variability (external variability) of surface physical conditions. Therefore large scale application requires a methodology for discrimination of physiographically uniform landscape units, as a basis (1) for adequate application of surface energy balance and microwave emission models and (2) for selection of control areas for the collection of ground truth. Such a discrimination can be done most efficiently using Landsat-TM data (visible + NIR channels) and standard image analyses techniques.

## 5. ON THE INTEGRATED USE OF REMOTELY SENSED SIGNATURES

As explained in the text above, both the microwave and thermal infrared methods need some information on the vegetation biomass and its structure. For passive microwave emission this is primarily to model or estimate the transfer of the signal through the canopy, whereas for thermal infrared it forms an integral part of the process of momentum, heat and moisture exchange within the earth-atmosphere interface. This explains the necessity to integrate visible and near-infrared with both microwave remote sensing and thermal-infrared remote sensing.

Combination of microwave remote sensing and thermal infrared remote sensing also offers obvious advantages. Although thermal infrared remote sensing can be used to determine both evapotranspiration and the thermal inertia of the top layers of the soil, it is obvious that the sensitivity of the surface temperatures to variations in soil moisture content decreases with increasing LAI. Therefore an independent estimate of top soil moisture reduces the number of unknown parameters in the surface energy balance model and may therefore contribute to the accuracy of the estimation of evapotranspiration. The integrated use of the different signals is shown in Fig. 10.

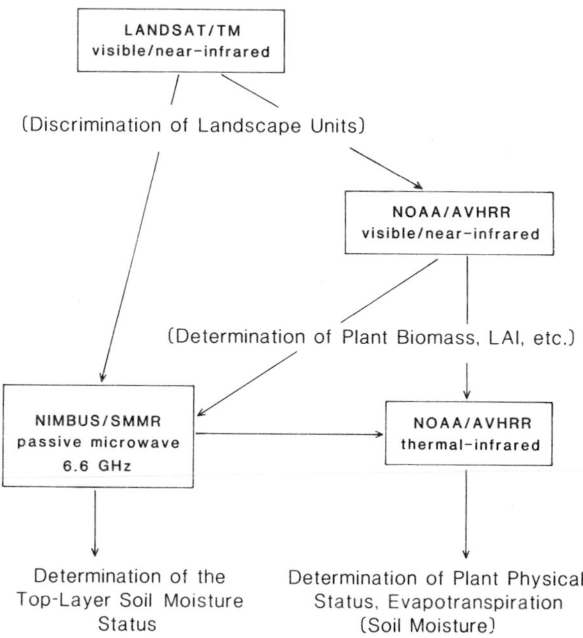


Fig. 10 Schematic representation of the integrated use of different types of remotely sensed signatures for monitoring the physical-hydrological status of the surface in (semi-)arid regions

## 6. APPLICATION OF THE INTEGRATED USE FOR BOTSWANA

The above described integration will be applied to Botswana using Landsat/TM, NOAA/AVHRR and Nimbus/SMMR data. An overview of specifications with respect to spatial and temporal resolution, applications and restrictions is given in table I.

Table I Overview of satellites and their specifications, used in an integrated approach for monitoring the physical-hydrological status of the surface in (semi-)arid regions.

| SATELLITE | | | | | SENSOR SPECIFICATONS | | | | | |
|---|---|---|---|---|---|---|---|---|---|---|
| Name | Orbit | Altitude | Identification | Description | Spectral domain | Wavelength/frequency | Spatial re-solution | Temporal resolution | Cloud Penetra-ting Yes | No. |
| LANDSAT | polar | 9.18 km | TM | Thematic mapper | visible | 0.4-0.8 μm | 30 m | 18 days | | * |
| | | | | | near infrared | 0.3-3.0 μm | 30 m | 18 days | | * |
| NOAA | polar | 8.35 km | AVHRR | Advanced very High Resolution Radio-meter | visible/near infrared | 0.50-0.68 μm | 1 km (LAC)/ | 12 hrs | | * |
| | | | | | | 0.73-1.1 μm | 4 km (GAC) | 12 hrs | | * |
| | | | | | thermal infra-red | 3.55-3.93 μm | 1 km (LAC) | 12 hrs | | * |
| | | | | | | 10.5-11.5 μm | 4 km (GAC) | 12 hrs | | * |
| | | | | | | 11.5-12.5 μm | | 12 hrs | | * |
| Nimbus | polar | 1100 km | SMMR | Scanning Multi-channel Micro-wave Radiometer | microwave | 37 GHz | 20 km | 1 day | * | |
| | | | | | | 21 GHz | ↓ | | * | |
| | | | | | | 18 GHz | | | * | |
| | | | | | | 10.7 GHz | | | * | |
| | | | | | | [illegible].6 GHz | 156 km | | * | |

(contin.)

| SATELLITE | | | APPLICATIONS |
|---|---|---|---|
| Name | Allows physically based modelling of the earth's physical status Yes | No | |
| LANDSAT | | * * | Classification of physiographically different landscape units (1) for selection of control <u>areas</u> for ground truth collection and (2) as a basis for regional scale application of remote sensing models |
| NOAA | | * * | Monitoring of biomass production, leaf area index (LAI), greenmass etc. |
| | * * * | | Thermal infrared modelling of surface energy balance processes to infer information on soil moisture status and heat fluxes |
| Nimbus | * * * * * | | Determination of soil moisture content in the topsoil by direct correlation and calibration or using radiative transfer models |

The approach will be tested for three areas of uniform vegetation characteristics around meteorological stations, where intensive ground truth sampling takes place. These test areas are approximately 120 km apart. In each area an intensive soil moisture monitoring program is going on within the framework of the Integrated Farming Pilot Project (IFPP) sponsored by the United Nations. Soil moisture is measured by neutron-probe. Each test area consists of 30 measurement sites in an area of 12 km long and 0.1 km wide. This soil moisture monitoring program provides unique data sets of ground truth, essential for studying the applicability of the above described approach as well as the study items described in section 1.

## 7. SUMMARY AND CONCLUSION

In this paper the potentialities of the integrated use of remotely sensed signatures are explained for monitoring the physical status of the earth's surface. The method will be used to estimate the spatial and temporal variability of soil moisture content, biomass development and evapotranspiration and will be applied to Botswana. Although the research program focusses at a final practical application, it has a series of scientific aspects and problems which require fundamental solutions. In this paper these scientific aspects have been summarised.

The program will run concurrently with an ongoing SE-Botswana groundwater recharge project (GRES-DGIS, 1987) with full mutual support and exchange of knowledge. It is therefore expected to contribute directly to a better understanding of the spatial variability of large scale soil moisture content and evapotranspiration. Both components play an important role for the assessment of water balance terms and thus may contribute to the better understanding of large scale recharge phenomena in semi-arid regions. Moreover, the program has several more scientific and practical applications, directly related to international programs and priorities (e.g. ISLSCP, HAPEX-MOBILHY).

## References


Asrar, G., Kanemasu, E.T., Miller, G.P. and Heiser, R.L. (1986) : Light interception and leaf area estimates from measurements of grass canopy reflectance. IEEE Transactions in Geosc. & Rem. Sens. vol. GE-24 (1), 76-82.

Asrar, G., Fuchs, M., Kanemasu, E.T. and Hatfield, J.L. (1984) : Estimating absorbed photosynthetic radiation and leaf area index from spectral reflectance in wheat. Agron. J. 76, 300-310.

Brutsaert, W. (1982) : Evaporation into the atmosphere. Reidel, Hingham, Massachusets.

Burke, H.K. & Schmugge, T.J. (1982) : Effects of varying soil moisture contents and vegetation canopies on microwave emissions. IEEE Trans. Geosci. Remote Sensing, GE-20 (3), 268-274.
Businger, J.A., Wijngaard, J.C., Izumi, IJ, and Bradley, E.F. (1971) : Flux profile relationships in the atmospheric surface layer. J. Atmos. Sci., 28, 181-189.
Camillo, P.J. & Gurney, R.J. (1986) : A resistance parameter for bare-soil evaporation models. Soil Science, vol. 141 (2), 95-105.
Camillo, P.J., Gurney, R.J. and Schmugge, T.J. (1983) : A soil and atmospheric boundary layer model for evapotranspiration and soil moisture studies. Water Res. Res. (19), 371-380.
Choudhury, B.J., Schmugge, T.J., Newton, R.W. and Chang, A. (1979). Effect of surface roughness on the microwave emission from soils. J. Geophys. Res. 84 (C9), 5699-5705.
Cihlar, J. & Ulaby, F.T. (1974) : Dielectric properties of soils as a function of moisture content. CRES Technical Report 177-47, Univ. of Kansas, Lawrence, Kansas.
Clough, S.A., Kneizys, F.X., Davies, R., Gamache, R. and Tipping, R. (1980) : Theoretical line shape for $H_2O$ vapor (application to the continuum). In : Deepak, A., et al. (1980). Atmospheric water vapor. Academic Press, London, 25-46.
Curran, P.J. (1980) : Multispectral remote sensing of vegetation amount. Prog. Phys. Geogr., 4, 315.
De Vries, J.J. (1984) : Holocene depletion and active recharge of the Kalahari groundwater - a review and an indicative model. J. Hydrol., 70, 221-232.
De Vries, D.A. (1975) : Heat transfer in soils. In : Heat and Mass Transfer in the Biosphere, Scipter, Washington, D.C. 1975.
Dyer, A. (1974) : A review of flux-profile relationships. Boundary Layer Meteorology, 7, 363-372.
Eagleson, P.S. (1970) : Dynamic Hydrology. McGraw-Hill, New York.
GRES-DGIS (1987) : Groundwater Recharge Research Program, funded by the Dutch Governmental Organization for International Cooperation.
Hatfield, J.L., Reginato, R.J. and Idso, S.B. (1983) : Comparison of long-wave radiation calculation methods over the United States. Water Res. Res., Vol. 19, 285-288.
Holben, B.N., Tucker, C.J. and Fan, C.J. (1980) : Spectral assessment of soybean leaf area and leaf biomass. Photogramm. Engin. and Remote Sensing, Vol. 45, 651.
Jackson, T.J., Schmugge, T.J. and O'Neill, P. (1984) : Passive microwave remote sensing of soil moisture from an aircraft platform. Remote Sensing of Env. (14), 135-151.
Jaskii, C.O., Townshend, J.R.G., Holben, B.N. and Tucker, C.J. (1985) : Analysis of the phenology of global vegetation using meteorological satellite data. Intern. Journal of Remote Sensing, Vol. 6 (8), 1271-1318.
Justice, C.O., Townshend, J.R.G., Holben, B.N. and Tucker, C.J. (1985) : Analysis of the phenology of global vegetation using meteorological satellite data. Intern. Journal of Remote Sensing, vol. 6 (8), 1271-1318.
Menenti, M. (1984) : Physical aspects and determination of

evaporation in deserts applying remote sensing techniques, ICW, Wageningen, the Netherlands.

NASA (1986) : Earth Observation System (EOS). Data and Information System volume II[a]. Technical memorandum 87777, NASA/GSFC.

Newton, R.W. and Rouse, J.W. (1980) : Microwave radiometer measurements of soil moisture content, IEEE Trans. Ant. Prop. AP-28, 680-686.

Owe, M., Choudhury, B.J. and Ormsby, J.P. (1986) : On large area variability in climate and climate-based soil moisture estimates. Intern Publ. NASA Godd. Space Flight Center, Hydrol. Services Branch, Greenbelt, MD.

Philip, J.R. & De Vries, D.A. (1957) : Moisture movement in porous materials under temperature gradients. EOS Trans. Geophys. Union 38, 222-228.

Raffy, M. & Baker, F. (1985) : An inverse problem occurring in remote sensing in the thermal infrared bands and its solutions. J. Geophys. Res. Vol. D3, No. 90, 5809-5819.

Raffy, M. & Becker, F. (1986) : A stable Interactive Procedure to obtain soil surface parameters and fluxes from satellite data. IEEE Transactions on Geosc. and Rem. Sensing, Vol. GE-24, No. 3, 327-333.

Raupach, M.R., Coppin, P.A. & Legg, B.J. (1986) : Experiments on scalar dispersion within a model plant canopy. Part I : The turbulence structure. Boundary-Layer Meteovol. 35, 21-52.

Rosema, A. (1975) : Simulation of the thermal behavior of bare soils for remote sensing purposes. In : Heat and Mass Transfer in the Biosphere, Scripta, Washington, D.C.

Rosema, A., Bijleveld, J.H., Reiniger, P., Tassone, G., Gurney, R.J. and Blyth, K. (1978) : Tellus, a combined surface temperature, soil moisture and evaporation mapping approach. Proc. Symposium on Remote Sensing (Manilla, the Philippines, 1978) Envir. Res. Inst. of Michigan.

Ross, J. (1975) : Radiation transfer in plant communities. In : Vegetation and the atmosphere, vol. 1 (Edited by : J.L. Monteith (London : Acad. Press), 13-52.

Schmugge, T. (1985): Remote sensing of soil moisture. In: Hydrological Forecasting, M.G. Anderson & T.P. Burt (Eds.), Wiley & Sons, 101-124.

Schmugge, T., O'Neill, P.E. and Wang, J.R. (1986) : Passive microwave soil moisture research. IEEE Trans. Geosc. & Remote Sensing, vol. GE-24 (1), 12-22.

Sellers, P.J. (1985) : Canopy reflectance, photosynthesis and transpiration. Int. J. of Remote Sensing, vol. 5 (8), 1335-1372.

Smith, J.A. (1983) : Matter-energy interaction in the optical region. In : Manual of Remote Sensing, Vol. I, Am. Soc. of Photogr., pp. 61-164.

Soer, G.J.R. (1980) : Estimation of regional evapotranspiration and soil moisture conditions using remotely sensed crop surface temperature. Remote Sensing of Envir., (9), 27-45.

Tucker, C.J., Holben, B.N., Elgin, J.H. and McMurtry, J.E. (1981): Remote sensing of total dry matter accumulation in winter wheat. Remote Sensing cf Environm. Vol. 11, 171.

Tucker, C.J., van Praet, C., Boerwinkel, E. and Gaston, A. (1983) : Satellite remote sensing of total dry matter accumulation in the Senegalese Sahel. Remote Sensing of Envir., Vol. 13, 461.

Ulaby, F.T., Razini, M., Fung, A.K., and Dobson, C. (1981) : The effects of vegetation cover on the microwave radiometric sensitivity to soil moisture. Remote Sensing Laboratory, Technical Report 460-6, University of Kansas, Lawrence, Kansas.

Ulaby, F.T., Moore, R.K. and Fung, A.K. (1982) : Microwave Remote Sensing. Vol. II, Addison-Wesley Publ. Comp., pp. 880 a.f.

Van de Griend, A.A. (1987) : A surface energy-balance model with a multi-layer canopy representation. NASA/Goddard Space Flight Center, Hydrol. Sciences Branch, Greenbelt, MD, Internal Report.

Van de Griend, A.A. and Camillo, P.J. (1986) : Estimation of soil moisture from diurnal surface temperature observations. Proc. IGARSS '86 Symposium, Zürich 8-11 September, 1986 (Ref. ESA-SP254), 1127-1230.

Van de Griend, A.A., Camillo, P.J. and Gurney, R.J. (1985) : Discrimination of soil physical parameters, thermal inertia and soil moisture, from diurnal surface temperature fluctuations. Water Res. Res. 21 (7), 997-1009.

Van de Griend, A.A. & Engman, E.T. (1985) : Partial area hydrology and remote sensing. J. Hydrol. (81), 211-251.

Wang, J.R. (1985) : Effect of vegetation on soil moisture sensing observed from orbiting microwave radiometers. Remote Sensing of Env. (17), 141-151.

---

[1]Institute of Earth Sciences (Dept. of Hydrogeology & Geographical Hydrology, and Dept. of Meteorology), Free University, Amsterdam, the Netherlands

[2]Goddard Space Flight Center (Hydrological Sciences Branch and Earth Resources Branch)

[3]Institute of Physics and Meteorology, Agricultural University, Wageningen, The Netherlands

# A PROPOSED STUDY OF RECHARGE PROCESSES IN FRACTURE AQUIFERS OF SEMI-ARID BOTSWANA

A. Gieske
University of Botswana, Faculty of Science
Private Bag 0022
Gaborone, Botswana

E. Selaolo
Geological Survey Department
Private Bag 14
Lobatse, Botswana

ABSTRACT. Determination of groundwater recharge in semi-arid regions by the classical water balance approach is of limited practical value, because in general the evapotranspiration and the groundwater discharge component are not directly measurable. Moreover, the storage of moisture in the unsaturated zone and the rates of infiltration along the various possible routes to the aquifer form important and uncertain factors. Linked in with plans to extend and upgrade the existing groundwater monitoring network of important aquifers in eastern Botswana, a thorough and long-term study is proposed, not only of moisture transport but also of dynamic aspects of groundwater replenishment. Classification of rainfall events will be combined with soil physical measurements, isotope analyses and study of solute transport.

## 1. INTRODUCTION

The eastern bushveld region of Botswana is largely of Precambrian origin (see paper of this workshop by De Vries & von Hoyer, 1987) and lies to the east of the main watershed which forms the margin of the Kalahari (see fig. 1). The pediplains of its flat denudational surface are only occasionally broken by some isolated hills and ridges of more erosion-resistant material.

The region drains by a number of ephemeral rivers into the Limpopo river system. The soils consist mainly of dark, red-brown, medium textured and ferruginous materials which in general favour rapid infiltration of rain water (Jennings, 1974). The vegetation that covers the area ranges from a fairly open bush and tree savanna on the sandveld to a more dense savanna on the hardveld (Timberlake, 1980).

Annual rainfall in eastern Botswana is approximately 500 mm with



*I. Simmers (ed.), Estimation of Natural Groundwater Recharge, 117–124.*

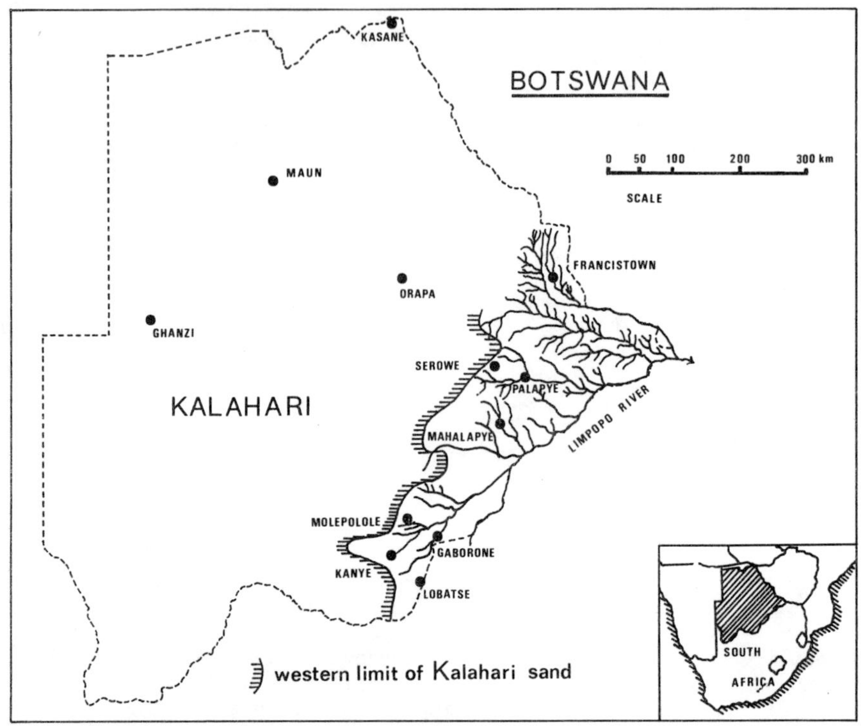


Fig. 1. Location and drainage system of eastern Botswana.

a variability of about 30 %. Potential evapotranspiration for short freely watered grass, is currently estimated as 1925 mm per year (P. Vossen, 1986, private communication). The actual evapotranspiration in field conditions, however, is not known.

In eastern Botswana aquifers have developed in basins which very often correspond to surface catchments and of which the fractured and permeable zones are normally characterized by depressions and covered with rock fragments and alluvial soils. Jennings (1974) started his investigations with the eastern aquifers of Lobatse, but later the emphasis of the research shifted towards the question of whether regular infiltration occurred through the sand cover of the Central Kalahari Basin. This resulted in a number of publications which have been discussed by De Vries & von Hoyer (1987). It appears that sporadic recharge may occur locally.

It was recognized by Jennings (1974) that the fracture aquifers of the eastern bushveld are different in that water can percolate very quickly through fractures, fissures and surface material to a depth where escape is no longer possible. Although little quantitative research on the problem of recharge in these eastern aquifers was done since the early 1970's, much effort was put into the search for new sources of groundwater as a consequence of the severe drought of the last five years and after long searches many high yielding aquifers were discovered.

In order to make a proper long-term assessment of these new sources, it was felt that more quantitative research into their recharge aspects was necessary and accordingly a three year research program will start in July 1987. In the following sections, the various problems encountered in the Botswana situation will be outlined together with some ideas for a possible approach.

## 2. CLIMATE, SOIL AND VEGETATION

Eagleson & Segarra (1985) have noted that in 'savannas the balance between pluvial inputs and evapotranspiration losses is so tight, that only in exceptionally wet years there is an excess of water that is lost through surface runoff or through deep infiltration'. De Vries & von Hoyer (1987) have discussed this balance for the Kalahari Basin and conclude that there is no convincing evidence for regular diffuse recharge.

In the eastern part of Botswana rainfall is higher, but with denser vegetation the same tight balance holds. However, fast infiltration of rainfall is possible through outcrop areas of fractured rock, through permeable zones of coarse alluvial material and through the sandbeds of ephemeral rivers. In section 3 some dynamic aspects of recharge in typical catchments will be discussed. This section will be limited to general remarks on the relationship between climate, soil and vegetation for a homogeneous, horizontal situation.

First, as mentioned in the introduction, the mean annual rainfall in the region is about 500 mm with a variability of 30%. Most of this rainfall occurs in short duration storms of high intensity, limited area and a marked seasonality. Very little has been done in Botswana on the analysis of individual rainfall events. Research has concentrated mainly on long-term periodicity of mean annual rainfall. Although 10 to 15 rainfall measuring stations were in operation around 1920, only very recently some close networks of rain gauges were established. Because recharge processes seem to occur within small catchment areas, it is important to gain more insight in the characteristics of individual rainfall events by establishing intensity-duration-frequency and depth-area-duration relations (Eagleson, 1978a).

Second, the actual evapotranspiration in field conditions has a typical interstorm character, determined by available moisture and soil sorptivity. It appears theoretically (Eagleson, 1978b) that evaporation under semi-arid conditions is similar for both vegetated and bare soils. No measurements have been made so far to verify this for the Botswana situation. Recently, the method described by Morton (1983) to calculate actual evapotranspiration in field conditions has been introduced by Buckley & Zeil (1984) and by Bons & van Loon (1985), but little is known about its accuracy in Botswana.

Another aspect that merits attention is the depth of the root zone. It seems that normally the thickness of this zone is much less than 10 m, but it has been observed that roots of trees particularly desert species, often go considerably deeper than that. Depths below 30 m are not exceptional (Jennings, 1974). The possibility that phreatophytes could reach the water table through fractures, introduces another unknown quantity in the water balance.

## 3. DYNAMIC ASPECTS OF RECHARGE IN SMALL CATCHMENTS

In eastern Botswana aquifers have developed in basins which often correspond to surface catchments. The fractured and permeable zones of these catchments are normally characterized by depressions and covered with alluvial soils of varying permeability. Good examples of this type of aquifer-catchment system can be found in the hills around Lobatse, where the aquifers are formed in fracture zones of dolomite, quartz and sandstone (Jennings, 1974 and De Vries & von Hoyer, 1987). Although this is certainly not the only type of aquifer that is found in eastern Botswana, it will be useful to describe one in order to elucidate the dynamic aspects of such systems. Moreover, the work by Jennings (1974) and Verhagen et al. (1974) who used water balance and environmental isotope methods in this area, can be used to give some figures.

The Lobatse West Basin is schematically illustrated in fig.2. Its catchment consists of hillslopes covered with vegetation and a densely vegetated valley floor, filled with permeable alluvial soil.

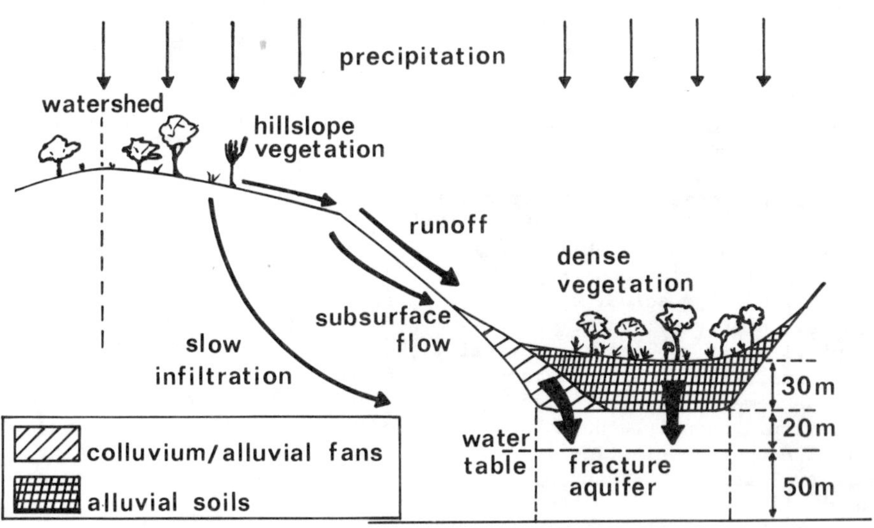


Fig. 2. Dynamics of infiltration in a hillslope catchment.

The soil which is about 30 m thick, covers a fractured aquifer. Against the base of the hillslope, colluvial/alluvial fans of coarse, highly permeable material are formed. Static water level is between 30 and 50 m below the surface. Abstraction from this basin has taken place since 1952. The recharge that takes place is made up of several components: direct diffuse infiltration of rain through the valley floor, infiltration of runoff from the hillslopes through the alluvial fans and possibly infiltration through the hills, although these consist of less pervious material. There is usually no appreciable surface outflow from the basin along the valley axis despite the considerable size of the entire catchment (about 10 $km^2$ ). Before abstraction started subsurface outflow might have occurred, because early observations noted the presence of perennial pools further down the valley.

On the basis of a tritium profile the diffuse recharge was estimated as 7% of the mean annual rainfall in the valley. The runoff from the hills was determined through extrapolated data from a weir located on the hillslope. For the wet season of 1970/71 this gave an infiltration of $1.6x10^5 m^3$ , which is two to three times as much as the direct infiltration. It was estimated further on the basis of the tritium content of the groundwater, that the transport of the water through the unsaturated zone of the valley floor takes 75 years, whereas it takes only 27 years through the alluvial fans.

Caution is needed to make a long-term recharge assessment of this system. For instance, the 1970/71 season with a precipitation of 579 mm, gave an infiltrated runoff of $1.6x10^5 m^3$ , while the 1971/72 season with 684 mm rain gave $3.8x10^5 m^3$ . This indicates that partial area concepts (see e.g. Kirkby, 1978) need to be used. Runoff depends on the characteristics of the catchment, the accumulated rainfall and the antecedent moisture. Together they determine subsurface stormflow, return flow and direct precipitation flow. It could be that in very wet years (e.g. the season 1966/67 with 900 mm) an exceptional amount of groundwater is added to the aquifer system, because the total runoff rises as a power function of the accumulated rainfall. The infiltration capacities of the alluvial fans and the valley floor would then be the limiting factors. Moreover, in very wet years temperatures will be lower and humidity higher thus decreasing evaporation and increasing ponding times. The amount of vegetation and consequently the evapotranspiration would increase of course, but to what extent the vegetation would be able to utilize the excess moisture is an open question.

Finally it appears that while the infiltration through hillslope runoff could in some cases be an exponential function of the accumulated rainfall, the same could be true for infiltration taking place through topographical collectors of areal rainfall such as riverbeds, sinkholes (Allison et al., 1985) and possibly some pans.

## 4. MEASUREMENT OF MOISTURE TRANSPORT THROUGH THE UNSATURATED ZONE

Simple surface water balance methods for the determination of infiltration below the root zone are not practical in a savanna situation because precipitation and evapotranspiration are well balanced. The difference of two fairly equal quantities each of which can only be determined to a certain accuracy, becomes completely unreliable. Recently however, coupled heat and moisture flux models (Camillo et al., 1983) have been developed which could in combination with high resolution remote sensing of the visible, infrared and microwave spectra (discussed elsewhere in this workshop) give much better insight into the energy and water balance. Meteorological ground stations must be set up which also monitor temperature and moisture as a function of depth to provide the necessary ground data for the area studied by satellites, allowing extrapolation to much larger areas at a later stage.

Movement of moisture through the unsaturated zone can in principle be determined by measuring soil moisture content and hydraulic gradient at various depths. In agricultural research in Botswana, soil moisture to a depth of 1.5 m is usually determined by gravimetric or neutron scattering methods whereas hydraulic gradients are measured with tensiometers. For the purpose of investigating infiltration below the root zone, neutron probes have not yet been used. Foster et al. (1982) and Jennings (1974) describe drillings where soil moisture profiles were determined with gravimetric methods. Such experiments could be made simultaneously with the insertion of aluminium tubes used for neutron probe techniques thus allowing calibration and continuous moisture measurements. At depths below the root zone and with moisture contents of less than 10%, tensiometers do not seem practical.

Over the past two decades experience has been gained with the analysis of environmental isotopes and chloride concentrations to determine recharge and transport through the unsaturated zone. As mentioned in section 2, tritium analysis in the Lobatse West Basin resulted in a diffuse recharge estimate of 7% of the mean annual rainfall (Jennings, 1974). Most of the later research concentrated on the question whether regular diffuse recharge occurs through the vast Kalahari sand cover. Verhagen et al. (1974) and Mazor et al. (1977) found definite signs of recharge using pumped samples from existing boreholes. Foster et al. (1982) conclude from combined measurements of tritium, soil moisture and chloride profiles in specially drilled boreholes that recharge would be unlikely through sand covers of more than 4 m. According to Mazor (1982) tritium profiles obtained from such boreholes provide "a minimum indication of recharge because coring does not sample water percolating quickly through fissures and zones of coarse material". Recently Sharma & Hughes (1985) concluded from a study of stable isotopes and chloride profiles in the deep homogeneous sands of western Australia that water could infiltrate through "preferred pathways", which might have contributed for about 50% to the infiltration below the root zone.

These considerations show that for the fractured and inhomogeneous aquifers of eastern Botswana where water can enter along various routes the problem of quantifying recharge through measurements becomes even more complex. In the following paragraphs some of the more pertinent problems are briefly discussed.

In a situation where as a consequence of sudden downpours the infiltration begins to accelerate through permeable collectors of rainfall, the concept of a simple stratification of successive years of moisture where transport is taken care of by some kind of piston displacement model, might no longer apply and fast gravity drainage along various routes could upset the pattern.

Possible vertical periodic movements of water and water vapour induced by thermal gradients might cause fractionation not only of stable isotopes but also of tritium (Foster et al. 1982) thus complicating the interpretation of the isotope analyses.

Regular measurement and monitoring of chloride in rain have only begun in 1983 (De Vries, 1986). The spatial distribution, seasonal variation and ratio between concentration in rain and dry deposition are not well known and to make accurate chloride water balances more data are needed. Chloride profiles could also be used in determining the depth of the root zone. Recent work by Sharma & Hughes (1985) has indicated that the end of this zone is marked by a maximum in the chloride concentration. This has yet to be verified for the Botswana situation. In any case, it is not as straightforward as it seems, for in the chloride profiles determined by Foster et al. (1982) the chloride maxima were mostly found in the top layers of the calcrete underlying the sand. It is important to include also stable isotope profiles in the analysis (Allison et al., 1985).

## 5. DISCUSSION

The problem of recharge in eastern Botswana has to be seen in the greater context of the savanna water and energy balance. The consequences of the great variability in rainfall and temperature on long-term equilibria between precipitation and vegetation have to be studied in more detail. It is hoped that high resolution remote sensing will provide support in this respect.

Second, in the heterogeneous situations encountered in eastern Botswana infiltration processes can be identified which are power functions of accumulated rainfall events. Easy to recognize are the processes related to topographical collector characteristics. The application of well-established partial area or variable source concepts means that study of depth-area-duration relations of rainfall must be combined with improved runoff (or run-in) observations in aquifer-catchment systems.

More difficult to analyze is the infiltration through thin soils of varying permeability into fissures and cracks of underlying rocks. It seems that some of these processes are also non-linear functions of accumulated rainfall for the simple reason that soil permeability increases with moisture content and when saturation has been reached, the rate of infiltration is determined by the height of the ponded water on the surface.

In the analysis of the complete system of the aquifer and its catchment, a careful study must be made of dynamic response of each of the elements in the system to possible successions of rainfall events. This can only be done when the whole range of soil physical, isotope and solute transport methods is applied to a few test areas over an extended period of time.

REFERENCES

Allison, G.B., Stone, W.J. and Hughes, M.W., 1985. ' Recharge in karst and dune elements of a semi-arid landscape as indicated by natural isotopes and chloride'. J. Hydrol., **76**: 1-25.

Bons, C.A. and van Loon, J.A.W.M., 1985. Water resources evaluation of a dolomite groundwater basin in S.E.Botswana. Report (unpublished), Institute of Earth Sciences, Free University of Amsterdam.

Buckley, D.K. and Zeil, P., 1984. 'The character of fractured rock aquifers in eastern Botswana'. Challenges in African Hydrology and Water Resources. Proc. of the Harare Symp., IAHS Publ. no. 144.

Camillo, P.J., Gurney, R.J. and Schmugge, T.J., 1983. ' A Soil and and Atmospheric Boundary Layer Model for Evapotranspiration and Soil Moisture Studies'. Water Resour. Res., **19**: 371-380.

De Vries, J.J., 1986. Groundwater data evaluation for the Nywane Basin. Report (unpublished), Geological Survey, Lobatse.

De Vries, J.J. and von Hoyer, M., 1987. 'Groundwater recharge studies in semi-arid Botswana - a review'. This Workshop.

Eagleson, P.S., 1978a. 'Climate, Soil and Vegetation. The Distribution of Annual Precipitation Derived from Observed Storm Sequences'. Water Resour. Res., **14**(5): 713-721.

Eagleson, P.S., 1978b. 'Climate, Soil and Vegetation. The Expected Value of Annual Evapotranspiration'. Water Resour. Res., **14**(5): 731-739.

Eagleson, P.S. and Segarra, R.I., 1985. 'Water-Limited Equilibrium of Savanna Vegetation Systems'. Water Resour. Res., **21**(10):1483-1493.

Foster, S.S.D., Bath, A.H., Farr, J.L. and Lewis, W.J., 1982. 'The likelihood of active groundwater in the Botswana Kalahari'. J. Hydrol., **55**: 113-136.

Jennings, C.M.H., 1974. The Hydrogeology of Botswana. Ph. D. Thesis, University of Natal, South Africa.

Kirkby, M.J. ed., 1979. Hillslope Hydrology. John Wiley & Sons, 389 pp.

Mazor, E., Verhagen, B.Th., Sellschop, J.P.F., Jones, M.T., Robins, N.E., Hutton, L. and Jennings, C.M.H., 1977. 'Northern Kalahari groundwaters : hydrologic, isotopic and chemical studies at Orapa, Botswana'. J. Hydrol., **34**: 203-234.

Mazor, E., 1982. 'Rain recharge in the Kalahari - a note on some approaches to the problem'. J. Hydrol., **55**: 137-144.

Morton, F.I., 1983. 'Operational estimates of areal evapotranspiration and their significance to the science and practice of hydrology'. J.Hydrol., **66**: 1-76.

Sharma, M.L. and Hughes, M.W., 1985. 'Groundwater recharge estimation using chloride, deuterium and oxygen-18 profiles in the deep coastal sands of western Australia'. J. Hydrol., **81**: 93-109.

Verhagen, B.Th., Mazor, E. and Sellschop, J.P.F., 1974. 'Radiocarbon and tritium evidence for direct rain recharge in the Northern Kalahari'. Nature, **249**, no. 5458, 643.

Verhagen, B.Th., 1984. 'Environmental isotope hydrology in southern Africa'. Challenges in African Hydrology and Water Resources. Proc. of the Harare Symp., IAHS Publ. no. 144.

# ESTIMATION OF NATURAL GROUNDWATER RECHARGE UNDER SAUDI ARABIAN ARID CLIMATIC CONDITIONS

Dr. M. J. Abdulrazzak, Dr. A. U. Sorman, Dr. O. Abu Rizaiza*
King Abdulaziz University, Department of Hydrology and Water Resources Management, P.O.Box 9034 and *Department of Civil Engineering, P.O.Box 9027, Jeddah - 21413, Saudi Arabia

**ABSTRACT** : The purpose of this study is to investigate the infiltration-recharge phenomena in a representative alluvial wadi system in order to determine the factors influencing it and subsequently formulate a groundwater recharge model. To achieve this goal a three kilometer experimental reach in one of the representative basins of the south-western part of Saudi Arabia was selected and instrumented with a data acquisition system to continuously monitor hydrological variables on rainfall, runoff, groundwater fluctuations, evapotranspiration, and soil temperature and moisture variation.

Analysis of data revealed different recharge responses at each well due to soil heterogeneity and moisture variation. The developed regression models show that the maximum flood hydrograph depth is the most important influencing factor affecting recharge. Also, factors such as rainfall depth and depth to water table affected the recharge rate.

## 1. INTRODUCTION

In the absence of dependable surface water supplies in arid countries such as Saudi Arabia, groundwater plays a significant role as a potential and dependable water resource. Optimal use and prevention of aquifer depletion requires reliable and precise information on natural recharge, and how it is affected by climatic conditions and soil parameters.

Consequently, to investigate the infiltration-recharge phenomenon in a typical arid zone, a wadi alluvial system in a representative reach in the Tabalah basin, located in the southwestern region of the country, has been selected. The drainage basin has been designated as one of the representative basins in the country where extensive data collection and analysis programs have been undertaken by the Ministry of Agriculture and Water (MAW) (3).

The hydrological network of the basin consists of one climatic station, 13 rainfall gauges, three runoff stations and five observation wells.



*I. Simmers (ed.), Estimation of Natural Groundwater Recharge, 125–138.*

## 2. EXPERIMENTAL SITE AND INSTRUMENTATION

In order to collect reliable information on the infiltration-recharge process and determine the influencing factors, the selected experimental reach in the lower Tabalah basin, has been instrumented along a three kilometer length, with one raingauge, three runoff stations, seven observation wells, a set of soil temperature and moisture sensors and evapotranspiration measuring units. These have been in operation since 1985. The location of the experimental reach in reference to the drainage basin is shown in Figure (1). A detailed layout map of the instrumented reach is shown in Figure (2). A reliable data logger system continuously retrieves and stores data according to a specified format which helped overcome the problems associated with the remoteness of the site and its accessibility during the flooding season. The data acquisition system consists of a battery operated electronic data logger and data pod units which had the desired features of accuracy combined with continuous recording into memory.

The logger stores soil temperature and moisture data on a storage pack with 64 KB memory from sensors installed at different depths. This data can be retrieved by computer through a reader. The data pods,which are equipped with microchips, can store uninterrupted hydrological data for rainfall, runoff and groundwater level during a two to four month period.

Both systems have been connected to the sensors and floating units to measure various hydrological parameters as shown on the flow diagram presented in Figure (3). The data pod recorder has either one or two sensors connected to it to measure water level fluctuations in the channel as well as in the observation wells, or for measuring evapotranspiration. On the other hand, the data logger unit, with 12 sensors extendable to 16, was used to measure soil moisture, soil temperature and stream water level. After each sensor was calibrated in the laboratory, it was installed properly on the site, in order to obtain accurate readings for subsequent computer recharge modeling. The instruments have been in operation for almost a year and their appropriateness under the Saudi Arabian climatic conditions has been tested. For more information, refer to reference (1).

The field survey of the study reach consisted of infiltration tests,along the alluvial wadi, and collection of representative soil samples. A geophysical survey was conducted to delineate the alluvial deposit depth and boundary. A cross-sectional survey was also made at each tentative well location and runoff station. Aquifer properties were determined by pump test and other laboratory methods.

## 3. DATA COLLECTION

The collection and analysis of field data for the experimental site began after the installation of the electronic equipment. Collected data as well as available records of rainfall, runoff, soil moisture and groundwater are currently being analyzed for modeling studies.

Data from the rainfall recorders near runoff station SW1 were

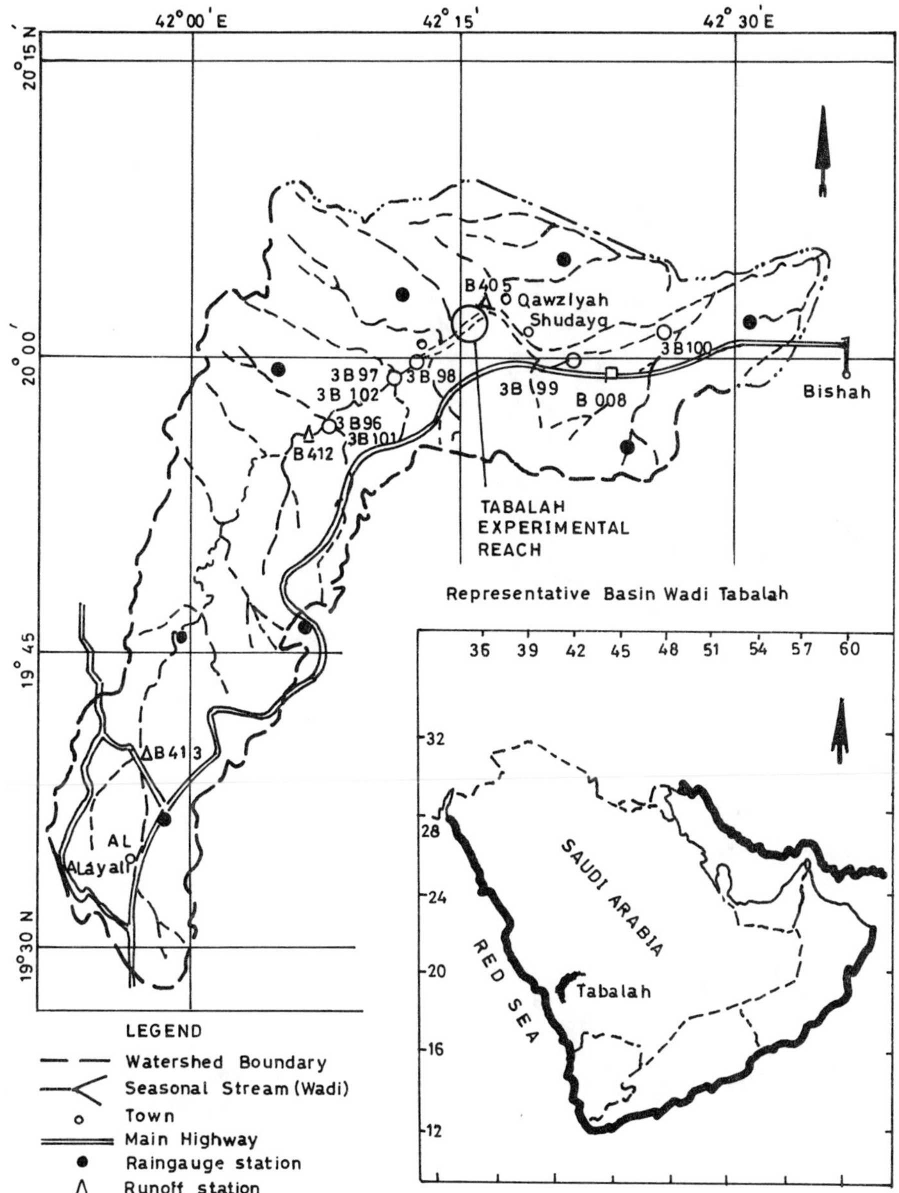


FIGURE (1). TABALAH EXPERIMENTAL REACH LOCATION MAP

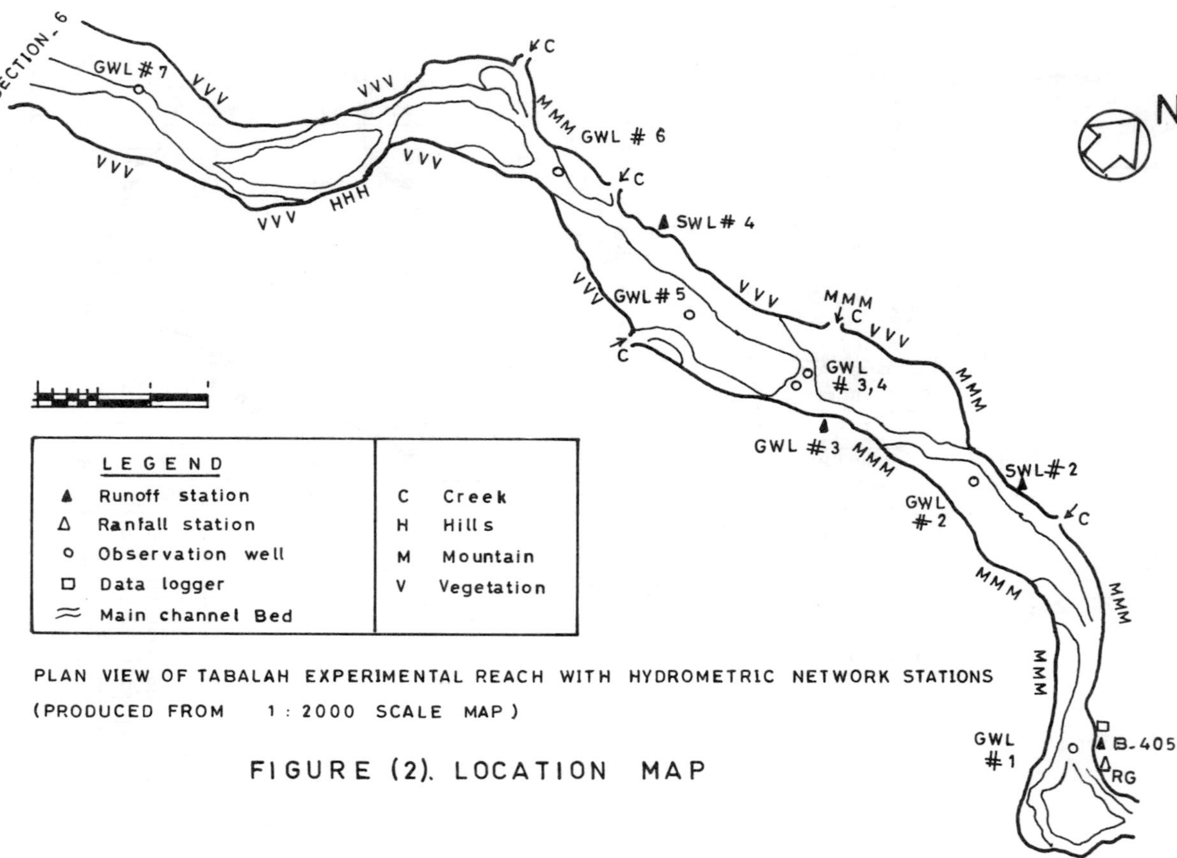


PLAN VIEW OF TABALAH EXPERIMENTAL REACH WITH HYDROMETRIC NETWORK STATIONS
(PRODUCED FROM 1 : 2000 SCALE MAP)

FIGURE (2). LOCATION MAP

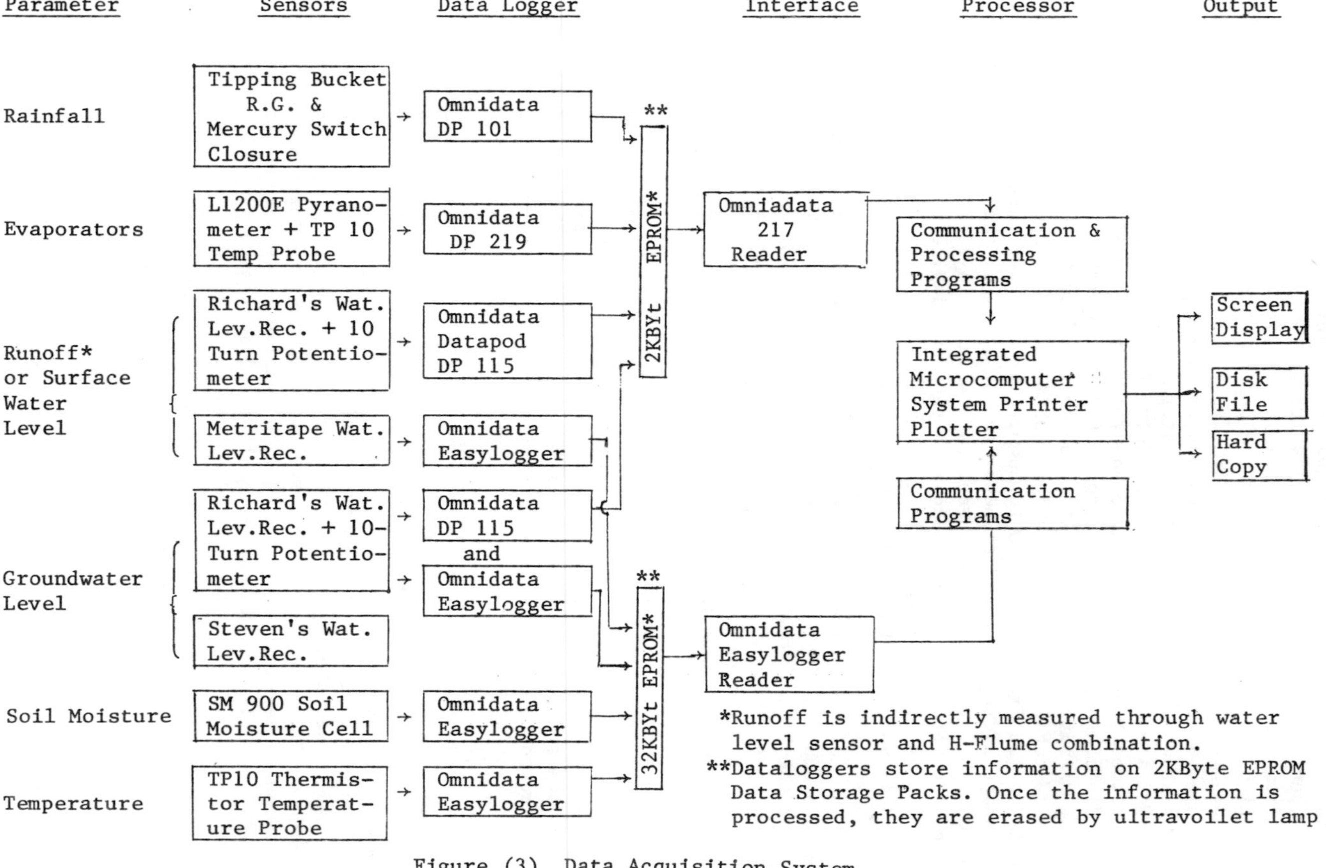


*Runoff is indirectly measured through water level sensor and H-Flume combination.
**Dataloggers store information on 2KByte EPROM Data Storage Packs. Once the information is processed, they are erased by ultravoilet lamp

Figure (3) Data Acquisition System

retrieved from the data storage module. Rainfall events during the rainy period from April through May 1986 were analyzed. In addition, data from other rainfall stations in the basin which have been in operation since 1983, were collected and the time-intensity distribution graphs were studied (2). The storm depth data is plotted for each recording station in order to show the areal storm isoheyets; a typical one is shown in Figure (4) for the storm on April 13, 1986. The area received a maximum precipitation depth of 25.8 mm, measured at station B227 between the hours of 18:45 and 20:15 pm on April 13, 1986.

Runoff stage graphs were collected through the automatic recorders at the two sites in addition to those operated by the Ministry of Agriculture (station SW1). Three major floods have been recorded in the six month period from March 1986 to August 1986. After correction was made for stage hydrographs due to scouring and silting, the resulting stage hydrograph for the analyzed event is presented in Figure (5). The rainfall hyetograph is also shown on the same day, obtained from the station installed by the research team. The analysis of the major flood events is summarized in tabular form and presented in Table I for the runoff station SW1.

The soil moisture data were collected from the set of moisture sensors at site G for the period April 10-13, 1986. The data is recorded on the logger system in bars, with the depths of soil ranging from 0.25mm to 1.75mm. The variation in the suction values showed the movement of the wetting front during the initial phase of the infiltration process. During the sensor calibration, the soil moisture-suction relationship is established for each sensor separately. Thus, the moisture content variation in the vertical soil horizon is determined during and after each flood has passed. A sample of data for the flood event of 10-13 April, 1986 is presented in Figure (6).

It was observed from the rainfall, groundwater rise and soil moisture data that the redistribution of soil moisture in the profile contributed towards groundwater rises. It took about 6-10 days for the steady state to be achieved. The data also identified the commencement of the infiltration and recharge processes.

Groundwater level recorders were installed on the wells and the data was recorded continuously. The groundwater table responded to each rainfall and flood event. An example of the data for April 13, 1986 is shown in Figure (7).

In order to compute the percolation rate and rate of recharge for modeling, the water level fluctuation of each well was analyzed as presented in Table (II). Analysis of water levels in wells No. 5 and 6, at 19:00 hours on April 13, were 2.88 and 3.67 meters from the soil surface, respectively. The water table began increasing in well #5 at 20:48 hours and in well #6 at 22:00 hours. Consequently, the percolation rates were estimated to be 1.6m/hr and 1.22m/hr for both wells. Also, it was noticed that the recharge rates vary from 3 mm/hr up to 63 mm/hr depending upon the depth of surface hydrograph, the areal depth of rainfall, as well as the initial depth of water table and the initial soil moisture content. Due to the heterogeneity of the soil formation at each well and soil stratification, wells 2 and 6 responded quickly when the initial moisture content was between 0.03 and 0.06%. This change

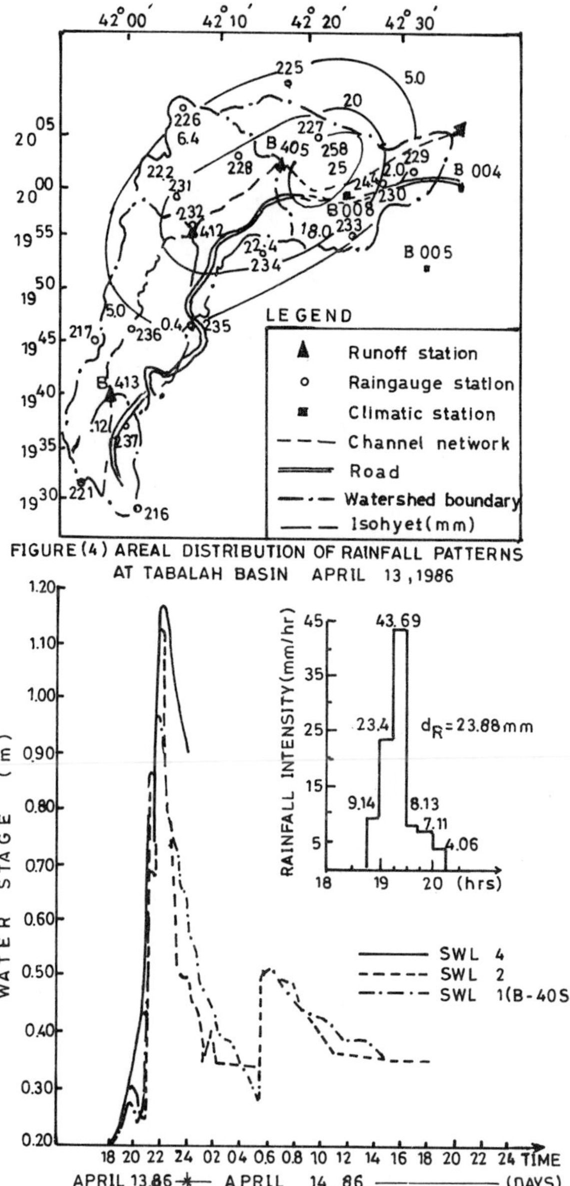


FIGURE (4) AREAL DISTRIBUTION OF RAINFALL PATTERNS AT TABALAH BASIN APRIL 13, 1986

FIGURE (5) SURFACE STAGE HYDROGRAPHS ON APRIL 13 AND 14, 1986

Table (I) Major Rainfall–Runoff Event (Tabalah Site)

| Date | Precip. depth (mm) | Maximum Rainfall | | Runoff depth (mm) | Max. Stage | | Corrected S.H. (m) | $Q_p$ $m^3/sec$ |
|---|---|---|---|---|---|---|---|---|
| | | rate mm/hr | time (hrs) | | height (m) | time (hrs) | | |
| 07 Apr 86 | 4.3 | 3.56 | 04:30 - 05:00 | 0.234 | 1.16 | 13:55 | 1.13 | 23.64 |
| 13 Apr 86 | 23.88 | 43.69 | 19:15 - 19:30 | 1.266 | 1.67 | 21:00 | 1.70 | 172.30 |
| 23 Apr 86 | 20.07 | 43.00 | 17:00 - 17:15 | 0.120 | 1.32 | 17:20 | 1.15 | 25.05 |
| 07 May 86 | 10.90 | 27.42 | 13:15 - 13:30 | - | 0.98 | 16:00 | 0.25 | - |

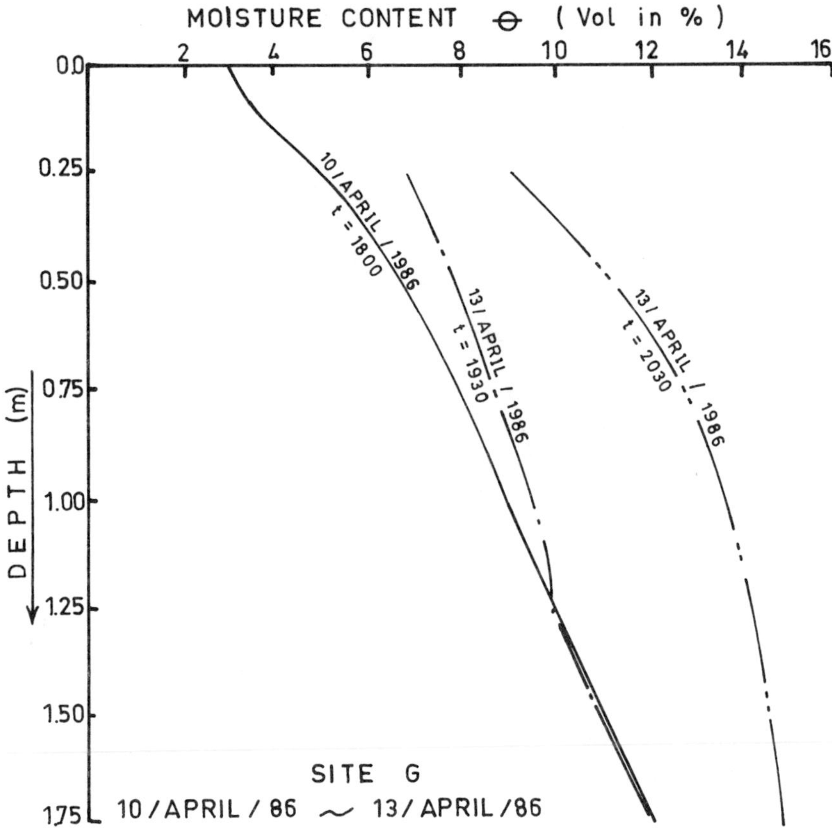


FIGURE (6 ). MOISTURE CONTENT VS DEPTH AT SITE G

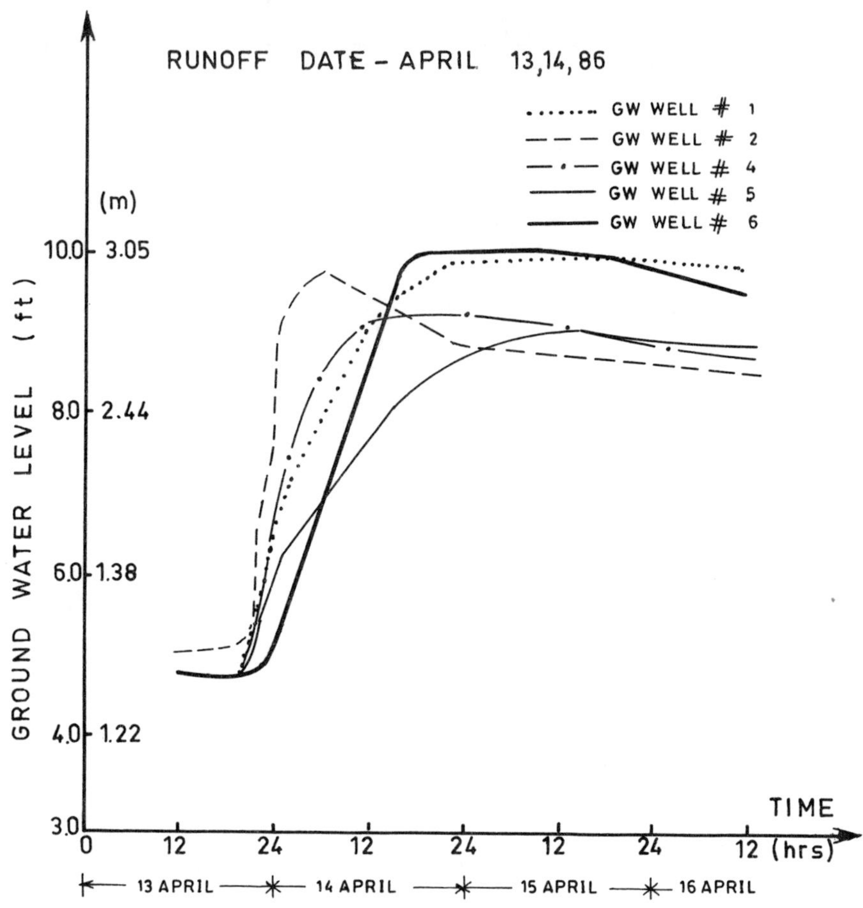


FIGURE (7). GROUND WATER RECHARGES AT WELL LOCATIONS.

Table (II) Ground Water Recharge Rates at the Wells

| | | GWL #1 | | GWL #2 | | GWL #4 | | GWL #5 | | GWL #6 | | Average |
|---|---|---|---|---|---|---|---|---|---|---|---|---|
| | | I | F | I | F | I | F | I | F | I | F | |
| Short Dump Data | CHN(ft) | 3.84 | 5.24 | 3.70 | 5.46 | 3.69 | 4.80 | 3.72 | 4.89 | 3.69 | 5.56 | |
| | Time(hrs) | $16^{33}$** | $16^{33}$ | $10^{07}$* | $07^{10}$ | $14^{33}$** | $14^{30}$ | $17^{00}$** | $17^{00}$ | $12^{78}$* | $07^{30}$ | |
| 07 Apr 86 | Rate(m/hr) | 0.0089 | | 0.0135 | | 0.0071 | | 0.0074 | | 0.0154 | | 0.0137 |
| | Water Rise (m) | 0.4267 | | 0.5364 | | 0.3380 | | 0.3870 | | 0.5820 | | 0.45 |
| | CHN(ft) | 5.13 | 9.98 | 5.05 | 9.79 | 4.77 | 9.31 | 4.78 | 9.03 | 4.69 | 10.02 | |
| | Time(hrs) | $20^{83}$ | $07^{93}$ | $18^{00}$ | $06^{30}$ | $20^{60}$ | $17^{43}$ | $18^{20}$ | $08^{30}$ | $20^{40}$ | $09^{50}$ | |
| 13 Apr 86 | Rate(m/hr) | 0.0420 | | 0.1175 | | 0.0665 | | 0.0340 | | 0.0044 | | 0.063 |
| | ΔH(m) | +1.48 | | +1.445 | | 1.384 | | +1.30 | | +1.625 | | 1.45 |
| | CHN(ft) | 8.61 | 9.53 | 8.16 | 9.47 | 8.35 | 9.15 | 8.21 | 9.05 | | | |
| | T(hrs) | $16^{33}$ | $16^{33}$ | $17^{00}$ | $20^{60}$ | $17^{03}$ | $14^{33}$ | $17^{00}$ | $09^{05}$ | | | |
| 23 Apr 86 | $R_t$(m/hr) | 0.01167 | | 0.1538 | | 0.0115 | | 0.0104 | | No Record | | 0.0122 |
| | ΔH(m) | +0.28 | | +0.40 | | 0.244 | | 0.253 | | | | 0.29 |
| | CHN(ft) | 8.33 | 8.57 | 8.19 | 8.39 | 8.43 | 8.65 | 8.19 | 8.33 | 7.94 | 8.18 | |
| | T(hrs) | $16^{33}$ | $12^{53}$ | $12^{80}$ | $15^{50}$ | $14^{23}$ | $14^{35}$ | $17^{00}$ | $17^{00}$ | $13^{50}$ | $16^{40}$ | |
| 07 May 86 | $R_t$(m/hr) | 0.0036 | | 0.0025 | | 0.0028 | | 0.0018 | | 0.0044 | | 0.003 |
| | ΔH(m) | +0.073 | | 0.67 | | 0.067 | | 0.043 | | 0.073 | | 0.065 |

* (24) hours lag
** (48) hours lag

was attributed to the texture of the alluvium channel profile, composed of particles of different sizes (1). The wells showed a total rise of 1.50-1.75 m in groundwater table during the month of April 1986 in which most of the rainfall-runoff activities occurred. Further data is expected to become available which will be used in determination of the influencing factors on natural recharge during the 1986-1987 rainy period.

It is clearly evident that soil moisture data is the most crucial in the interpretation of soil responses to recharge. Reliable estimates before and after rainfall-runoff events will contribute towards a more accurate estimate of surface runoff and groundwater recharge during model development studies under arid climatic conditions.

## 4. PRELIMINARY MODEL STUDIES

In order to study the influence of different hydrological and soil parameters on the rate of recharge, a regression analysis technique was applied to the collected data using one, or a combination of, causative factors on the dependent parameters such as:

A- maximum rise in the runoff stage hydrograph in m.

B- point rainfall depth near the reach and areal average depth of rainfall over the basin causing the runoff in mm.

C- unsaturated depth of the soil profile in m.

The results of the preliminary regression analysis studies are tabulated in Table (III) under four different model forms, three of which show the effect of each individual parameter on the groundwater recharge estimation in the linear regression type equation. In the fourth model, the multilinear regression equation is used in selecting the combination of the two most predominant factors. The model studies clearly indicated that maximum flood hydrograph depth is the most important factor affecting recharge. The area coverage rainfall depth compared to the single station records also showed high correlation in explaining the recharge phenomena. Finally, the unsaturated depth of soil proved to be the third most important factor affecting the recharge rate.

## 5. DISCUSSION OF RESULTS

The analysis of collected data provided valuable insight on the infiltration-recharge mechanisms in the alluvial system and the influencing factors. Considering the erratic nature of the rainfall, the response of the basin to rainfall events, and the initial moisture content of the soil, different recharge responses were experienced without the consistency expected in an arid region. Also, because of the limited number of observed flood events in the year 1986, few factors were used in the early stage of model development. Only two of these factors were studied in multilinear form, and the terms which are considered in this study were concerned mainly with potential head. The constant term in the regression equation reflects the effects of unsaturated moisture

Table (III) Regression Models for Groundwater Recharge

| Model # | Independent Model Parameter | Unit | Regression Terms | | | Corelation Coefficient |
|---|---|---|---|---|---|---|
| | | | LRA | LRB=B1 | LRB=B2 | COR = R |
| 1 | Max. Runoff Depth ($\sigma H$) | (m) | -0.200 | 0.989 | - | 0.9996 |
| 2a | Areal Rainfall Depth ($Pd_a$) | (mm) | -0.004 | 0.102 | - | 0.8770 |
| 2b | Point Rainfall Depth ($Pd_p$) | (mm) | -0.050 | 0.042 | - | 0.600 |
| 3 | Unsaturated Soil Depth | (m) | -0.655 | 0.454 | - | 0.515 |
| 4 | ($\sigma H + Pd_a$) | | -0.198 | 0.884 | 0.015 | |

suction head which is a function of soil moisture deficit ($\theta$ sat - $\theta$ residual) and intrinsic soil characteristics such as saturation, hydraulic conductivity, and porosity. The model will be further calibrated with the addition of more data from the study reach and the entire experimental basin.

## ACKNOWLEDGEMENTS

The authors would like to express their appreciation to the King Abdulaziz City for Science and Technology (KACST) for sponsoring the project for the years 1984-1987. Thanks are also extended to the Ministry of Agriculture and Water (MAW) for providing supplementary hydrologic data and publications, without which this study would not have been possible. The encouragement provided by the Faculty of Meteorology, Environment, and Arid Land Agriculture, KAU is gratefully acknowledged.

## REFERENCES

1. King Abdulaziz University KACST Research Project "Estimation of Natural Groundwater Recharge", Progress Reports 1,2,3; 1985,1986.
2. Ministry of Agriculture and Water: "Rainfall and Runoff Data", Hydrological Publications 93, 95, 98, 101, 103, 105; 1981, 1984, 1985.
3. Saudi Arabian, Dames and Moore, "Representative Basins Study", Interim Reports, 1, 2, 3; 1983, 1984, 1985.

# SOLUTE PROFILE TECHNIQUES FOR RECHARGE ESTIMATION IN SEMI-ARID AND ARID TERRAIN

W.M. Edmunds, W.G. Darling and D.G. Kinniburgh
Hydrogeology Research Group
British Geological Survey
Crowmarsh Gifford
Wallingford
Oxfordshire, OX10 8BB, U.K.

ABSTRACT. Conventional methods for recharge estimation have limitations when applied to arid and semi-arid regions; the use of tritium profiles is also not always applicable. Unsaturated zone solute profiles, using a reference solute such as chloride, offer an alternative technique. Sampling may be undertaken by percussion drilling, augering or from dug wells; the methods developed are described and examples discussed. Recharge estimates using chloride profiles from Cyprus (420 mm mean annual rainfall) are in good agreement with results estimated from tritium profiles and indicate a mean annual recharge of around 50 mm/year. In Central Sudan (180 mm mean annual rainfall), good agreement was found between adjacent unsaturated zone chloride profiles and these indicated a net annual direct recharge via interfluve areas of around 1 mm/year. It is concluded that solute profiles offer a cheap and effective tool for estimating direct recharge in porous lithologies of semi-arid regions and also for investigating recharge history, providing input data for chloride are available. In more arid regions, however, a component of discharge may occur during hyperarid episodes. Further validation of moisture composition using stable isotope techniques is required under such conditions.

## 1. INTRODUCTION

In many arid and semi-arid regions of the world it is not clear whether sustainable replenishment of groundwater is currently taking place. Many groundwater schemes exploiting deep aquifers are either in operation or under consideration and some of these are inadvertently exploiting fossil resources or other limited resources. This exploitation is often at the expense of shallow unconfined resources, developed by traditional methods over long periods, and in tune with the delicate natural water balance. It is in this context that improved techniques for accurate recharge estimation and a better understanding of recharge processes are vitally needed.

A common indirect method of evaluating recharge involves consideration of the difference between rainfall and estimates of evapotranspiration, taking into account surface runoff and vegetation



*I. Simmers (ed.), Estimation of Natural Groundwater Recharge, 139–157.*

cover. Recharge estimates using this water balance approach rely on small differences between two large numbers (rainfall, evaporation) both of which present severe measurement problems in arid and semi-arid zones.

The use of environmental (thermonuclear) tritium to investigate recharge rates through the unsaturated zone has been widely and successfully used in temperate zones (Smith *et al.*, 1970) and has also been applied in semi-arid and arid zones, e.g. in Australia (Allison and Hughes, 1974), India (Sukhija and Shah, 1975), Saudi Arabia (Dincer *et al.*, 1974), and Cyprus (Edmunds and Walton, 1980). Although a proven tool for recharge estimation, tritium suffers from several disadvantages:

(i) tritium is not conservative in behaviour and is lost from the system by evaporation and transpiration;

(ii) the relatively short half life (12.3 yr) limits the long term usefulness of the method, especially at the present when atmospheric levels are declining;

(iii vulnerability to contamination during sampling and processing, a factor which is enhanced in remote areas and at low total moisture levels;

(iv) analysis is highly specialised and costly;

(v) quantitative studies are difficult to achieve since it is difficult to determine a tritium mass balance;

(vi) there are not many tritium measurements of pre-1970 rainfall and so the input of tritium is rather poorly defined (especially when taken in conjunction with (i) above).

Alternative techniques are therefore desirable. Solute profiles offer one possible approach. Unlike tritium, chloride and perhaps some other solutes are not subject to the restrictions outlined above; in particular, there should be no loss during evaporation and 'conservative' behaviour should occur. The use of chloride mass balance in recharge estimation was demonstrated by Eriksson (1976) in India using regional groundwater samples. This technique is excellent for providing a minimum estimate of recharge but additions of solutes from various non-atmospheric sources may take place, limiting its usefulness. The use of unsaturated zone solute profiles in recharge estimation has been developed by Allison and Hughes (1978), Allison *et al.* (1985), Sharma and Hughes (1985) in Australia and by Kitching *et al.* (1980) and Edmunds and Walton (1980) in Cyprus. The objective of this paper is to describe the approach and methods developed during the BGS research programme in Cyprus and subsequently in Sudan. Results from each study are then used to illustrate the potential as well as some of the limitations of the solute profile technique.

## 2. MODEL FOR THE USE OF SOLUTES TO ESTIMATE RECHARGE

A working hypothesis for the use of solutes to evaluate the direct component of recharge is illustrated in Figure 1.

The input of solutes to the aquifer depends initially on the total atmospheric fallout per unit time, made up of rainfall ($F_p$) and the net dry deposition ($F_d$) fluxes. Both the rainfall amount (P) and local composition of the total deposition ($F_p + F_d$) may be determined for a given site, although the regional variation in both quantities must be considered if recharge estimation is required for large areas.

Solutes will be deposited on and transported through the upper soil during the rainy season at rates depending on the rainfall intensity. These solutes will undergo concentration as a result of evapotranspiration (E). Solutes may be removed from solution by plant uptake, by mineral precipitation, or by adsorption. Similarly solutes may be released by decay of dead plant material, by mineral dissolution or by desorption. In the absence of a knowledge of these ancilliary fluxes, only those constituents for which there is no <u>net</u> release or storage by the soil or rock matrix may be used. Chloride is the solute that most frequently and most conveniently meets these requirements.

Nutrient cycling by plants may affect solute movement (including that of chloride) on an annual basis, but, in stable landscapes, the amounts removed annually by plant uptake are balanced by the amounts released by plant decomposition, i.e. a steady state should have been achieved. This assumes that there are no additions of the solute in fertilisers or permanent removal by crop harvesting (including the export of grazing animals).

The solute concentrations in the soil or in the upper unsaturated zone will vary seasonally or annually depending upon the intensity of the moisture flux due to the incident rainfall and evapotranspiration. Complex movement of solutes both upwards and downwards may take place in response to water movement,which in turn depend upon the prevailing water potential gradients. A 'zero flux plane' (ZFP) exists (Wellings and Bell, 1980) which effectively separates moisture and solutes moving upwards (evapotranspiration) from that moving downwards (drainage). The position of the ZFP will shift seasonally between the surface and a depth of several metres and may, in some places and at certain times, be coincident with the water table in which case discharge may occur. Its position will also vary spatially in response to root development. However complex the soil moisture distribution might be in the soil zone, therefore, the transfer of moisture/solutes at depth will be a relatively straightforward process. Under conditions of recharge a maximum depth can thus be defined at which a net, steady state, moisture and solute transfer should take place towards the water table. The amount of solute crossing the ZFP would be expected to vary in relation to antecedent rainfall over one or more seasons and some oscillation in the solute profile would then occur. A detailed discussion of the transmission of solutes across the ZFP is given in Wellings and Bell (1980). The average composition of interstitial water in this profile

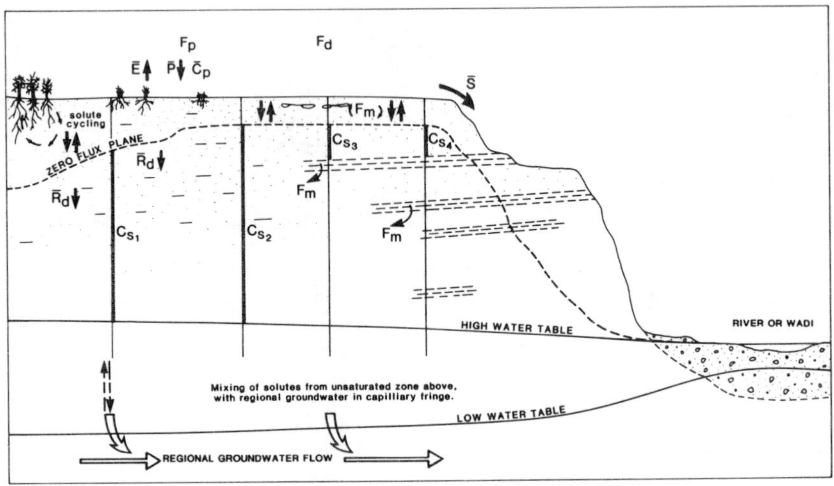


Figure 1. Schematic representation of solute movement and recharge via the unsaturated zone. The symbols are defined in the text. Sections of unsaturated zone profiles $C_{s1-4}$ useful for direct recharge estimation are shown in solid lines.

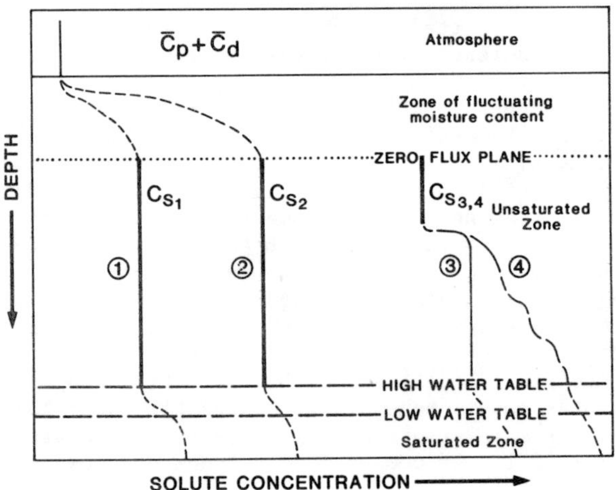


Figure 2. Idealised solute concentrations developed during percolation in profiles $C_{s1-4}$. Steady-state sections are indicated by solid line.

($C_s$) will, under steady state conditions, be proportional to the concentration factor, P/(P-E), assuming no loss of solute to minerals and that the water and 'inert' solutes are transported at the same rate.

In the steady state, the water balance equation can therefore be given by

$$\bar{R}_d = \bar{P} - \bar{E} - \bar{S} \tag{1}$$

where $\bar{R}_d$ is the direct recharge flux and $\bar{S}$ is the surface runoff flux; the bars indicate time- and space-averaged quantities. Providing surface runoff is negligible ($\bar{S} \sim 0$), this leads to

$$\bar{R}_d = \bar{P} - \bar{E} \tag{2}$$

Similarly the solute balance is given by

$$\bar{F}_p + \bar{F}_d = \bar{F}_s + \bar{F}_m \tag{3}$$

where $\bar{F}_p$ and $\bar{F}_d$ are the average precipitation and net dry deposition fluxes (= input), respectively, and $\bar{F}_s$ and $\bar{F}_m$ are the net steady state output fluxes in the drainage water and the net flux of solute precipitated or adsorbed by minerals (dissolution and desorption give a negative flux), respectively. $\bar{F}_s$ is given by the output water flux multiplied by the solute concentration (appropriately averaged), i.e.

$$\bar{F}_s = \bar{R}_d \, \bar{C}_s \tag{4}$$

where $\bar{C}_s$ is the average concentration of the reference solute in the below-ZFP water. If we assume $\bar{F}_m = 0$, then Equations (3) and (4) combine to give

$$\bar{F}_p + \bar{F}_d = \bar{R}_d \, \bar{C}_s \tag{5}$$

or on rearranging

$$\bar{R}_d = \frac{(\bar{F}_p + \bar{F}_d)}{\bar{C}_s} = \frac{(\bar{P}\bar{C}_p + F_d)}{\bar{C}_s} \tag{6}$$

Hence the amount of direct recharge can be estimated from a knowledge of the volume-averaged concentration of the reference solute in the rainfall ($\bar{C}_p$) and in the deep interstitial water ($\bar{C}_s$), the long term average annual precipitation ($\bar{P}$), and the net dry deposition flux of the reference solute ($\bar{F}_d$). Note that if $\bar{F}_d = 0$, then the fraction of the rainfall contributing to direct recharge is simply given by the ratio $\bar{C}_p/\bar{C}_s$.

To recapitulate, the steady state model is subject to certain assumptions:

(1) since there is a time lag ($\Delta t$) in solute input to the unsaturated zone and its output to the saturated zone, it must be assumed that no major climatic change has occurred over this period;

(2) that there have been no external, e.g. fertiliser, additions nor recent changes in atmospheric pollution;

(3) that there is no net change in storage of the 'reference'solute above or below the ZFP, either by (a) plants or animals; or (b) by mineral precipitation/dissolution or adsorption/desorption. In the first instance this assumption should be valid if there have been no significant natural vegetation changes or changes in agricultural practices.

In principle, it may be possible to use as a reference any solute that is not released by weathering or removed by precipitation; even interacting solutes, e.g. cations on clays, in theory, may be able to be used since the quantity of exchangeable ions are often effectively constant and need not lead to a net change in storage. In practice the most inert solutes are likely to prove the most reliable and for this reason chloride has been used here for the recharge calculation although $SO_4$, $NO_3$ and $SEC_{25}$ (specific electrical conductance at 25°C) have also been considered. It is possible that only a restricted portion of the unsaturated zone profile may be usable for recharge estimation (Figure 1). For example, the presence of certain lithologies, e.g. residual marine bands, may release chloride. The possible development of solute profiles is summarised in Figure 2. Profiles 1 and 2, represent steady state drainage under different vegetation/soil conditions where evapotranspiration rates differ. Profiles 3 and 4 represent two possible cases where solute compositions have been modified by reaction and/or changes in storage. Only the upper part of profiles 3 and 4 would be of value in recharge calculations and, in certain reactive lithologies, no steady state profile may be developed at all.

The drainage compositions, $C_{s1-2}$, would be expected to be similar to those encountered at the surface of the water table. However the composition of the saturated flow will have been modified by the incoming lateral flow with higher salinity.

Therefore, water table samples taken from shallow wells are unlikely to be reliable for accurate recharge estimates. However, since chloride is not lost during drainage and saturated flow, theshallow groundwater chemistry, or river baseflow, can always be used in areas of active recharge to derive a minimum figure for total recharge (Eriksson, 1976) and could be more widely used in regional water balance studies.

## 3. METHODOLOGY

Sampling of the unsaturated zone has so far been carried out using three different methods depending on the lithological nature of the terrain and the logistics. As the project has developed, the simpler techniques have proved to be the more effective.

### 3.1 Dry percussion drilling

In Cyprus, profiles of up to 30 m were obtained by a wireline percussion technique, drilling dry, using a claycutter with a one-way valve. Metal casing was advanced every metre to avoid contamination. Samples were bulked every 20 cm and collected in polythene bags; drilling progress was typically 4 m per working day. This technique was attempted in Sudan but was unsuccessful due to the more indurated and clay-rich nature of the terrain, as well as logistical problems.

### 3.2 Power augering

The principal technique used in Sudan involved using a Land Rover-mounted, petrol driven lightweight auger with 2-inch diameter flights. Samples were obtained to a maximum of 10 m by this method which proved the limit in this terrain (sandy-clays overlying Cretaceous sandstone). The sampling method adopted involved collection of comminuted sandy material delivered in sequence at the surface; each metre of material was then quartered and stored in polythene bags. Slight heating of samples occurred during drilling which might have caused water vapour loss but this was not considered to be significant. Isotope measurements were made on these samples and no fractionation was observed.

### 3.3 Dug well sampling

In order to take advantage of traditional construction methods, a dug well sampling programme was organised in Sudan. Initially, side wall sampling of a recently dug well at 25 cm intervals to depths of 30 m or greater was carried out using both a portable electric drill and an adze. However this technique proved to be operationally difficult and ran the risk of prior loss of moisture and capilliary migration of solutes to the side wall. This technique was abandoned in favour of the 'low technology' solution which is now being routinely used. Well-digging teams collect fresh material from the well at predetermined intervals and/or at the end of each day's shift. Samples are collected in glass fruit bottling jars with rubber seals. Supplementary samples have also been obtained from shallow pits and excavations (up to 2 m).

Samples obtained by any of the above techniques have been used for soil moisture and solute determinations; however only percussion drilling and dug well sampling methods are suitable for isotopic determinations (including tritium). Typically 50 g samples of moist disaggregated sediment are dispersed and elutriated in 30 ml aliquots of distilled, demineralised water for at least an hour, with occasional stirring; the optimum period is determined for each lithology to ensure effective elutriation. Supernatant solutions are then filtered (0.45 μm) prior to analysis.

Moisture content and $SEC_{25}$ have generally been determined in the field and in this way, it is possible to make preliminary interpretations on soil moisture conditions and recharge estimations using $SEC_{25}$. Laboratory analysis was carried out for Cl and $NO_3$ using automated colorimetry; sulphate and cations were analysed if required by ICP-OES. Moisture was extracted for stable isotope determination by azeotropic distillation with toluene. In nearly all instances, agreement between duplicates was within ±2 °/oo for $\delta^2H$ and ±0.2 °/oo for $\delta^{18}O$; duplicates falling outside this range were repeated until satisfactory. Isotopic analysis was carried out by conventional mass spectrometry, namely the reaction of 10 µl water with heated Zn shot for $\delta^2H$ and the equilibration of 5 ml water with $CO_2$ of known isotopic composition for $\delta^{18}O$. Some tritium analyses of sandy material were carried as an independent check of the results from the solute profiles. Water samples were obtained by vacuum distillation.

## 4. RESULTS

### 4.1 Cyprus

Recharge investigations were carried out on the Akrotiri peninsula in the extreme south of the island of Cyprus. Drilling was carried out in unconsolidated deposits of Recent age comprising mainly fine grained sands, where the unsaturated zone was at least 10 m thick. These studies were linked with a long term lysimeter study of recharge (Kitching <u>et al</u>., 1980). Vegetation varied from grassland to forest; twelve boreholes were drilled in an area of 6 $km^2$ in locations representative of all land use types. The sites drilled represented undisturbed conditions with some animal grazing but with no inputs of artificial fertilisers. Of the twelve boreholes drilled, eight provided profiles to depths (15 m - 29 m) suitable for interpretation (Figure 3), the others being abandoned due to drilling breakdowns of various kinds. A detailed discussion of the results is given in Kitching <u>et al</u>. (1980), Edmunds and Walton (1980) and Edmunds (1983), but a summary of the essential features is given here.

(1) Below a certain depth, chloride concentrations reach a relatively constant value. This depth (generally 2-3 m) is interpreted as the effective zero flux plane for solutes.

(2) At depth, a nearly constant interstitial water chloride concentration, $C_s$, is developed in all profiles with the possible exception of AK 7. There are fluctuations about this mean value but there are no significant trends with depth. Indeed, the amount of variation about the mean value gives an indication of the reliability that can be placed on the recharge estimate derived from a profile. The mean chloride concentrations also vary from profile to profile and this is thought to reflect a genuine spatial variation in the rate of recharge.

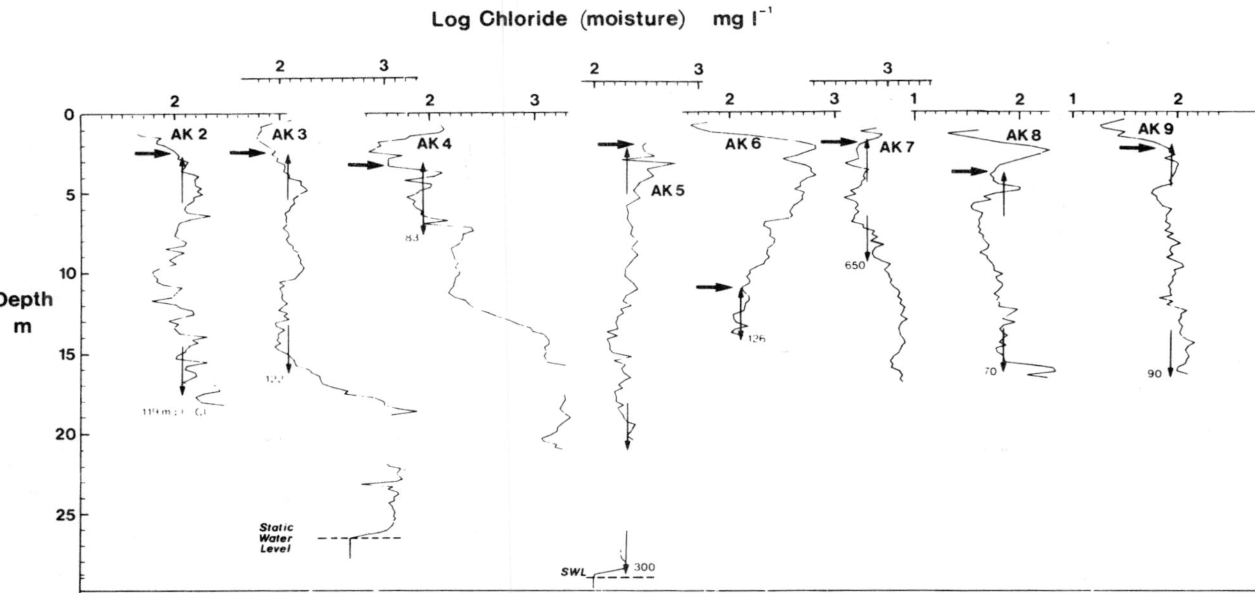


Figure 3. Chloride profiles (log Cl) from eight boreholes drilled at Akrotiri, Cyprus. Vertical arrows indicate the profile length over which steady-state conditions are considered to exist. Horizontal arrows indicate the likely maximum depth of the zero flux plane (ZFP). Mean chloride concentrations for the intervals are given in $mgl^{-1}$.

Table 1. Mean annual rainfall (3 yr and 25 yr) for stations at Akrotiri (Cyprus) with the average chloride deposition (wet and dry) over the three year period 1977-80.

| Year | Rainfall (mm) | Mean total deposition of chloride (kg/ha/yr) |
|---|---|---|
| AKROTIRI (RAF) | | |
| 1977/78 | 455 | 55 |
| 1979/79 | 246 | 43 |
| 1979/80 | 506 | 77 |
| AKROTIRI VILLAGE | | |
| 1977/78 | 426 | 66 |
| 1978/79 | 235 | 57 |
| 1979/80 | 564 | 96 |
| MEAN VALUE | 406 (25 yr) | 66 (3 yr) |

Table 2. Data used from Figure 3 for recharge estimation ($R_d$)*. Comparative results from tritium profiles, $R_d$ ($^3H$), are given for reference

| Borehole | AK 2 | AK 3 | AK 4 | AK 5 | AK 6 | AK 7 | AK 8 | AK 9 |
|---|---|---|---|---|---|---|---|---|
| Profile Interval(m) | 2.4-17.3 | 2.4-16.1 | 3.2-6.9 | 2.0-28.3 | 10.9-14.0 | 1.7-9.4 | 3.7-16.0 | 2.0-16.5 |
| Profile Length(m) | 14.8 | 13.7 | 3.7 | 26.3 | 3.1 | 7.7 | 12.3 | 14.3 |
| $C_s$ (mg/l) | 119 | 122 | 83 | 200 | 126 | 650 | 70 | 90 |
| log $C_s$ | 2.076 | 2.086 | 1.920 | 2.301 | 2.100 | 2.813 | 1.845 | 1.954 |
| $R_d$ (Cl) (mm/yr) | 56 | 55 | 80 | 33 | 53 | 10 | 94 | 74 |
| $R_d$ ($^3H$) (mm/yr) | 52 | 53 | N/A | 22 | N/A | N/A | 62 | 75 |

* $C_s$ is the mean interstitial water chloride concentration estimated by elutriation of samples from the profile intervals indicated in Figure 3.

(3 Steady state conditions are found to persist to the water table (e.g. AK 5) or to the maximum depth drilled (AK 8, 2, 9). In other boreholes there is an indication of chloride addition from the aquifer; in AK 3 there is an abrupt increase in chloride which coincides with a marine horizon (shelly gravels) and the profile below this cannot be used. In AK 4 there is a suggestion of continuous uptake of chloride through the succession, and only a limited portion of the profile has been used.

(4) Distinct peaks are apparent within the steady state sections of some profiles (e.g. AK 2), which indicate that homogenisation by dispersion does not take place. The peaks of AK 2 and possibly other profiles have an apparent periodicity of less than 1 m and are considered to represent cyclic, mainly annual, inputs of solutes from the soil zone (Edmunds and Walton, 1980).

Those sections of the solute (chloride) profiles that were used in recharge calculations are indicated in Figure 3. They represent an integrated record of recharge over a period of time ranging from 1978 back to the late 1940's.

The 25 yr mean rainfall value for Akrotiri (406 mm) was used for calculations and the 3 yr mean chloride deposition flux including that of dry deposition is given in Table 1. The 3 yr mean chloride deposition flux was used in conjunction with the 25 yr rainfall value in the recharge calculations. There is some fluctuation in the chloride deposition both over the period of investigation and between the two stations. This is reflected in the total rainfall figures. The use of long term means in this technique is advantageous since the profile itself represents the average input over a number of years. The recharge estimates for the eight boreholes are given in Table 2.

At all eight locations, recharge was estimated to be in the range 10-94 mm/yr. The variability is consistent with changes in topography and vegetation cover between sites with the highest values occurring in areas of sparse vegetation and low recharge (~10 mm) taking place where bush vegetation exists. The results compare well with the independent recharge estimates obtained from tritium peak analysis.

### 4.2 Sudan

Estimation of direct recharge was carried out as part of a wider study of the recent and palaeorecharge history and water resources of central Sudan. Profiles were obtained by augering and from dug wells at interfluve sites in the vicinity of Abu Delaig, 100 km ENE of Khartoum. Abu Delaig lies on a wadi system draining into Wadi el Hawad. The interfluve areas (Figure 4) generally comprise sandy, colluvial clay sediments probably of Quaternary age, overlying the basal Nubian Sandstone (Cretaceous). Profiles were therefore obtained from indurated sediments including cemented sandstones in contrast to the unconsolidated deposits found in Cyprus.

This area is on the Sahel margin and renewable sources of groundwater are principally found in the Nile valley where there is

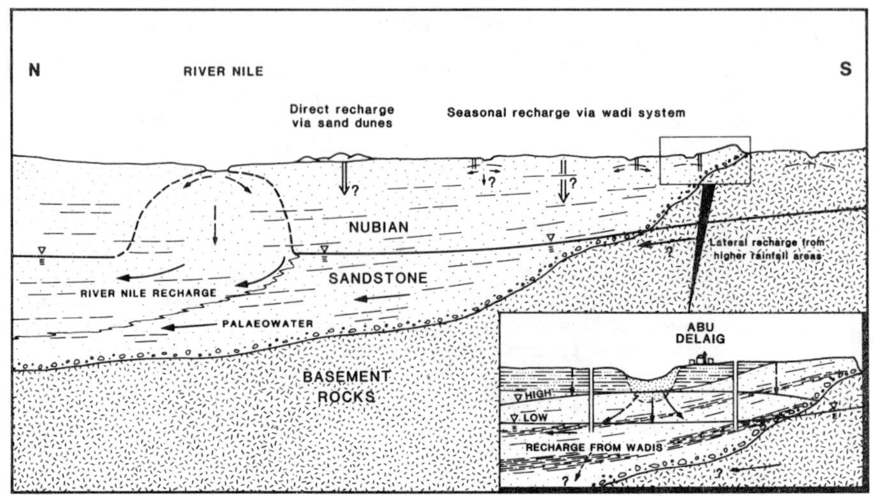


Figure 4. Schematic cross section of the Wadi Hawad area showing the situation of Abu Delaig and relative position of recharge profiles on either side of the wadi system. The only significant current recharge sources are from the River Nile and via the wadi system.

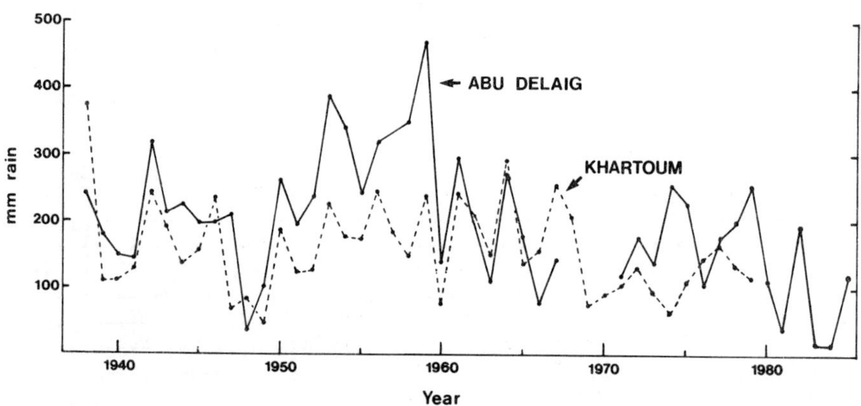


Figure 5. Annual rainfall at Abu Delaig and at Khartoum between 1938-1985.

groundwater recharge from the River Nile and along the wadi lines. Isotopic and geochemical studies demonstrate that much of the water being exploited from the deeper Nubian aquifer is palaeowater, recharged during the mid-Holocene (Darling et al., 1987). There appears to be little or no recharge via the wadi system to the deep regional aquifer at the present day.

Rainfall records are available for Abu Delaig from 1938 to the present day and the annual rainfall is compared with that of Khartoum in Figure 5. Over the 29-year period, 1938-1967, the mean annual rainfall at Abu Delaig was 225 mm but in the period 1971-1985 the corresponding value was only 154 mm. The period of study, 1982-1985, was the most arid on record with only about 15 mm recorded in both 1983 and 1984. Rainfall was sampled for chemical analysis during each of the years 1982-1985 and the mean concentrations of chloride are summarised in Table 3. These values represent bulk deposition during the rainy season and it is assumed that during the dry season there is no input to or removal of chloride from the surface. A value of 5 $mgl^{-1}$ Cl has therefore been adopted for $\bar{C}_p$, although, as discussed below, this may not be valid as a long term average.

Table 3. Chloride concentrations of rainfall (mean annual average) at Abu Delaig, Sudan, for the period 1982-1985. Results in $mgl^{-1}$.

| Year | Total rainfall | Mean Cl (weighted average) |
|---|---|---|
| 1982 | 192 | 6.1 |
| 1983 | 15.5 | 10.6 |
| 1984 | 15.2 | 5.6 |
| 1985 | 118 | 2.4 |

Volume weighted 4 year average: 4.9 $mgl^{-1}$

Four solute profiles from the unsaturated zone at Abu Delaig illustrate the results for Sudan (Figure 6) which are described in detail elsewhere (Edmunds et al., in prep.); hole A is a hand dug well and G, M, Q are auger holes. All profiles were obtained in November 1983 some three months after the very poor rainy season.

The moisture contents, plotted on a dry weight basis, range from 2% to 11% and strongly depend upon lithology, with highest values occurring in clay-rich horizons. The overall solute distribution is given by $SEC_{25}$ which reflects the total mineralisation of the moisture profile. The chloride and nitrate profiles are also illustrated. Results are expressed as elutriate concentrations and then as calculated pore water concentrations after correcting for the dilution during elutriation. After correction, it was found that estimated pore water $SEC_{25}$ values range from 2190 to 79430 $\mu S\ cm^{-1}$, equivalent to a total mineralisation from about 1300 $mgl^{-1}$ to values rather above sea water concentrations. These corrected $SEC_{25}$ values should only be

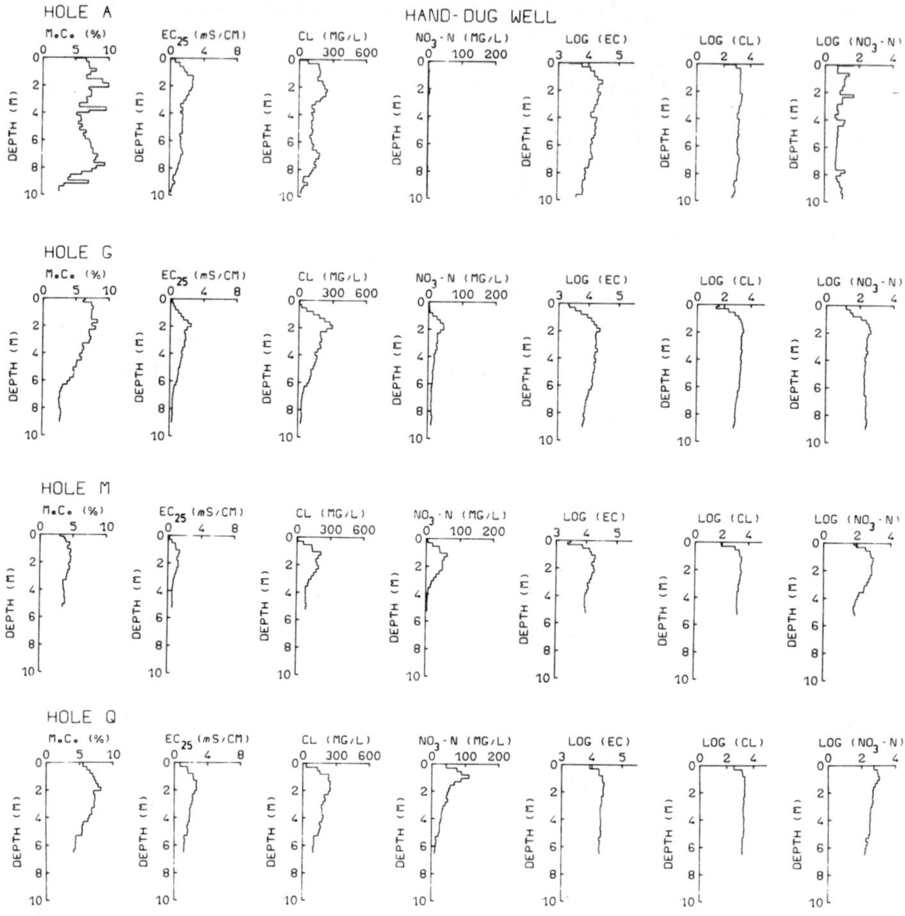


Figure 6. Moisture and solute profiles for one dugwell (A) and three augers at Abu Delaig, Sudan. Data shown are: moisture content M.C. in per cent (dry weight basis); specific electrical conductance ($EC_{25}$) of the elutriate; chloride concentration of the elutriate (Cl); nitrate concentration for the elutriate ($NO_3$-N); log of specific electrical conductance of soil moisture after correction for moisture content (LOG EC); log chloride concentration of the soil moisture after correction for moisture content (LOG CL); log concentration of nitrate in soil moisture after correction for moisture content (LOG NO3).

Table 4. Mean chloride concentrations in interstitial waters from profiles at Abu Delaig (Sudan) and the corresponding estimates of direct recharge. Mean chloride concentrations were obtained after correction for the dilution resulting from elutriation.

| Profile | Interval (m) | Mean Chloride | | No. of Meas. | $R_d$(mm) |
|---|---|---|---|---|---|
| | | mg/l | log (mg/l) | | |
| A | 1.0 -9.80 | 1288 | 3.110 | 39 | 0.78 |
| B | 1.05-2.85 | 1738 | 3.240 | 8 | 0.58 |
| C | 1.05-5.00 | 2239 | 3.350 | 20 | 0.45 |
| D | 2.24-4.20 | 2228 | 3.348 | 8 | 0.45 |
| E | 0.80-6.95 | 783 | 2.894 | 25 | 1.28 |
| G | 2.05-8.80 | 1675 | 3.224 | 27 | 0.60 |
| H | 0.80-3.30 | 1208 | 3.082 | 10 | 0.83 |
| J | 0.15-2.80 | 1120 | 3.049 | 8 | 0.89 |
| K | 0.80-2.30 | 971 | 2.987 | 6 | 1.03 |
| M | 0.80-5.30 | 1936 | 3.287 | 18 | 0.52 |
| N | 1.55-1.80 | 3936 | 3.595 | 22 | 0.25 |
| P | 0.55-3.00 | 173 | 2.238 | 10 | 5.78 |
| Q | 0.80-6.55 | 1845 | 3.266 | 23 | 0.54 |
| S | 1.30-4.30 | 1663 | 3.221 | 9 | 0.60 |

interpreted in a semi-quantitative sense because of the lack of strict proportionality between $SEC_{25}$ and concentration. Nevertheless, they give a good indication of the general trend in mineralization of the interstitial water down a profile.

Chloride concentrations generally mirror $SEC_{25}$ indicating that Cl is an important constituent of the total mineralisation; sulphate is also present as a major solute, particularly in the upper section of the profiles (Edmunds et al., in prep.). Nitrate is also an important constituent of profiles G, M and Q on the southern side of the wadi with highest concentrations up to 1000 $mgl^{-1}$ $NO_3$-N in pore solutions. This is in contrast to hole A on the northern side of the wadi where background levels closer to 10 $mgl^{-1}$ $NO_3$-N are found. The variation in nitrate concentrations is not yet fully explained but is likely to relate to different vegetation or cultivation on either side of the wadi; it is unlikely therefore that in this location nitrate could form a suitable reference solute for recharge estimation.

Chloride concentrations increase with depth over the first 1-2 m and below this reach near constant values. They may be interpreted as typical recharge profiles with a zone of fluctuating moisture content above a zero flux plane at about 2 m depth. The absence of any salt accumulation indicates a net recharge for these profiles and a steady-state portion of several metres can be observed in each profile. None of these profiles reaches the water table but data are available from several dug wells in the vicinity where much lower chloride concentrations are observed (in the range 40-60 mg Cl $l^{-1}$).

The main chloride concentrations have been calculated for the steady-state sections of all 14 profiles taken from an area of some 6 $km^2$ of interfluve terrain (Table 4). The interstitial chloride concentrations range from 1000 to nearly 4000 mg/l and there is some correspondence between concentration and site location. The highest value (N) occurs in a profile taken from a sandstone ridge above the main wadi where some surface run-off might have occurred. The lowest value (K) is from a relatively clay-rich profile from flat lying ground. Using steady-state interstitial water chloride concentrations ($C_s$), a mean chloride concentration in rainfall ($\bar{C}_p$) of 5 $mgl^{-1}$ and the long-term average rainfall ($\bar{P}$), values for $R_d$ are calculated. Excluding the wadi profile, these lie in the range 0.25-1.28 mm/yr. A regional long term average ($\bar{R}_d$) of 0.72 mm/yr may be calculated for the direct component of recharge via interfluve areas in this area of Sudan.

## 5. DISCUSSION

Good agreement exists between estimates of direct recharge obtained using chloride interstitial water concentrations and tritium peaks contained in unsaturated zone profiles from sands in Cyprus, where the mean annual rainfall is around 410 mm/yr. Realistic, if very low, values for direct recharge were derived from chloride profiles in an arid area of Sudan where mean annual rainfall is around 180 mm/yr.

It is apparent that under favourable conditions, chloride alone can be used with very simple procedures as a tool for recharge estimation and that many of the difficulties inherent in the use of tritium are avoided. Although no direct comparisons are available within the present study, the advantages of the chloride profile technique over a conventional water balance approach are immediately apparent. Soil physics theory may be needed to interpret the profiles and their evolution, but no detailed physical data are required apart from moisture content and bulk density.

A main advantage of the chloride profile technique is that results are derived in part from long term data, compared with direct measurements of flow in the unsaturated zone where results are typically obtained on an annual basis. The solutes in the unsaturated zone act as a totalising rain gauge with storage of moisture over decades or centuries. Profiles therefore can act as a means of obtaining long term averages of recharge and also as a means of detecting climatic change (Edmunds and Walton, 1980; Allison et al., 1985). In exceptional circumstances, where the data are not smoothed by dispersion it may be possible to discern annual increments in the profile. This can be seen in the case of the chloride profile AK 2 (Figure 3) where individual annual peaks are visible (Edmunds and Walton, 1980); using the $SEC_{25}$ profile from the same borehole gives a good correlation with the oscillations in rainfall since the 1940's.

The chloride profiles from Sudan, indicating much lower net recharge rates, indicate that water is stored in the unsaturated zone for much longer periods. Using a value for porosity of 30%, saturation at 25% and a recharge rate of 1 mm/yr, a 25 m profile can be storing water which has been in transit for about 2000 years. The shape of some profiles from Sudan indicates that slightly less saline water is found with increasing depth which may be a reflection of slightly higher recharge rates in earlier centuries. An alternative explanation could be that there is also a component of upward movement of water vapour in these profiles and this is suggested by isotopic results ($\delta^{18}O$, $\delta^{2}H$) from some of the profiles in Sudan (Darling et al., 1987). In particular, the isotopic composition of the deepest moisture is similar to that of lower salinity water contained in the saturated zone. However, although discharge (loss of water vapour) may be occurring during certain periods net recharge is indicated on the basis of the chloride profiles. The profiles might therefore represent the complex result of both processes, with recharge only activated during episodes of higher rainfall. Resolution of this problem requires further data from similar areas especially from those with higher mean annual rainfall.

Just as important as measuring and obtaining a reliable figure for the mean interstitial water chloride concentration, is the reliable estimation of the atmospheric inputs. This is particularly difficult in most semi-arid regions where long-term records of the chloride composition of rainfall and dry deposition are practically non-existent. In the present study, reliable data have been obtained for total solute deposition over periods up to 4-years, and as indicated, the rainfall in Sudan during this study was well below the long term average. Longer runs of data would improve the reliability of the recharge estimates. There is also the problem that rainfall samples can contain dust and that both rain and dry deposition are therefore likely to be sampled together during the rainy season. The effective rainfall chemistry used for recharge estimate purposes in Sudan is rather higher in chloride than that expected from its continental position. In this study it has been assumed that, during the rainy season all chloride is washed into the profile, and that dust is 'on the move' during the dry season; thus net deposition is insignificant except during the rainfall episodes when sampling of bulk deposition took place.

## 6. CONCLUSIONS

Solute profiles provide a potentially important technique for investigating the unsaturated zone of semi-arid and arid regions and for estimating recharge. The advantages of the method may be summarised.

1. it is a quantitative method which overcomes most of the limitations and errors of conventional techniques in semi-arid and arid terrains;

2. chloride is the most suitable reference solute for recharge estimation and may be used independently, i.e. without supporting evidence such as tritium;

3 recharge estimates derived from the chloride mass balance approach reflect averages of rainfall inputs over tens, hundreds or even a few thousands of years;

4. local differences in recharge due to vegetation, or soil type or slope may be confirmed. Integration of solute profile information with other data (e.g. remote sensing) may provide a means of obtaining improved regional estimates of recharge;

5. sampling and analysis are relatively straightforward, cheap and unsophisticated and adaptable to remote areas underlain by unconsolidated formations and also many consolidated porous media.

The principal limitations must also be recognised. The method requires a representative value for solute inputs; a 3 or 4 year spatially-averaged value may be adequate but ideally there should be supporting evidence to indicate that this value is representative of a much longer time period. There is a strong case for a network of rainfall chemistry, even chloride only, stations in subtropical regions which would enable wider use of solute balance studies to be made.

## 7. ACKNOWLEDGEMENTS

The British Overseas Development Administration are thanked for their financial support to carry out this work. We also wish to thank colleagues in ACSAD (Damascus) and the National Administration for Water (Khartoum) for their help in many ways. This paper is published with the permission of the Director, British Geological Survey, Natural Environment Research Council.

## 8. REFERENCES

ALLISON G.B. and HUGHES M.W. 1974. Environmental tritium in the unsaturated zone: estimation of recharge to an unconfined aquifer. Proc. Symp. Isotope Techniques in Groundwater Hydrology, **Vol I,** 57-72. IAEA Vienna.

ALLISON G.B. and HUGHES M.W. 1978. The use of environmental chloride and tritium to estimate total recharge to an unconfined aquifer. Aust. J. Soil Res., **16,** 181-95.

ALLISON G.B., STONE W.J. and HUGHES M.W. 1985. Recharge in karst and dune elements of a semi-arid landscape as indicated by natural isotopes and chloride. J. Hydrol. **76,** 1-25.

DARLING W.G., EDMUNDS W.M., KINNIBURGH D.G. and KOTOUB S. (in press). Sources of recharge to the basal Nubian Sandstone aquifer, Butana area, Sudan. Proc. Symp. Isotope Hydrology 1987. IAEA Vienna.

DINCER T., AL MUGRIN A. and ZIMMERMAN U. 1974. Study of the infiltration and recharge through the sand dunes in arid zones with special reference to the stable isotopes and thermonuclear tritium. J. Hydrol., **23,** 79-105.

EDMUNDS W.M. 1983. Use of geochemical methods to determine current recharge and recharge history in semi-arid zones. Proc. IIeme symposium arabe sur les ressources en eau, Rabat 1981. pp 282-301. Arab Centre for Semi Arid Zones and Dry Lands (ACSAD), Damascus.

EDMUNDS W.M. and WALTON N.R.G. 1980. A geochemical and isotopic approach to recharge evaluation in semi-arid zones - past and present. 'Arid Zone Hydrology: Investigations with Isotope Techniques', pp 47-68. IAEA Vienna.

EDMUNDS W.M., DARLING W.G. and KINNIBURGH D.G. (in prep.). Recharge to the shallow aquifer system at Abu Delaig, Sudan.

ERIKSSON E. 1976. The distribution of salinity in groundwaters in the Delhi region and recharge rates of groundwater. Interpretation of Environmental Isotope and Hydrochemical Data in Groundwater Hydrology, IAEA Vienna. pp 171-178.

KITCHING R., EDMUNDS W.M., SHEARER T.R., WALTON N.R.G. and JACOVIDES J. 1980. Assessment of recharge to aquifers. Hydrol. Sci. Bull., **25,** 217-235.

SHARMA M.I. and HUGHES M.W. 1985. Groundwater recharge estimation using chloride, deuterium and oxygen-18 profiles in the deep coastal sands of Western Australia. J. Hydrol, **81,** 93-109.

SMITH D.B., WEARN D.L., RICHARDS H.J. and ROWE P.C. 1970. Water movement in the unsaturated zone of high and low permeability strata by measuring natural tritium. 'Isotope Hydrology 1970', pp 73-87, IAEA, Vienna.

SUKHIJA B.S. and SHAH C.R. 1975. Conformity of groundwater recharge rate by tritium method and mathematical modelling. J. Hydrol. **30,** 167-178.

WELLINGS S.R. and BELL J.P. 1980. Movement of water and nitrate in the unsaturated zone of Upper Chalk near Winchester, Hants, England. J. Hydrol., **48,** 119-136.

RECHARGE ESTIMATION FROM THE DEPTH-DISTRIBUTION OF ENVIRONMENTAL CHLORIDE IN THE UNSATURATED ZONE - WESTERN AUSTRALIAN EXAMPLES

M.L. Sharma
Senior Principal Research Scientist
CSIRO
Division of Groundwater Research
Wembley, Western Australia. 6014.

ABSTRACT. This paper presents methods for computing long-term average recharge rates to groundwater from the analyses of the depth-distribution of environmental chloride in the unsaturated zone. A one-dimensional equation, incorporating convective and diffusion terms, describing steady-state transport of non-reactive solutes, is solved in an inverse fashion, enabling computation of the depth-distribution of vertical water flux density. This and other simplified models were applied to interpret the observed chloride profiles beneath native vegetation in two coastal regions of Western Australia. The presence of preferred pathways for water movement has been suggested for both regions. In the sandy coastal region, the overall recharge was some 15% of the annual precipitation (775 mm $yr^{-1}$). Over 50% of this recharge occurs through preferred pathways. In the profiles of the south-western region, the majority of recharge occurs through preferred pathways since the estimated recharge through the soil matrix is negligible (<0.5% annual rainfall). The limitations of using a simple one-dimensional model for such complex lateritic profiles are discussed.



*I. Simmers (ed.), Estimation of Natural Groundwater Recharge, 159–173.*

## 1. INTRODUCTION

Reliable estimation of rates of water addition to an aquifer is basic to the assessment of groundwater resource potential so that efficient long-term management schemes can be developed to avoid any adverse environmental consequences. Furthermore, knowledge of recharge rates is crucial in assessing aquifer contamination and for developing efficient prevention plans.

Despite the recognition of the requirements for recharge data, there is a paucity of readily applicable methods for estimating natural recharge rates. This is particularly so in arid and semi-arid regions where recharge is not only a small proportion of precipitation but also the absolute recharge rates are relatively very small. In these regions the use of the traditional water balance method, in which other components of the water balance are measured or estimated indirectly and recharge is computed by the difference, will usually be of inadequate accuracy while the uncertainty in recharge estimation is likely to be magnified.

As an alternative, tracer techniques are possibly more appropriate for arid regions. In these water movement through the profile to the aquifer is followed by either natural or applied tracers. Under many semi-arid regions environmental chloride of oceanic origin has been found to be a good tracer. The presence of chloride in the profile is considered an imprint of water movement, and the observed depth-distribution of chloride concentration can be used in the inverse fashion for computing recharge. In this paper we present the theoretical basis of interpreting chloride profiles for computing vertical water flux density as a function of profile depth. Application of this technique in computing recharge rates is demonstrated in two regions of Western Australia where important problems related to groundwater management are being encountered.

## 2. BACKGROUND AND METHODS

### 2.1. Choice of a Tracer

An environmental tracer most suitable for following the movement of water must be highly soluble, relatively inert (conservative), foreign to the system and not substantially taken up by vegetation. In addition it should be available in relatively large quantities to be measured with reliability and ease. Chloride ion satisfies most of these criteria and therefore is considered a suitable tracer, particularly in coastal areas where large quantities of aeolian chloride are deposited from the winds carrying precipitation and dry salts from over the ocean (Carroll, 1962; Hingston and Gailitis, 1976; Menger and Eisenreich, 1983). Such chloride has been used for estimating areal recharge based on its concentration in the saturated zone (Eriksson and Khunakasen, 1969; Kitching *et al*., 1980) as well as in estimating localized recharge by considering its concentration in the unsaturated zone (Allison and Hughes, 1978; Peck et al., 1981; Sharma et al., 1983; Sharma and Hughes, 1985).

## 2.2. Theory and Methods

The use of chloride as an environmental tracer for computing recharge is based on the consideration of conservation of mass of the solute. Computations are usually based on the assumption that chloride is a conservative tracer, and that unsaturated/saturated soil systems have attained a steady state or at least a quasi-steady-state. The steady-state, one-dimensional vertical flux (J) of a non-sorbing solute in a non-dispersive and relatively inert system is given (Bresler, 1981; Peck et al., 1981) by:

$$J = -D\theta \frac{\partial C}{\partial z} + Cq + S \qquad (1)$$

where D is the diffusion coefficient of the solute, $\theta$ is the volumetric water content, C is the solute concentration, z is the depth (measured positive downward), q is the vertical flux of water and S is the source/sink term for the solute. Under many conditions S can be assumed negligible, specially at depths below the root zone. For deep profiles, which may store several years' recharge, a quasi-steady-state can be assumed and Equation (1) (with S=0) can be rearranged and solved for computing the steady-state vertical water flux as a function of depth as:

$$q = \frac{1}{C} \left[J + D\theta \frac{\partial C}{\partial z}\right] \qquad (2)$$

In the subsequent discussion the application of Equation (2) is demonstrated. When the contribution of diffusion can be assumed negligible, this equation simplifies to:

$$q = \frac{J}{C} \qquad (3)$$

Assuming that $Cl^-$ is derived from precipitation alone, its annual input flux (J) is equated to $\bar{C}_p\bar{P}$, where $\bar{P}$ is long-term average annual precipitation and $\bar{C}_p$ is long-term average $Cl^-$ concentration in precipitation. Thus for practical purposes, long-term average annual recharge rate (R) below the root zone (where S ~ 0) can be calculated from:

$$R = \frac{\bar{C}_p \bar{P}}{\bar{C}_z} \tag{4}$$

where $\bar{C}_z$ is the average $Cl^-$ concentration of the draining water, and this can be estimated from the depth-average $Cl^-$ concentration of the unsaturated zone below the maximum rooting depth.

### 2.3. Regions Used for Recharge Estimation

The simplified, one-dimensional steady-state models, described above, have been used for computing long-term average vertical water flux in deep (>15 m) unsaturated profiles. For demonstration purpose we use examples of chloride profiles measured beneath native vegetation, which has remained relatively undisturbed over a very long period (>100 yrs), and thus assumptions of steady-state can be considered valid.

The examples are from two coastal regions (south-western region, northern coastal region) of Western Australia (Fig. 1), where estimation of recharge in relation to land use is considered very important from the water resources management point of view. The climate of both areas is Mediterranean with distinct summer and winter seasons. The majority (>80 per cent) of the rain falls during the winter period (May to October) and is the source of possible recharge. The rainfall in the south-western lateritic region declines from a maximum of 1300 mm $yr^{-1}$ near the Darling Scarp to about 600 mm $yr^{-1}$ some 50 km inland (Fig. 1). In the north coastal region which extends only 30 km inland from the coast, rainfall ranges from 600 to 800 mm $yr^{-1}$. Potential evaporation (Class A pan) is high (1600 to 1800 mm $yr^{-1}$), and far exceeds the rainfall during most part of the year with the exception of the winter months (June to August).

The land formation and vegetation characteristics of the two regions differ considerably. These and water management issues are briefly discussed below.

2.3.1. North Coastal Sandy Plain. In this region, the 'Gnangara Mound' is a major groundwater reservoir, which is being used increasingly for supplying water to Perth, Western Australia. In recent years, the Mound has supplied up to 42% of the public water supply for a population of about 1 million. Although, the overall net recharge rate for the Mound has been reasonably well estimated to be about 11% of the annual precipitation of about 800 mm $yr^{-1}$ (Allen, 1981), there is considerable variability in recharge due to variation in vegetation, topography and soil type (Sharma *et al.*, 1983).

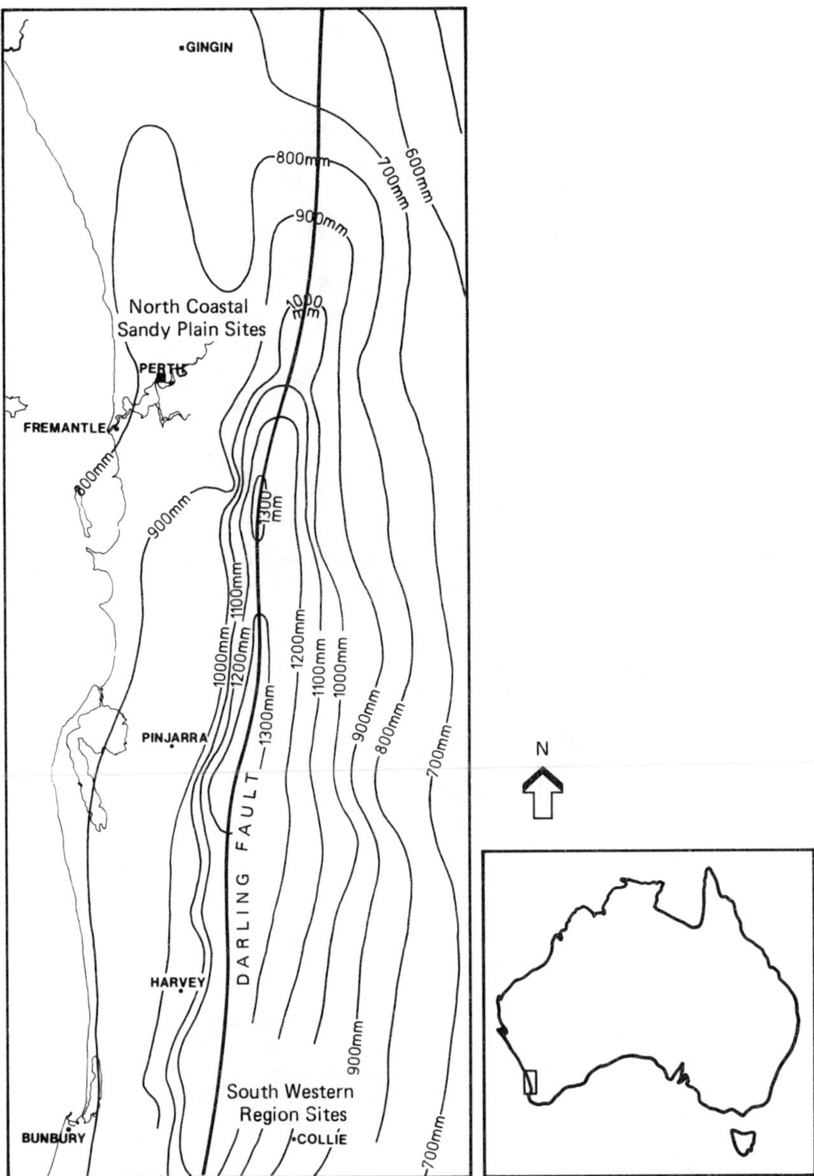


Fig. 1. Location of coastal regions of Western Australia being investigated.

One of the major changes in land use has been the replacement of native Banksia woodland by pine plantations (Butcher and Havel, 1976; Carbon _et al_., 1982). This is believed to have affected recharge. However, the reduction in recharge rate due to replacement of Banksia woodland by pines has not been adequately quantified. Of course, such modification will not only be influenced by the age, structure and density of the original and planted vegetation, but also by factors such as soil type and topographical position in the landscape.

The soil systems consist of sand dunes, which are aeolian in origin. They form an upper crust over a relatively shallow (up to 90 m deep) unconfined aquifer. These deep (5-50 m) sands are relatively inert and infertile and do not contain any $Cl^-$ of pedogenic origin. The use of chloride as a tracer in this region is therefore considered attractive (Sharma _et al_., 1983).

2.3.2. _South-western Region_. The development of agriculture after removal of native Eucalyptus vegetation over the last 100 years or so has caused salinization of water resources in the region (Peck _et al_., 1983). Of particular interest is the degradation of surface water resouces in the region, and this is putting increasing pressure on exploitation of the groundwater resources of the 'Gnangara Mound' near Perth.

In this region lateritization has given rise to deeply weathered duplex profiles. The upper gravelly surface soil layer is 2-8 cm thick with a saturated hydraulic conductivity ($K_s$) of about 2 m $d^{-1}$, and overlays a deep (up to 40 m) kaolinitic clayey subsoil with a much lower hydraulic conductivity ($K_s = 10^{-3}$ m $d^{-1}$). This subsoil stores a large quantity of salts which are believed to be aeolian in origin (Sharma _et al_., 1980).

Replacement of perennial, deep-rooted native vegetation (mainly _Eucalyptus marginata_ and _E. calophylla_) by annual, shallow-rooted agricultural species, causes increased recharge through the deep lateritic profiles, and leaches salts stored from the profile. Because of its economic significance, several studies have been carried out to quantify the effect of change in land use on various components of the hydrological cycle (e.g. Peck and Williamson, 1987; Sharma _et al_., 1987; Williamson _et al_., 1987.)

2.4. Sampling and Data Analysis

The data presented here are based on soil core samples collected mechanically using a drilling rig. The soil cores were sectioned for depth intervals and oven dried for determination of soil water content. Then a predetermined ratio of soil:water mixtures were prepared from which water was extracted.

Chloride concentrations of soil water extracts from the sandy Swan Coastal Plain soils were determined colorimetrically (Sharma and Hughes, 1985), while for lateritic soils of the south-western region

$Cl^-$ was estimated from measured electrical conductivity of extracts based on pre-determined correlations (Peck *et al.*, 1981; Johnston, 1987).

Average deposition of oceanic chloride falling in precipitation and dry fallout were estimated from the relationships established by Hingston and Gailitis (1976). The chloride fall decreases from west to east with distance from the ocean. Therefore depending on the location of the sampling site, a separate chloride input was computed. With the assumption of steady-state in solving the solute transport equations (1 to 3), the output of chloride flux will be the same as the input, and therefore J was equated to $\bar{C}_p\bar{P}$.

For all the profiles q(z) was computed by solving Equation (2) for the observed data sets C(z) and $\theta(z)$. Since $\partial C/\partial z$ needs to be smooth for the computation of q(z), a cubic spline was fitted to the experimental data of chloride concentrations as a function of depth, and $\partial C/\partial z(z)$ was computed from the smooth curve.

In solving Equation (2) for the lateritic profiles, the diffusion coefficient $D(\theta)$ was computed (Peck *et al.*, 1981) using the empirical relationships published by Porter *et al.*, (1960), while for the sandy soils with very low water holding capacity a constant value of D ($=0.01\ m^2\ d^{-1}$) was used. For these soils, the computations are negligibly affected by the wide range of possible D values (Sharma, 1986).

## 3. RESULTS AND DISCUSSION

### 3.1. North Coastal Sandy Plain

Measured chloride concentration of soil water with the fitted cubic spline and the corresponding volumetric water content as a function of depth for two profiles (within 100 m distance) measured beneath native vegetation are presented in Fig. 2. Approximate locations and average chloride concentrations of the groundwater are also shown. In general, there is only a small change in C with depth down to about 10 m, then C increases and approaches a rather large value which remains relatively constant to a considerable depth. It then declines and approaches $\bar{C}_G$, which is much lower than the value of C in the bulk of the profile. The corresponding q(z) for the two profiles (Fig. 3) show that q becomes relatively constant below 10 m, but it increases as the saturated zone is approached. Dynamics of soil water profiles observed for the area (Fig. 3) confirm that 10 m can be regarded as the maximum depth of water extraction for the native vegetation. Thus the values of q computed below the root zone (>10 m) should be interpreted as the average recharge rate through the soil matrix. The difference between the two profiles is considered as spatial variability, reasons for which have been discussed elsewhere (Sharma, 1986).

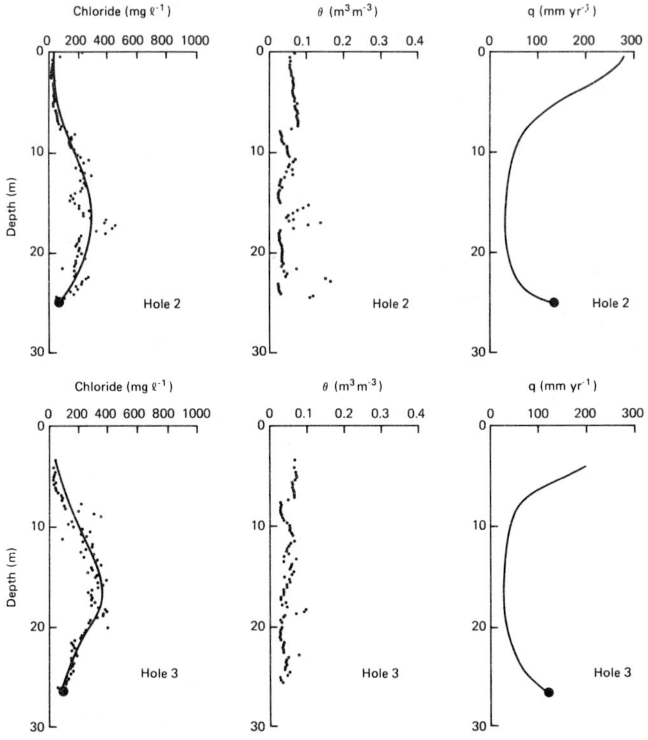


Fig. 2(a) Measured chloride concentration (C) and soil water content (θ), and the cubic spline fitted to the chloride data for borehole #2 beneath a native Banksia woodland, Swan Coastal Sandy Plain, Western Australia. Long-term average vertical water flux (q) was computed using Eqn (2), where $J = \bar{C}_p \bar{P}$, $\bar{C}_p = 12$ mg $L^{-1}$ and $\bar{P} = 775$ mm $yr^{-1}$.

(b) Same as 2(a) for borehole #3 from an adjacent location.

Considering several plausible hypotheses for the overall lower $\bar{C}_G$ values compared to C in the bulk of the profile below 10 m Sharma and Hughes (1985) concluded that the most convincing explanation is the presence of preferred pathways for water movement. In such systems some portion of infiltrated water will pass to the groundwater with minimal interaction with the soil matrix.

A simplified bimodal system of water movement is suggested for such recharge mechanism. The recharge occurs through the soil matrix ($R_m$) as well as through the preferred pathways ($R_p$), and $\bar{C}_G$ represents the average $Cl^-$ concentration of the total drained water. Thus the total recharge (R) for the system can be estimated by Equation

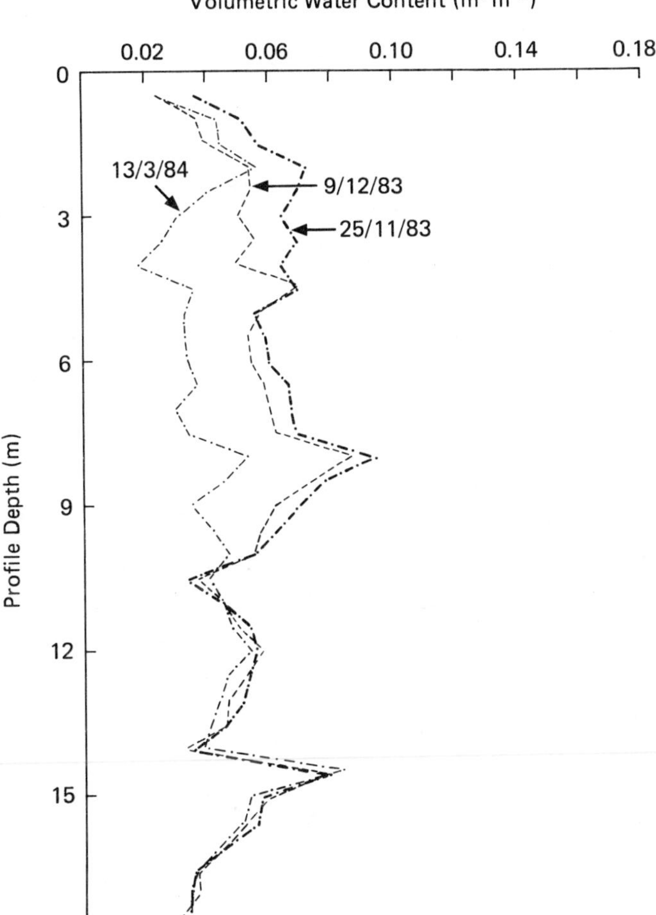


Fig. 3. Dynamics of soil water in the profile after the soil water storage was renewed by a heavy rainfall event. The soil water contents were measured by a neutron moderation technique, beneath a Banksia woodland located in the proximity of borehole #2 (Fig. 2(a)).

(3) with $C_G$ as the value of $C_z$. Provided $C_G$ is not unduly influenced by the chlorinity of surrounding groundwaters, R can be partitioned into $R_m$ and $R_p$ i.e.,

$$R_m = R \left(\frac{C_G - C_p}{C_z - C_p}\right) \tag{5}$$

$$R_p = R\left(\frac{\bar{C}_z - \bar{C}_G}{\bar{C}_z - \bar{C}_p}\right) \tag{6}$$

where $\bar{C}_G$ is the depth-average $Cl^-$ concentration of the shallow (~1 m deep) groundwater, and $\bar{C}_z$ is the depth-average $Cl^-$ concentration in the profile below the root zone and above the groundwater (average C for the profile excluding the first 10 m and the last 2 m). Such computations suggest that the total recharge values for the two profiles (Fig. 2 a,b) were respectively 129 mm $yr^{-1}$ and 178 mm $yr^{-1}$. Of these only 37 mm $yr^{-1}$ and 28 mm $yr^{-1}$ respectively were through the matrix, the remainder occurred through preferred pathways. It is clear that over 50% of the total recharge occurs through preferred pathways. Based on these computations it can be shown that these sandy profiles store some 10-15 yrs of the recharge occurring through the soil matrix.

## 3.2. South-western Lateritic Region

Profiles for C and θ, and the computed q for the two boreholes measured beneath a native Eucalyptus forest from (a) the high rainfall (1150 mm $yr^{-1}$) area and from (b) the low rainfall (800 mm $yr^{-1}$) area, presented in Fig. 4 a,b clearly show the effect of rainfall on the chloride accumulation in the profile. Analyses showed that the diffusion contributed significantly in the transport of $Cl^-$ particularly at depths where $\partial C/\partial z$ is large. In profile A, diffusion contributed up to 22 percent of the convection term, while in profile B, its contribution was of similar magnitude to the convection term (Peck *et al*., 1981). In general, q is high at shallow depths (<2 m) and it decreases with depth, approaches a constant value for some intermediate depth then increases sharply and attains a somewhat constant value at depths close to the watertable. In the intermediate profile zone, q is rather low (approximately 4.5 mm $yr^{-1}$ and 0.4 mm $yr^{-1}$ for A and B profiles) and increases respectively to 50 mm $yr^{-1}$ and 3.5 mm $yr^{-1}$ at the watertable. In these profiles groundwater recharge through the soil matrix is only 0.35% and 0.05% of the annual rainfall.

The value of q below the root zone should be regarded as the recharge rate. However, for these native perennial forests, the effective root depth cannot be defined reliably (Carbon *et al*., 1980; Sharma *et al*., 1982). An analysis of the seasonal dynamics of soil water storage in similar profiles indicates that the largest changes occur within the upper 4 m of the profile (Sharma *et al*., 1987) although water extraction by plant roots extends to depths of 6 m and beyond (Fig. 5).

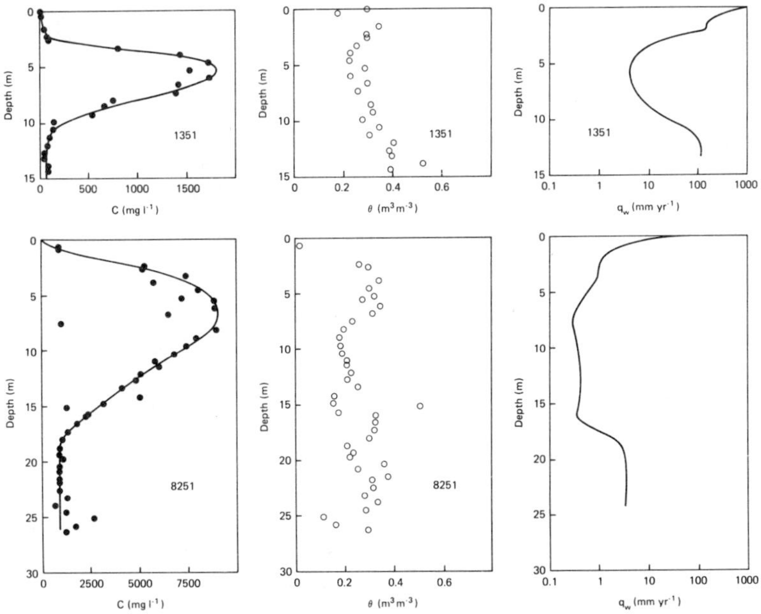


Fig. 4(a) Measured chloride concentration (C) and soil water content (θ), and the cubic spline fitted to the chloride data for a borehole beneath a native Eucalyptus forest in a relatively high rainfall zone of south-western Australia. Long-term vertical water flux (q) was computed using Eqn (2), where $J = \bar{C}_p \bar{P}$, $\bar{C}_p = 9.3$ mg L$^{-1}$, $\bar{P} = 1150$ mm yr$^{-1}$.

(b) Same as Fig. 4(a) for a borehgole in a relatively low rainfall zone of south-western Australia. Here $C_p$ = ~4 mg L$^{-1}$, P = 800 mm yr$^{-1}$.

Considerably higher $Cl^-$ concentrations (lower q) at intermediate depths probably indicate the maximum depth of root activity for water extraction. However, substantial increases in q below these depths cannot be explained easily. After considering several possible reasons, Peck et al., (1981), concluded that the most likely reason for this is the presence of preferred pathways imbedded in the soil matrix, and these facilitate injection of water of low chlorinity into the profile at depths close to the water table. Johnston et al., (1983) experimentally demonstrated the presence of preferred pathways in a lateritic profile in a 1200 mm yr$^{-1}$ rainfall region. These consist primarily of root channels. In a similar rainfall region, Johnston (1987) has further shown the presence of small-scale localized recharge throughout the unsaturated zone with an estimated q of 50 to 100 mm yr$^{-1}$, while most of the other surrounding area has q values below 5 m, of 2.2 to 7.2 mm yr$^{-1}$.

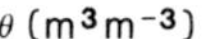


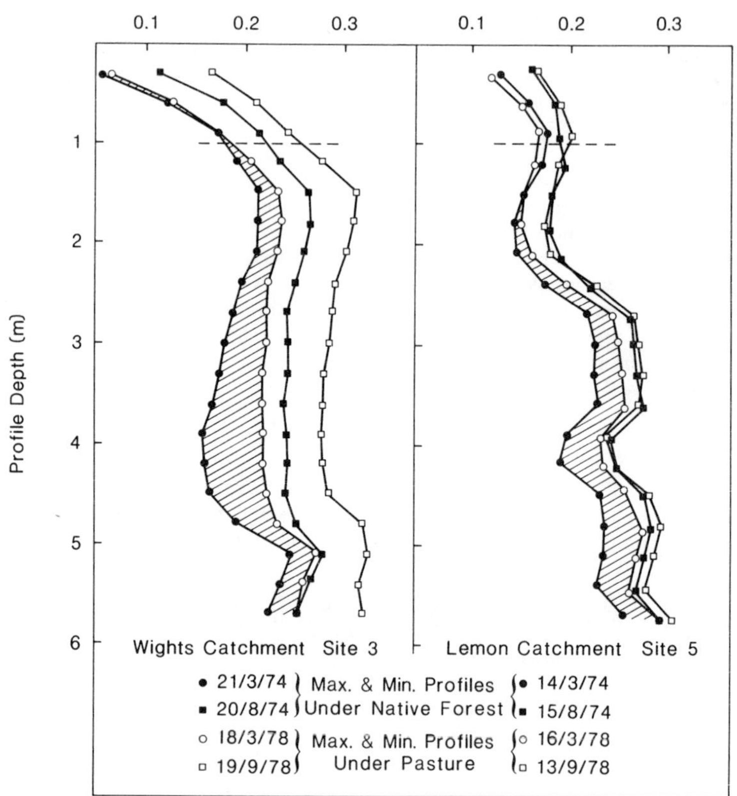


Fig. 5. Soil water profiles at (a) a location in proximity of borehole #1351 (rainfall = 1150 mm $yr^{-1}$), and another (b) at a location in the proximity of borehole #8251 (rainfall = 800 mm $yr^{-1}$). Shaded areas show minimum water extraction by the native vegetation. (For more explanation see Sharma et al., 1987.)

In comparison with the sandy coastal plain profiles, analyses of lateritic profiles are much more complex, firstly because the maximum root depth cannot be defined, and secondly because of recognized vertical layering. Thus the assumption of uniform vertical transport of water and $Cl^-$ through the soil profile cannot be justified. This is particularly the case for profile 'A' for which water flux of the order of 145 mm $yr^{-1}$ must occur through preferred pathways. Limitations of the analysis should be recognized. One of the major limitations is that in the one-dimensional analysis, the lateral flow of water and $Cl^-$ above the clayey subsoil layer is ignored, while other evidence suggests that this component may induce a substantial loss of $Cl^-$ (Sharma *et al*., 1980; Stokes and Loh, 1982). Furthermore, the analysis could be improved by incorporating the effect of dispersion (Johnston, 1987).

## 4. SUMMARY

Based on the conservation of mass, a one-dimensional vertical solute transport model for steady-state conditions was applied to computing vertical water flux density from the chloride profiles measured beneath native perennial vegetation. This chloride originates from the ocean and is deposited with the precipitation and is redistributed in the soil profile by water movement.

Analysis of data sets of profiles from two coastal areas of Western Australia, indicated that water movement through the soil profiles is not steady and uniform, and that water flow through preferred pathways contributes significantly to groundwater recharge. It is estimated that the majority of recharge occurs through preferred pathways in structured duplex lateritic profiles. Even in "apparently uniform" unstructured coastal sands more than 50% of recharge is contributed by this process.

## 5. REFERENCES

Allen, A.D. 1983. Ground water resources of the Swan coastal plain, near Perth, Western Australia. In B.R. Whelan (ed.) *Ground water resources of the Swan Coastal Plain*. CSIRO, Division of Land Resources Management, Wembley, Western Australia. pp. 29–47.

Allison, G.B. and Hughes, M.W. 1978. The use of environmental chloride and tritium to estimate total recharge to an unconfined aquifer. *Aust. J. Soil Res*. **16**, 181–195.

Bouwer, H. 1978. Groundwater hydrology. McGraw-Hill Book Company, New York.

Bresler, E. 1981. Transport of salts in soils and subsoils. *Agric. Water Manage*. **4**, 35–82.

Butcher, T.B. and Havel, J.J. 1976. Influence of moisture relationships on thinning practice. *New Zealand J. Forest Sci*. **6**, 158–170.

Carbon, B.A., Bartle, G.A., Murray, A.M. and Macpherson, D.K. 1980. The distribution of root lengths and the limits to flow of soil water to roots in a dry schlerophyll forest. *For. Sci*. **26**, 656–664.

Carbon, B.A., Roberts, P.J., Farrington, P. and Beresford, J.D. 1982. Deep drainage and water use of forests and pastures grown on deep sands in a Mediterranean environment. *J. Hydrol*. 55, 53–64.

Carroll, D. 1962. Rainwater as a chemical agent of geological processes - A review. USGS-WSP-1535-G, U.S. Dept. of the Interior. Superintendent of Documents, U.S. Government Printing Office, Washington, D.C. 18p.

Eriksson, E. and Khunakasen, V. 1969. Chloride concentrations in ground water, recharge rate and rate of deposition of chloride in the Israel coastal plain. J. Hydrol. **7**, 178-197.

Hingston, F.J. and Gailitis, V. 1976. The geographic variation of salt precipitated over Western Australia. Aust. J. Soil Res. **14**, 319-335.

Johnston, C.D. 1987. Hydrology and Salinity in the Collie River Basin, Western Australia. 7. Preferred water flow and localized recharge in a variable regolith. J. Hydrol. (in press).

Johnston, C.D., Hurle, D.H., Hudson, D.R. and Height, M.L. 1983. Water movement through preferred paths in lateritic profiles of the Darling Plateau, Western Australia. CSIRO Australia, Division of Groundwater Research. Technical Paper No. 1, p. 1-34.

Kitching, R., Edmunds, W.M., Shearer, T.R., Walton, N.R.G. and Jacovides, J. 1980. Assessment of recharge to aquifers. Hydrol. Sci. Bull. **25**, 217-235.

Munger, J.W. and Eisenreich, S. 1983. Continental-scale variations in precipitation chemistry., Environ. Sci. Technol. **17**, 32A-41A.

Peck, A.J., Johnston, C.D. and Williamson, D.R. 1981. Analyses of solute distributions in deeply weathered soils. Agric. Water Manage. **4**, 83-102.

Peck, A.J., Thomas, J.F. and Williamson, D.R. 1983. Effects of Man on Salinity in Australia. Water 2000 Consultant Report Vol. 8 (Australian Government Publishing Service, Canberra, 1983) pp. 78.

Peck, A.J. and Williamson, D.R. 1987. Hydrology and Salinity in the Collie River Basin, Western Australia. 3. Effects of forest clearing on groundwater. J. Hydrol. (in press).

Porter, L.K.. Kemper, W.D., Jackson, R.D. and Stewart, B.A. 1960. Chloride diffusion in soils as influenced by moisture content. Soil Sci. Soc. Am. Proc. **24**, 460-463.

Sharma, M.L., Williamson, D.R. and Hingston, F.J. 1980. Water pollution as a consequence of land disturbance in southwest of Western Australia. In P.A. Trudinger, M.R. Walter, and B.J. Ralph (ed.) Biochemistry of ancient and modern environments. Aust. Acad. Sci., Canberra p. 429-439.

Sharma, M.L., Johnston, C.D. and Barron, R.J.W. 1982. Soil water and groundwater responses to forest clearing in a paired catchment study in Southwestern Australia. The First National Symposium on Forest Hydrology, Melbourne, May 11-13, The Institution of Engineers, Australia. Preprints of Papers p. 118-123.

Sharma, M.L., Farrington, P. and Fernie, M. 1983. Localized groundwater recharge on the "Gnangara Mound", Western Australia. International Conference on Groundwater and Man, Sydney, Australia. Vol. 1: The Investigation and Assessment of Groundwater Resources. pp. 293-302.

Sharma, M.L., and Hughes, M.H. 1985. Groundwater recharge estimation using chloride, deuterium and oxygen-18 profiles in the deep coastal sands of Western Australia. J. Hydrol. **81**, 93-109.

Sharma, M.L., Barron, R.J.W. and Williamson, D.R. 1987. Hydrology and Salinity in the Collie River Basin, Western Australia. 2. Soil water dynamics of lateritic catchments as affected by forest clearing for pasture. J. Hydrol. (in press).

Sharma, M.L. 1986. Groundwater recharge along a hillslope on the Coastal Plain of Western Australia, estimated by a natural chemical tracer. AWRC Conference on Groundwater Systems Under Stress, Brisbane 1986. pp. 37-46.

Stokes, R.A. and Loh, L.C. 1982. Streamflow and solute characteristics of a forested and deforested catchment pair in Southwestern Australia. The first National Symposium on Forest Hydrology, Melbourne, May 11-13, The Institution of Engineers, Australia. Preprints of Papers p. 60-66.

Williamson, D.R., Stokes, R.A. and Ruprecht, J.J. 1987. Hydrology and Salinity in the Collie River Basin, Western Australia. 1. Response of input and output of water and chloride to clearing for agriculture. J. Hydrol. (in press).

# NATURAL RECHARGE MEASUREMENTS IN THE HARD ROCK REGIONS OF SEMI-ARID INDIA USING TRITIUM INJECTION - A REVIEW

R.N. Athavale
International Crops Research Institute for the Semi-Arid Tropics (ICRISAT), Patancheru P.O.,
Andhra Pradesh 502 324,
India

R. Rangarajan
National Geophysical Research Institute (NGRI),
Hyderabad,
Andhra Pradesh 500 007,
India

ABSTRACT Hard rocks cover about 66 % of the land area of India. They are mainly located in the semi-arid tropical belt characterised by seasonal (monsoonal) precipitation. The main rock types are Granites and Basalt and the corresponding soil types are Alfisols and Vertisols.

Natural recharge to phreatic aquifers of this region has been estimated in the case of several large basins (area 500 Sq. Km. and more) and two watersheds (area around 50 Sq. Km.) using the Tritium injection technique. Tritium was injected at representative sites before onset of monsoon and the displacement of Tritium peak and variation of moisture content in vertical soil profiles collected in post-monsoon period were used for determining spot values of recharge due to precipitation. The average recharge values show a range from almost nil to about 100 mm depending upon the soil type, temperature, rainfall amount and pattern, hydrogeological conditions etc.

The data for various basins is presented and evaluated. Some general features emerging from these measurements, conducted during the last ten years, are recapitulated.



*I. Simmers (ed.), Estimation of Natural Groundwater Recharge, 175–194.*

## INTRODUCTION:

The Semi-arid tropics are areas where monthly rainfall exceeds Potential Evapotranspiration for 2 to 7 months in a year and the mean monthly temperature is above 18 degree C. The areas with 2 to 4.5 wet months are called dry semi-arid tropics and those with 4.5 to 7 wet months are called wet-dry semi-arid tropics. The semi-arid tropics (SAT) comprise all or part of 49 countries on 5 continents. The total area is around 19.6 million Sq. Km.supporting a population of more than 700 million people of the 49 countries.

The Indian subcontinent has the largest semi-arid tropical area of any of the developing countries and represents about 10% of the total SAT. But on a regional basis, the largest geographical areas lie in Africa (66%) and Latin America (19%). India has by far the largest total population in the SAT, more than 400 million people or 55% of the total.

Out of a total area of 2 million Sq.Km.of SAT in India, about 0.7 million Sq.Km.is covered by vertic soils derived from Deccan Trap Basalts and about all the rest is covered by Alfisols or related soils. Practically all of the SAT region of India is underlain by Granites, Basalts and indurated Pre-cambrian sediments. The undependable monsoonal rainfall is characterised by storm events and gaps and the farmer has to depend on supplemental irrigation from dug wells or shallow borewells for sustainance of his crops.

The phreatic zone is the principal aquifer in SAT region. It is located in the weathered mantle or overburden and the fracture zones in hard rocks underneath, extending down to a depth of about 40m.

In view of the importance of ground water in supplemental irrigation in SAT agriculture, it is necessary that the exact quantum of annual replenishment (recharge) to the ground water reserves in river basins or watersheds is evaluated, so that the safe yield is determined and the resource is properly managed and equitably distributed amongst the users.

## NATURAL RECHARGE MEASUREMENT USING INJECTED TRITIUM TECHNIQUE:

The primary source of recharge is deep percolation of a fraction of the rainfall, after the soil profile is saturated. The secondary sources of recharge are:

i) Percolation along streams, canals and lake beds

ii) Return flow of irrigation water, derived from surface or subsurface sources and

iii) Artificial recharge through spreading at suitable sites or injection in wells.

Measurement of the component of recharge accrued through direct percolation of precipitation has been carried out in several basins and watersheds of India over last ten years. The results of these measurements, carried out in hard rock covered regions of SAT India, by using the injected Tritium technique, are recapitulated in this paper.

Tritium (H ), a radioactive isotope of Hydrogen, is a soft Beta emitter, having mass 3 and a half-life of 12.26 years. It exists in the form of the water molecule and as such it is an ideal tracer for studying groundwater movement. The use of Tritium, as a tracer in recharge measurement, involves injection of Tritiated water at a certain depth in the unsaturated zone of the soil column, and study of the vertical movement that this tracer undergoes during the next hydrological cycle. The application of artifical Tritium tracer in recharge measurement is based on an assumption of the piston-flow model for water movement in the unsaturated zone. The piston flow model, proposed by Zimmerman et al (1967 a and 1967 b), and by Muninch (1968 a and 1968 b), assumes that the percolating soil moisture moves downward in discrete layers, and any addition of a fresh layer of moisture at the surface, would push down an equal amount of water immediately below and so on, till the last such layer in the unsaturated zone will be added to the saturated regime or the water table.

FIELD AND LABORATORY PROCEDURE:

In the Tritium injection technique, the moisture at certain depth in the soil profile is tagged with Tritiated water. The tracer moves downward along with the infiltrating moisture due to subsequent precipitation or irrigation. A soil core is collected from the injection site after a certain interval of time and the moisture content and tracer concentration are measured from various depth intervals. The displaced position of the tracer is indicated by the peak in its concentration. The peak may be broadened because of other factors such as diffusion, irregularities in water input and streamline dispersion. The centre of gravity of the profile is assumed to correspond to the displaced position of the tagged layer. Moisture content of the soil column, between injection depth and displaced depth of the soil core, is the measure of recharge to ground water over the time interval between injection of Tritium and collection of soil profile.

The various basins in which recharge measurements have been carried out are shown in Fig. 1. Tritium injections at representative sites in each basin were made before the onset of monsoon, i.e., in general in the first week of June. Selection of injection sites was made on the basis of geology, soil type, topography, drainage pattern and also approachability.

The number of injection sites was decided on the basis of the area and also on the feasibility of completing field and laboratory work within a specified time.

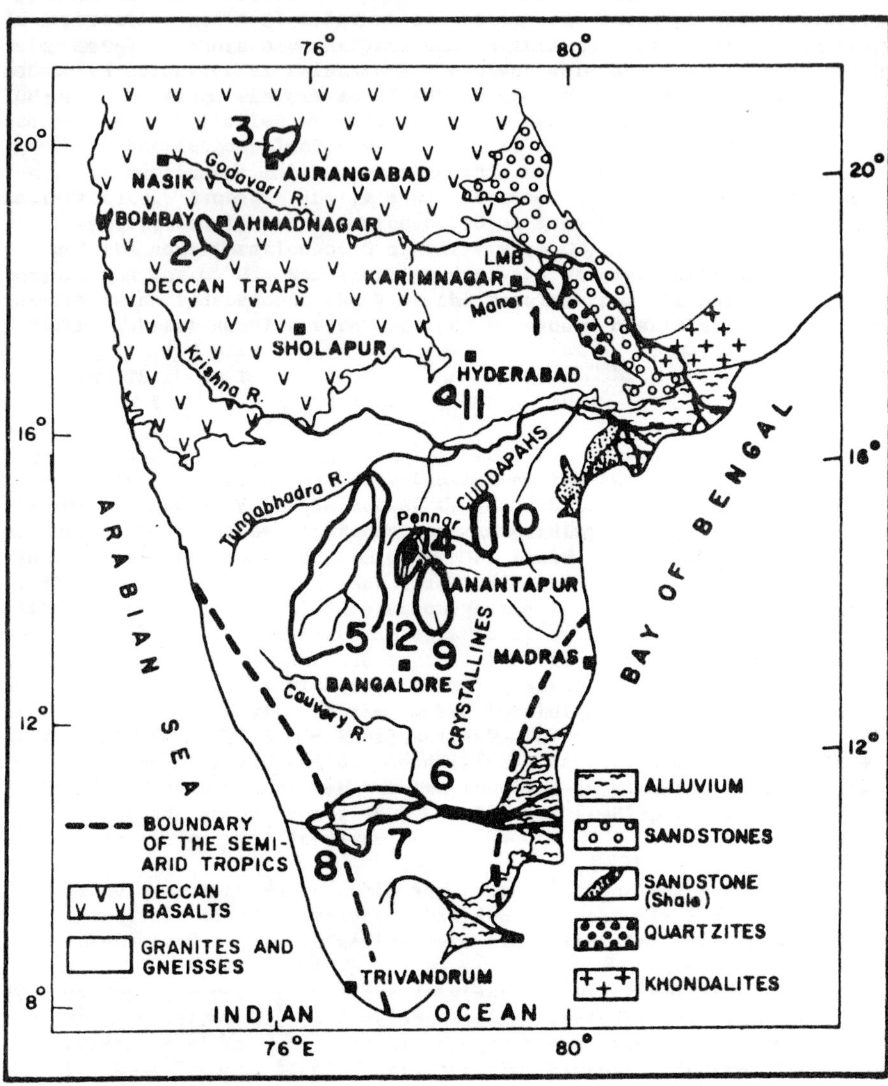


Fig. 1. Geological map of semi-arid South India showing recharge measurement basins.

At any injection site, a relatively flat patch of land was selected, sufficiently distant from big trees, but near important land marks such as milestones and electrical poles. The site sometimes had a thin cover of grass or shrubs. In cases where a non-agricultural site was not available, a ploughed or unploughed farm plot, having no facility for well or canal irrigation, was selected.

The depth of injection varied from basin to basin but within the range of 60 cm to 80 cm. 2.5 ml of Tritiated water, having an activity of 10 Micro-Curie per ml was used in each injection. Five injections (Total 125 Micro Curie of H ) were made at each site. A 1.25 cm diameter hole upto desired depth was made with a drive rod and the hole was backfilled with local soil after Tritium injection. Four injections were placed symmetrically on the circumference of a circle having a diameter of 10 cm, and the fifth injection was made at the centre of the circle. Each injection site location was precisely determined through triangulation so that it could be found easily again for vertical collection of soil profiles at the end of monsoon. Soil profiles having a depth interval of 10 cm or 20 cm were generally collected in November-December down to maximum depth of 3m. Duplicate injections were made at a few sites for studying reproducibility. A second collection from a few injection points was made at the end of the hydrological cycle, to study the effect of evaporation and transpiration during the post-monsoon lean period.

In the laboratory, the mositure content in about 25 g of soil, taken from each sample, was determined on a torsion balance heated by an infrared lamp upto 105 degree centigrade. The rest of the sample was subjected to partial vacuum distillation. 4 ml of the distillate was mixed with 10 ml of scintillator solution and the Tritium activity was counted using a liquid scintillation spectrometer. These field and laboratory procedures are described in detail by Athavale et al (1978 and 1980).

The tritium activity of each 10 cm or 20 cm depth interval of a profile was plotted against depth along with moisture content. Grain size analysis of the soil core material was carried out.

## RESULTS AND DISCUSSION:

The results of recharge measurements carried out in various basins in the SAT region of South India are presented in Table 1. Some of the results are already published by Athavale et al 1980, (for Sr.No.1 in Table 1), 1983 a (for Sr. No.2 and 3) and 1983 b(for Sr. No. 4). The work in Vedavati basin (Sr. No. 5) and in Noyil, Ponnani and Vattamalaikarai basins (Sr. No. 6,7,8) was carried out at the request of the Central Ground Water Board of Govt. of India and the results have been communicated to them. Rangarajan and Ramesh Chand (1987 C) have carried out recharge measurements for Chitravati and Kunderu basins.

Table 1. Mean values of precipitation recharge for hard rock covered basins in semi-arid India.

| Sl. No. | Basin/Watershed name | Coordinates (Lat & Longt) | Main rock types | Area (Sq. Km.) | Year of measurement | No. of recharge measurements | Average rainfall (mm) | Mean recharge (mm) |
|---|---|---|---|---|---|---|---|---|
| 1. | Lower Maner | 18° 05' -18° 42'N<br>79° 32' -80° E | Sand Stone, shale, quartzite, granite | 1,600 | 1976 | 26 | 1,250 | 100 |
| 2. | Godavari-Purna | 19° 45' -20° 15' N<br>75° 10' -75° 50' E | Basalt | 1,091 | 1980 | 24 | 652 | 56 |
| 3. | Kukadi | 18° 45' -19° 10' N<br>74° 10' -74° 40' E | Basalt | 1,153 | 1980 | 19 | 612 | 46 |
| 4. | Marvanka | 14° 15' -15° N<br>77° 15' -77° 47' E | Granite, greiss, schist | 2,044 | 1979 | 19 | 550 | 42 |
| 5. | Vedavati | | | | | | | |
| | (1) West Suvarnamukhi | 13° 15' -13° 45' N<br>76° 30' -76° 45' E | Granite, gneiss schist | 958 | 1978 | 18 | 565 | 39 |
| | (2) Lower Hagari | 14° 45' -16° N<br>76° 45' -77° 30 E | Granite, gneiss, schist | 3,679 | 1978 | 44 | 565 | 6.5 |
| 6. | Noyil | 10° 54' -11° 19' N<br>76° 39' -77° 56' E | Granite, gneiss, schist | 3,420 | 1979 | 21 | 715 | 69 |
| 7. | Vattamalaikarai | 10° 52' -11° 52' N<br>77° 15' -77° 45' E | Granite, gneiss, schist | 512 | 1979 | 2 | 460 | 61 |
| 8. | Ponnani | 10° 15' -11° N<br>76° 15' -77° 15' E | Granite, gneiss, schist | 3,973 | 1979 | 9 | 1,320 | 61 |
| 9. | Chitravati | 13° 35' -14° 50' N<br>77° 30' -78° 15' E | Granite, gneiss, schist | 6,100 | 1981 | 48 | 615 | 25 |
| 10. | Kunderu | 14° 30' -16° N<br>77° 45' -79° E | SST, Shale, LST, quartziti, granite, gneiss | 8,650 | 1982 | 45 | 615 | 29 |
| 11. | Aurepalle Watershed (Mahaboobnagar) | 17° 47' -17° 53' N<br>78° 33' -78° 43' E | Granite | 64 | 1984<br>1985 | 15<br>12 | 563<br>583 | 32<br>17 |
| 12. | Manila Watershed (Anantapur) | 14° 30' -14° 36' N<br>77° 38' -77° 44' E | Granite & gneiss | 40 | 1986 | 25 | 390 | 24 |

Note : Serial numbers correspond with numbers in Fig. 1

TABLE: 2 RECHARGE DUE TO MONSOON PRECIPITATION IN AUREPALLE WATERSHED DURING 1984 AND 1985

| 1984 | | | 1985 | | |
|---|---|---|---|---|---|
| Sr. No. | Site | Recharge (in mm) | Sr. No. | Site | Recharge (in mm) |
| 1. | Near Konapur | -2 | 1. | Near Konapur | 5 |
| 2. | Akuthotapally - Konapur | -15 | 2. | Near Lambaditanda | 22 |
| 3. | South of Akuthotapally | 39 | 3. | Akuthotapally-Konapur | 8 |
| 4. | Dodlapahad | 10 | 4. | Akuthotapally-Konapur | 19 |
| 5. | Singampally II | 43 | 5. | North of Akuthotapally | 4 |
| 6. | Nallavarpally | 70 | 6. | Aurepalle | 17 |
| 7. | Aurepalle-Chinnamadgul | 103 | 7. | Singampally | 33 |
| 8. | Singampally I | 78 | 8. | South of Singampally | 39 |
| 9. | Reddipuram } Duplicate | 16 | 9. | Turkalkunta | 25 |
| 10. | Reddipuram } Duplicate | | 10. | Reddipuram-Chinnamadgul | 8 |
| 11. | Chandrayanpalli-Kalkunda | 37 | 11. | Reddipuram-Kalkunda | 15 |
| 12. | Kalkunda NE | 8 | 12. | Near Irven | 12 |
| 13. | Ajalapuram-Chinnamadgul | 5 | | | |
| 14. | Near Ajalapuram | 31 | | | |
| 15. | Irven | 24 | | | |
| | Total 14 sites mean | 31.9 | | Total 12 sites mean | 17.3 |

As can be seen from this table, the area of these basins varies from 500 Sq.Km to 8500 Sq.Km. On an average, each recharge measurement site represents an area of 50 to 200 Sq.Km.

Recently some detailed studies on two small watersheds (Sr. No. 11 and 12 in Table 2) have been carried out. In the case of these watersheds the density of recharge measurement was one site per 2 to 4 Sq.Km.

Recharge measurement in the case of each basin has been carried out for only one season. This has been necessary from the point of logistic considerations and coverage of a large region. In the case of the Aurepalle and Manilla watersheds (Sr. No.11 and 12) the measurements have been carried out for two monsoons (Rangarajan et al, 1987 a , 1987 b). Results of recharge measurements for Aurepalle watershed for 1984 and 1985 are reported here in detail as an illustration. In the case of Manilla watershed, 1985 results are presented and the samples collected in November 1986 are being processed.

RECHARGE MEASUREMENTS IN AUREPALLE WATERSHED:

Tritium injection studies were carried out in the Aurepalle watershed during two hydrological cycles, 1984 and 1985. This watershed, covering an area of approximately 64 Sq.Km.is located in granites and has a cover of alfisols. The weathered zone thickness varies from a few meters upto 20 meters, with an average value of about 8-10m. Recharge measurements at 15 sites injected in June 1984, gave a mean value of 31.9 mm, which is 6.2% of the 1984 annual rainfall of 563mm. The input to ground water system, due to direct percolation of precipitation, was estimated as about 2 Million cubic Meters(MCM).

Recharge measurements were repeated in the 1985 hydrological cycle. The mean value of recharge measured at 12 sites was found to be 17mm, which is only 3% of the 1985 total rainfall of 583mm.

Fig. 2 shows a drainage map of Aurepalle watershed and Tritium injection sites for 1984 and 1985. Typical Tritium Vs. depth profiles in the case of four injection sites from this watershed are presented in Fig. 3. The recharge data for 1984 and 1985 is presented in Table 2.

EFFECT OF RAINFALL PATTERN ON RECHARGE:

Infiltration and runoff due to a precipitation event depend upon the amount and intensity of rainfall as well as the antecedent moisture condition of the soil profile. The cumulative recharge due to a monsoon season, made of several precipitation events, is also considerably dependant on the characteristics of the events as well as their spacing. This is well brought out in the study of precipitation recharge over two hydrological cycles 1984, and 1985, in the case of the Aurepalle watershed.

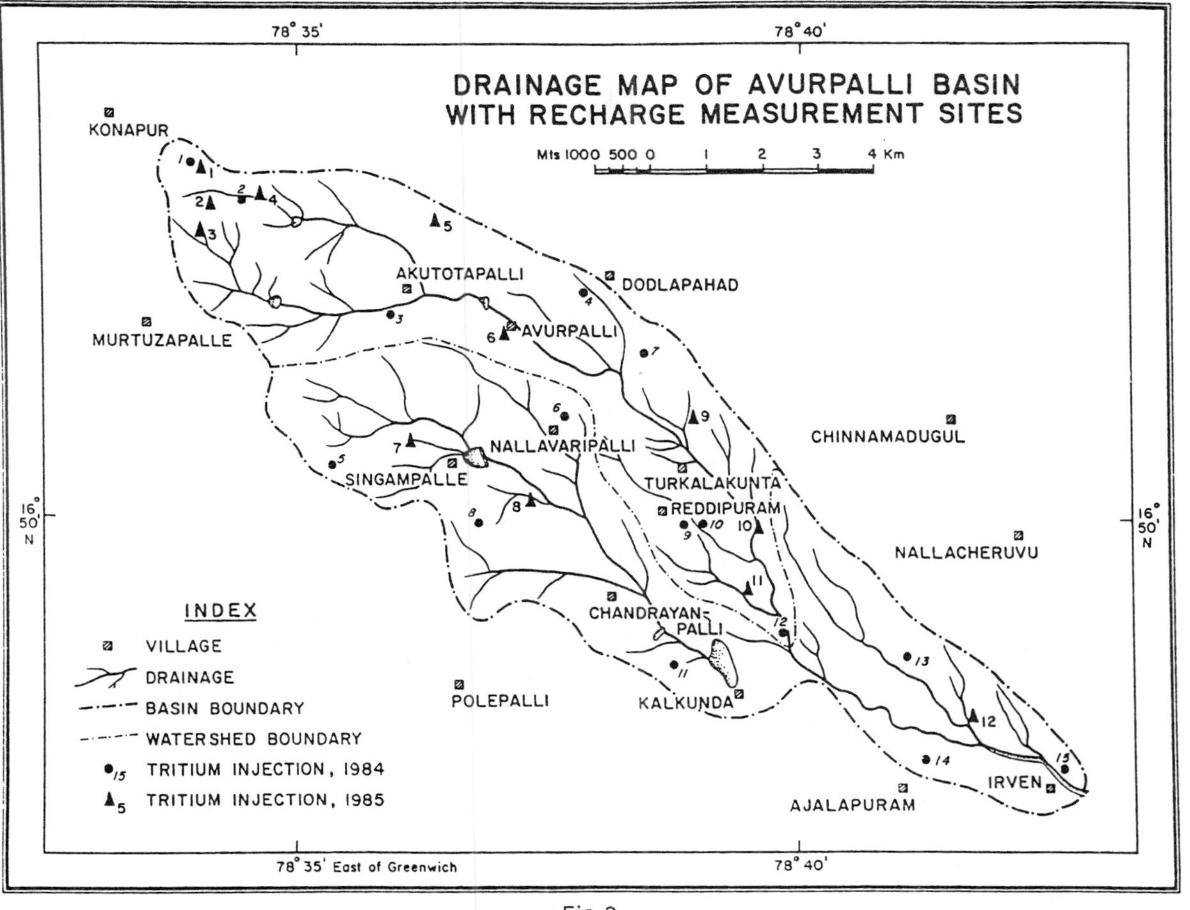

DRAINAGE MAP OF AVURPALLI BASIN
WITH RECHARGE MEASUREMENT SITES
Mts 1000 500 0 1 2 3 4 Km
78° 35'
78° 40'
16° 50' N
78° 35' East of Greenwich
KONAPUR
MURTUZAPALLE
AKUTOTAPALLI
DODLAPAHAD
AVURPALLI
NALLAVARIPALLI
SINGAMPALLE
TURKALAKUNTA
REDDIPURAM
CHINNAMADUGUL
NALLACHERUVU
CHANDRAYAN-PALLI
KALKUNDA
POLEPALLI
AJALAPURAM
IRVEN
INDEX
VILLAGE
DRAINAGE
BASIN BOUNDARY
WATERSHED BOUNDARY
TRITIUM INJECTION, 1984
TRITIUM INJECTION, 1985


Fig. 2

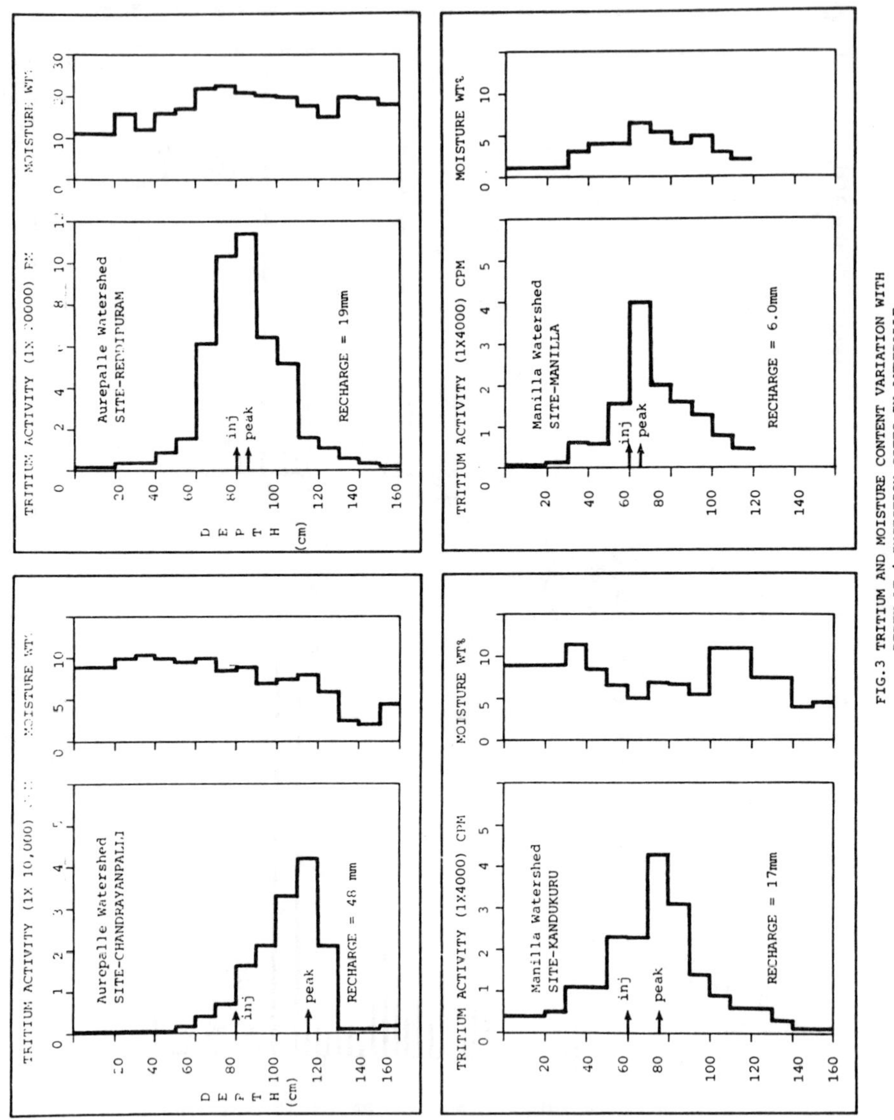


FIG.3 TRITIUM AND MOISTURE CONTENT VARIATION WITH DEPTH AT 4 INJECTION SITES IN AUREPALLE WATERSHED

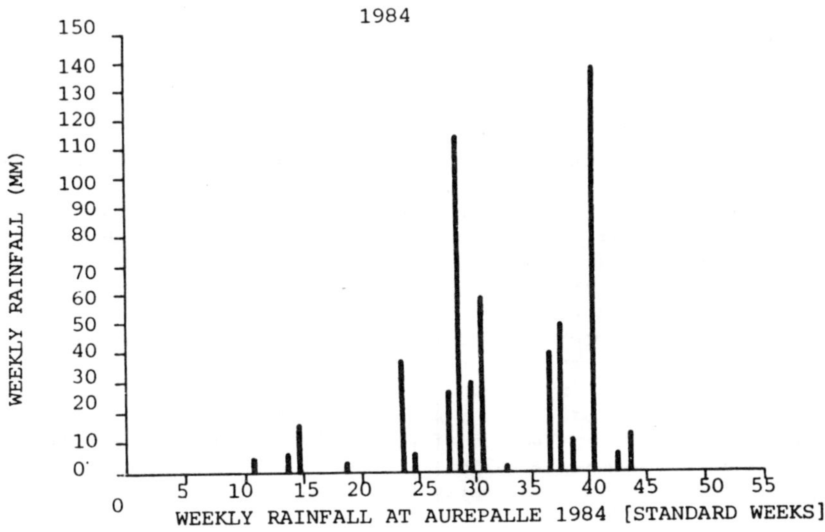


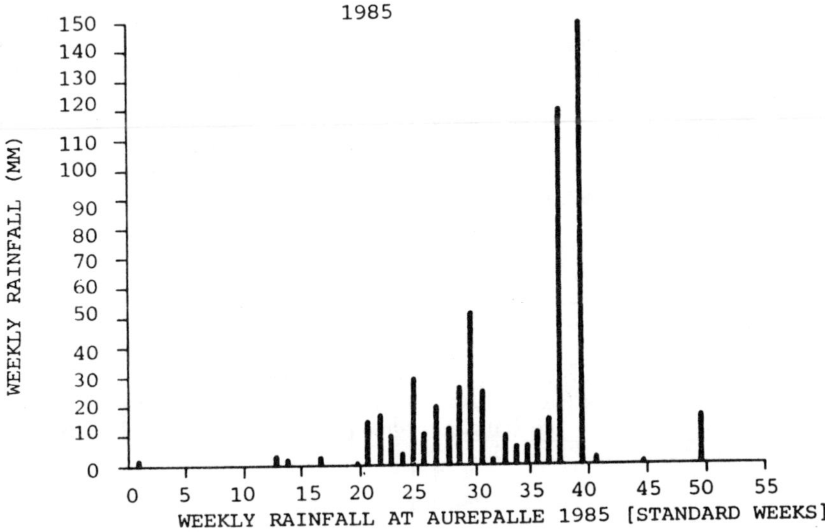


FIGURE 4 RAINFALL PATTERN IN AUREPALLE BASIN

The annual rainfall values for 1984 (563 mm) and for 1985 (583 mm) are almost equal but there is a great deal of difference in the rainfall pattern of the two consecutive years, as can be seen from figure 4. The 1984 monsoon is characterised by two major storm events, occurring during the 29th and 41st standard weeks. These events resulted in a rainfall of 115.6 mm and 139.6 mm respectively and accounted for nearly 45 % of the annual rainfall. The mean recharge value for 1984 was 32 mm. In the case of the 1985 monsoon also, two storm events, having intensities of 120.1 mm and 149.4 mm, account for 46 % of the total rainfall. However, these events occurred during the 38th and 40th standard week, a situation giving rise to higher runoff than the 1984 storm events, which were separated by about 11 weeks. Thus the average recharge due to the 1985 monsoon (17 mm) is lower than that due to 1984 (32 mm) by about 50 %, as a result of the rainfall pattern. Similarly, the recharge would be more than the 32 mm average figure, if the same amount of precipitation is more evenly distributed than in 1984.

This case illustrates the need for generating recharge data in a small watershed for a period of about 5 years and for preparing a rainfall pattern:recharge model, which would then be able to predict the recharge directly from rainfall data.

EFFECT OF GRAIN SIZE ON RECHARGE VALUES:

A mechanical grain size analysis of the soil material collected beneath the injection depth was carried out in the case of all basins. The soil was soaked in water in a beaker for about a day. The wet sample was first washed with a standard sieve of size 60 µm so that all particles less than 60 µm in size and representing the silt-clay fractions were removed. The remaining sample was dried in an oven and further sieved into the sand (60 µm to 2 mm mesh) and gravel (> 2 mm mesh) fractions (Athavale et al, 1978).

The recharge values of individual sites were then plotted against the average sand content (volume percent) below the depth of injection. An example of such a study for 26 sites in the Lower Maner Basin (Sr. No. 1 of Table 1) is shown in Figure 6. A best fit curve drawn through the points has one asymptote at sand content of about 20 % and another asymptote at about 70 %. A more or less linear correlation between recharge values and sand content seems to be valid for two separate segments of the curve. One of these is in the sand percentage range of 20 - 40 and the other in the range 55 - 65.

DIFFUSION OF INJECTED TRITIUM ACTIVITY IN SOIL:

In order to study the extent of lateral spreading of injected Tritium activity in the soil due to diffusion, an experiment was conducted in clayey soils at one site in Lower Maner Basin (No. 1 in Fig. 1 and Table 1). Soil cores for this study were obtained from the point where Tritium was originally

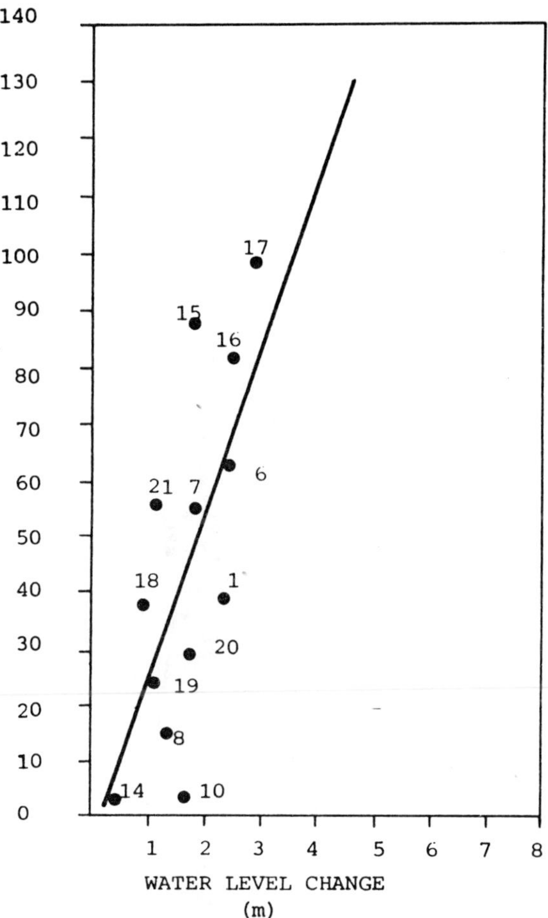


FIGURE 5 RELATION BETWEEN TRITIUM RECHARGE VALUES AND W.L. FLUCTUATION IN DUG WELLS OF MARVANKA BASIN

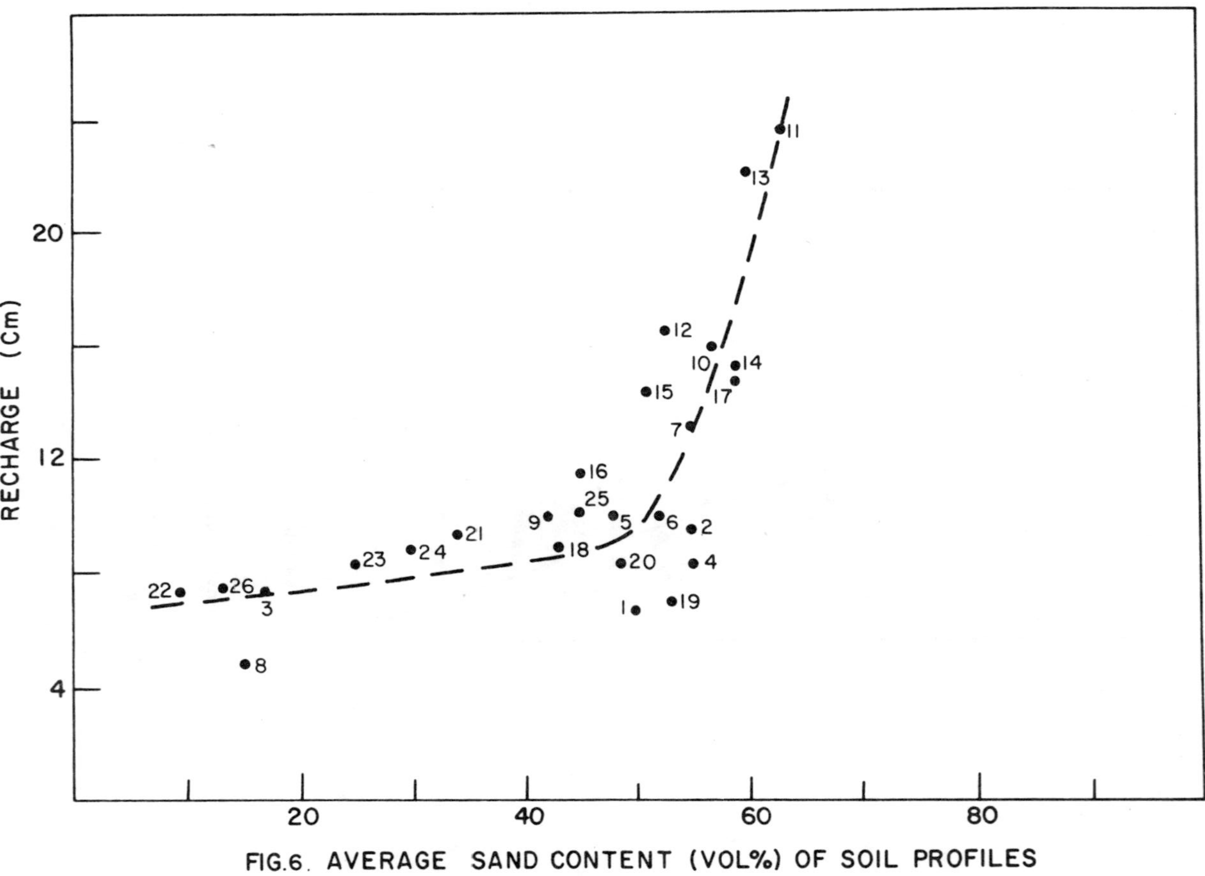


FIG.6. AVERAGE SAND CONTENT (VOL%) OF SOIL PROFILES

injected and also from points at distances of 0.5, 1.0 and 2.0 m from the injection point. The tracer was injected in April, 1976 and samples were collected in July, 1977 (Athavale et al., 1978). Tritium activity was present up to a lateral distance of 1.0 m, while it was beyond the detection limit at the distance of 2.0 m. The cumulative Tritium activity of profiles at distances of 50 cm or 100 cm was found to be 0.26 and 0.005 times the activity at the injection point:

The values fit into the exponential law of diffusion having the relation

$$C(r,t) = Co.\ C^{\frac{-r2}{4DT}} \quad \text{where}$$

$c(r,t)$ = Activity at distance r at time t

Co = Activity at the centre

D = Diffusion constant for the soil

For the experiment lasting for 15 months (t = 450 days) the diffusion constant for clayey soil at side Dhanwada was found to be $1.21 \times 10^{-5}\ Cm^{2} .\ Sec^{-1}$. This is within the range of reported values for diffusion rates in clays.

## VARIABILITY OF RECHARGE RATES:

The recharge rate at a site is decided by many factors. Besides the site specific factors such as temperature (demands of evapotranspiration) and soil permeability, there are certain dynamic factors such as the amount of rainfall and also the rainfall pattern, as is already discussed in the case of Aurepalle watershed. The local water table depth also influences the rate of recharge and in cases where the water table is very shallow, negative recharge values are obtained because exfiltration is the dominant mode of water transfer.

On a basin scale, there is a regional systematic variation in recharge values with the values showing a general declining trend as one moves from the upper reaches to the lower reaches. This trend is similar to that for the amount of water table fluctuation (Fig. 5). The second factor on a regional scale is the spatial distribution of rainfall. For example, in the 1984 rainy season, the total rainfall varied from 401 mm at Kalkonda (Fig. 2) to 563 mm at Aurepalle at a distance of 8 Km. The third and probably the most important factor operating on a basin/watershed scale is soil heterogeneity.

In addition to these natural factors of both static and dynamic nature, resulting in variability of recharge values, the experimental error in estimation of recharge values also adds to the scatter of data. In the case of the Tritium injection method, we have measured the reproducability error as ± 10%. Availability of such estimates of experimental error in the case of other methods used for recharge estimation is desirable.

The discussion above indicates that there is no justification for presenting the average recharge value, in the strict physical sense of the term. There exist a range of discrete (site specific) values and also a regional trend which gives rise to "variability". It is, however, still necessary to present the recharge values as averages for the practical purpose of intercomparison of groundwater potential of different basins/watersheds.

The cumulative effect of the various factors leading to scatter of recharge values can also, at best, be expressed as standard errors of the average value. In the case of Aurepalle Watershed (Table 2) the values are 31.9 ± 33.0 and 17.3 ± 11.0.

GENERAL OBSERVATIONS ON RECHARGE MEASUREMENTS IN HARD ROCK REGIONS OF SEMI-ARID INDIA:

Table 1 presents data on deep percolation or recharge in the case of 10 basins and two watersheds, covered by Vertic soils and Alfisols and underlain by Basalts, Granites, Gneisses, Schists, Quartzites, Sandstones and Shales. The annual rainfall values range from 390 mm to 1250 mm and the average recharge values from 6.5 mm to 100 mm. A broad correlation between the rainfall and recharge amounts exists although this correlation is greatly influenced by other factors such as soil type, topography and hydrogeological conditions, in addition to the annual effect of rainfall pattern, as shown in the case of the Aurepalle watershed.

The following other observations can also be made from Table 2.

(1) Although the Vertic soils have less primary permeability than Alfisols, the average recharge values for Vertic and Alfic soils, receiving approximately same quantity of rainfall, are comparable. This is probably because vertic soils develop a secondary transient high permeability, duc to their property of shrinking and swelling. The vertic soils develop a network of cracks in the dry season preceding the monsoon and allow rapid infiltration through these cracks until they swell and form a relatively impermeable layer.

(2) Comparison of the Tritium injection recharge measurement method with other methods has been carried out in the case

of several basins, in both qualitative and quantitative terms. In general, a broad agreement between the results from different methods is seen.

Specific cases of such comparison are noted below:

**o Water level fluctuation:**

A qualitative correlation between the recharge rate measured by using Tritium tracer technique and the amplitude of water level fluctuation in nearby dug wells is noticed. This is illustrated in the case of the Mar Vanka basin through Fig. 5 (Athavale et al, 1983 b).

o Hydrometeorological **data:**

Ground water recharge to the Neon sub-basin, forming a part of the Betwa basin, has been carried out under an Indo-British project on evaluation of ground water resources (An onymous, 1980). The Neon sub-basin has an area of 1085 Sq.Km. and is covered by vertic soils underlain by Deccan Trap basalt flows. The rainfall in this basin is comparable to that in the Godavari-Purna and Kukadi basins (Sr. No. 2 and 3, Table 1). The average annual input to the ground water system in Neon sub-basin, due to 1977 monsoon, was calculated as 32 MCM to 36 MCM. This estimate is in good agreement with the estimates of 31.9 MCM and 35.4 MCM obtained by Athavale et al (1983 a) for the Kukadi and Godavari-Purna basins respectively.

In a companion paper (Muralidharan et al, 1987) presented in this workshop, an agreement between recharge estimates arrived at from climatic water balance studies, and Tritium injection studies is shown in the case West-Suvarnamukhi basin.

**o Recharge estimate from regional ground water model:**

In a companion paper appearing in these proceedings it is shown that a good agreement exists between the annual recharge or safe-yield values for the West-Suvarnamukhi basin, calculated from a regional ground water model based on water level fluctuation data for 1978, and by using the injected Tritium technique for the same year (Muralidharan et al, 1987).

**o Hydrogeological Estimates:**

Recharge estimation through hydrogeological studies involves determination of specific yield (sy) through pumping tests. A product of the maximum annual water level change ($\Delta h$) in the phreatic aquifer and the specific yield, gives a spot value of recharge. A good agreement between the hydrogeological estimate and Tritium injection estimate of average recharge was observed in the case of the Godavari-Purna basin (Athavale et al, 1983 a) and the Marvanka basin (Athavale et al, 1983 b).

Thus, a broad agreement between the recharge estimates obtained by various methods is seen. The recharge estimates given in Table 2 are considered to be 'low' by some hydrogeologists. It may, therefore, be mentioned that a higher average value of recharge of 210 mm has been found by Datta et al (1973) for the Gangetic plains area characterised by higher rainfall, sandy soils having high permeability and a climatic water balance favourable for deep percolation. Also one of us (RR) has found recharge values of 300 to 500 mm in the case of Neyveli basin, located in Tamilnadu state.

(3) In these studies, conducted over a period of 10 years, it has been ascertained, through special experiments, that the reproducibility of recharge measurements using Tritium tracer technique, is within 10%.

(4) There is no significant difference between recharge values obtained for two adjacent injection sites, one of which was collected at the end of the monsoon (November-December) while the other was collected before the beginning of next year's hydrological cycle (i.e. in May-June, 1980). This observation is valid only in cases where winter rainfall is insignificant and suggests that there is no significant evaporation loss of moisture from depths below 60 cm.

(5) The recharge values in a basin show a wide range. In a few cases negative recharge values, indicating exfiltration in place of infiltration, are also recorded in areas having shallow water table.

## CONCLUSIONS:

The Tritium injection technique is found to be a comparatively quick and reliable method in estimating ground water recharge due to precipitation having a peaked monsoonal character in the Semi-arid Tropical Region. These recharge values are in good agreement with those obtained using various other techniques. Recharge values in a basin or a watershed are greatly influenced by the pattern of seasonal rainfall. There is, therefore, a need for conducting recharge measurements for a period of about five years in representative watersheds covered by Aflisols and by Vertisols, in order to prepare a data based model for direct determination of recharge from the rainfall record for any spe cified year. Systematic recharge measurements over large areas would provide a useful basis for optimal utilisation of the replenishable but limited ground water reserves of semi-arid tropical regions. The recharge measurements will also form a primary data base for artificial recharge programs which would be needed for meeting the increasing demand for ground water by farmers of the semi-arid tropical regions.

## REFERENCES

An onymous (1980)-An interim report on summary results of ground water resource evaluation studies in Upper Betwa river basin, India", Tech. paper No. 10, Indo-British Betwa Ground Water Project, Central Ground Water Board, Government of India, pp: 105-107.

Athavale, R.N., Murthy, C.S., and Chand, R. (1978) "Estimation of recharge to the phreatic aquifers of Lower Maner basin by using the tritium injection Method", Tech. Report no. GH9-NH1, National Geophysical Research Institute, Hyderabad.

Athavale R.N., Murti C.S. and Chand R. (1980)," Estimation of recharge to the phreatic aquifers of Lower Maner basin using the Tritium injection method", Journal of Hydrology, Vol.45, pp: 185-202.

Athavale R.N., Ramesh Chand and Rangarajan R. (1983) a). "Ground water recharge estimates for two basins in the Deccan trap basalt formation". Hydrological Sciences Journal, 28, 4, 12/1983, pp: 525-538.

Athavale R.N., R. Chand, Murali D.M., Muralidharan D. and Murti C.S.(1983) b), "Measurement of ground water recharge in Marvanka basin, Anantapur district". Proc. of Seminar on Assessment of Ground water resurces, Vol. III, Central Ground water board, New Delhi, pp: 275-290.

Datta, P.S., Goel, P.S., Rama, and Sangal, S.P., (1973), "ground water recharge in Western Uttar Pradesh", Proc. Ind. Acad.Sc., Vol. LXXVIII, No. 1, pp 1-12.

Munnich, K.O., (1968 a) "Moisture movement measured by isotope tagging", Guide book on Nuclear techniques in hydrology, IAEA, Vienia, pp 112-117.

Munnich, K.O., (1968 b) "Use of Nuclear techniques for the determination of ground water recharge rates", Guide book on Nuclear techniques in hydrology, IAEA, Vienna, pp. 191-197.

Muralidharan D.,Athavale R.N., and Murti C.S., (1987) "comparison of recharge estimates from injected Tritium technique and regional hydrological modelling in the case of a granitic basin in semi-arid India", in Proceedings International Workshop on "Estimation of natural recharges to Ground Water (with special reference to arid and semi-arid regions"), Antalya, Turkey, March 7-15, 1987.

Rangarajan R., Hodlur G.K., Athavale R.N., and Rao T.G. (1987 a), "Estimation of Recharge to aquifers of Aurepalle watershed, Mahaboobnagar district by tritium tagging method", (Tech. report:National Geophysical Research Institute, Hyderabad).

Rangarajan R ., Hodlur G.K., Athavale R.N.,and Rao T.G. (1987 b), "Estimation of recharge to aquifers of Manila watershed, Anantapur district," by tritium tagging method (Tech. report:National Geophysical Research Institute).

Rangarajan R., and Ramesh Chand (1987 c)," Groundwater recharge to Kunderu and Chitravati basins of Rayalaseema region of A.P.",(Tech. Report, National Geophysical Research Institute, Hyderabad).

Zimmermann, U., Ehhalt, D, and Munnich, K.O., (1967 a) "Soil water movement and Evapotranspiration: changes in the isotope composition of water", Isotopes in hydrology, IAEA, Vienna.

Zimmermann, U., Munnich, K.O. and Roether, W., (1967 b) "Downward movement of soil mositure traced by means of hydrogen isotopes", Geophy. Monograph No.11, American Geophysical Union, Washington, pp. 28-36.

COMPARISON OF RECHARGE ESTIMATES FROM INJECTED TRITIUM TECHNIQUE AND REGIONAL HYDROLOGICAL MODELLING IN THE CASE OF A GRANITIC BASIN IN SEMI-ARID INDIA

D. Muralidharan, and
C.S. Murti
National Geophysical Research Institute (NGRI),
Hyderabad,
Andhra Pradesh 500 007,
India

R.N. Athavale
International Crops Research Institute for the
Semi-Arid Tropics (ICRISAT),
Patancheru P.O.,
Andhra Pradesh 502 324,
India

ABSTRACT. The injected Tritium technique was used for estimation of recharge from 1978 monsoon precipitation to the phreatic aquifers of the West Survarnamukhi sub-basin, forming a part of the Vedavati basin and having an area of 958 Sq. Km., covered with granite, gneisses, and schists. Recharge was measured at 20 sites, of which 6 were in schists and the rest in gneisses. Displacement of Tritium tracer peak in the unsaturated Alfisol zone was used for measuring the spot values of recharge which varied from 0 mm to 127 mm. The average recharge calculated from spot values was found to be 39.2 mm or 8.5 % of the seasonal rainfall. The input to the groundwater regime of the basin, calculated by using the Thiessen polygon method works out to be 43.8 Million Cubic Meters (MCM).

A regional groundwater model of the Vedavati basin has recently been prepared by Sridharan et al. (1986). Based on water level fluctuation data for the period November 1977 to November 1978, these authors have estimated the annual recharge for various parts of the Vedavati basin. The annual recharge or safe yield value for West-Survarnamukhi basin, obtainable from the model, is 42.5 mm. Thus, an agreement between recharge values estimated from Tritium injection in the unsaturated zone, and from a model calibrated with water level fluctuations in the saturated zone is observed in the case of West Survarnamukhi basin. A similar agreement with an average recharge value of 34.4 mm, estimated from a climatic water balance study, is also noteworthy.



*I. Simmers (ed.), Estimation of Natural Groundwater Recharge, 195–204.*

INTRODUCTION

The Vedavati river basin located in the central part of semi-arid southern India, covered with hard rocks and having an area of 22400 Km ., was studied by the Central Ground Water Board, during 1975-1980, for evaluation of ground water resources. As a part of the Vedavati basin project, the West Survarnamukhi sub-basin, located in the upper reaches of the Vedavati basin, was selected for intensive hydrogeological and water balance studies (Fig.1). This sub-basin is drained by the western branch of the Suvarnamukhi, a tributary of the Vedavati, and has an effective irrigation area of 958 Sq.Km.

The West-Survarnamukhi sub-basin is covered by Pre-Cambrian granites, granite gneisses and schists (Fig.2). These rocks are traversed by dolerite dykes, pegmatite veins and quartz reefs. Gneissic terrains have well defined systems of jointing, with steep vertical dips (Anonymous, 1977).

The sub-basin exhibits an undulating topography with an elevation ranging from 700m to 900m and has a dendritic drainage pattern. Red gravelly soils cover the uplands and red loamy soils occur in the plains. Patches of lateritic soils are present in the South-Eastern part of the sub-basin. The rainfall is mostly due to monsoons and is confined to the months of June to September, which contribute about 70% of the total precipitation. The annual average precipitation recorded at the Chiknayakanhalli hydrometeorological station is 580 mm and the mean monthly temperature shows a variation from 22 degree centigrade to 29 degree centigrade.

Exploitation of groundwater in the sub-basin is mainly through dug wells tapping the phreatic aquifer. The depth of weathering in the granitic and gneissic rocks ranges from 1.5m to 7.5m and in schists from 2 to 13m.

RECHARGE MEASUREMENTS IN THE WEST-SUVARNAMUKHI BASIN:

Recharge to the phreatic aquifers of the West-Suvarnamukhi basin, due to the monsoon precipitation of 1978, was measured using the Artificial Tritium Injection Technique. Tritium was injected at 20 sites, well distributed over the basin, as shown in Figure 2. Six of the sites were located over schist rocks and the rest over granite gneisses. Tritium injections were made in June 1978, before the onset of the monsoon. At each injection site, 125 micro-curies of Tritium, contained in 12.5 ml of water, was injected at the bottom of a cluster of 5 holes, having a depth of 80cm in the soil profile. The cluster of holes occupies a diameter of 10cm., with each hole having diameter of 1.25 cm. The holes were back-filled with soil and the injection site was pinpointed with reference markers and triangulation. Vertical soil profiles, down to a depth of 2 to 2.5 m, were collected from each site, after the cessation of monsoon season, with a sampling depth interval of 20 cm. The sample was weighed and packed in polythene bags in the field.

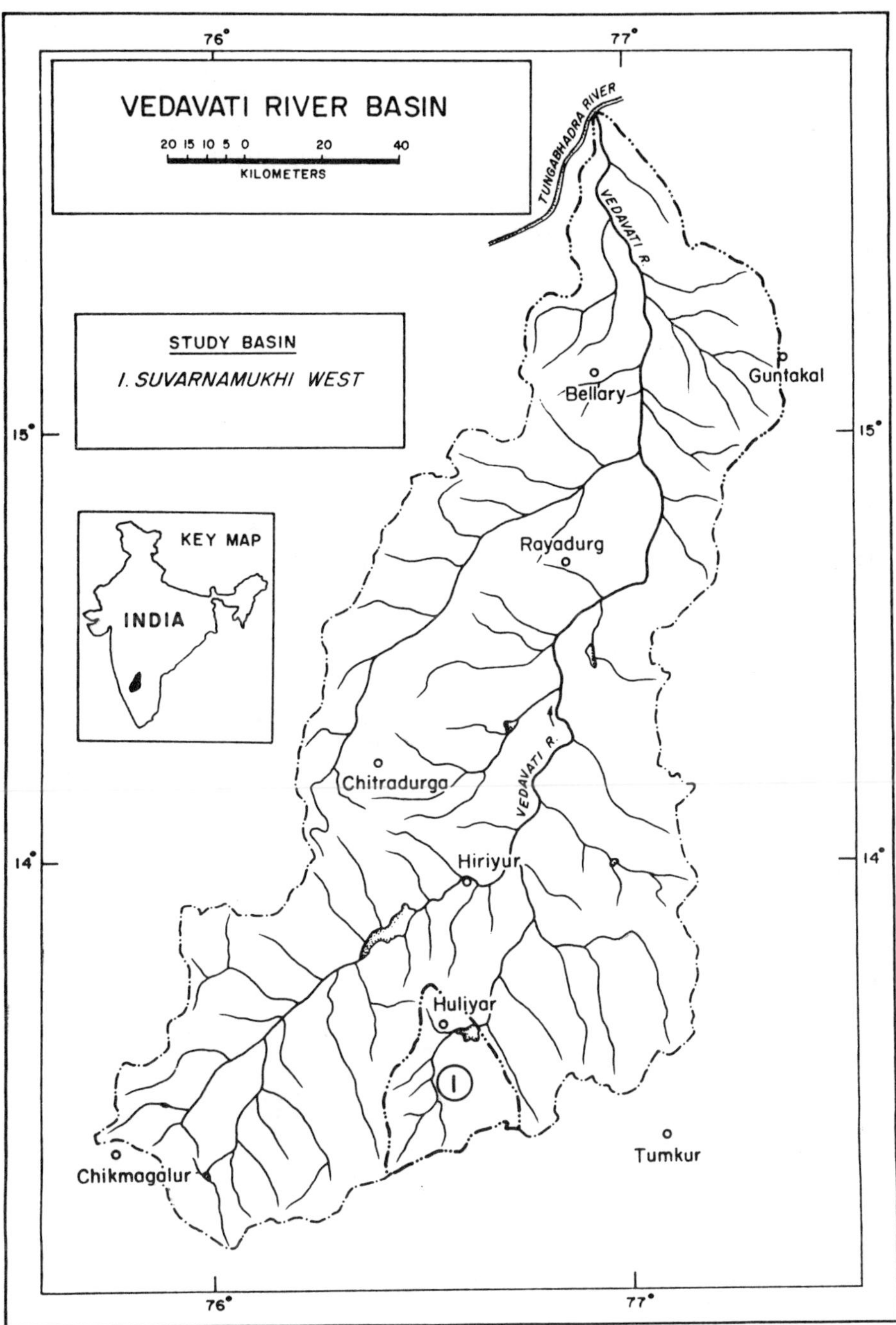
VEDAVATI RIVER BASIN
20 15 10 5 0 20 40
KILOMETERS
STUDY BASIN
I. SUVARNAMUKHI WEST
KEY MAP
INDIA
TUNGABHADRA RIVER
VEDAVATI R.
Bellary
Guntakal
Rayadurg
Chitradurga
Hiriyur
Huliyar
Chikmagalur
Tumkur
76°
77°
15°
14°

FIGURE:1

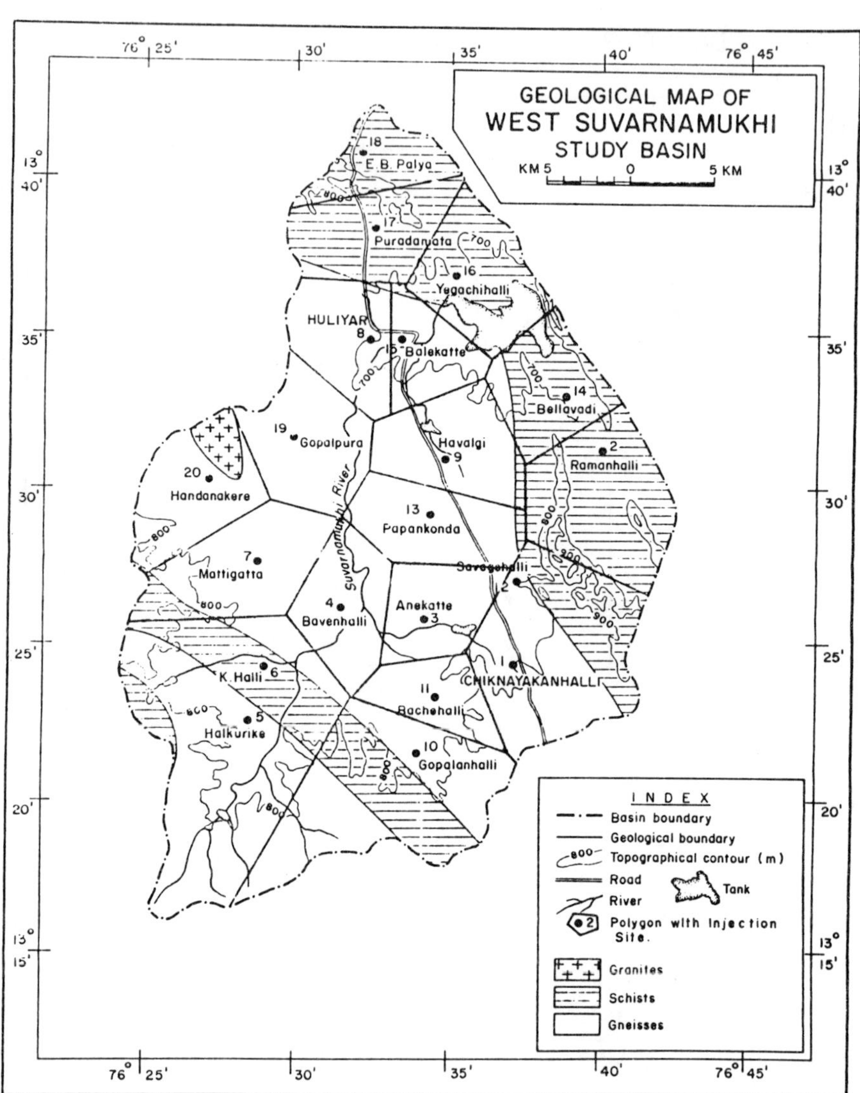

GEOLOGICAL MAP OF
WEST SUVARNAMUKHI
STUDY BASIN
KM 5 0 5 KM
E.B. Palya
Puradamata
Yegachihalli
HULIYAR
Balekatte
Bellavadi
Gopalpura
Havalgi
Ramanhalli
Handanakere
Suvarnamukhi River
Papankonda
Mattigatta
Savegehalli
Anekatte
Bavenhalli
K. Halli
CHIKNAYAKANHALLI
Bachehalli
Halkurike
Gopalanhalli
INDEX
Basin boundary
Geological boundary
Topographical contour (m)
Road
River
Tank
Polygon with Injection Site.
Granites
Schists
Gneisses


FIGURE: 2

Each sample was reweighed in the laboratory. About 25 gm of soil from each sample was used for determination of the moisture content, using an Infra-red torsion balance. Moisture from the rest of the soil sample was extracted through partial vacuum distillation. 4ml of the distillate, representing moisture from each 20 cm interval, was used in determining the Tritium activity. The water sample was mixed with 10 cc of the liquid scintillator 'Instagel' in a low potash vial and the vial was kept overnight for dark-setting. The Tritium activity of each vial sample was counted on a manually operated liquid scintillation spectrometer model LSS-20, supplied by the Electronic Corporation of India.

Details of the field and laboratory procedures are described elsewhere (Athavale et al, 1980).

## RESULTS AND DISCUSSION:

The variation of Tritium content and weight percentage of moisture was plotted against depth in the case of the soil profile collected at each site. Four such typical depth-profiles are presented in Fig.3. The spot value of recharge for each site was calculated by first determining the centre of gravity of the Tritium vs depth profile. The mean displacement of the tracer was taken as the distance between the injection depth and the depth of the centre of gravity. The recharge was calculated from the moisture concentration, tracer displacement and wet bulk density. The recharge values for 18 out of the 20 sites are presented in Table 1. At one site (Halkurki-No. 5), the soil profile could not be collected, as a boulder was found located just below injection depth. The results for another site (Savegahalli - No. 12) had to be ignored as the field was irrigated. The arithmatic mean of the remaining 18 recharge values is 39.2 m.

The average monsoonal rainfall, recorded at three rain gauging stations in West-Suvarnamukhi basin, for the period June-November, 1978, was 463 mm. The mean recharge of 39.2 mm, thus amounts to 8.5% of the seasonal precipitation. Taking the effective infiltration area of the sub-basin as 958 sq. km., the average input to the phreatic aquifers, due to direct infiltration of a fraction of the monsoonal precipitation, then works out to be 39.2 mm X 958 Sq.Km = 37.6 Million Cubic Meters (MCM).

## RECHARGE CALCULATION USING THE THIESSEN POLYGON METHOD:

In the Thiessen polygon method, the non-uniformity in the distribution of raingauges is accounted for by giving a weighting factor to each gauging station. The weighting factor is expressed as the ratio of area surrounding the gauge over the total area. The average precipitation is the sum of the gauge data, each multiplied by its weighting factor. The polygons

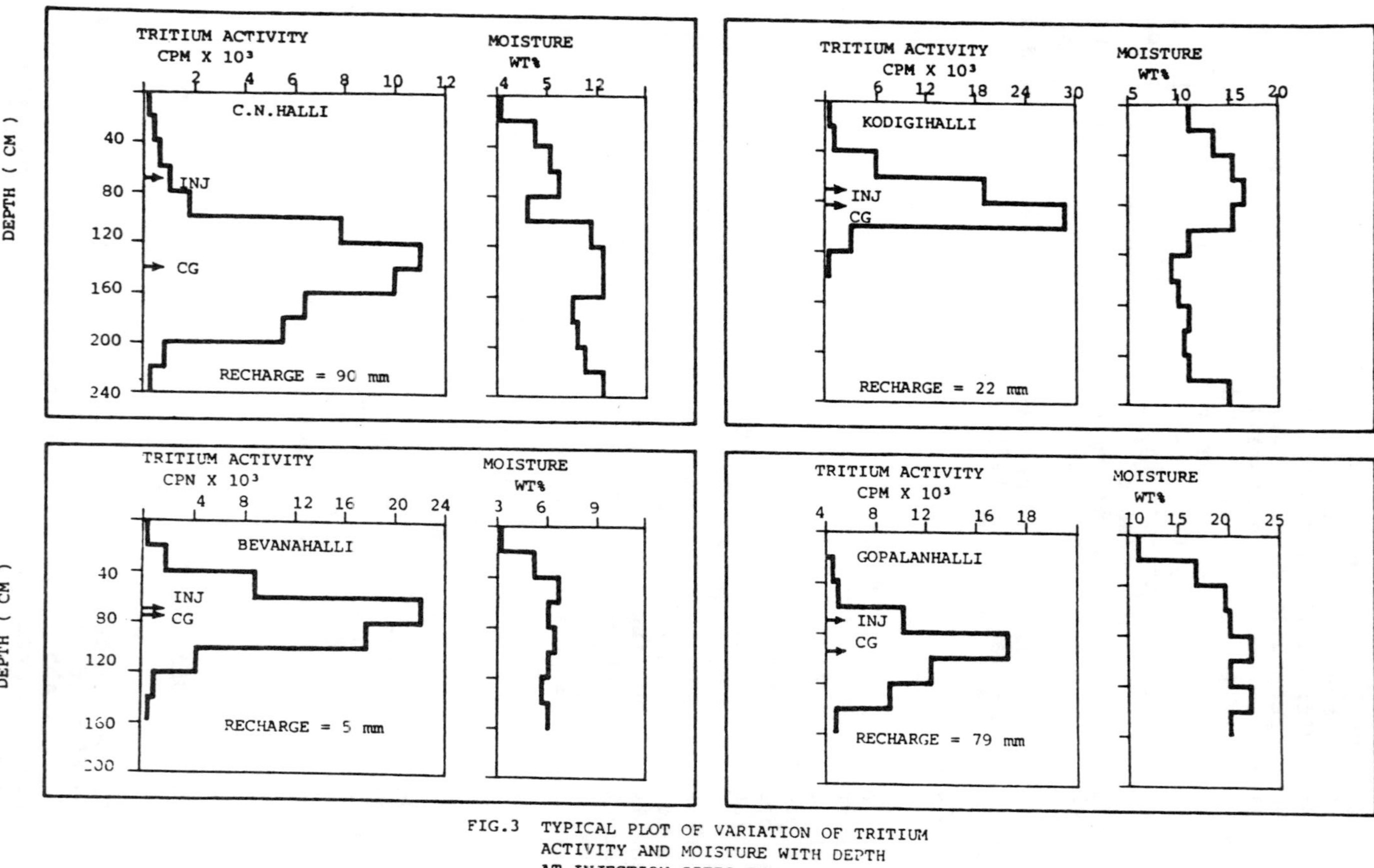


FIG.3 TYPICAL PLOT OF VARIATION OF TRITIUM ACTIVITY AND MOISTURE WITH DEPTH AT INJECTION SITES IN WEST-SUVARNAMUKHI BASIN

**TABLE - I**

Recharge due to 1978 monsoon precipitation in West Suvarnamukhi Vasin

| Sl. No. | Site Name | Recharge in mm | Polygonal area in Sq. Kms. | Net Recharge in each polygon in MCM |
|---|---|---|---|---|
| 1. | C.N. Halli | 90 | 89 | 8.01 |
| 2. | Ramanhalli | 127 | 68 | 8.63 |
| 3. | Annekatte | 22 | 35 | 0.77 |
| 4. | Bevenahalli | 5 | 40 | 0.20 |
| 5. | Halkurki | - | - | - |
| 6. | Kodigihalli | 22 | 140 | 3.08 |
| 7. | Mattigadda | 22 | 48 | 1.06 |
| 8. | Huliyar | 30 | 40 | 1.20 |
| 9. | Havalgiri | 11 | 50 | 0.55 |
| 10. | Gopalanhalli | 79 | 87 | 6.87 |
| 11. | Bachihalli | 3 | 26 | 0.08 |
| 12. | Savegahalli | - | - | - |
| 13. | Papakonda | 86 | 39 | 3.35 |
| 14. | Bellavadi | 18 | 47 | 0.84 |
| 15. | Balakate | 2 | 28 | 0.06 |
| 16. | Yegachihalli | 109 | 50 | 5.45 |
| 17. | Puradamata | 32 | 44 | 1.40 |
| 18. | E.B. Palya | 0 | 31 | - |
| 19. | Gopalpura | 22 | 52 | 1.14 |
| 20. | Handankare | 26 | 44 | 1.14 |
| | Arithmatic Average | 39.2 | Total 958 | 43.83 |

Note : Serial numbers in Table correspond to site numbers in Figure 2.

surrounding each gauge are obtained by erecting bisectors which are orthogonal to the segments joining adjacent stations.

We have used the Thiessen polygon approach in the case of the recharge values. The results from the polygon method are also presented in Table 1. The total 1978 monsoonal recharge obtained by this technique works to be 43.8 MCM. The reproductibility error in each of the spot values of recharge determined by using the Tritium injection techniques is within 10% (Athavale et al, 1980). We can therefore, conclude that the total recharge values of 37.6 MCM and 43.8 MCM, calculated from arithmetic mean and the Thiessen polygon method, are in reasonable agreement. This is possibly because the 20 injection sites were fairly well distributed over the sub-basin.

COMPARISON WITH RECHARGE ESTIMATES USING OTHER METHODS:

1. Sridharan et al (1986) have prepared a regional groundwater model of the Vedavati basin, using a leaky aquifer concept. This model was calibrated using observation well water level data for the period November, 1977 to November 1978, and aquifer parameter data from long duration pump-tests as well as short duration air-tests,conducted by the Central Ground Water Board. These authors divide the entire basin area into 96 blocks on the basis of hydrogeological characteristics. Block nos. 70 and 71 represent the West Suvarnamukhi basin in this model.

   The annual recharge due to precipitation was calculated by Sridharan et al (1986) by calibrating the model for the water level data for Nov. 1977 to November, 1978 and after assuming that the net recharge due to precipitation in non-monsoon months is zero. Thus the model recharge values are effectively for the same period as are the injected Tritium values i.e. over the monsoon months of June to November, 1978. The values of net annual recharge (safe yield), estimated from the model, for blocks 70 and 71 representing the West Suvarnamukhi sub-basin are 37 mm and 48 mm respectively, giving a mean value of 42.5 mm, which agrees very well with the average recharge value of 39.2 mm obtained by the Tritium injection technique, for the year 1978, which was a year with normal rainfall.

2. Radhakrishna and Duba (1973) have carried out an evaluation of ground water resources of Karnataka state. Most of the Vedavati basin drainage area is located in this state. These authors have worked out a climatic water balance for thirteen districts in Karnataka and have arrived at an approximate estimate of annual recharge to ground water.

   West Suvarnamukhi basin is located within the Tumkur district where the annual average rainfall is 687.9 mm. Radhakrishna and Duba (1973) estimate that about 5% i.e. 34.4 mm of the total rainfall contributes to annual recharge. This figure is in good agreement with the average recharge value of 39.2 mm obtained from Tritium injection studies.

CONCLUSIONS:

(1) The average recharge to phreatic aquifers of West Suvarnamukhi basin, estimated by using Tritium injection technique, was about 40 mm or 8.5% of the seasonal rainfall.

(2) The mean recharge value, calculated through arithmetic averaging of the 18 spot values, agrees with the mean value obtained after using Thiessen polygons.

(3) In the case of the West Suvarnamukhi sub-basin the annual recharge or safe yield values estimated by other workers from a calibrated regional ground water model are found to be in close agreement with those obtained by using the Artificial Tritium Injection Technique.

A similar agreement with recharge estimates derived from climatic water balance calculations is also observed.

ACKNOWLEDGEMENTS:

Estimation of recharge, using the Tritium injection method, was carried out by us at the suggestion of Dr. A. Achyutrao, Director of the Vedavati basin project. We are thankful to him and other members of the Vedavati project team of the Central Ground Water Board, for help and advice in field-work and also for providing rainfall and other hydrological data.

## REFERENCES

Anonymous, (1977) - Hydrology and Ground Water balance of the Vedavati river basin in parts of Karnataka and Andhra Pradesh - an interim report on concept, methodology and tentative findings - A technical report, Central Ground Water Board, New Delhi, pp 1-183.

Athavale, R.N., Murti, C.S. and Chand R., (1980) - Estimation of recharge to the phreatic aquifers of the lower Maner basin, India, by using the Tritium injection method. Journal of Hydrology, Vol. 45, pp 185-202.

Radhakrishna, B.P., and Duba Dusan, (1973) - Ground water resources evaluation in Mysore state, in Proceedings International Symposium on Development of Ground Water Resources, Madras, India, Vol. 3, pp: V.63 to V.73.

Sridharan, K., Lakshmana Rao, N.S., Mohankumar, M.S. and Ramesam, V., (1986) Computer model for Vedavati ground water basin. Part 2. Regional model. Sadhana, Vol.9, part 1, pp. 43-55.

# STUDIES ON NATURAL RECHARGE TO THE GROUNDWATER BY ISOTOPE TECHNIQUES IN ARID WESTERN RAJASTHAN, INDIA

H. Chandrasekharan, S.V. Navada*, S.K. Jain*,
S.M. Rao* and Y.P. Singh**,
Nuclear Research Laboratory, IARI, New Delhi 110012, INDIA

(*BARC, Bombay 400085, **Defence Laboratory, Jodhpur 342001)

ABSTRACT. Studies on natural recharge to the groundwater is an important parameter for careful assessment of groundwater potential in arid regions. In this paper, the use of both environmental ( $^{2}H$, $^{18}O$ and $^{14}C$) and artificial ( $^{3}H$ and $^{60}Co$) isotopes in understanding the nature of recharge and recharge condition of groundwater in certain parts of arid Western Rajasthan are presented and discussed in the light of results of other investigations.

Investigations carried out for three years since 1982 indicate that the natural recharge to the groundwater in the investigated areas may occur once in several years either by direct infiltration of local intense precipitation or through river channels from episodic flash floods or remotely from distant outcrops.

## INTRODUCTION

Groundwater is an important natural resource and needs to be developed for agricultural and domestic purposes in any water scarce areas. It can be compared with the money in the bank and, therefore, can be used whenever required. But at the same time, groundwater, unlike the money in the bank, does not show a negative balance in overdrafting. It only indicates that the withdrawals exceed the deposits (in other words, recharge) and the balance is declining. Such overdrafting or rather over-exploitation of the groundwater is common in the arid Western Rajasthan, due to erratic and low rainfall and lack of suitable surface water resources. Hence, in order to have a knowledge of how much of draft of groundwater may be permitted on perennial basis without causing any deleterious effects on the underground reservoir, the natural recharge to the groundwater has to be evaluated periodically.

To date, we have a number of methods to compute and study the natural recharge. They may be broadly described under three main groups thus.

1. Hydrogeological Methods
   a. Water Table Methods
   b. Water Balance Methods



*I. Simmers (ed.), Estimation of Natural Groundwater Recharge, 205–220.*

2. Soil Physical Methods
   a. Lysimeter Methods
   b. Moisture Probe Techniques
3. Isotope Tracer Methods
   a. Environmental/Injected Radioisotope Methods.
   b. Environmental/Injected Stable Isotope Methods.

Details of the first two groups can be had from a number of text books and literature (1,2,3 and 4). Isotope Tracer Methods using both environmental isotopes ( $^{3}H$, $^{14}C$, $\delta^{2}H$ and $\delta^{18}O$) as well as injected radioisotopes could be employed for studying groundwater recharge in arid regions.

Environmental $^{3}H$ and $^{14}C$ are useful in determining the period of recharge to the groundwater. If a particular groundwater contains $^{3}H$ significantly above 10 TR (1 TR is the activity of tritium equal to 3.2 p Ci/litre), then it has been clearly recharge since 1952. Large quantities of $^{3}H$ were injected into the atmosphere in the years 1952-63 by atmosphere H-bomb tests. However due to scarcity of rains in arid areas and small infiltration rates through the unsaturated zone, the shallow groundwaters in such areas may not show any significant $^{3}H$ indicating that modern recharge is negligible. Environmental $^{14}C$ could be used to date very old waters; that is, paleowaters upto age of about 30,000 years. A number of hydrological investigations (5,6,7,8 and 9) have been carried out in arid zones mostly in the Sahara using isotope techniques and it has been observed that the deep groundwater is generally very old. The $^{14}C$ content indicates an age greater than 20,000 years B.P. These paleowaters could have been recharged through distant outcrop areas and the source of recharge could be identified using their stable isotope contents (deuterium and oxygen-18). The hydraulic interconnections between aquifers; that is, whether the shallow aquifer is contributing to the deeper aquifer or vice versa could also be studied using stable isotopes.

Injected radiotracers are useful in estimating infiltration of local precipitation for recharging the shallow aquifers. Direct recharge to the groundwater has been studied extensively in the country and elsewhere (10,11 and 12) using injected tritium tracer. A gamma tracer; for example, Cobalt-60 in the form of a cynide complex could be used for this purpose and has the advantage of insitu measurement of tracer movement. The technique has been employed in some parts of Maharashtra (13) and Rajasthan.

In this paper, results of investigations on natural recharge to the groundwater using isotope tracer methods carried out with the above background are presented and discussed.

## MATERIALS AND METHODS

Radioactive and stable isotope investigations were carried out in different places of Western Rajasthan (table 1). The general background of these areas are detailed elsewhere (14); however, we present below a brief summary of the same.

Table 1. Isotope Investigations for Studying Natural Recharge in Western Rajasthan, INDIA.

| S.No. | Location | Isotopes used |
|---|---|---|
| 1. | Siwana | $^{3}H$, $^{60}Co$ (A) |
| | | $^{2}H$, $^{18}O$ (E) |
| 2. | Raital | $^{3}H$, $^{60}Co$ (A) |
| 3. | Bhadrajun | $^{3}H$, $^{60}Co$ (A) |
| 4. | Jodhpur | $^{3}H$, $^{60}Co$ (A) |
| 5. | Jalore and adjacent areas | $^{2}H$, $^{3}H$, $^{14}C$, $^{18}O$ (E) |

E = Environmental

A = Artificial

Table 2. The General Geological Succession of the Investigated Areas

| Era | Period | Lithology |
|---|---|---|
| Quarternary | Recent to Subrecent | Windblown sand and Younger Alluvium<br>Older Alluvium |
| Early Palaeozoic | Vindhayan System | Limestone<br>Sandstone |
| | Malani Suite | Volcanics<br>Jalore and Siwana Granite<br>Erinpura Granite |
| Purana (Algonklan) | | Quartzite, feldspathic grits,<br>Biotitic schist limestone<br>and calcium silicate rocks |
| Archean | Aravalli system | Slate, phyllite and Mica schist<br>with intercalated quartz |

Pre Aravalli Gneisses

Table 3. Groundwater Occurrence in the Investigated Areas

| Type of Aquifer | Occurrence of Groundwater | Mean Discharge(lph) | Quality of Water |
|---|---|---|---|
| Younger Alluvium<br>Older Alluvium | Pore spaces below water table | 40,000<br>36,000 | Potable<br>saline & potable |
| Vindhyan (Sandstone and limestone) | Fractures and to a limited extent through joints, bedding planes and solution channels. | 31,000 | Potable |
| Granite (Erimpura Jalore and Siwana) | Weathered and joint planes | 15,000 | Fresh and brackish |

## Climate

The climate in general is characterised by extremes of temperature, low and erratic rainfall, large variations in relative humidity and high evapotranspiration. The year can be divided into four distinct seasons, namely, Winter (December-February), Summer (March-June), Monsoon (July-September) and Post-Monsoon (October-November). The mean annual rainfall is around 300 mm of which 90 per cent occurs due to the southwest monsoon. The mean maximum and minimum temperatures vary from $25^0$ C to $43^0$ C and from $5^0$ C to $28\ ^0$C respectively. During summer dust storms (andhi) and dry hot dust wind (loo) occur frequently. The evapotranspiration is of the order of 2000 mm per year.

## Geological and Hydrogeological Features

Pre-Aravalli gneisses are the oldest rocks encountered in the investigated areas (table 2). It appears that the area has been stable throughout the quarternary and the tertiary time. Occurrence of the groundwater is controlled by the physical character and structures of the above geological formations. Porosity and permeability of these formations depend on the presence of primary and secondary openings. Extension of these openings' continuity and interconnection controls the groundwater occurrence and movement. It has been found that the presence of primary openings is nil. Hence, groundwater occurs here mainly in the secondary porosities like fractures, joints, fissures, etc. The principal source of recharge is rain and hence water table fluctuates accordingly. The approximate yield of wells and groundwater quality are given in Table 3. In general salinity in groundwater decreases with depth.

## Isotope Investigations

Radiotracers ($^3$H and $^{60}$Co) and variations in stable isotope ratios $\delta^2$ H and $\delta^{18}$ O were used to study the natural recharge due to rain. The estimation of recharge was done on the basis of the soil moisture movement of a tagged layer by a radiotracer (10,11 and 12). Details of injection, quantity, strength, etc. are given in Table 4. The layout of injection in a field is shown in figure 1. After analysing the samples collected four times (in January 1983, February 1984, July 1984 and February 1985) respective histograms of tracer profiles are shown in figures 4,5,6 and 7. The C.G. or the peak of the tracer profile represents the displacement of the tracer. The soil moisture contents were determined by the gravity method and the in-situ wet bulk density ( $\gamma_W$) of each site by sand logging method and presented in Table 4.

Two tubewells were sampled for $^{14}$C analysis and a number of tube wells and dugwells were sampled for $^3$H and stable isotopes $\delta^2$ H and $\delta^{18}$O analyses. The wells sampled are indicated in figures 2 and 3. The measurement of $\delta^2$ H and $\delta^{18}$ O were carried out using a 602E micro mass spectrometer after preparing the samples using carbondioxide (15) and zinc (16). The $^3$H analysis was carried out by electrolytic enrichment followed by liquid scintillation counting using a LKB, LS counter. The $^{14}$C analysis was carried out by gas proportional counting at NGRI, Hyderabad. The results are presented in Table 5.

Table 4. Details of Radiotracer Injection, Sampling & Results

| Location | Depth of Injection (cm) | Total Activity ($\mu$ ci) | | Displacement (cm) of the tracer after last sampling | | Mean | | Mean Fractional Recharge $\left(\frac{\text{Recharge}}{\text{Rainfall}} \cdot 100\right)$ (%) |
|---|---|---|---|---|---|---|---|---|
| | | $^{60}Co$ | $^{3}H$ | $^{60}Co$ | $^{3}H$ | Moisture content % ($m_w$) | Bulk Density ($\gamma_w$) | |
| Siwana | 60 | 400 | 100 | 0 | 10 | 7.8 | 1.38 | 0.8 |
| Raital | 60 | 200 | 100 | Disturbed | 10 | 5.8 | 1.31 | 1.9 |
| Bhadrajun | 70 | 100 | 100 | 10 | 0 | 26.0 | 1.32 | 1.2 |
| Jodhpur | 70 | 320 | 100 | 38 | 35 | 13.0 | 1.52 | 13.0 |

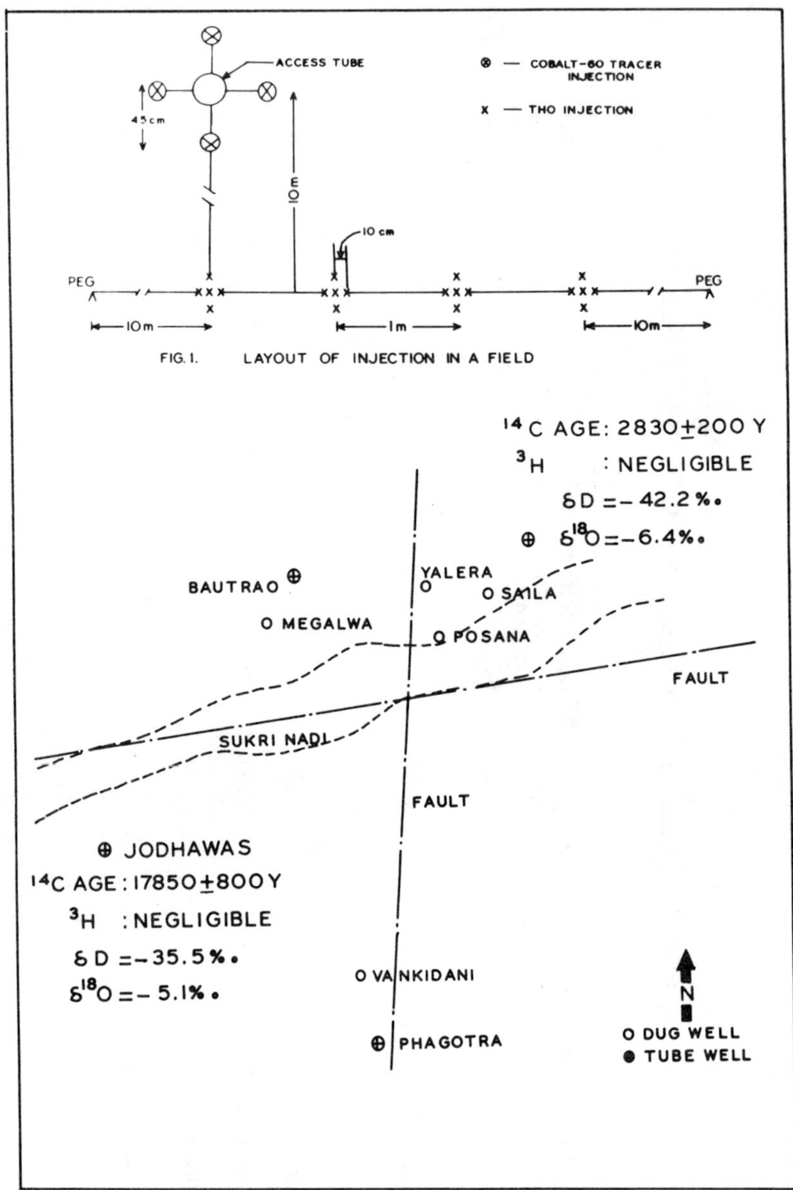


FIG.2. SAMPLING POINTS OF SAMPLES COLLECTED FROM JALORE AREA

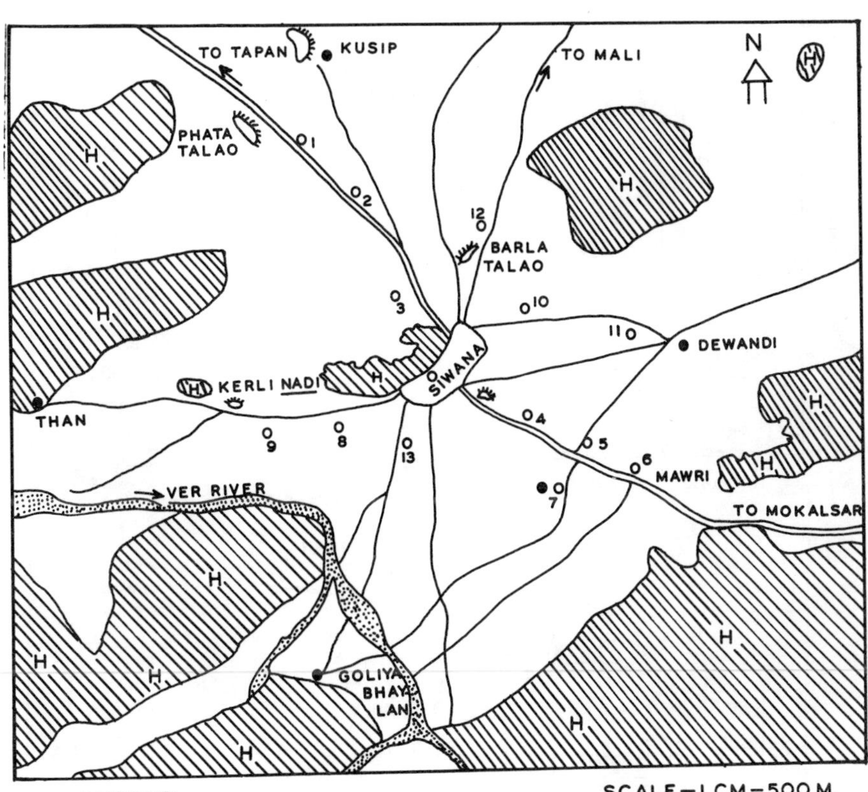


FIG. 3. RADIOTRACER INJECTION POINT AND DUG WELL LOCATION

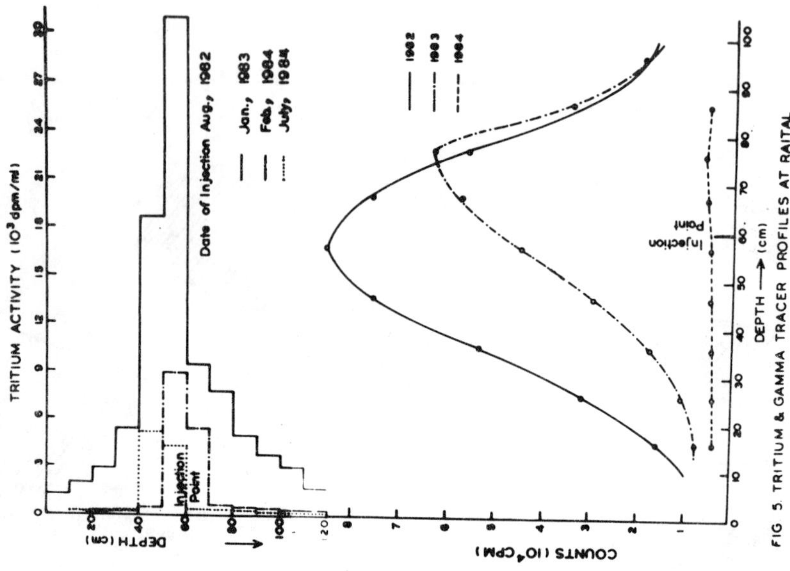


FIG 5. TRITIUM & GAMMA TRACER PROFILES AT RAITAL

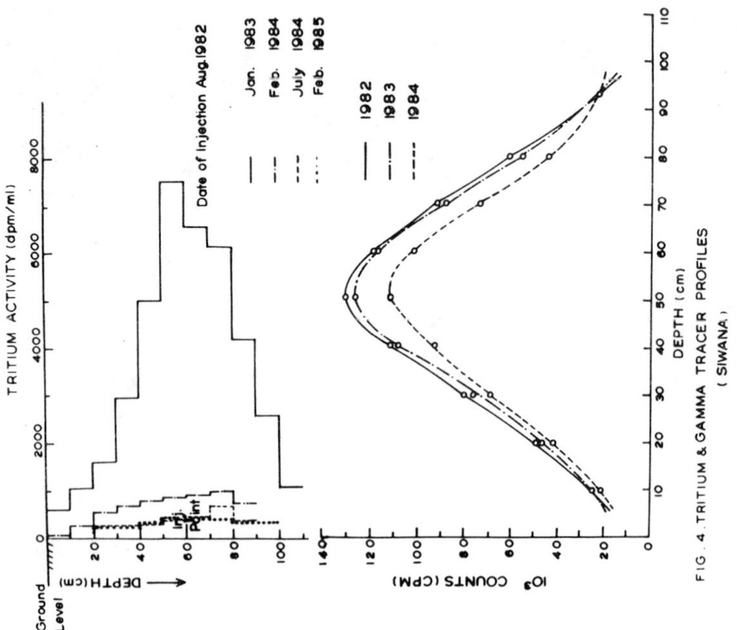


FIG. 4. TRITIUM & GAMMA TRACER PROFILES (SIWANA)

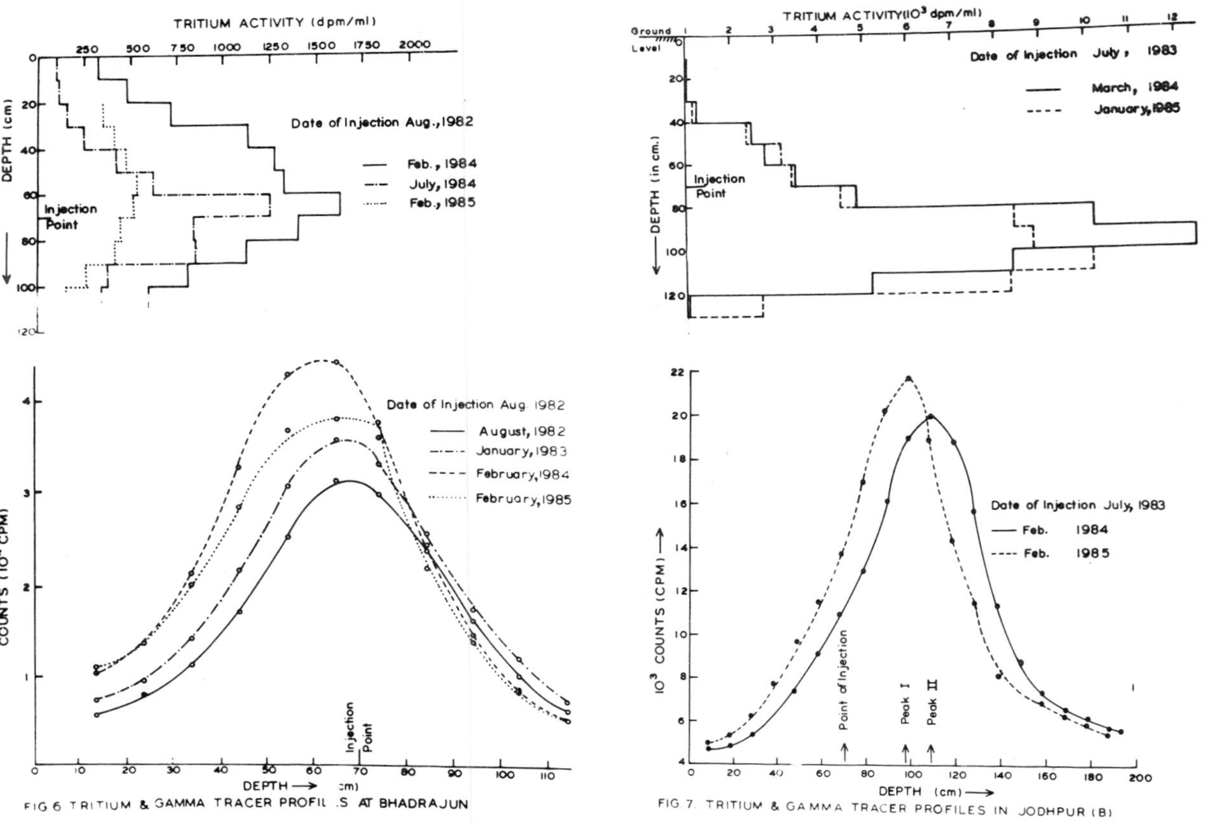


FIG 6 TRITIUM & GAMMA TRACER PROFIL.S AT BHADRAJUN

FIG 7 TRITIUM & GAMMA TRACER PROFILES IN JODHPUR (B)

Table 5. Environmental Isotopic Results of Water Samples from Jalore and Siwana Areas

(W. Rajasthan, India)

| Location | Depth (m) | $\delta\,^{2}H$ (±2‰) | $\delta\,^{18}O$ (±0.2‰) | $^{3}H$ (±1 TU) | Remarks |
|---|---|---|---|---|---|
| | | | | | Uncorrected Carbon-14 age: |
| Tube Wells | | | | | |
| Jodhawas | 266 | -35.3 | -5.1 | 3 | 17850 ± 800 |
| Sayla | 175 | -42.2 | -6.4 | 3 | 2830 ± 200 |
| Bautra | - | -41.8 | -5.8 | - | |
| Magalwa | - | -51.6 | -6.8 | - | |
| Phagotra | 150 | -50.2 | -6.7 | - | |
| Dug Wells | | | | | |
| Vankidani | 30 | -51.9 | -7.2 | 4 | |
| Bautra | 50 | -43.5 | -6.5 | 3 | |
| Valera | 43 | -43.5 | -6.5 | 3 | |
| Sayla | 15 | -55.4 | -7.3 | - | |
| Posana | 20 | -50.4 | -7.0 | 3 | |
| Kusip | 30 | -41.0 | -6.09 | - | |
| Siwana | 25 | -48.3 | -7.62 | - | |
| Siwana | 24 | -46.3 | -6.41 | - | |
| Mawri | 27 | -41.4 | -6.19 | - | |
| Mawri | 30 | -36.7 | -5.59 | - | |
| Siwana | 28 | -43.2 | -6.18 | - | |
| Siwana | 30 | -32.6 | -6.32 | - | |
| Siwana | 27 | -41.4 | -7.29 | - | |
| Meli | 33 | -36.8 | -6.76 | - | |
| Siwana | 22 | -40.5 | -5.71 | - | |
| Siwana | 22 | -34.1 | -6.39 | - | |
| Siwana | 25 | -43.4 | -6.69 | - | |

## RESULTS AND DISCUSSIONS

Analysis of tritium and cobalt-60 tracer profiles (figs. 4,5,6 and 7) indicate that there is very little displacement of the tracer in three of the four places. The corresponding recharge to the groundwater works out to about 1% in Siwana, Raital and Bhadrajun and 13% in Jodhpur of the rainfall occurred during the period of investigations. Geo-electrical investigations conducted at Siwana (17) indicate that clayey zones located at varying depths may hinder the rainfall recharge. On the other hand, the irrigated area in Siwana increased from 12.3 per cent in 1970-71 to 41.6 per cent in 1982-83 (18) which means that the exploitation of groundwater has increased more than three times. This is possible if the groundwater reservoir in Siwana region is recharged by some means. To understand the recharge process further, variations in stable isotope ratios ($\delta^2$ H and $\delta^{18}$O) of well water samples were analysed. It has been found that the $\delta$s are highly depleted in relation to the corresponding rainfall values of -14.0‰ and -2.25‰ . This indicates that the groundwater recharge of the present day rainfall is negligible. These depleted values are comparable to the corresponding values of rainfall in a relatively more humid area like Delhi ($\delta^2$ H = -51.2‰ and $\delta^{18}$ O = -7.65‰ (19)). Thus, it is observed that $\delta$ s of well water samples around Siwana and the $\delta$ s of precipitation in a more humid region are nearly comparable thereby indicating that the groundwater available today was mainly recharged by the precipitation occurring years back when the Thar Desert was under a more humid and cooler period than the present. This does not, however, rule out the possibility that the groundwater may also be recharged once in several years either by direct infiltration of local intense precipitation or by episodic floods.

The tritium and Cobalt-60 tracer profiles at Raital indicate a lot of difference. During the first three samplings there was a movement of about 10 cm (based on C.G. positions) for tritium whereas cobalt-60 has moved so fast that its peak/C.G. could not be located in the third samplings. On enquiry, it was confirmed that the movement of the cobalt tracer was due to a local problem and not due to rain.

At Bhadrajun, both tracers did not move since the date of injection in spite of rainfall of 226 mm, 832 mm and 248 mm respectively in 1982, 1983 and 1984. The tracer profiles at different samplings indicate that the gamma counts are increasing with time for a particular depth indicating that cobalt-60 tracer moves towards the access tube most probably due to gravity.

At Jodhpur B area, both tracers behave in a similar way. The fractional recharge works out to 13% for tritium and 14% for cobalt-60 tracers at the end of the last sampling (Table 4). The investigations clearly indicate that the rainfall recharge does take place in sandy soils of Western Rajasthan, but it depends on the distribution of rainfall with time. The source and period of recharge to the deep aquifers in Jalore district have been studied using environmental isotopes ($^3$H, $^{14}$C, $\delta^2$H and $\delta^{18}$O). The $\delta^2$H and $\delta^{18}$O values fall on a line which has a slope of 8 similar to the meteoric line and an interception of 4‰ (Fig. 8). Similar lower intercept compared to present day meteoric waters with intercept of +10% has been observed in paleowaters sampled in some parts of Sahara and has been attributed to a higher humidity of over 90 per cent over the oceans during the period when the recharge took place (16).

The artesian flowing well in Jalore district (Jodawas) which is located south the Jowai river and the suspected fault has an uncorrected carbon-14 age of 17,850 years and negligible $^3H$ content showing that they are paleowaters. The stable isotopic composition $\delta^2 H$ = -35.3‰ and $\delta^{18} O$ = -5.1‰ is more depleted compared to the present day precipitation (average values $\delta^2 H$ = -15‰ and $\delta^{18} O$ = -2.5‰ ). Hence these waters were possibly recharged remotely during a more cooler and humid period. The source of recharge to this well is being studied by sampling for $^{14}C$, $\delta^2 H$ and $\delta^{18} O$ from some more wells tapping only the tertiary aquifer. The Ec value of 5650 μ mhos/cm and presence of $H_2S$ gas in the sample show that this is in the discharge area beyond the redox barrier. The deep well at Sayla has water sample showing a $^{14}C$ uncorrected age of 2830 $\pm$ 200 years. This well taps both tertiary as well as quarternary alluvium and is near the river course. The sample is possibly a mixture of ancient and young water. The stable isotopic contents of this well as well as other deep wells are depleted compared to the present day precipitation. The dug wells sampled also show depleted stable isotopic composition similar to deeper aquifer indicating possible interconnection between the two. Most of them have negligible $^3H$ content showing that these wells have negligible contribution from the present day precipitation.

The depleted stable isotopic compositions in both Jalore and Siwana areas show that recharge took place during a period more cool than the present. Paleontological studies (20) show that at periods beyond 10,000 years B.P. and even at a period of 3000-5000 years B.P. moist and cooler conditions might have prevailed over the Rajasthan desert. From 3000 years onwards the indications are that progressively dry conditions prevailed. Hence most of the recharge of the wells might have occurred during the last pluvial period.

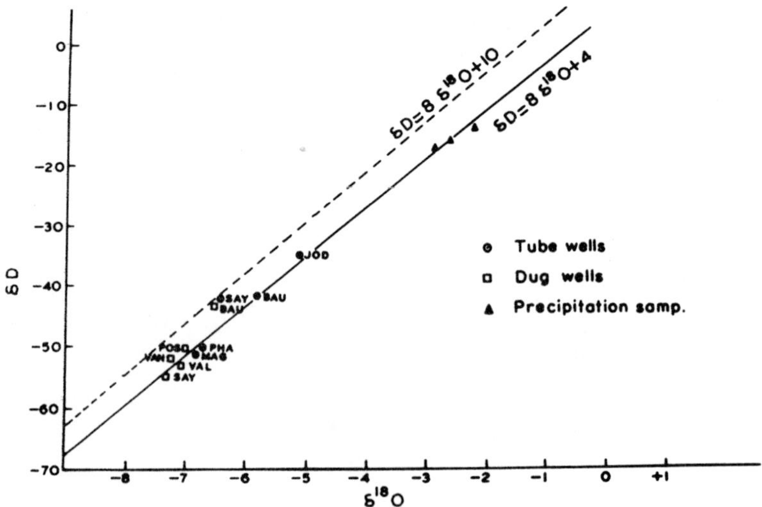


FIG.8. δD-δ18O RELATIONSHIP OF SAMPLE FROM JALORE AREA

## CONCLUSIONS

The injected radiotracer studies show that most of the arid areas have very little contribution towards groundwater from local precipitation with few exceptions. The natural recharge in Siwana region worked out to about 1% of the rainfall from 1982 to 1985 and at Jodhpur it is around 13% of the rainfall from 1983 to 1985. The environmental isotope study shows that the artesian flowing well and other deep dug/tube wells have waters which are ancient and recharged during a more moist and cool period than the present. The deep wells were possibly recharged remotely through distant outcrop areas. The shallow and deeper aquifers appear to be interconnected as evidenced by their stable isotopic compositions.

## ACKNOWLEDGEMENTS

Authors are very thankful to the Directors of Central Arid Zone Research Institute, Jodhpur and Defence Laboratory, Jodhpur for providing the required facilities. Grateful thanks are due to Dr. P.N. Tiwari, Project Director, Nuclear Research Laboratory, IARI, New Delhi for encouragements. We thank Shri B.L. Paliwal and Dr. C.S. Doshi of Rajasthan Groundwater Department, Jodhpur for their help. Thanks are also due to Dr. Sukhija of NGRI, Hyderabad for help in $^{14}C$ analysis of well samples from Jalore area.

## REFERENCES

1. B.S. Sai, C.S. Murti and M.S. Bhalla 1982 'Evaluation of Rainfall Contribution to Groundwater Recharge in Crystalline Aquifers' of Karimnagar District, Andhra Pradesh; Geoviews, Vol. X No. II, pp. 459-489.
2. Kitching R. 1982 'Construction and Operation of a Large Undisturbed Lysimeter to Measure Recharge to the Chalk Aquifer', England; Journal of Hydrology, 58, pp 267-277.
3. Chandrasekharan H. 1978 'Geohydrological Studies of the Chikkahagari Basin', Ph.D. Thesis, Department of Geology Bangalore University, Bangalore.
4. Walton W.C. 1970 Groundwater Resources Evaluation, Mcgraw Hill Book Company.
5. Munnich K.O. and Vogel J.C. 1962 'Untersuchungen and Phuvielen Wassern der Ost - Sahara', Geol. Rundsh. 52, 611.
6. Gonfiantini R., Dincer T. and Derekey A.N. 1974 'Environmental Isotope Hydrology in Honds Region, Algeria', Isotope Techniques in Groundwater Hydrology, Proc. Symp. Vienna, Vol. I, pp. 293.
7. Edmunds W.M. and Wright E.P. 1979 'Groundwater Recharge and Paleo Climate in the Sirte and Kufra Basins, Libya', J. Hydrology, 40, 215.
8. Dray M., Gonfiantini R. and Zuppi G. 1983 'Isotopic Compositions of Groundwater in Southern Sahara - Paleo Climates and Paleo Waters' - Proc. Adv. Group Meetings, Vienna (IAEA).

9. Haynes C.V. and Haas H. 1980 'Radiocarbon Evidence of Holocene Recharge of Groundwater in Western Egypt' Radiocarbon, 22, 705.
10. Zimmermann U. et al 1966 'Tracers to Determine Movement of Soil Moisture and Evapotranspiration', Science 152, 346.
11. Datta P.S., Goel P.S., Rama and Sangal S.P. 1973 'Groundwater Recharge in Western U.P.', Proc. of the Indian Academy of Sciences Vol. LXX VII Sec. A No.1.
12. At havale R.N., Murthy C.S. and Chand R. 1980 'Estimation of Recharge to the Phreatic Aquifers of the Lower Maner Basin, India by Using the Tritium Method', J. Hydrology, 45, pp. 195-202.
13. Nair et al 1973 'Groundwater Recharge Studies in Maharashtra Development of Isotope Techniques and Field Experience Isotopes in Hydrology', Proc. Symp. Neuherberg, IAEA, Vienna 803.
14. Chandrasekharan H. 1985 'Studies in Groundwater Hydrology Using Geo-electrical and Nuclear Techniques' Ph.D. Thesis (submitted) Jodhpur University, p.224.
15. Epstein S. and Mayeda T. 1953 'Variations of $^{18}$O Content of Waters from Natural Sources' Geochim, et Cosmochin Acta, pp 213-214.
16. Merlivat L. and Jourd J. 1979 'Global Climatic Interpretation of $\delta^2$H -- $\delta^{18}$O relation for Precipitation', J. Geophys. Research, 84 5029.
17. Chandrsekharan H. and Sharma M.L. 1985 'Geo-electrical Investigations for Groundwater on a Granitic Region Around Siwana, Western Rajasthan', Geophysical Research Bulletin, Vol. 23, No. 4.
18. Chandrasekharan H. and Balakram 1985 'Groundwater - A Parameter in Determining Landuse Pattern in Siwana Region, Western Rajasthan', The Indian Geographical Journal, Vol. 60, No. 1, pp 1-8.
19. Bahadur J., Gupta R.K. and Saxena R.K. 1977 Hydrology Review Jan. - Apr. pp. 40-52.
20. Bryson R.A. and Barreis D.A. Bull. Amer. Soc Boston, 48 , 3, 136.

# GROUNDWATER RECHARGE ESTIMATION
# (Part 2)
# NUMERICAL MODELLING TECHNIQUES

# NUMERICAL AND CONCEPTUAL MODELS FOR RECHARGE ESTIMATION IN ARID AND SEMI-ARID ZONES

K.R.Rushton
Civil Engineering Department
University of Birmingham
Birmingham B15 2TT
England

ABSTRACT. Recharge estimation can be based on a wide variety of models which are designed to represent the actual physical processes. This paper considers the direct estimation of recharge using soil moisture balance models, recharge due to losses from irrigation schemes, the influence of the unsaturated zone and recharge due to rivers and other sources of run-off. A study of the water table fluctuation method indicates that indirect methods often provide unreliable estimates. Several examples are included to demonstrate that, by identifying and representing the flow mechanisms, realistic estimates of recharge can be made.

## 1. INTRODUCTION

Any model of a recharge system involves a decision about the likely flow mechanisms. Once these mechanisms have been defined, calculations can be carried out to estimate the recharge.

Before identifying the probable flow mechanisms it is essential to examine the field evidence carefully. It is dangerous to assume that, because a successful method has been devised to estimate the recharge in one locality, the same method can be used in a situation which appears to be similar. The recharge mechanism at a particular location can depend on a wide variety of factors which will be discussed further below; if an important factor is not included in the recharge model, the results may be very misleading. For example, models may ignore the difference between the potential recharge from the soil zone and the actual recharge which enters the aquifer. These two quantities may differ due to the influence of the unsaturated zone or alternatively the aquifer may be unable to accept the potential recharge.

This paper reviews a number of alternative approaches to the modelling of recharge mechanisms in arid and semi-arid areas. Examples are chosen to illustrate the wide range of alternative approaches. Some of the models are sophisticated and have large data requirements; others are relatively simple and concentrate on the dominant features which control the quantity of water that reaches the main aquifer. In all these models, the important criterion is that



*I. Simmers (ed.), Estimation of Natural Groundwater Recharge, 223–238.*

they represent the essential features of the flow mechanisms.

## 2. IMPORTANT FEATURES IN RECHARGE MECHANISMS

In the Introduction, reference was made to the fact that the recharge mechanism depends on many important features. Certain of the important features are listed below.

At the land surface:
- topography
- precipitation: magnitude, intensity, duration, spatial distribution
- run-off, ponding of water
- cropping pattern, actual evapotranspiration

Irrigation:
- nature of irrigation scheduling
- losses from canals and water courses
- application to fields, land preparation, losses from fields

Rivers:
- rivers flowing into the study area
- rivers leaving the study area
- rivers gaining water from or losing water to the aquifer

Soil zone:
- nature of the soil, depth, hydraulic properties
- variability of the soil, spatially and with depth
- rooting depth in soil
- cracking of soil on drying out or swelling due to wetting

Unsaturated zone between soil and aquifer
- flow mechanism through unsaturated zone
- zones with different hydraulic conductivities

Aquifer
- ability of the aquifer to accept water
- variation of aquifer conditions with time

This list is far from complete but it does illustrate that a large number of features can have a significant effect on the recharge mechanisms. The list also suggests that it may be necessary to use a series of models to represent the different features. These could include:

-a catchment model to simulate conditions on the surface
-a model of certain features of the irrigation system
-a soil water balance model to represent the soil zone
-an unsaturated flow model for the unsaturated zone
-a regional groundwater flow model for the aquifer conditions.

Reference will be made to a limited number of models to illustrate the general approach.

## 3. SOIL MOISTURE BALANCE MODEL

Soil moisture balance models have proved to be valuable in estimating the potential recharge. In a standard soil moisture balance calculation the volume of water required to fully saturate the soil is expressed as an equivalent depth of water and is called the soil moisture deficit. The change in soil storage any one day is expressed as

$$\Delta S = P - RO - EA \qquad (1)$$

where P is the precipitation
RO is the run-off and
EA is the actual evapotranspiration.
One condition that is enforced is that if the soil moisture deficit is greater than a critical value, called the root constant, evapotranspiration will occur at a rate less than the potential rate. The magnitude of the root constant depends on the crop, the stage of crop growth and the nature of the soil (see for example Rushton and Ward 1979).

When the change in soil storage, $\Delta S$, is positive the soil moisture deficit is reduced. On other days when there is no rainfall, $\Delta S$ is negative and the evaporative demands are taken from soil storage leading to an increase in the soil moisture deficit. In terms of recharge, the important condition is when the soil moisture deficit is sufficiently small that the quantity of water available to change storage is greater than the soil moisture deficit. After reducing the soil moisture deficit to zero, the excess water will either pond on the ground surface, or the soil becomes free draining and the excess water is transmitted downwards as potential recharge.

When applying this method to estimate the recharge for a catchment, the calculation should be repeated for areas with different

precipitation and evapotranspiration,
crop type and
soil type.

More detailed studies using soil physics techniques have shown that this approach contains many simplifications. In particular it infers that there is a fully saturated zone and an unsaturated zone with an abrupt change between the two. Field experiments indicate that the volumetric water content in the unsaturated zone varies in a complex manner (Wellings and Bell 1982). Furthermore the soil moisture deficit approach assumes that recharge only occurs when the soil is fully saturated and becomes free draining yet in practice some vertical drainage does occur at other times. Nevertheless, the annual potential recharge estimated using the soil moisture balance technique is similar in magnitude to that predicted by more accurate methods although the distribution throughout the year is somewhat different.

Although this technique was developed for temperate regions, it can be used for semi-arid regions in which there are distinct rainy and dry seasons. One of the important features is that the wilting of the crop can be represented. When wilting occurs due to a lack of moisture in the soil, there is little further change in the moisture conditions in the soil provided that cracking of the soil does not occur. It is essential to adequately represent the soil conditions during the dry season to provide a reasonable value for the soil moisture deficit at the onset of the rainy season. This will indicate what volume of water is required to saturate the soil before any recharge can occur.

When the soil moisture balance method is applied in semi-arid areas it is found that the recharge estimates may differ significantly between years that have similar annual rainfalls. If much of the rainfall occurs as short intense precipitation, the soil moisture deficit can be satisfied with the excess rainfall becoming recharge. However, when the rainfall is distributed fairly uniformly over several months, the soil may never reach saturation so that little recharge occurs (Khan 1986). These findings have been supported by observations of water table fluctuations which indicate that the annual recharge should not be estimated as a constant fraction of the annual precipitation.

## 4. LOSSES FROM IRRIGATION

Losses from irrigation schemes frequently provide a greater contribution to recharge than rainfall. Potential recharge due to losses from fields can be estimated using the soil moisture balance technique described in the previous section, but losses from canals and flooded rice fields require a more detailed analysis.

The efficiency of canal irrigation schemes, which is defined as the water actually used by the crop divided by the water released from the source, is often low with the losses from the main canals and the distribution canals each equalling about 20% of the water released from the source works (Bos and Nugteren 1978). Although some of these losses occur due to evaporation, the major part of the canal losses are due to seepage to the underlying aquifer.

Losses from canals are usually estimated as a function of the wetted perimeter with no account taken of the aquifer into which losses are occurring (Ahmad 1974). However, a detailed investigation by Bouwer (1969) demonstrated that the losses depend primarily on the hydraulic conductivity and the nature of the boundaries to which the flow is occurring. Bouwer considered two sets of boundary conditions; one a highly permeable zone at some distance beneath the canal and the other an impermeable base with a lateral fixed head boundary, Fig.1. Using numerical models he obtained a series of design curves. A more recent study by Wachyan and Rushton (1987) has considered a further boundary condition of an underlying zone of lower hydraulic conductivity through which water can move vertically. This is more representative of the situation in many practical problems. Figure 2 shows the form of the flow patterns for this situation.

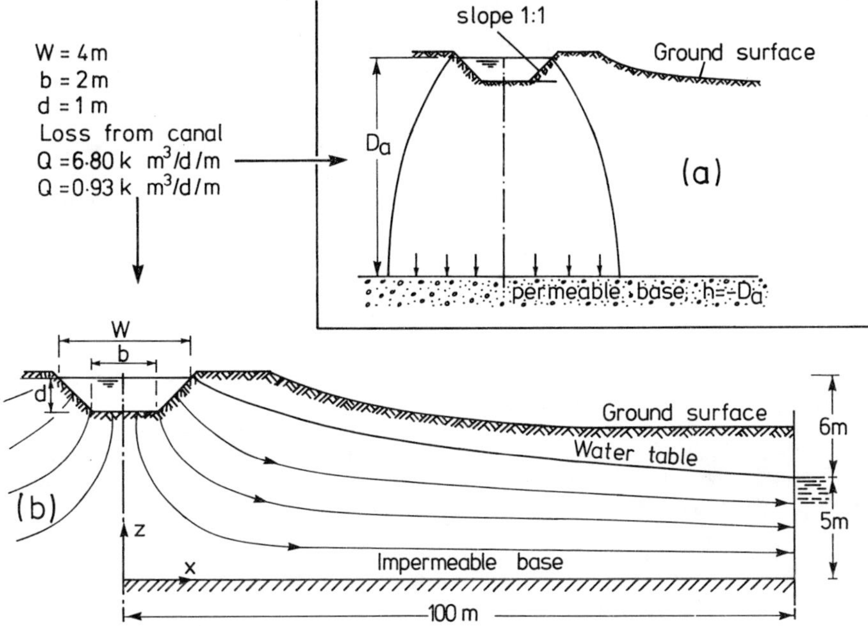


Figure 1. Losses from canals (a), to a highly permeable zone, (b) to a lateral boundary for an aquifer with an impermeable base.

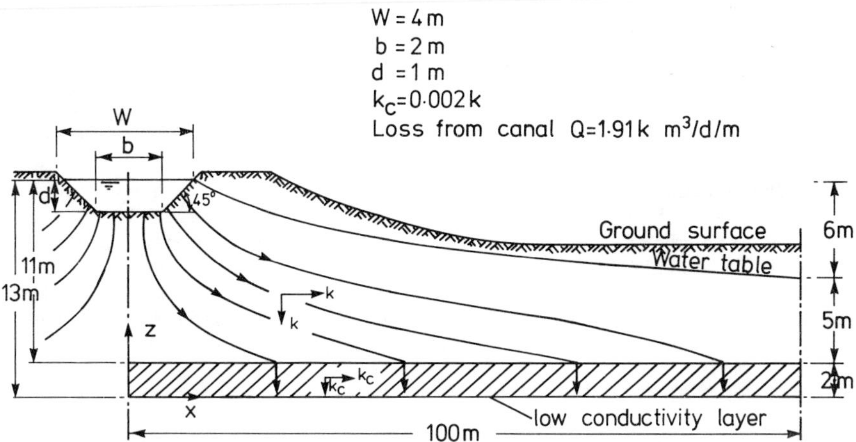


Figure 2. Losses from canal through underlying zone of lower conductivity.

The results of Fig.2 refer to a small canal having a base of 2m and sides inclined at $45^{o}$; the depth of water is 1.0m. According to the usual approach the loss for each square metre of wetted perimeter is 0.1 $m^3/d$; therefore for each metre length of canal the loss is 0.48 $m^3/d$.

Adopting the approach of Bouwer, flow occurs from the canal to lateral fixed head boundaries which for this example are assumed to be 100m on either side of the canal with constant heads 6m below canal level; the loss equals 0.93k $m^3/d/m$. For a typical sandy soil with some clay the hydraulic conductivity, k, is approximately 1.0 m/d and therefore the loss would equal 0.93 $m^3/d/m$.

The example represented by Fig.2 refers to flow from the canal to a layer with hydraulic conductivity 0.002 times that of the overlying zone positioned 10m below the base of the canal; there is no lateral fixed head boundary. When k = 1.0 m/d for the overlying zone, the loss is 1.91 $m^3/d/m$. This loss is roughly three times the usual estimate based on the wetted perimeter.

It is often suggested that these losses can be reduced to very low values by the lining of canals. Experience does not always support this view. Further model solutions representing this situation show that if the canal lining is 99% perfect,the losses are about 72% of the value if no lining is present (Wachyan and Rushton 1987). This result refers to an average hydraulic conductivity of about 1.0 m/d; for aquifers with higher hydraulic conductivities the canal lining is usually more effective.

Model simulations have also proved to be useful in investigating the losses from flooded rice fields. Detailed examination of field results from many countries (Walker and Rushton 1986) have shown that the net losses from the fields are equivalent to about 20 mm/d which is more than twice the evaporative losses. Numerical models have indicated that the main source of the losses is flow through the bunds (small banks between fields). Although the beds of the field are puddled to reduce the downwards percolation to small values, the puddling does not continue under the bunds. The water can, therefore, move through the bunds to the underlying aquifer; indeed the bunds are very efficient hydraulic systems for transmitting water from the surface to the aquifer.

These examples show that losses from irrigation schemes can be a major source of recharge. The magnitude of the recharge depends on the source of the water and the nature of the underlying zones, therefore it is not acceptable to say that the flow to an aquifer due to irrigation is a fixed percentage of the flow in an irrigation system. Instead, detailed studies are necessary into the different components of the losses.

## 5. MOVEMENT THROUGH THE UNSATURATED ZONE

From a soil moisture balance or a study of losses from canals or rice fields an estimate can be made of the potential recharge. In many situations, this water then has to move through the unsaturated zone until it reaches the water table. The actual recharge reaching the water table may be substantially less than the potential recharge due to the influence of the unsaturated zone.

A striking example of the difference between potential and actual recharge has been identified in the Mehsana alluvial aquifer in Gujarat, India (Rushton 1986a). Due to heavy exploitation, a steady but significant decline has occurred in the water table; presently it is more than 20m below ground level. In recent years, canal irrigation has spread to part of the Mehsana area and losses from the canals, distributaries and fields can be observed. It was anticipated that these losses would lead to a recovery in the water table. This has not occurred; the water table position is largely unchanged and waterlogging is observed in the vicinity of certain canals. This is the result of the actual recharge to the water table being far less than the potential recharge from the irrigation canals due to the restricted flow through the unsaturated zone arising from the presence of low conductivity layers.

Flow conditions within the unsaturated zone are far more complex than the flow mechanisms in a saturated aquifer. The main difficulty is that there are three parameters,

the moisture content,
the matric potential (or fluid pressure)
the hydraulic conductivity,

which are inter-related. Furthermore, the relationship between these parameters is very sensitive; a change in the volumetric water content of 5% often corresponds to a change in the hydraulic conductivity by two or more orders of magnitude.

A number of numerical models have been devised and used successfully to represent the vertical flow conditions in the unsaturated zone between the soil surface and the water table. Reference should be made to Bouwer (1978), Hillel (1982), Singh and Saini (1986) and Watson (1986).

The appropriate equation for one dimensional vertical flow in the unsaturated zone is

$$\frac{\partial \theta}{\partial t} = \frac{\partial}{\partial z}\left(k(\psi)\frac{\partial \psi}{\partial z}\right) + \frac{\partial}{\partial z}\left(k(\psi)\right) \qquad (2)$$

where $\theta$ is the volumetric water content,

$k$ is the hydraulic conductivity ($LT^{-1}$),

$\psi$ is the matric suction potential (L)

$t$ is the time (T) and

$z$ is the vertical ordinate (L).

Both the volumetric water content, $\theta$, and the hydraulic conductivity, $k$, are functions of the unknown potential, $\psi$. The dependence of $\theta$ and $k$ on $\psi$ is determined from experiments; there is the added complication of hysteresis in the relationships depending on whether the soil moisture is increasing or decreasing. Provided that suitable techniques are used, numerical solutions can be obtained for equation (2); many valuable insights into the flow in the unsaturated zone have been gained.

Most of the numerical models of flow in the unsaturated zone have been concerned with relatively homogeneous properties. However, the zone between the soil and the main aquifer often consist of layers of sands, silts and clays; the saturated hydraulic conductivities of these layers differ by at least three orders of magnitude. In such a situation, an approximate estimate of the vertical flow can be gained by considering the zone of lowest hydraulic conductivity.

Figure 3 illustrates a situation of a sandy clay zone in the unsaturated region between the soil zone and the water table. Due to the low hydraulic conductivity of this sandy clay layer (say 0.001 m/d) vertical flow from the surface is likely to cause perched water tables. Beneath the sandy clay zone there is a more permeable region through which water passes rapidly. As a first approximation, it can be assumed that the pressure in this sandy layer is close to atmospheric and therefore, for the dimensions recorded in the diagram, the vertical hydraulic gradient is

$$\partial h/\partial z = -3.9/3.0 = -1.3$$

When multiplied by the vertical hydraulic conductivity, $k_z = 0.001$ m/d, the vertical velocity downwards through the clay layer is 1.3 mm/d even though the potential recharge in the vicinity of canals and rice fields may be greater than 5 mm/d.

This calculation contains many simplifying assumptions. Nevertheless, since it represents the dominant feature that is restricting vertical flow through the unsaturated zone, it will provide a reasonable approximation to the actual recharge.

Bouwer (1978) has presented another approximate model which can be used to represent infiltration and the movement of the wetting front during intermittent irrigation. The model refers to the case where water ponds on the soil surface and it predicts the rate at which infiltration can occur and the wetting front move downwards. Assuming a sharp front, a modified form of the Green and Ampt equation leads to the expression for the vertical infiltration rate

$$v_i = k\,(H_w + L_f - h_{cr})/L_f \tag{3}$$

where $k$ = hydraulic conductivity of the wetted zone

$H_w$ = depth of water above the soil

$L_f$ = depth of wetting front (see Fig.4)

$h_{cr}$ = critical pressure head for soil wetting.

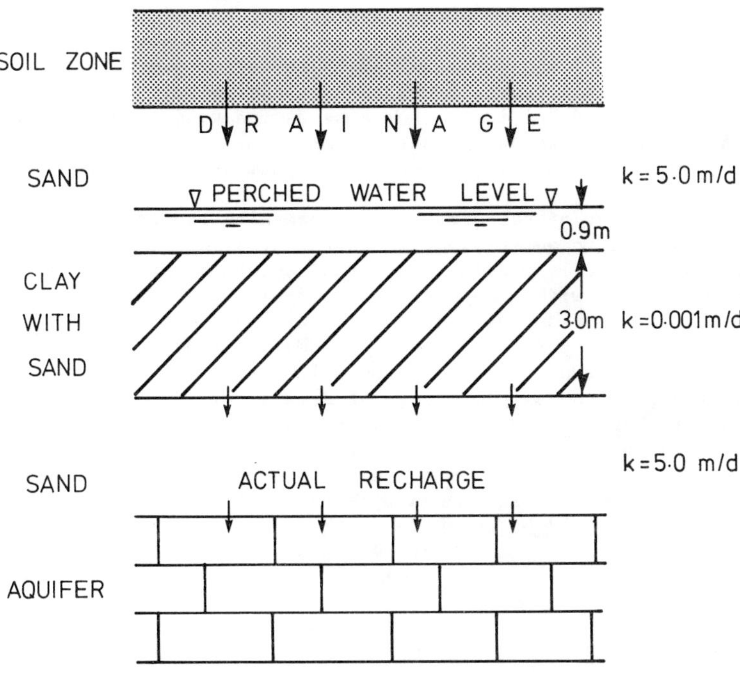


Figure 3. Effect of a low permeability zone on flow to underlying aquifer

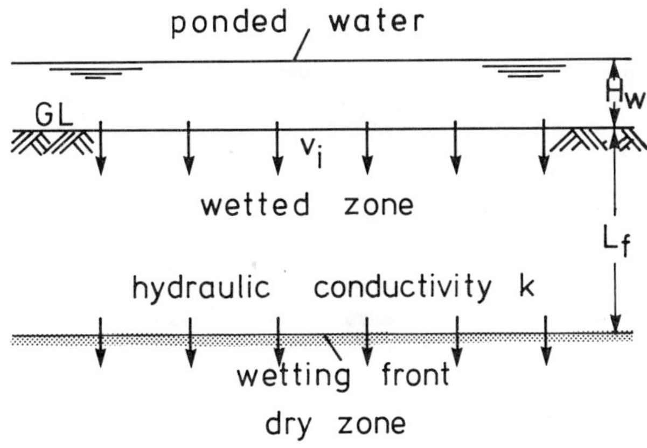


Figure 4. Movement of wetting front in homogeneous aquifer

The term, $h_{cr}$, is equivalent to an extension of the saturated zone to represent the capillary fringe. The hydraulic conductivity used in these calculations is the resaturated hydraulic conductivity and is approximately half of the normal hydraulic conductivity.

Equation (3) refers to a Darcy velocity; the actual velocity of the wetting front equals $v_i$ divided by the effective porosity. As the depth of the wetted front, $L_f$, increases the velocity, $v_i$, tends to k, the hydraulic conductivity. For a hydraulic conductivity of 0.5 m/d the infiltration rate approaches 0.5 m/d from initially higher values. If the porosity is 0.25, the vertical velocity of the wetting front tends to 2.0 m/d.

In the light of the discussion concerning the complex conditions in the unsaturated zone, it is clear that many simplifying assumptions are incorporated in this model. Nevertheless, Watson and Awadalla (1985) make comparisons with a more accurate model and show that in many situations the predictions of equation (3) are approximately correct although there is a tendancy to over-estimate the infiltration rate.

The discussion in this section will be closed by considering the implications of the simple models of Figs.3 and 4. The situation depicted in equation (3) and Fig.4 indicates that infiltration can occur at a rate greater than or equal to the hydraulic conductivity. However, if the wetting front reaches a zone of lower hydraulic conductivity such as the clay layer of Fig.3, equation (3) no longer applies. On the other hand, it would be wrong to restrict the infiltration according to the method of Fig.3 since water collects above the clay layer and continues to move slowly through the clay layer for long after the infiltration event.

## 6. INFILTRATION DUE TO SURFACE FLOWS

Even when annual precipitation is low, rainfall may occur as occasional events of high intensity. These high intensity storms often lead to river flows especially when some of the precipitation occurs over impervious regions. These river flows may subsequently pass over a more pervious region and part may infiltrate into the aquifer. This mechanism can be described qualitatively but the quantative assessment of the recharge is difficult because it depends on so many factors.

There is little clear guidance in the literature as to how this form of recharge should be estimated. Reports of specific areas indicate how estimates were made but the methods are usually site specific. Furthermore, they often rely on a limited correlation between rainfall and another parameter such as the water level at some location within the aquifer.

In the absence of any general models, some of the features which should be considered are listed below. They may indicate the general form that the method of estimation should follow, but having made a direct estimate,an independent check should be made to ensure that the estimates are reasonable.

Features of River Flow:

magnitude and duration of flow passing over the aquifer
intervals between events
silt content of the water.

Features of the River Bed:

dimensions and slope of the river bed
hydraulic conductivity of the bed and sides of the channel
degree of saturation before onset of flow
evaporation from the river bed when river flow ceases.

Sub-surface Features:

presence of clay layers in the unsaturated zone
current moisture content of the unsaturated zone
depth of water table in the aquifer
hydraulic properties of the aquifer
ability of the aquifer to accept water.

In the light of the above features the assessment of likely recharge should be considered in four stages.

(1) Determination of the quantity of water that the aquifer can accept. Small sand or gravel aquifers may have been filled with water during a previous high flow event so that no more water can be accepted. On the other hand, the hydraulic conductivity of an aquifer may be low with the result that the water which infiltrates along the line of the river cannot move away laterally at a sufficient rate thereby limiting the actual recharge. An understanding of the aquifer conditions is therefore an essential first step in estimating the recharge.

(2) The next stage is to estimate the vertical flow which can occur through the unsaturated zone. This was discussed in the previous section where it was emphasised that it is necessary to consider the interaction between effects such as infiltration rates, movement of the wetting front and the influence of zones of low hydraulic conductivity.

(3) Having estimated how much water can pass into and through the unsaturated zone and how much water can be accepted by the aquifer, it is necessary to examine flow mechanism from the river into the aquifer. Initially water will flow vertically downwards from the river bed into the unsaturated zone. Whilst the wetting front is moving downwards, Fig.5(a), water leaves from the full width of the saturated river bed. If the river flow is the result of a short storm then the vertical flow is the predominant mechanism but if the river flow continues for several days, the flow conditions will change to those shown in Fig.5(b). When horizontal flow becomes significant, most of the flow occurs from the edges of the river bed with little flow occurring from the middle of the river; this result is deduced from numerical modelling of a river section (Rastogi 1983). The presence of silt deposits in the bed and sides of the river can reduce the magnitude of the infiltration but rarely to less than half of the value when no silt is present.

(4) The final stage involves an examination of the river flows to ascertain the time period for which the river flow is large enough to meet the infiltration potential of the river -aquifer system. In certain instances the volume of water available far exceeds the infiltration potential but in other situations the river flow is too small and a drying up of the river will occur. The manner in which the river dries up provides important information about the river-aquifer interaction mechanism.

The above discussion indicates that a wide variety of simple 'models' are required in estimating the flow from an intermittent river into an aquifer. Often it is suggested that the mechanisms are too complex and no realistic attempts are made to estimate the recharge,yet acceptable estimates can usually be made by following an approach similar to that described in this section. Additional information such as changes in well hydrographs can be used to confirm that the estimates are of the right order.

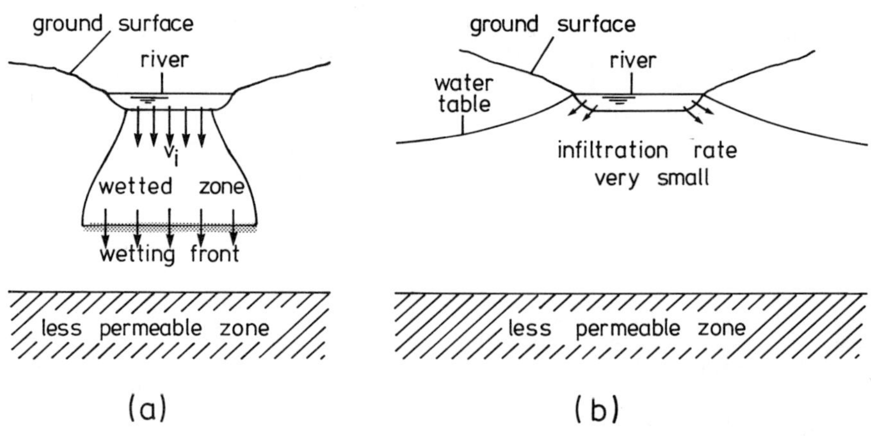


Figure 5 Infiltration from rivers (a) Conditions shortly after start of river flow (b) Conditions after river flow has continued for several days.

## 7. WATER TABLE FLUCTUATION METHOD

Since all the techniques described above require adequate data and appropriate models, an alternative approach is to use the indirect method of deducing the recharge from the fluctation of the water table. Provided that there is a distinct rainy season with the remainder of the year being relatively dry, the rise in the water table during the rainy season is used to estimate the recharge. A method used in India (Anon 1983) will be used to illustrate the approach.

The basic assumption is that the rise in the water table during the rainy season is primarily due to the rainfall recharge. It is recognized that other factors such as pumping or irrigation during the rainy season do have an influence. If the rise in water table is $\Delta s$, the rainfall recharge, $R_r$ is estimated as

$$R_r = S_y \Delta s + Q_a - R_i \tag{4}$$

where $S_y$ is the specific yield,

$Q_a$ is the abstraction during the rainy season divided by the study area (L/T),

$R_i$ is the return flow due to any irrigation which occurs during the rainy season (L/T).

Though at first sight this method is attractive, there are a number of factors which mean that it can lead to unreliable results. One of the fundamental inadequacies is the implicit assumption that every inflow or outflow is uniformly distributed over the area. This may be approximately true for the rainfall and even for the return flow from irrigation but it is rarely true for the abstraction from the aquifer. When pumping is reduced or ceases during the rainy season, a redistribution of groundwater heads occurs so that part of the observed increase in water level may be due to normal well recovery.

Another important question is whether abstraction from both shallow and deep wells should be included in the term $Q_a$. Abstraction from shallow wells will have an effect on the water table whereas deep wells continue to draw water from the phreatic surface by a form of vertical leakage even when they are not pumping for a number of months (Rushton 1986b).

The estimation of recharge using equation (4) is also critically dependent on the value of the specific yield. The earlier discussion on unsaturated flow indicated that changes in the volumetric water content are gradual and cannot be represented by the concept of the soil being either fully saturated or completely unsaturated with the volumetric water content changed by $S_y$. Since the water table fluctuation occurs in the partially saturated zone, it is difficult to determine the appropriate value of the specific yield.

The significance of the specific yield determined from a pumping test is equally open to question. For example Nwankwor et al (1984) estimated the specific yield of a clean unconsolidated sand;

(a) using pumping tests with type curve analysis from which they obtained values of the specific yield of 0.07 to 0.08,

(b) using a volume balance method for the same pumping test the specific yield varied from 0.02 at 0.25 hours to 0.25 at 65 hours,

(c) by draining a column of sand the specific yield equalled 0.3.

For materials containing clay the rate of change in specific yield is very much slower. In another study Sophocleous (1985) presents a detailed analysis of conditions in the unsaturated zone following infiltration and shows that the water table appears to rise faster due to the influence of the capillary fringe. The rate of rise is more than six times the value obtained by dividing the infiltration rate by the specific yield.

All these uncertainties and particularly the difficulty over the definition of the specific yield means that the simple model postulated by equation (4) should not be used. However this does not mean that the change in elevation of the water table can never be used for recharge estimation. The long term response of the water table multiplied by the specific yield determined as in (c) should equal the long term difference between the recharge and the outflows from the aquifer system. Consequently the long term change in water table elevation can be used as a check on recharge estimates.

## CONCLUSIONS

The aim of this contribution has been to indicate how numerical models can be used to estimate the magnitude and distribution of the recharge which actually reaches an aquifer. Perhaps the greatest advantage of a model is that it requires the investigator to identify the important mechanisms. First attempts are likely to provide an incomplete model, either because the model is not representing the important features or because some of the parameter values are incorrect. Further attempts are essential and they usually indicate what modifications need to be made to the model and what additional field information is required. Since the recharge is often the most important quantity in a groundwater resource estimation, considerable time and effort must be spent both in the field and in the office in modifying and refining the methods of estimation.

This paper has reviewed only a selection of alternative models; other important models have not been included. As the usefulness of models for recharge estimation becomes more widely recognized, further improved models will be developed. These will not necessarily be more sophisticated or require a more extensive data base. The greatest need is to learn how to identify the important features which influence the recharge processes.

## REFERENCES

Ahmad,N. 1974. Groundwater resources of Pakistan. Ripon Printing Press, Lahore. 295pp.

Anon. 1983. Ground Water Estimation Methodology. Report of the Ground Water Estimation Committee, Central Ground Water Board, Ministry of Irrigation, New Delhi.

Bos, M.G. and Nugteren,J. 1978. 'On Irrigation Efficiencies' Int.Inst. for Land Reclamation and Improvement, Wageningen, The Netherlands. pp138.

Bouwer,H. 1969. Theory of seepage from open channel. Advances in Hydroscience, Vol.5, Academic Press, New York, 121-172.

Hillel,D. 1982. Introduction to Soil Physics, Academic Press. 364pp.

Khan. 1986. Inverse problem in Ground Water: Model Application. Ground Water, 24, 39-48.

Nwanker,G.I.,Cherry,J.A. and Gillham,R.W. 1984. A comparative study of specific yield determinations for a shallow sand aquifer. Ground Water, Vol.22, 764-772

Rastogi,A.K. 1983. Numerical solutions of confined and unconfined aquifer interactions with partially penetrating rivers. Ph.D. Thesis University of Birmingham,U.K. 309pp.

Rushton,K.R. 1986(a). Surface water-groundwater interaction in irrigation schemes. Conjunctive Water Use (Proceedings of the Budapest Symposium, July 1986) IAHS Publ.No.156. 17-27.

Rushton,K.R. 1986(b). Vertical flow in heavily exploited hard rock and alluvial aquifers. Ground Water, 24, 601-608.

Rushton,K.R. and Ward,C. 1979. The estimation of groundwater recharge. Journal of Hydrology, 41, 345-361.

Singh,S.R. and Saini,A.K. 1986. A two-dimensional finite element model for saturated-unsaturated flow. Regional Workshop on Groundwater Modelling, WRTDC, University of Roorkee, India, 21-42.

Sophocleous,M. 1985. The role of specific yield in ground-water recharge estimations: a numerical study.. Ground Water, 23, 52-58.

Wachyan,E. and K.R.Rushton. 1987. Water losses from irrigation canals. Accepted for publication in Journal of Hydrology.

Walker,S.H. and K.R.Rushton. 1986. Water losses through the bunds of irrigated rice fields interpreted through an analogue model. Agricultural Water Management, 11, 59-73.

Watson,K.K. 1986. Numerical analysis of natural recharge to an unconfined aquifer. Conjunctive Water Use, IAHS, Publ.No.156, 323-333.

Watson,K.K. and Awadalla. 1985. Comparative study of Green and Ampt analysis for a falling water table in a homogeneous sand profile. Water Resources Research, 21,8, 1157-1164.

Wellings,S.R. and Bell,J.P. 1982. The physical controls of water movement in the unsaturated zone, Q.J.Eng.Geol. London, 15, (3) 235-241.

# METHODS FOR ESTIMATION OF NATURAL GROUNDWATER RECHARGE DIRECTLY FROM PRECIPITATION - COMPARATIVE STUDIES IN SANDY TILL

Per-Olof Johansson
Department of Land Improvement and Drainage
Royal Institute of Technology
S-100 44 Stockholm
Sweden

ABSTRACT. Six different methods for estimation of natural groundwater recharge directly from precipitation were tested and compared in a sandy till area in southeastern Sweden. A one dimensional soil water flow model was tested against observed groundwater levels. The fit between simulated and observed groundwater levels was shown to be rather insensitive to displacements between evapotranspiration and groundwater recharge. Applying a single soil moisture reservoir method, recharge had to be allowed even when a moisture deficit existed in order to correctly reproduce the dynamics revealed as groundwater level fluctuations. The estimation made from groundwater level fluctuations and a specific yield value was not satisfactory. Comparisons of chloride deposition and concentration in spring discharge gave promising results for studies of relative areal variability of recharge. Spring discharge measurements and a catchment area model, calibrated against them, gave valuable information of total recharge quantities, which could be used for comparisons with the other methods.

The study clearly demonstrated the need for comparative studies with several methods, since all estimations suffered from substantial uncertainty.

## 1. INTRODUCTION

The methods available for estimation of groundwater recharge directly from precipitation could conceptually be divided into inflow, aquifer response and outflow methods according to where the studies are concentrated (Johansson, 1983). Applying inflow methods, it is presumed that the water movement in the unsaturated zone is vertical and the results are not directly assigned to an area but could be considered as point values. Lysimeter measurements, tracers, soil moisture budget models and one dimensional soil water flow models are examples of inflow methods. The aquifer response in water quantity or chemical composition may be used for estimation of the groundwater recharge,



I. Simmers (ed.), Estimation of Natural Groundwater Recharge, 239–270.

e.g. transformation of groundwater level changes to amounts of water by using the specific yield concept, determination of the recharge necessary to maintain the groundwater levels (inverse modelling) or comparison of chloride deposition and concentration in groundwater. Using outflow methods groundwater recharge and groundwater discharge are put equal. Determination of the groundwater portion in stream discharge, measurements of spring discharge and continuous withdrawal from wells are possible methods. The outflow methods give integrated groundwater recharge for an area. A correct determination of this area is crucial.

When estimating groundwater recharge it is essential to proceed from a good conceptualization of different recharge mechanisms and their importance in the study area. The choice of methods should, besides this conceptualization, be guided by the objectives of the study, available data and possibilities to get supplementary data. Of course economy also must be considered. Since all available methods are afflicted with substantial uncertainty, it is desirable to apply more than one method based on independent input data.

The objectives of the present study were to test and compare different methods for estimation of groundwater recharge and its areal and temporal variations and to obtain information on groundwater recharge in a sandy till area.

## 2. DESCRIPTION OF THE STUDY AREA

The field investigations were performed in an area around Emmaboda in southeastern Sweden (Fig. 1). The studies were concentrated to some springs and their catchment areas and to the groundwater observation tubes belonging to the groundwater networks of the Geological Survey of Sweden (SGU).

The geology of the area was earlier described by Knutsson (1962, 1966 and 1971) and here only some main features will be given. The whole area is situated above the highest shore line and sandy till is the dominating soil type. The till cover is on average 4 m deep and underlain by granitic bedrock. The moraine landscape may be divided into drumlin and hummocky terrain (Knutsson, 1971). The drumlin terrain consists of hills of basal till, formed around rock cores and oriented in the direction of the glacial movement. The till is mostly sandy-silty and very compact. Lenses of sorted sand and gravel within the till can be continuous over rather long distances. The lower parts of the study area are dominated by hummocky terrain. Here rock outcrops are scarce and the orientation of the hills is irregular and not influenced by the bedrock morphology. The till is mostly sandy and rather loose. The lenses of sorted material are often short and unconnected. Small fens are common. The study area is mainly covered by coniferous forest.

In this part of Sweden the precipitation is comparatively low. Average monthly precipitation and air temperatures are shown in Table I. The study area is located within the catchment area of the

River Lyckebyån. The mean annual runoff for the catchment area (785 $km^2$) is approximately 200 mm.

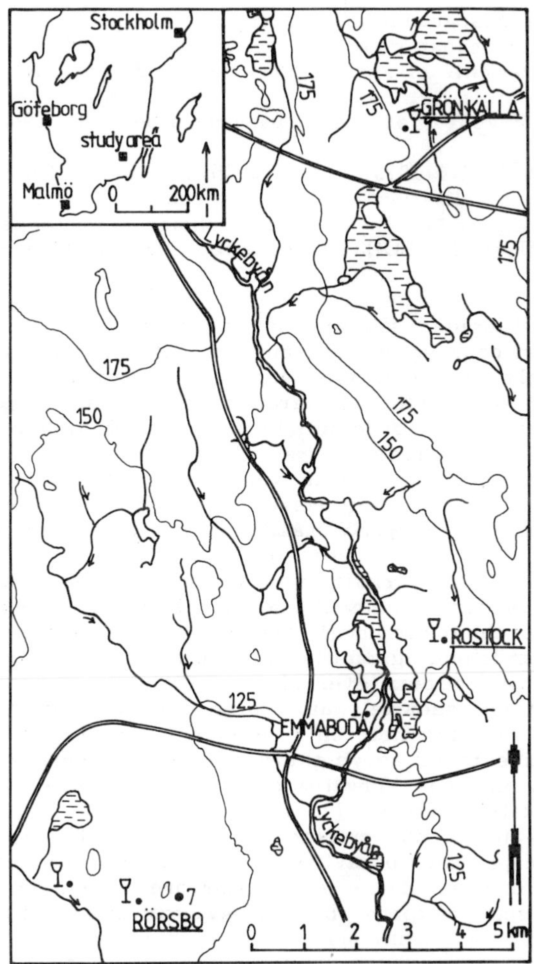


Figure 1. Map of the central part of the study area. (Ỵ. spring; • groundwater level observation tube belonging to the networks of SGU).

TABLE I. Average monthly precipitation (uncorrected) and air temperatures from the stations Rörsbo and Växjö respectively, run by the Swedish Meteorological and Hydrological Institute (SMHI) (see Fig. 5). (Based on data from 1968-1985 and 1951-1980 respectively.)

| | J | F | M | A | M | J | J | A | S | O | N | D | Y |
|---|---|---|---|---|---|---|---|---|---|---|---|---|---|
| P (mm) | 42 | 27 | 35 | 34 | 44 | 47 | 64 | 45 | 54 | 50 | 59 | 48 | 546 |
| T (°C) | -2.8 | -3.0 | 0.4 | 4.6 | 10.3 | 14.7 | 15.8 | 15.2 | 11.3 | 7.1 | 2.6 | -0.8 | 6.4 |

## 3. SOME FUNDAMENTAL GEOHYDROLOGICAL FEATURES OF THE STUDY AREA

Sandy till is a well graded soil with low porosity and permeability. The water retention capacity is good and the capillary rise and transport is considerable. These properties are quite different in the upper part of the soil profile compared to the conditions deeper down, due to climatic and biological factors (Fig. 2 a, b).

The above-mentioned features make the groundwater flow highly transient. It also implies that the unsaturated and saturated zone can hardly be treated separately, but must be dealt with as one system. Groundwater levels have a yearly maximum in spring after snowmelt and a minimum in autumn. The depth to the water table varies between 0.5-4.0 m, with an annual variation of 2.0-2.5 m in recharge areas.

The infiltration capacity in the recharge areas exceeds the rainfall or snowmelt intensity, with few exceptions. Surface runoff does not occur over longer distances but primarily due to topographical conditions, it may appear over shorter distances especially in connection with intense snowmelt and frozen ground. This may cause an areal variation in infiltration in a smaller scale. Temporary saturated horizons due to underlying low permeable layers may also cause lateral flow. Little is known about the significance of macropores in the percolation process.

Most of the water is not transported over long distances in the till and the landscape can be divided into recharge and discharge areas of different orders. The size of these areas may vary during the year. The drainage is often poor and swampy areas occur.

## 4. METHODS APPLIED FOR ESTIMATION OF GROUNDWATER RECHARGE

Six methods, representing all three groups from the conceptual division into inflow, aquifer response and outflow methods, were chosen for tests and comparisons. The methods, some fundamental input data and way of calibration are schematically presented in Table II. A detailed description of the methods and adaptations to the conditions in the study area are given below.

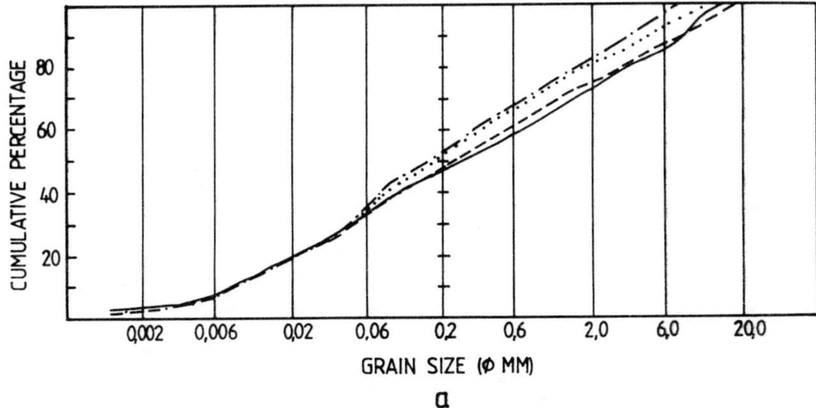


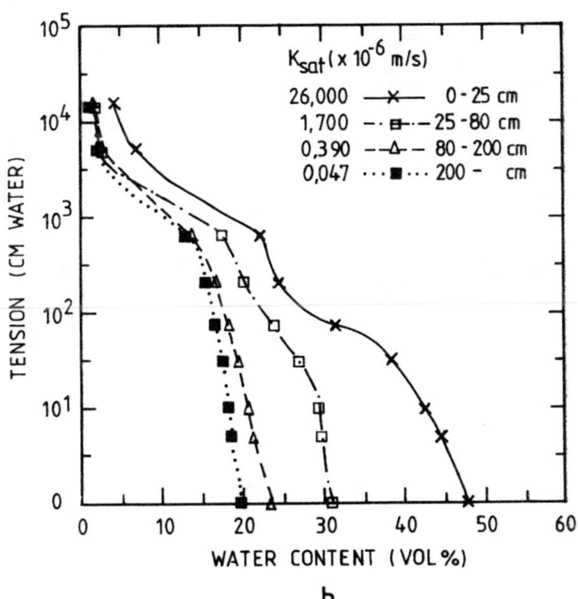


Figure 2 a. Grain size distribution for a soil profile in the Grön källa area. (—— 0-25 cm; —·— 25-80 cm; ---- 80-200 cm; ......... 200- cm).

b. Soil moisture retention properties and saturated hydraulic conductivities.

TABLE II. Tested and compared methods for estimation of groundwater recharge.

| METHOD/MODEL | CATEGORY | NEED FOR INPUT DATA | | CALIBRATION |
|---|---|---|---|---|
| | | Climatical | Soil moisture and groundwater | |
| One dimensional soil water flow model (SOIL) | inflow | prec., temp., wind speed rel. humidity | soil water retention properties, hydraulic cond., groundwater outflow | measured groundwater levels |
| Soil moisture budget model | inflow | prec., temp., wind speed, rel. humidity | size of soil moisture reservoir, soil moisture-recharge relation | soil water flow model |
| Groundwater level fluctuations | aquifer response | | groundwater levels, specific yield | |
| Chloride concentration | aquifer response | prec., wet and dry deposition of chloride | concentration of chloride in groundwater | |
| Spring discharge | outflow | | spring discharge, size of catchment area | |
| Catchment area model (PULSE) | outflow | prec., temp., potential evapotranspiration | size of soil moisture reservoir, soil moisture-recharge relation, outflow from the groundw. reservoir | spring discharge |

### 4.1 One dimensional soil water flow model

A one dimensional soil water model, with appropriate submodels for the boundary conditions, gives possibility to estimate groundwater recharge. This type of model was presumed to be advantageous for studies of recharge processes and dynamics, but also to give possibilities to study areal variability in recharge.

The soil water and heat model used is called SOIL. The model is described in detail by Jansson and Halldin (1980) and Jansson and Thoms-Hjärpe (1986). Here only the main features and the adaptations to the studied problem will be presented.

The soil water flow equation, expressed in the units employed in this study, reads

$$\delta\theta/\delta t = -\delta/\delta z\left[k(\delta S/\delta z + 1\right] - s \qquad (1)$$

where $\theta$ = volumetric water content ($cm^3/cm^3$); t = time (min); k = unsaturated conductivity (cm/min); S = soil moisture tension (cm water), positive in unsaturated soil; z = depth (cm), positive downward and s = sink term (l/min). The equation is solved by an explicit forward finite difference method. The water content - tension relationship is treated by a modified form of the analytical function given by Brooks and Corey (1964) and possible hysteresis effects are ignored. The unsaturated hydraulic conductivity is calculated from the water retention curve and a measured value of the saturated hydraulic conductivity as proposed by Mualem (1976). For the upper boundary condition submodels are available for precipitation, interception, snow dynamics and evapotranspiration. The snowmelt and refreezing are calculated by a heat budget approach. The potential evapotranspiration is calculated by the Penman-Monteith equation,on a daily basis, from climatical and vegetation data. At high soil moisture tensions the actual evapotranspiration is reduced. In this application the lower boundary is a no-flow boundary. A lateral groundwater outflow is introduced in the form of a sink. The outflow ($q_{gr}$) is governed by the equation

$$q_{gr} = q_1 \max(0,(z_1 - z_{sat})/z_1) + q_2 \max(0,(z_2 - z_{sat})/z_2) \qquad (2)$$

where $q_1$, $q_2$ = maximum peak and base flow respectively; $z_{sat}$ = level where the soil moisture tension is nil and $z_1$, $z_2$ = level where peak and base flow respectively ceases. These parameters are obtained from calibrations.

The heat flow is calculated as the sum of conduction and convection, using the general heat flow equation. Soil frost treatment is based on a function for freezing point depression and an analogy between freezing-thawing and drying-wetting capillary processes.

Groundwater recharge was defined as the accumulated flow across the compartment boundary immediately above the groundwater table. The model was applied in two different areas. A long term study (16 years) was performed in the Rörsbo area (at SGU:s groundwater level observation tube No. 7), where the groundwater was shallow, 0-3 m below the ground surface (Johansson, 1987). The other application, which will be presented here, was made in the catchment area of Grön källa (Fig. 3), where the water table was deeper (2-6 m) and where also a direct comparison with several other methods was possible.

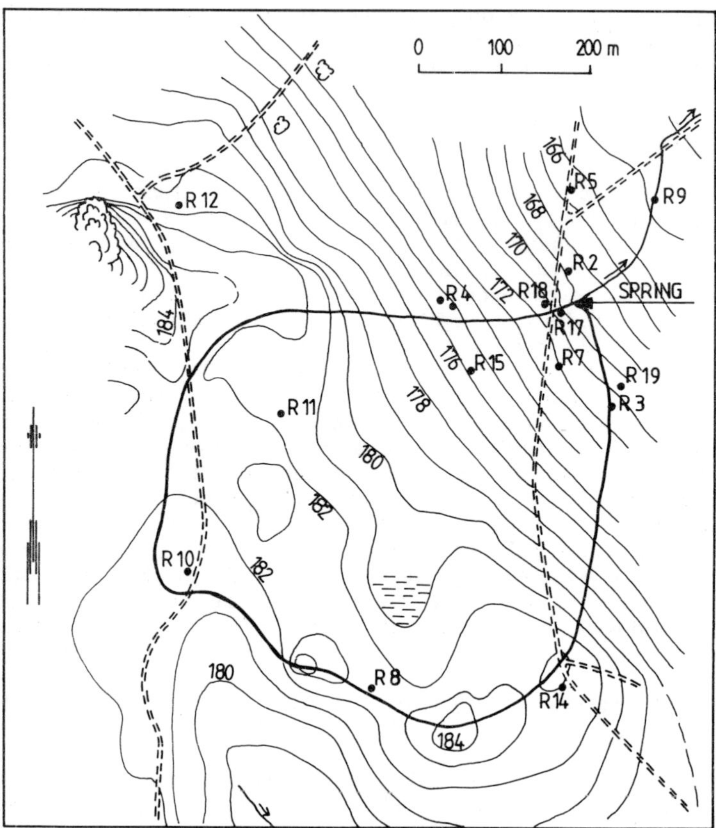


Figure 3. Topographical map of the catchment area of Grön källa.

Precipitation data were obtained from SMHI 's meteorological stations in Rörsbo, Lessebo and Gullaskruv (Fig. 5). The measured precipitation was increased by 7 percent for rain and 21 percent for snow, compensating for evaporation, wetting and wind losses. The other climatical input data needed were taken from the meteorological station in Växjö (50 km NW of the study area).

Soil moisture retention properties and saturated hydraulic conductivities, at four different depths down to 2.50 m below ground, were obtained from laboratory analyses of undisturbed soil samples (Fig. 2b). The model was tested against groundwater level fluctuations in observation tube No. 10, located close to the water divide.

Depth to groundwater table and groundwater level fluctuations are almost identical in the whole area except for a very small part close to the spring.

The vegetation at observation tube No. 10 is dominated by pine forest.

## 4.2. Soil moisture budget model

The most commonly used model of the unsaturated zone, for estimation of groundwater recharge, is a single reservoir model (see e.g. Penman, 1949; Grindley, 1967 and 1969; Eriksson and Johansson, 1978; Howard and Lloyd, 1979 and Rushton and Ward, 1979). Conceptually the size of the reservoir is meant to represent the amount of water contained in the root zone between field capacity and wilting point. At every time step, precipitation minus evapotranspiration is added to the reservoir. When the reservoir is full, the surplus will go to groundwater recharge. The evapotranspiration takes place at potential rate until a certain soil moisture deficit is reached. The actual evapotranspiration then decreases and reaches zero when the reservoir is empty.

Some earlier comparative studies suggested an underestimation of the groundwater recharge calculated with this model and an inability to represent the recharge dynamics as reflected in groundwater level fluctuations (Smith et al, 1970; Fox and Rushton, 1976 and Kitching et al, 1977). Rushton and Ward (1979) tested different ideas of by-pass flow to the groundwater zone when the reservoir was not full.

Results from the model are of very limited value without calibration and validation, since besides the substantial uncertainty in input data (precipitation and potential evapotranspiration) the model parameters do not have a direct physical representation which can be measured in the field.

Considering the soil water retention and flow properties in the study area, and the groundwater level fluctuating from immediately below ground surface and almost down to the bedrock, it seemed not possible to determine a single size reservoir with the physical representation outlined above. However, it was considered to be interesting to test and compare this commonly used method against the simulations by the one dimensional soil water flow model and against observed groundwater level fluctuations.

The soil moisture budget model was applied in the Rörsbo area, where simulations with the one dimensional soil water flow model had been performed earlier and tested against groundwater level fluctuations (Johansson, 1987). The SOIL simulations gave good fits between simulated and observed groundwater levels for an average groundwater recharge between 134 and 197 mm/year for the studied period. The soil moisture budget model was calibrated and compared with the SOIL simulation giving the highest recharge for the period 1970-1984. The same climatical input data were used.

The structure of the soil moisture budget model programme was taken from Nilsson (1983). The relation between potential and actual evapotranspiration was given by

$$AE = PE(S/SFC)^{C} \tag{3}$$

where AE = actual evapotranspiration; PE = potential evapotranspiration; S = actual soil moisture storage; SFC = maximum soil moisture storage and C = coefficient obtained from calibration. The snow routine was based on a degree-day approach. The concepts of no groundwater recharge before the soil moisture reservoir was full and allowing a fraction of rainfall or snowmelt for recharge also when a deficit existed, were both tested. The fraction forming recharge depended on the deficit (Fig. 4). The latter concept is often adopted in runoff modelling (see e.g. Bergström, 1976).

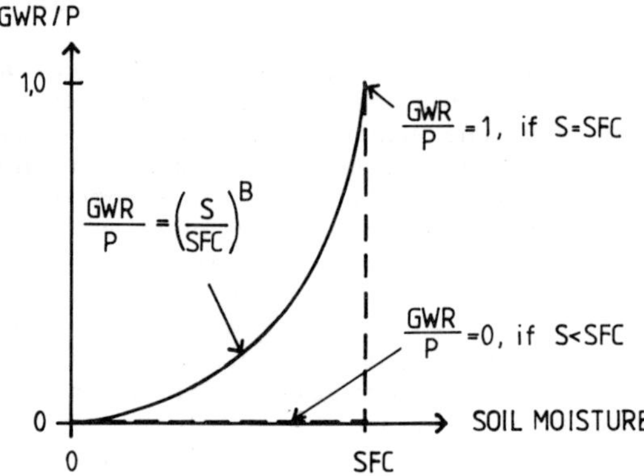


Figure 4. Different concepts for the relation between groundwater recharge and water content in the soil moisture reservoir (GWR = groundwater recharge, P = rainfall and snowmelt); S = actual water content in the soil moisture reservoir; SFC = maximum water content in the soil moisture reservoir; B = coefficient obtained from calibration).

## 4.3. Groundwater level fluctuation method

The groundwater table and its fluctuations are functions of groundwater recharge, water yielding properties, transmissivity and geometry of the aquifer. The groundwater recharge (GWR), during the time period $\Delta t$, can be written

$$GWR = \Delta h \times SY \times A + Q \times \Delta t \qquad (4)$$

where $\Delta h$ = change in groundwater level; SY = specific yield; A = area and Q = net groundwater flow. The groundwater flow can be calculated by analytical or numerical methods. These solutions require information on aquifer parameters and boundary conditions. In this study a technique is adapted, which uses recessions derived from observations of decreasing groundwater levels when no recharge is supposed to occur. This technique is frequently used (see e.g. Einsele, 1975; Lemmelä, 1976; Karanth and Srinivasa Prasad, 1979; Olsson, 1980 and Soveri, 1985) and was described in detail and adapted for a till area by Johansson (1987).

The shape of the recession curve depends on water yielding properties, transmissivity and geometry. Vertical heterogeneities in the soil profile may give this curve a complicated form. If the groundwater table is shallow, the recessions may be influenced by evapotranspiration and frost penetration. To get the unaffected recession curve, caused only by groundwater flow, periods must be found when these processes are insignificant. The distance between the actual groundwater level and the groundwater level calculated from the recession curve, for every time interval, can then be summed up and multiplied by the specific yield to get the recharge. The specific yield will only be a constant when the soil profile is homogeneous and the water table is deep enough to give an equilibrium water content in the uppermost part of the profile which is the same as at free drainage.

The method is attractive since groundwater level observations often are available. The method could also give information of temporal and areal variations. Only observation tubes at or close to the water divide should be used to minimize the influence of lateral groundwater flow.

The results from an application in the Rörsbo area were earlier described (Johansson, 1987). The results were discouraging, which was thought mainly to depend on the shallow groundwater table and on inhomogeneities in the soil profile within the range of groundwater level fluctuations. Here the method is applied in the Grön källa area, where the groundwater table is deeper and the soil water yielding properties within the range of groundwater level fluctuations are more homogeneous, which should be favourable.

The groundwater level observations in Grön källa have been carried out since autumn 1983. The unaffected recession for the 14 days measuring interval was determined from observations during the stable winter 1984-1985. Depending on the short period of measurements, it was not possible to extract different recession rates for different depths. The specific yield was defined as the water released between tensions 0 and 100 cm of water, and was extracted from the pF-analyses of the layer extending from 2 m below ground and downwards. The groundwater level varied between 2.32 and 5.67 m below ground.

### 4.4. Chloride concentration method

In many areas, atmospheric deposition is the only significant source of chloride in groundwater. The adsorption of chloride in soil is negligible and the vegetation uptake is small. If the chloride ion is considered to be conservative, it should be possible to determine the groundwater recharge by comparison of wet and dry deposition and the concentration in groundwater (Eriksson and Khunakasem, 1969; Eriksson, 1976 and Eriksson, 1985, pp. 153-166). However, some investigations of small forested watersheds with low dry deposition (Andersson-Calles and Eriksson, 1979 and Rosén, 1982) gave outputs lower than the input from wet deposition. In spite of considerable uncertainties in measurements of the wet deposition, the question arose if not a small amount of chloride was stored in the vegetation (Rosén, 1984).

If the assumption of chloride as a conservative ion is accepted, the groundwater recharge is given by

$$GWR = D/C \tag{5}$$

where GWR = groundwater recharge (mm/year); D = wet and dry deposition ($mg/m^2/year$) and C = concentration in groundwater (mg/l). A proper areal mean value of groundwater recharge, from a number of point values of chloride concentration in groundwater, is obtained from

$$GWR = D/N \sum_{k=1}^{N} (1/C_k) \tag{6}$$

where N = number of measurements and $C_k$ = $Cl^-$-concentration in sample k (Eriksson, 1976).

The method is convenient, fast and cheap. The drawback is the uncertainty in the determination of the wet and especially the dry deposition. The method is most attractive in areas with high evapotranspiration, where the infiltrating water gets highly concentrated. An error in input data gives here a small absolute error in estimated recharge (see e.g. Jacks and Sharma, 1982). The problem with errors in input is also less accentuated in inland areas, where the dry deposition is low.

It was presumed that in the present study, the obtained absolute recharge values should be marred by unacceptable uncertainty but that interesting information of relative areal variability could be gathered.

Since the study area is located above the highest shore line and the content of chloride is neglible in the granitic rock, atmospheric deposition could be regarded as the only source of chloride in those parts not directly affected by human activities. No significant local sources of atmospheric chloride were present. The deposition could be assumed to be long distance transported chloride originating mainly from sea spray.

The deposition data were taken from stations run by the Department of Meteorology (MISU) at University of Stockholm (Fig. 5).

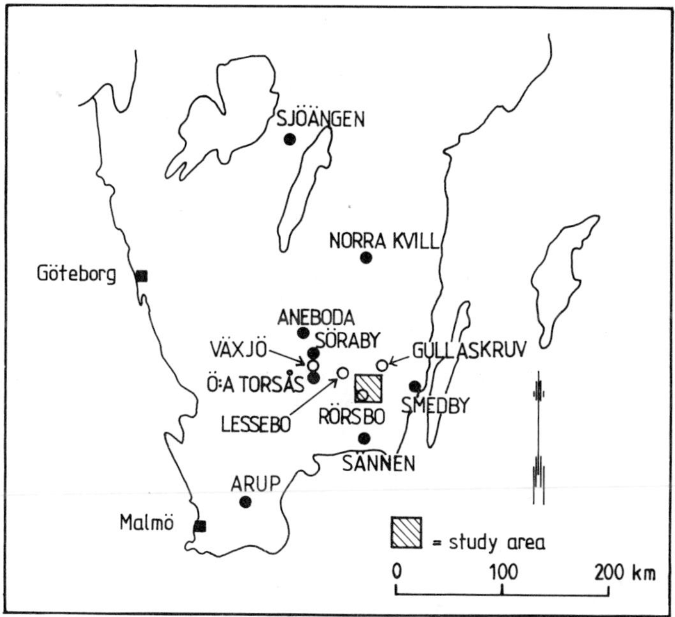


Figure 5. Location of SMHI's climatic stations (o) and MISU's observation stations for wet deposition and aerosol concentration (●) used in this study.

Different stations were used for different time periods due to changes in the observation networks and in local conditions at the measuring sites. For the wet deposition, the stations in Östra Torsås/Söraby and Smedby were used for the period before 1981. July 1981 - June 1982 measurements were made in Rörsbo within a project run by MISU. From 1983-1985, measurements were available for the stations Sännen, Aneboda och Norra Kvill. Different collecting devices were used but were constantly open and the water was monthly accumulated in bottles connected to the collectors. Parallel sampling during shorter periods by constantly open collectors and collectors only open when precipitation occured gave approximately 10 percent higher values for the former (Granat, personal communication). SMHI's regular networks were used to get the amount of precipitation (stn. Rörsbo, Lessebo and Gullaskruv).

Data of aerosol concentrations of chloride were available only from a few places. The aerosol concentration was estimated in two different ways. In the regular networks, run by MISU, the aerosol content was collected on a filter. Due to reactions on the filter, chloride was lost. From other experiments $Na^+$-concentration was known to be equivalent to the $Cl^-$-concentration. The $Cl^-$-concentration was estimated from the $Na^+$- concentrations 1983-1985 in Arup and Sjöängen, but also data from other stations were used to get an idea of the gradients. A special study of the aerosol concentration was made in Sjöängen 1977 by using a cascade impactor (Lannefors et al. 1983). Here particles were differentiated by size, which meant that the problem with reactions on filters was eliminated, therefore the $Cl^-$-concentrations could be directly measured.

Springs, located in coniferous forest and not directly affected by human activities, were chosen to get the $Cl^-$-concentration in groundwater. Three different studies were performed. A long term study, 1970-1984, was based on SGU's measurements of the chemical composition of the discharge of a spring in Rörsbo (2-5 analyses/year). Comparisons on long term averages could be made with the simulations by SOIL. The second study dealt with the Grön källa area for the period 1981-1985. $Cl^-$-concentration measurements were carried out 1980 (Jacks and Knutsson, 1981) and 1983-1986 (35 analyses). Finally, to study areal variability of groundwater recharge, discharge from 9 springs located within a radius of 15 km were sampled 1980, 1985 and 1986 (3-4 analyses/spring). The chloride deposition was supposed to be the same in all the catchment areas, which was reasonable since the topographical differences were small and all the springs were located in forested areas.

Other chemical constituents were analysed to determine indications of contamination.

### 4.5. Spring discharge method

Spring discharge may be used as a measure of the groundwater recharge provided that the changes in groundwater storage are known. For longer periods, the difference in storage may be neglected. An integrated value of the groundwater recharge in the catchment area will be obtained. The determination of the catchment area will be crucial and the spring outlet has to be well-defined.

The springs in moraine areas are usually relatively small, having a mean annual discharge of less than one litre per second. It was realized that it would be a difficult task to find springs with well-defined catchment areas within the study area, but if possible interesting information on variations in groundwater recharge in different types of terrain could be obtained. If it was impossible to determine the catchment areas with an acceptable accuracy for quantification of recharge, the studies might still be valuable for analyses of groundwater dynamics, runoff forming processes and water pathways.

From a total of 25 springs surveyed in the field, five springs located in coniferous forest and with comparatively well-defined catchment areas were chosen. The springs were chosen to represent both hummocky and drumlin terrain. Special attention was paid to find well-defined outlets and discharge areas which were more or less just a point. V-notch weirs were build and the catchment areas were determined based on topography.

Continuous recording measurements were performed in two of the springs as well as manual measurements every second day for some period of time. Due to comparatively slow responses it appeared satisfactory with weekly measurements of discharge. For example the difference in runoff volume, between measurements every second day and every eighth day, was for the springs of Grön källa and Rostock less than one percent during March-May 1985.

A preliminary evaluation after two years of measurements indicated that the determination of the catchment areas, based on topography only, was not satisfactory. The studies were then concentrated to Grön källa and Rostock. Grön källa is a comparatively large spring in drumlin terrain while the Rostock spring is small and located in hummocky terrain. Approximately 20 groundwater level observation tubes were installed to obtain better determinations of the catchment areas (Fig. 3 and 6). The catchment areas for Grön Källa and the Rostock spring were determined to 0.194 and 0.0164 $km^2$ respectively. The catchment area of Grön källa is completely covered with coniferous forest while in Rostock approximately 15 percent is grazing land.

4.6. Catchment area model

Conceptual runoff models often contain the kind of unsaturated zone model described under "Soil moisture budget model". The idea was that calibration of such a runoff model against the spring discharge should mean a way of testing the total groundwater recharge quantities obtained as an output from the unsaturated zone submodel. It was understood that the test of the obtained groundwater recharge was not very strict, especially for the dynamic behavior, since the simulated discharge was heavily dependent on the choice of the outflow parameters for the groundwater reservoir. An areally integrated value of the recharge was obtained.

The model used was PULSE (Bergström and Sandberg, 1983) (Fig. 7). The snow routine is based on the degree-day approach and the soil moisture reservoir is of the type where recharge also takes place when the reservoir is not full (Fig. 4). Evapotranspiration is supposed to take place at potential rate until a certain deficit in the soil water reservoir is reached. It is then reduced linearily to zero when the soil moisture reservoir is emptied. The outflow from the groundwater reservoir is controlled by four recession coefficients activated at different levels. Capillary upflow from the response

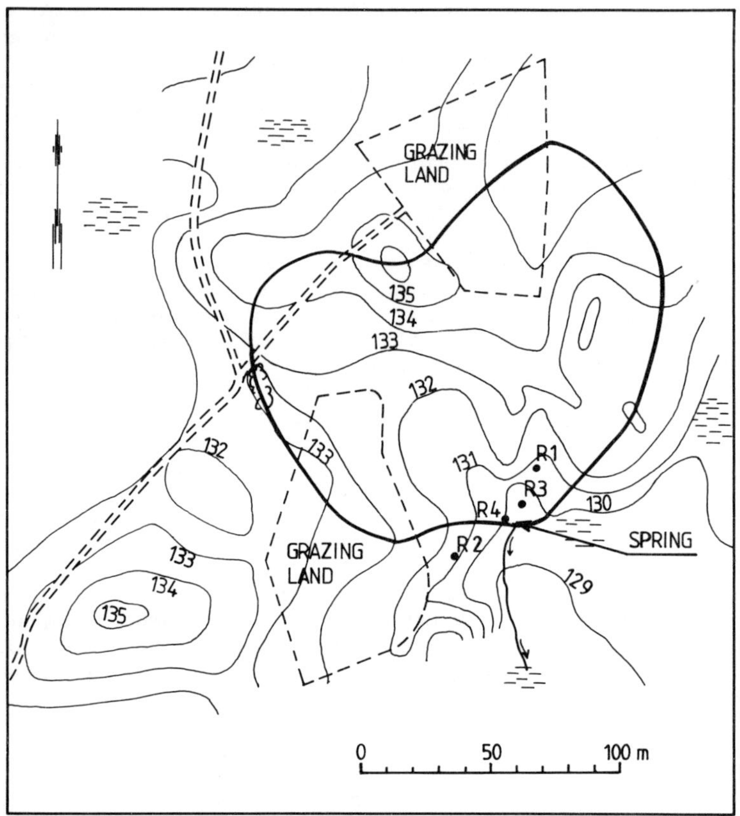


Figure 6. Topographical map of the catchment area of the Rostock spring.

function to the soil moisture reservoir is simulated as a linear function of the deficit in the reservoir. Net groundwater recharge is calculated as gross groundwater recharge minus capillary upflow.

The model was applied for the catchment areas of the springs Grön källa and Rostock. The simulations were performed on a daily basis. Precipitation data were taken from Rörsbo, Lessebo and Gullaskruv for Grön källa but from Rörsbo only for Rostock. Average values of monthly potential evapotranspiration were taken from Wallén (1966) and were based on Penman's equation.

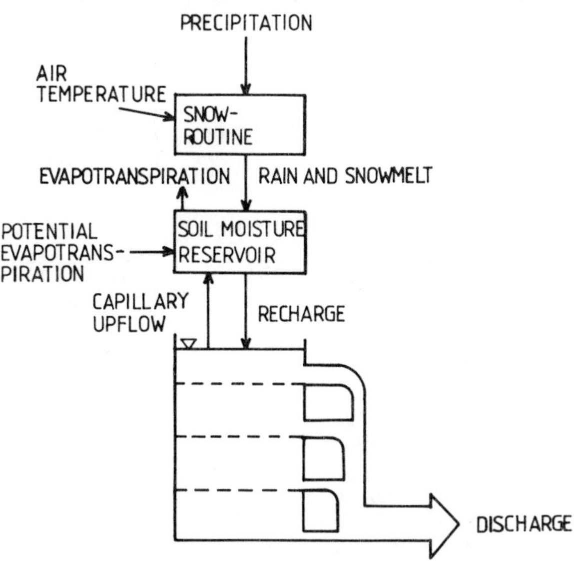


Figure 7. Schematic structure of the PULSE model (after Bergström and Sandberg, 1983).

## 5. RESULTS

### 5.1. One dimensional soil water flow model

Since observed groundwater levels were available only for Oct. 1983-Nov. 1985, the whole period was used for calibration of the groundwater outflow parameters and tests of the evapotranspiration parameters. The groundwater outflow was calibrated separately for the winter periods, when the influence of evapotranspiration could be supposed to be negligible. An increase of the air-entry values, deduced from the pF-analyses, from 2 to 10 cm for the two bottom layers was necessary to get responses in the simulated groundwater level in acceptable agreement with the observations. The response of the simulated level was still somewhat slow and seemed to give greatly accentuated peaks during spring-time. An increase of the saturated hydraulic conductivity with a factor 5 and 10 for the two bottom layers respectively was tested and gave faster responses and less accentuated peaks. Earlier simulations (Johansson, 1987) had revealed that simulated groundwater levels were rather insensitive to displacement between evapotranspiration and groundwater outflow, and therefore sensitivity tests were made. Three simulations will be discussed (A-C), where the last one was an example of a simulation with increased saturated hydraulic conductivity values for the two

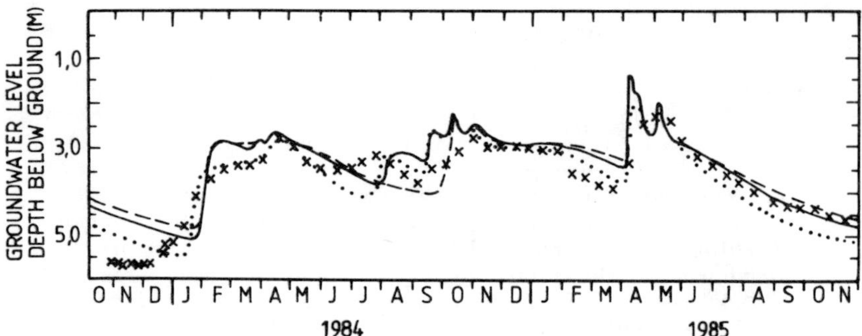


Figure 8. Simulated (SOIL) and observed groundwater levels for the period, Oct. 1983 - Nov. 1985 at groundwater level observation tube No. 10 at Grön källa (---sim A; —— sim B; ....... sim C; x observations).

bottom layers (Fig. 8). The correlation coefficients were 0.88, 0.89 and 0.92. For comparison with other methods the period 1981-1985 was simulated. Simulations A, B and C gave mean annual recharges of 162, 179 and 213 mm respectively. The average corrected precipitation was 679 mm/year. Yearly sums and seasonal variations of recharge were calculated (Fig. 9). Contrary to the simulations in the Rörsbo area (Johansson, 1987), where the groundwater table was more shallow, no upward flow from the groundwater zone occurred in any of the simulations. This meant that it was easier to separate between effects of evapotranspiration and groundwater runoff on the simulated groundwater level.

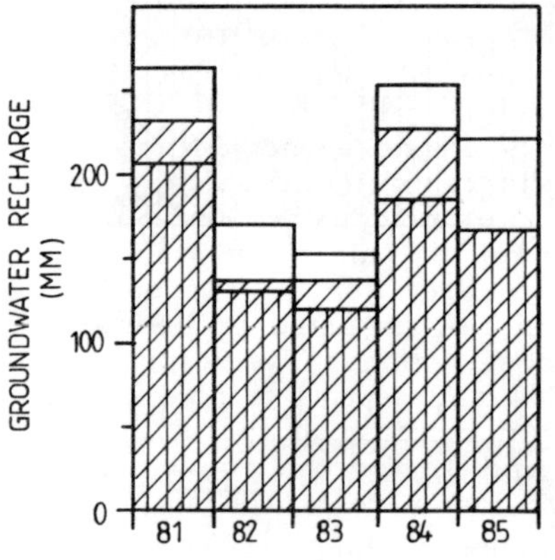


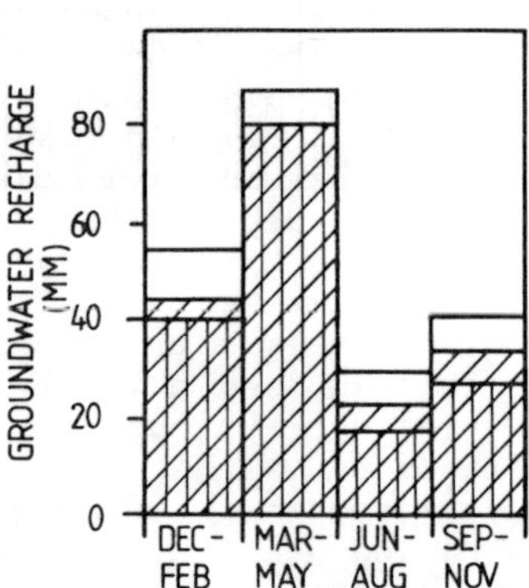


Figure 9. Simulated (SOIL) annual groundwater recharge and seasonal variations at Grön Källa during 1981-1985 (▥ sim A; ▨ sim B; ▭ sim C).

5.2. Soil moisture budget method

The concepts with no groundwater recharge before the soil moisture reservoir was full and allowing a fraction of rainfall or snowmelt for recharge also when a deficit existed, were both tested. Several tests, with different sizes of the reservoir and the coefficient determining the fractional recharge, were performed. Results from one simulation based on each concept will be discussed, both giving approximately the same average annual recharge, for 1970-1984, as the SOIL simulation giving the highest recharge. Using the "no recharge before full reservoir" concept, the reservoir had to be about 150 mm, while 250 mm gave the best result together with a value of 5.0 for the coefficient B for the other concept. The average annual recharge was 194 and 195 mm respectively, compared to 192 mm for the SOIL simulation.

Studying annual and average monthly recharge, the simulation with recharge also when a reservoir deficit existed showed a better agreement with the SOIL simulation (Fig. 10 a,b). Still substantial deviations were present for single years. The largest deviations, 25-30 percent, appeared for the years 1976-1977. No systematics in the deviations coupled for example to dry and wet years could be observed.

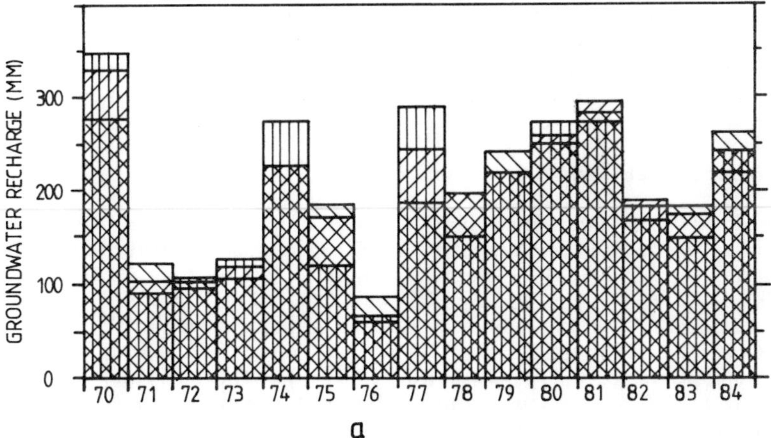


Figure 10a. Annual groundwater recharge, 1970-1984, at Rörsbo (SGU's observation tube No. 7) simulated by two different variants of the soil moisture budget method and by the one dimensional soil water flow model (SOIL) (▥ soil moisture budget method with no recharge when reservoir deficits existed; ▨ soil moisture budget method with a fractional recharge also when deficits existed; ▧ SOIL).

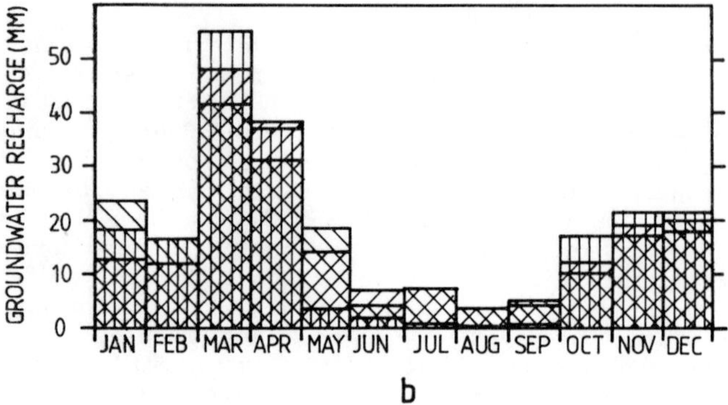


Figure 10b. Monthly averages of groundwater recharge for the period 1970-1984 (symbols as above).

Looking closer to the dynamics, it was obvious that groundwater recharge during the summer was underestimated when no reservoir deficit was accepted. No recharge was simulated in situations when rising groundwater levels were observed. Recharge in autumn was also delayed compared to rising groundwater levels and sometimes completely missed before winter. Both summer recharge and autumn recovery was better reproduced in the simulations where recharge was allowed also when the soil moisture reservoir was not full.

5.3. Groundwater level fluctuation method

The groundwater level recession, caused only by groundwater flow when no groundwater recharge occured, was from observations during the stable winter 1984-1985 determined to 17 cm/14 days. The specific yield, taken from the pF-analyses, was 0.03. The total net groundwater recharge, for the whole period of groundwater level observations Oct. 30, 1983 - Nov. 14, 1985, was calculated to 275 mm. The SOIL simulations gave 355, 395 and 459 mm for the same period.
The distribution in time was quite different compared to the results obtained from the SOIL simulations (Fig. 11).

5.4. Chloride concentration method

Average values of $Cl^-$-concentration in precipitation for the two stations Östra Torsås/Söraby and Smedby were calculated by Malmer and Wallén (1980), for the periods 1955-1962 and 1954-1975 respectively, to 1.0 and 0.8 mg/l. The measurements in Rörsbo Jul. 1981 - Jun. 1982 gave an average of 0.6 mg/l (Granat, personal communication). The average for the three stations Sännen, Aneboda and Norra Kvill, where data are available 1983-1985, was 0.8 mg/l (unpublished data from the MISU/PMK networks).

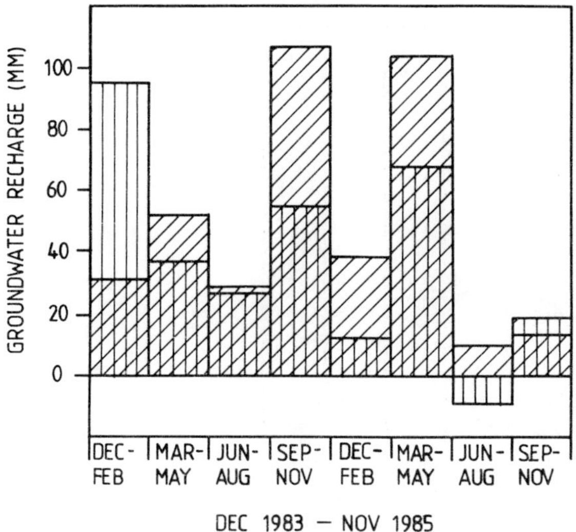


Figure 11. Groundwater recharge, Dec. 1983 - Nov. 1985, at groundwater level observation tube No. 10 at Grön Källa calculated by the groundwater level fluctuation method (▥) and comparison with SOIL simulation B (▨).

The cascade impactor measurements in Sjöängen during 1977 (Lannefors et al., 1983) gave a $Cl^-$-concentration in the air of 190 ng/m$^3$. If the $Na^+$-concentrations from MISU's regular measurements in Arup and Sjöängen 1983-1985 (unpublished) were transferred to equivalent amounts of $Cl^-$, the aerosol concentration in the study area could be estimated to 350-500 ng/m$^3$. Lannefors et al. (1983) estimated the deposition velocity of $Cl^-$ to 1.8 cm/s at the Sjöängen site with aerosol distribution present there. This velocity gives an annual dry deposition of approximately 110 and 200-280 mg/m$^2$ respectively in the study area if the above-mentioned aerosol concentrations are used.

The uncertainties involved in estimations of deposition are substantial. Uncertainty factors for the wet and dry deposition values could roughly be estimated to 1.3 and 3.0 respectively (Granat, personal communication).

For the further calculations, the $Cl^-$-concentration in precipitation was put to 0.8 mg/l and three values for the dry deposition used (110, 200 and 280 mg/m$^2$year).

For Rörsbo, 1970-1984, the corrected precipitation used in the one dimensional soil water flow simulations was 591 mm/year, resulting in a wet deposition of approximately 470 mg/m$^2$/year. The different estimations of the dry deposition gave a total deposition of 580, 670 and 750 mg/m$^2$year. The average $Cl^-$-concentration in the Rörsbo spring 1970-1984 (50 analyses) was 5.7 mg/l (st.dev. 1.3). The

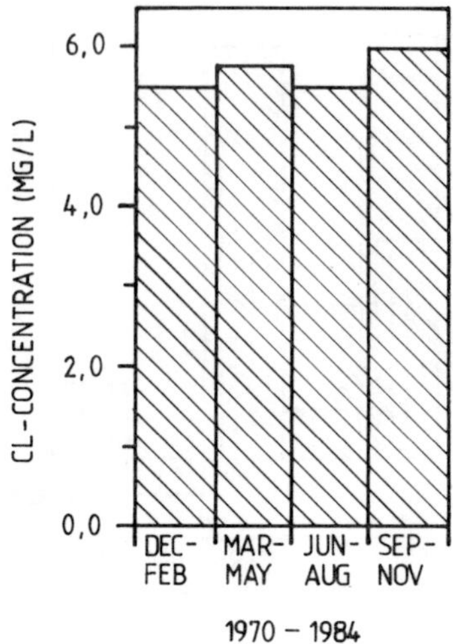


Figure 12. Seasonal variations of the chloride concentration in the spring discharge of the Rörsbo spring (1970-1984).

seasonal variations were small (Fig. 12). The mean annual groundwater recharge corresponding to the different deposition values were 102, 118 and 132 mm. The three one dimensional soil water flow simulations gave groundwater recharges ranging from 134 to 197 mm corresponding to depositions between 760 and 1 110 mg/m$^2$/year. If the estimated wet deposition was supposed to be correct the dry deposition then should correspond to 58 to 131 percent of the wet deposition.

For the Grön källa area the period 1981-1985 was considered. The corrected annual mean precipitation was 679 mm. The total deposition was, with the assumptions from above, 640, 730 and 810 mg/m$^2$/year. The average concentration in the spring discharge was 7.0 mg/l (st. dev. 0.4), giving a recharge of 91, 104 and 116 mm respectively for the different dry deposition assumptions.

The average $Cl^-$-concentrations in the 9 springs measured 1980, 1985 and 1986 (3-4 analyses) varied between 4.3 and 8.5 mg/l. The lowest value originated from a spring, where the catchment area probably was dominated by bedrock outcrop and very thin soil layers and was therefore not representative for the study area. The spring with the highest concentration had a catchment area which was difficult to determine but there was a risk that part of it could have been contaminated by grazing cattle. If these two extreme springs were excluded, the average concentration varied between 5.0 and 7.2 mg/l with a mean value of 6.2 mg/l. If the highest deposition, estimated from the measurements, was used the groundwater recharge in the catchment areas of these springs should vary between 113 and 162 mm.

5.5. Spring discharge method

The average annual spring discharges for Grön källa and Rostock, for the period Oct. 1981 - Sep. 1985, were 178 and 123 mm respectively. Differences in precipitation may partly explain the lower discharge in Rostock. Rostock is situated close to the precipitation station in Rörsbo while the distance from Grön källa to the stations in Rörsbo, Lessebo and Gullaskruv are approximately the same. The measured annual precipitation in Rörsbo was 39 and 68 mm lower than in Lessebo and Gullaskruv respectively.

The evenly distributed seasonal discharge in Grön källa, compared to Rostock, was an effect of the larger groundwater reservoir (Fig. 13).

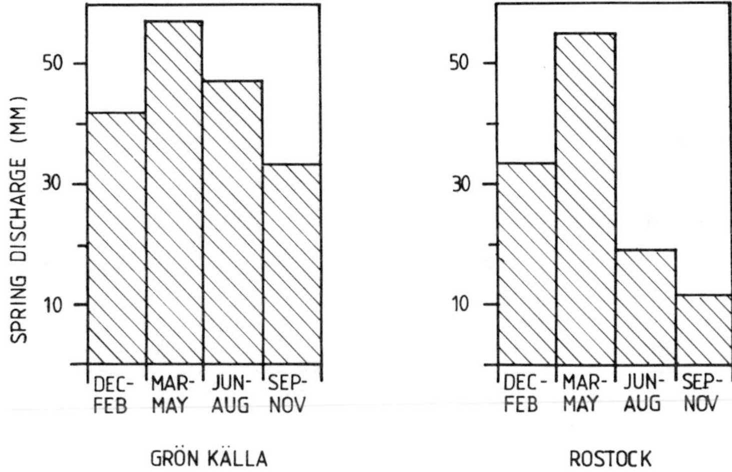


Figure 13. Sesonal distribution of the spring discharge of Grön Källa and the Rostock spring (Dec. 1981 - Nov. 1985).

5.6. Catchment area model

Since only four years of spring discharge measurements were available, the whole period was used for calibration. A relatively good agreement was obtained between simulated and observed discharge for the two springs (Fig. 14 a,b). The correlation coefficients were 0.84 and 0.86 for Grön källa and Rostock respectively and the accumulated differences between simulated and observed discharge were -27 and +38 mm for the whole period. A generel feature for the Grön källa simulation was too fast recessions during winter compared to the observations. The difference between winter and summer recessions depends on the effects of evapotranspiration and the interaction

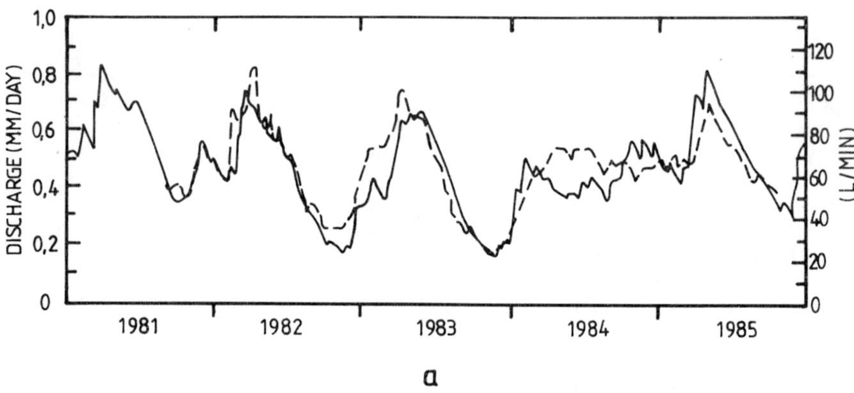


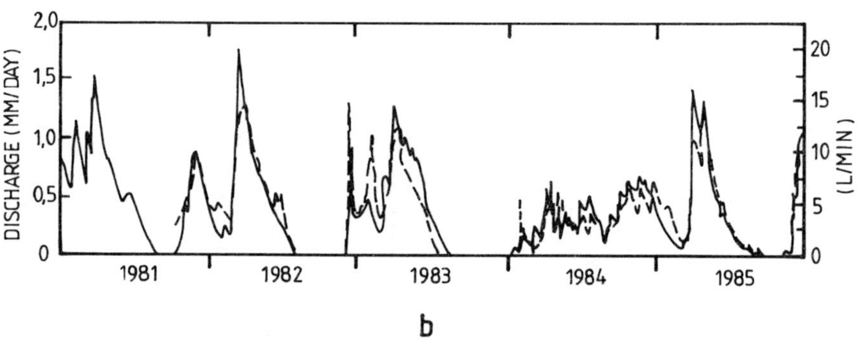


Figure 14. Simulated (PULSE) and observed spring discharge at Grön Källa and Rostock, respectively ( ——— simulated; ---- observed).

between the unsaturated zoneand the groundwater zone. The lack of agreement between simulated and observed discharge during the winter periods indicates that this interaction is not described in an acceptable way.

The average annual groundwater recharge, 1981-1985, was calculated to 178 mm for Grön källa and 134 mm for Rostock. The difference in average groundwater recharge is partly explained by lower precipitation in the Rostock simulation, which was based on data from the station in Rörsbo while an average of the stations in Rörsbo, Lessebo and Gullaskruv was used in the Grön källa simulation.

The average monthly recharge in Grön källa estimated by PULSE was compared to the SOIL simulation giving approximately the same total recharge for the period 1981-1985 (Fig. 15). The total amount of

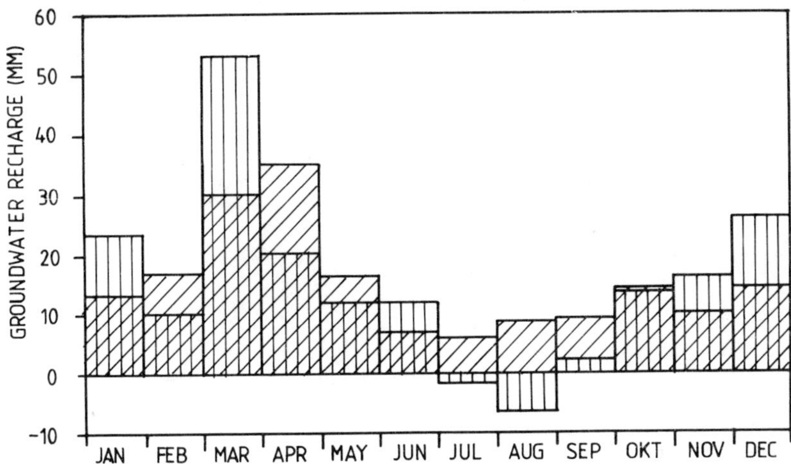


Figure 15. Average monthly groundwater recharge in Grön Källa estimated by PULSE (▥) and SOIL (▨) at Grön Källa, 1981-1985.

recharge in March and April was approximately the same but the distribution between the months was quite different. The snowmelt usually appeared in the end of March and the beginning of April and therefore the timing of the snowmelt was crucial. This partly explained the difference but also the tendency of SOIL to give delayed recharge sometimes, compared to recharge as reflected in raising groundwater levels, was of importance. PULSE gave almost no recharge during summer-time. Average recharge was negative for July and August.

An interesting test of the specific yield value used in the groundwater level fluctuation method was obtained by multiplying the water content in PULSE'S groundwater reservoir with the specific yield value. Corrections for the absolute level were also necessary. A comparison to the observed groundwater level indicated that the measured specific yield value could be used to simulate the ground water level fluctuations, and that a single constant value was enough in this case (Fig. 16). The deviations between observed and calculated groundwater level reflected the inability of the model to reproduce the discharge correctly (see Fig. 14a).

## 6. DISCUSSION AND CONCLUSIONS

The study clearly demonstrated the difficulties involved in groundwater recharge estimation. All the estimations suffered from uncertainties in input data as precipitation, potential evapotranspiration,

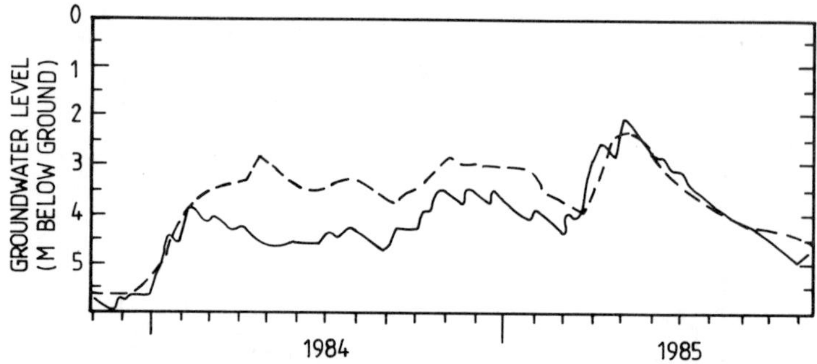


Figure 16. Comparison between simulated groundwater levels by PULSE and observations (—— simulated; ---- observed).

soil water retention and conductivity properties, wet and dry deposition and catchment area determination. The need for comparative studies with different methods based on independent input data, was obvious.

It is impossible to give general recommendations to use certain methods. Before doing that, the objective of the study must be specified in detail. The wanted resolution in time is an important criterion when a method is chosen. In scientific studies the interest may vary from estimation of instantaneous recharge to long time averages. In groundwater resource studies in humid climate, the interest usually varies from seasonal recharge to averages over several years. An attempt to classify the applied methods from this criterion is made in Table III. The character of the results from an areal aspect is also indicated. Geohydrological features and available data are of course other restrictions in the choice of method.

The problem with uncertainties in input data is clearly illustrated in the three model studies (SOIL, the soil budget model and PULSE). If the hypothesis of no surface runoff in recharge areas is accepted, net groundwater recharge is equal to precipitation minus evapotranspiration. The error in the input of precipitation, due to errors in point measurements and representativity of these measurements, may be considerable. As pointed out earlier the evapotranspiration is calculated in different ways in the models. The most physically correct treatment of the evapotranspiration is performed in SOIL. The potential evapotranspiration is calculated on a daily basis with Penman-Monteith's equation and reductions are performed due to high soil moisture tensions for different soil layers individually. Studying annual groundwater recharge, evapotranspiration is probably the most critical input. It should be desirable to get better estimations of the evapotranspiration from independent information of the controlling parameters connected to the vegetation, mainly the surface resistance. If the interest is concentrated to instantaneous recharge or events precipitation and soil properties are more important.

TABLE III. Classification of the applied methods for estimation of groundwater recharge according to the resolution in time of their results. Dotted line indicate point values of groundwater recharge and solid line an areally integrated value.

TIME SCALE

| METHOD/MODEL | Instantaneous | Events | Monthly | Seasonal | Annual | Long time average |
|---|---|---|---|---|---|---|
| One dimensional soil water flow model (SOIL) | - - - - - | - - - - - | - - - - - | - - - - - | - - - - - | - - - - - |
| Soil moisture budget model | - - - - - | - - - - - | - - - - - | - - - - - | - - - - - | - - - - - |
| Groundwater level fluctuations | - - - - - | - - - - - | | | | |
| Chloride concentration | | | | | | ——— |
| Spring discharge | | | | | —— | ——— |
| Catchment area model (PULSE) | ——— | ——— | ——— | ——— | ——— | ——— |

The water balance element which could be measured with the best accuracy is runoff, therefore it is of great value if it is possible to calibrate this type of model simulations against runoff. The spring discharge measurements performed in the study were valuable from this aspect. The problem with these results was the determination of the catchment areas. In the Grön Källa area the determination from the topography was considerably improved by installation of groundwater level observation tubes. In Rostock the topographical conditions were more favourable. In both cases, there was also a risk for by-pass flow under the measuring weirs.

The SOIL simulations calibrated against groundwater levels, were rather insensitive to displacements between evapotranspiration and groundwater flow. The need for better data on soil water properties, e.g. air-entry values, hydraulic conductivities especially at low tensions, were obvious. Simultaneous measurements of water content and tension are probably necessary to calibrate the model. An indication of which of the performed SOIL-simulations that gave the most probable total amount of recharge was obtained by comparison with the spring discharge measurements. It is fully recognized that the results obtained from the spring discharge measurements were areally integrated values, but as pointed out earlier the conditions within the catchment area of Grön källa are very homogeneous. It might also have been possible to calibrate SOIL simultaneously against both groundwater levels and spring discharge.

The results from the soil moisture budget model are of very limited value if not calibrated against other methods, but gave in the long term study in the Rörsbo area an annual and seasonal distribution of recharge rather similar to the SOIL simulation giving the same total recharge. Recharge had to occur, also when the soil moisture reservoir was not full, to reproduce the recharge dynamics as reflected in groundwater level changes.

The groundwater level fluctuation method gave results deviating from those obtained by the other methods at Grön källa. It seemed only possible to get qualitative information from this technique. It was, however, interesting to see that the groundwater level could be simulated in an acceptable way by multiplying the daily recharge calculated by PULSE with the constant specific yield value which was used in the recharge estimation made by the groundwater level fluctuation method.

If the total values of groundwater recharge obtained from the spring discharge measurements at Grön källa, approximately 180 mm/year, were accepted the chloride concentration method could also be calibrated. The total atmospheric deposition needed to get this recharge was 1 260 mg/m$^2$/year, which was considerably more than estimated from measurements. If the wet deposition was correctly estimated (545 mg/m$^2$/year) the dry deposition accordingly must be 715 mg/m$^2$/year or 131 percent of the wet deposition. In the study of areal variability of chloride concentration in spring discharge, this total deposition should give groundwater recharge values between 175 and 250 mm/year. The areal variation of evapotranspiration should, according to this, be about 75 mm between the different small catchments, all located in till and covered by coniferous forest. It must be pointed out that the forest stand is of considerable importance for the final $Cl^-$-concentration in groundwater. A mature stand catches more chloride in the form of dry deposition and has a higher evapotranspiration than a young one. Thus there is an additative effect on the $Cl^-$-concentration. The limited investigation carried out here needs to be developed. Better time series of the chloride deposition and the chloride content and its variations in the spring discharge are needed and should be coupled to discharge measurements. Correctly designed, such an investigation could give interesting information on areal variations in evapotranspiration which could be coupled to vegetation and hydrogeological conditions.

In the PULSE-model simulations, groundwater recharge was in principle calculated in the same way as in the soil moisture budget model but a calibration of the total amount of recharge was obtained from the spring discharge measurements. The representation of the unsaturated and saturated zones as two separate systems, however, complicates a physical interpretation of results obtained in sandy till areas, where the groundwater is shallow and the annual variations are of the same magnitude as the thickness of the overburden. The capillary upflow parameter could perhaps be made more sophisticated, but the limitations in the conceptualization could not be overcome. For studies of recharge processes a physical model is necessary.

Of the models applied in this study, the SOIL-model has the most physically correct description of important processes like snowmelt, evapotranspiration and soil water flow, which means that it has the best potential for studies of recharge processes and further development. The sensitivity of the calculated recharge values to variations in soil water retention and conductivity will be major interest when discussing areal representativity of obtained results.

## ACKNOWLEDGEMENTS

I am indebted to Prof. Gert Knutsson, Dept. of Land Improvement and Drainage, who initiated the project, gave untiring support during the work and valuable comments on the manuscript. Prof. Gunnar Jacks and Lena Maxe, MSc, at the same department, took part in the field investigations, gave ideas for the work and valuable comments on the manuscript. Dr. Per-Erik Jansson, at the Swedish Agricultural University, introduced me to the SOIL-model, supported me in the modelling and gave valuable comments on the manuscript. Dr. Sten Bergström at SMHI, placed the PULSE-model at my disposal, helped me to get started and gave valuable comments on the manuscript. Dr. Lennart Granat, at MISU, was very helpful discussing the chloride deposition. The groundwater level data was obtained free of charge from SGU. The work was mainly financed by the Swedish Natural Sciences Research Council.

## REFERENCES

Andersson-Calles, U.-M. and Eriksson, E., 1979. 'Mass balance of dissolved inorganic substances in three representative basins in Sweden'. Nordic Hydrol., 10: 99-114.

Bergström, S., 1976. Development and application of a conceptual runoff model for Scandinavian catchments. Dept. of Water Resources Engineering, Lund Institute of Technology/University of Lund.

Bergström, S. and Sandberg, G., 1983. 'Simulation of groundwater responses by conceptual models - three case studies'. Nordic Hydrol., 14:71-84.

Brooks, R.H. and Corey, A.T., 1964. Hydraulic properties of porous media. Hydrology paper No. 3, Colorado State Univ., Fort Collins, 27 pp.

Einsele, G., 1975. 'Eichung von Grundwasser-Ganglinien zur Bestimmung der Grundwasserneubildung und des Grundwasserabflusses'. Z. Dtsch. Geol. Ges., 126:293-315.

Eriksson, E., 1976. 'The distribution of salinity in groundwaters of the Dehli region and recharge rates of groundwater'. In: Interpolation of environmental isotope and hydrochemical data in ground water hydrology. IAEA, Vienna, pp. 171-177.

Eriksson, E., 1985. Principles and applications of hydrochemistry. Chapman and Hall, 183 pp.

Eriksson, E. and Khunakasem, V., 1969. 'Chloride concentrations in groundwater, recharge rate and rate of deposition of chloride in the Israel Coastal Plain'. J. Hydrol. 7:178-197.

Eriksson, E. and Johansson, S., 1978. Gotlands grundvattenresurser. Avd för Hydrologi, Uppsala Universitet, 13 pp.

Fox, I.A. and Rushton, K.R., 1976. 'Rapid recharge in a limestone aquifer'. Groundwater, 14:21-27.

Granat, L., 1986. Dept. of Meteorology, University of Stockholm (personal communication).

Grindley, J., 1967. 'The estimation of soil moisture deficits'. Meteorolog. Mag., 96:97-108.

Grindley, J., 1969. 'The calculation of actual evaporation and soil moisture deficits over specified catchment areas'. Meteorol. Off., Bracknell. Hydrol. Mem., No. 38.

Howard, K.W.F. and Lloyd, J.W., 1979. 'The sensitivity of parameters in the Penman evaporation equations and direct recharge balance'. J. Hydrol., 41:329-344.

Jacks, G. and Knutsson, G., 1981. Känsligheten för grundvattenförsurning i olika delar av landet (förstudie). Kol-Hälsa-Miljö; Teknisk Rapport, No. 11, 190 pp.

Jacks, G. and Sharma, V.P., 1982. 'Hydrology and salt budget in two tributaries to Cauvery river, India'. French-Swedish seminar on Hydrology of arid zones, Lund, Sweden. Swedish Natural Sciences Research Council - French-Swedish Research Society, Report No. 41, pp. 155-165.

Jansson, P.-E. and Halldin, S., 1980. Soil water and heat model. Technical description. Swedish Coniferous Forest Project, Technical Report No. 26, Uppsala, 81 pp.

Jansson, P.-E. and Thoms-Hjärpe, C., 1986. 'Simulated and measured soil water dynamics of unfertilized and fertilized barley'. Acta Agric. Scand., 36:162-172.

Johansson, P.-O., 1983. Metoder för bestämning av grundvattenbildningens storlek - en litteraturstudie (English summary). Trita-Kut 1030. Inst. för kulturteknik, Kungl. Tekn. Högskolan, Stockholm, 31 pp.

Johansson, P.-O., 1987. 'Estimation of groundwater recharge in sandy till with two different methods using groundwater level fluctuations'. (J. Hydrol., 90:183-198).

Karanth, K.R. and Srinivasa Prasad, P., 1979. 'Some studies on hydrologic parameters of groundwater recharge in Andhra Pradesh'. J. Geol. Soc. India, 20: 404-414.

Kitching, R., Shearer, T.R. and Shedlock, S.L., 1977. 'Recharge to Bunter Sandstone determined from lysimeters'. J. Hydrol., 33:217-232.

Knutsson, G., 1962. 'Algutsbodatraktens geologi'. Algutsboda Sockenbok I, Nybro.

Knutsson, G., 1966. 'Grundvatten i moränmark'. Svensk Naturvetenskap: 236-249.

Knutsson, G., 1971. 'Studies of ground-water flow in till soils'. Geologiska Föreningens i Stockholm Förhandlingar, 93:1-22.

Lannefors, H., Hansson, H-C. and Granat, L., 1983. 'Background aerosol composition in southern Sweden - fourteen micro and macro constituents measured in seven particle size intervals at one site during one year'. Atmospheric Environ., 17:87-101.

Lemmelä, R., 1976. 'Waterbalance in sandy areas'. Nordic Hydrological Conference, Reykjavik, 4 pp. (photocopy).

Malmér, N. and Wallén, B., 1980. Wet deposition of plant mineral nutrients in southern Sweden. Medd., No. 43. Växtekologiska institutionen, Lunds Universitet, 28 pp.

Mualem, Y., 1976. 'A new model for predicting the hydraulic conductivity of unsaturated porous media'. Wat. Resour. Res., 12: 513-522.

Nilsson, T., 1983. Dokumentation av subrutin för beräkning av potentiell grundvattenbildning. Avd. för Hydrologi, Uppsala Universitet, 7 pp. (photocopy).

Olsson, T., 1980. Ground-water-level fluctuations as a measure of the effective porosity and ground-water recharge. The Geological Survey of Sweden, Report No. 21, Uppsala, 46 pp.

Penman, H.L., 1949. 'The dependence of transpiration on weather and soil conditions'. J. Soil Sci., 1:74-89.

Rosén, K., 1982. Supply, loss and distribution of nutrients in three coniferous forest watersheds in central Sweden. Reports in forest ecology and forest soils, 41. Dept. of Forest Soils, Swedish University of Agricultural Sciences, 70 pp.

Rosén, K., 1984. 'Hydrokemiska budgetstudier - Med utgångspunkt från små avrinningsområden i skogsmark'. In: Vattnet i det terrestra ekosystemet. Swedish IHP, Report No. 58:111-118.

Rushton, K.R. and Ward, C., 1979. 'The estimation of groundwater recharge'. J. Hydrol., 41:345-361.

Soveri, J., 1985. Influence of meltwater on the amount and composition of groundwater in quaternary deposits in Finland. Publ. of the Water Research Inst. No. 63, Helsinki, 92 pp.

Smith, D.B., Wearn, P.L., Richards, H.J. and Rowe, P.C., 1970. 'Water movement in the unsaturated zone of high and low permeability strata by measuring natural tritium'. In: Symposium on the Use of Isotopes in Hydrology, IAEA, Vienna, pp 73-87.

Wallén, C.C., 1966. 'Global solar radiation and potential evapotranspiration in Sweden'. Tellus, 18:786-800.

# THE PRINCIPLES OF INVERSE MODELLING FOR ESTIMATION OF RECHARGE FROM HYDRAULIC HEAD

H. Allison
Principal Research Scientist
CSIRO
Institute of Energy and Earth Resources
Division of Groundwater Research
Private Bag, P.O., Wembley, Western Australia, 6014.

ABSTRACT. The Boussinesq equation permits, in principle, estimation of recharge for steady groundwater flow by taking derivatives of hydraulic head and transmissivity functions. The obstacles for doing this are as follows. Taking derivatives of the spatially-distributed data, even when they are known at every point, leads to numerical instability. Hydraulic heads are always measured with inaccuracies. Differentiating this "noisy" data leads to large errors in recharge estimation. It is shown that by using a special modification of the Boussinesq equation it is possible to overcome all three difficulties simultaneously. The methods of recharge estimation proposed are suitable for taking spatial derivatives from "noisy" measurements in other fields of Science and might be useful in Remote Sensing for edge enhancement.

## 1. INTRODUCTION

Estimation of groundwater recharge is of practical importance, particularly in the arid zones and in places where soil salinity is a problem (Sharma, 1979)(Peck, et al., 1981).

There are two principal methods of recharge estimation: (a) 'from above' - by analysis of water moving downwards through the unsaturated zone of soil, and, (b) 'from below' - by inferring the recharge from water-table changes. This second approach was considered by Freeze (1983) to be the most "straightforward way of estimating groundwater recharge". This is the approach considered in this article. It is called an "Inverse Problem" because, contrary to the "Forward or Direct Problem", where recharge is postulated known and hydraulic heads computed, it is the recharge estimate which is computed from field measurements of hydraulic head. According to classification of Kisiel and Duckstein (1976), this is the "Inverse Problem Type IV". Starting from the pioneering work of Emsellem and de Marsily (1971), considerable understanding of the Inverse Problem Type I, that is, estimation of aquifer parameters, has been achieved (see reviews by Allison and Peck, 1985 and by Dietrich (1986).



*I. Simmers (ed.), Estimation of Natural Groundwater Recharge, 271–282.*

Recharge estimation by an inverse method has not yet achieved the share of attention it deserves and in this article an attempt is made to rectify the situation.

We start from formulation of this Inverse Problem, then describe some naive approaches to solve it and difficulties encountered, and explain the origin of difficulties. After cursory analysis of theoretical background, we describe the proper approaches to the solution of this Inverse Problem and concentrate on a general method which seems to be the most theoretically sound and which also opens new computational avenues.

Interestingly, some of the known methods used for taking spatial derivatives in remote-sensing for image inhancement (edge detection), such as splines and spatial filtering, follow as particular cases from the method proposed in this article. However, the method described here has advantages over the Wiener filtering as the proposed method does not require spatial stationarity of variables.

## 2. THE PROBLEM

The spatially distributed and time-varying recharge $q(x,y,t)$ enters as a free term into the parabolic second-order partial differential equation which describes groundwater flow in a non-homogeneous isotropic two-dimensional aquifer:

$$- S\frac{\partial h}{\partial t} + \frac{\partial}{\partial x}\left(T\frac{\partial h}{\partial x}\right) + \frac{\partial}{\partial y}\left(T\frac{\partial h}{\partial y}\right) = q(x,y,t) \tag{1}$$

where $h$ is the hydraulic head [L], $T(x,y)$ is transmissivity $[L^2T^{-1}]$, $S$ is the dimensionless storage coefficient and $q$, in units $[LT^{-1}]$, is a recharge rate per unit area of the aquifer domain.

It is seen from (1) that should one possess the analytical expressions for $h$ and $T$, determination of $q$ would have been a trivial exercise of calculus in computing the derivatives in equation (1). Recognising that taking the derivatives of experimentally obtained $h$ is an unreliable procedure one might prefer to bypass the differential equation (1) and to find $q$ from the integral equation (1a).

$$h(x,y,t) = \int H(x,\xi,y,\eta,\ t-\tau)\, q(\xi,\eta,\tau)\, d\xi\, d\eta\, d\tau \tag{1a}$$

where $H$ is a Green function.

An example of using the integral equation (1a) for recharge estimation can be found in Hunt, (1983)(pp.217-221). The difficulty with the integral representation (1a) is that the Green function $H$ is known only in some simple cases. Nevertheless, the integral representation has useful theoretical properties and the steady-state version of (1a) will be used below to demonstrate that rather different recharges $q_1$ and $q_2$ can produce nearly indistinguishable (in the "least squares sense") hydraulic heads.

The problem of recharge estimation is thus ill-posed for two reasons: it does not necessarily have a unique solution within a given accuracy of computation, and it is sensitive to errors - 'noise' in the hydraulic head.

In this article the methods for solving the recharge estimation problem for a steady-state case are considered; the transient case will be reported elsewhere.

## 3. ILL-POSEDNESS OF RECHARGE ESTIMATION

The steady-state counterparts of equations (1) and (1a) will be from now on considered; firstly, because they provide a mathematically tractable case, but also they are important practically, when either slowly varying in time recharge is considered, or when recharge is more or less cyclic with a period of a year.

In nature, recharge is the cause of changes in hydraulic head; the physically natural description of this process is, therefore, the integral representation (3). We denote thus the integral operator $A$ ; the more familiar differential operator which acts upon the hydraulic head, is denoted $A^{-1}$ as inverse to $A$ .

$$q(x,y) = A^{-1}h(x,y) = \frac{\partial}{\partial x}\left(T\frac{\partial h}{\partial x}\right) + \frac{\partial}{\partial y}\left(T\frac{\partial h}{\partial y}\right) \tag{2}$$

$$h(x,y) = Aq(x,y) = \int H(x,\xi,y,\eta)q(\xi,\eta)\,d\xi\,d\eta \tag{3}$$

To understand the roots of ill-posedness of recharge estimation, assume that the integral operator $A$ possesses a complete series of eigenfunctions $\{\psi_n\}$ and eigenvalues $\{\lambda_n\}$ , ordered so that $\lambda_1 > \lambda_2 > \lambda_3 \cdots$ and $\lambda_n \to 0$ as $n \to \infty$ . Introducing the projection operator $P(\lambda_n)$ onto the space spanned by $\psi_n$ , one has the following representations:

$$h = \sum_n b_n\psi_n = \sum_n P(\lambda_n)h \tag{4a}$$

$$q = \sum_n c_n\psi_n = \sum_n P(\lambda_n)q \tag{4b}$$

$$Aq = \sum_n \lambda_n c_n\psi_n = \sum_n \lambda_n P(\lambda_n)q \tag{4c}$$

By comparing (4a) with (4c) it follows:

$$c_n = \frac{b_n}{\lambda_n} \quad \text{and, from (4b):}$$

$$q = \sum_n \frac{b_n}{\lambda_n}\psi_n = \sum_n \frac{1}{\lambda_n} P(\lambda_n)h \tag{5}$$

Let us now consider two heads, $h_1$ and $h_2$ ; the difference between which is small in the least-squares sense, so that $||h_1 - h_2||^2 = \varepsilon^2$ . The difference in their projections

$P(\lambda_n)h_1$ and $P(\lambda_n)h_2$ can be then taken to be $\varepsilon$ . From (5), the corresponding difference in $q_1$ and $q_2$ is:

$$q_1 - q_2 = \sum_n \frac{\varepsilon}{\lambda_n} \tag{6}$$

and with $\lambda_n \to 0$ as $n \to \infty$ the deviation in recharge estimate can be large, particularly large for small $\lambda_n$ , which corresponds to high spatial frequency.

Hence, the claims made often in hydrological literature that the computer model is good if it reproduces "as closely as possible the observed hydraulic head values" (Aboufirassi and Marino, 1984, p.127) should be viewed with scepticism, because quite different recharges $q_1$ and $q_2$ may produce indistinguishable hydraulic heads.

Turning now to the Boussinesq differential equation (2) one can immediately expect difficulties, associated with attempts to find $q(x,y)$ by "simply taking derivatives" of $h(x,y)$ . To start with, $h(x,y)$ is never known in its entirety; at best one has scattered observation wells, only rarely arranged in a rectangular grid. Interpolation by 'kriging' may produce the kriged surface, which is discontinuous in its second derivatives. The kriged function may thus be simply not in the domain of analytical definition of the differential operator $A^{-1}$ of equation (2). The crude forceful use of the finite difference schemes of the type:

$$T_i \frac{h_2 + h_3 + h_4 + h_5 - 4h_i}{a^2} = q_i \tag{7}$$

where $i$ is the central node number, $h_2$ to $h_5$ are heads of adjacent nodes $q_i$ - recharge estimate in this node with $T_i$ being a local transmissivity at node $i$ , and $a$ is the width of grid interval, may lead to large errors in recharge estimations, particularly if the fine grid, provided by abuse of kriging is thought to improve the accuracy of recharge estimation.

One can see what errors the fine grid may induce in recharge estimation by considering the following two-dimensional mesh of nodal lines at the grid width $a = \frac{\ell}{n}$ with a small amplitude $\varepsilon$ , simulating a standard deviation in hydraulic head, considered as an "input" $\bar{h}$ to the equation (2):

$$\bar{h}(x,y) = \varepsilon \sin\frac{n\pi x}{\ell} \sin\frac{n\pi y}{\ell} , \tag{8}$$

where $\ell$ is the dimension of the aquifer, taken here equal in both and $x$ and $y$ direction. The magnitude of error in recharge estimation, from (8) and (2) with uniform $T$ is, upon taking second derivatives, equal to:

$$|\text{error in } q| = \varepsilon T \frac{\pi^2 n^2}{\ell^2} = \frac{\varepsilon T \pi^2}{a^2} \tag{9}$$

Thus small grid width $a$ contributes as its squared value to error in recharge estimation, while in the forward problem the smaller the grid, the better the accuracy of solution.

From the above consideration there follows a practical rule: when estimating recharge by use of (7), start from estimating the minimal permissible grid cell size, $a$, using equation (9) as an estimate of error.

The above difficulties are typical of the so-called "Ill-posed problems". Careful approach to such problem can considerably improve the crude recharge estimations obtained by using the equation (7).

## 4. CORRECT APPROACHES TO THE ILL-POSED PROBLEM OF RECHARGE ESTIMATION

The difficulties with equations (2) and (3) have their explanation in the following theorem:

Theorem: The existence and boundedness of the inverse operator $A^{-1}$ is a necessary and sufficient condition for the inverse problem to be correctly posed, that is to possess only one solution and be insensitive to small variations in input data. (Formulation here has been deliberately stripped of finer mathematical details; for rigour see Allison, 1979).

Both conditions of the theorem are violated in recharge estimation: the operator $A^{-1}$ as given by equation (2) is unbounded (the error in recharge as given by (9) grows infinitely as $a \to 0$) and the operator may not exist in the sense that hydrualic head $h(x,y)$ may not possess the second derivatives.

In addition to explaining the source of difficulties one encounters in recharge estimation the theorem also hints on the ways in which the problem can be remedied and made well-posed. One approach is to modify the operator $A$ so that the inverse operator will satisfy the demands of the theorem.

The modification proposed by Phillips (1962) consisted in the introduction of an artificial parameter $\alpha > 0$, so that instead of the integral equation of the first kind (eq.3) one has a stable equation of the second kind:

$$(\alpha I + A)q_{\alpha} = h \tag{10}$$

The approximate solution $q_{\alpha}$ of (10) is, of course, not the true recharge $q$, because the equation (10) has been modified from equation (3), but it is possible, in principle, to choose such a value of $\alpha$ that the "trade-off" can be achieved between stability of solution $q_{\alpha}$ and its reasonably small departure from $q$. The more general approach by Tikhonov and Arsenin (1977) introduces for the stable approximate the solution

$$q_{\alpha} = (\alpha I + A^*A)^{-1}A^*h \tag{11}$$

One can see immediately the relationship of (11) to the method of Ridge Regression in statistics, (Hoerl and Kennard, 1970) where a small positive quantity is added to the diagonal of ill-conditioned least-square estimation matrix. The analogy with the Ridge Regression becomes even more striking upon comparing the projection solution (12) of equation (11) with the previous equation (5):

$$q_\alpha = \sum_n \frac{b_n}{\lambda_n + \alpha} \psi_n = \sum_n \frac{1}{\lambda_n + \alpha} P(\lambda_n) h \qquad (12)$$

The Ridge Regression modification of the eigenvalues (equation 2.5 in Hoerl and Kennard) is immediately recognisable in the right side of equation (12). The Ridge Regression eigenvalues thus are beginning to have physical meaning: they may be thought of as the eigenvalues of the elastic membrane (a drum), the thickness of which varies in space according to $T(x,y)$ of the equation (2).

The projection solution (12) has its limitations, requiring existence of eigenvalues, but it gives a hint on how to tackle the problem with the unbounded operator $A^{-1}$: namely, instead of the spectral function $\lambda_n^{-1}$ which tends to infinitely large values as $\lambda_n \to 0$ (this, actually, is the reason, why the operator $A^{-1}$ is unbounded), the finite approximation $(\lambda_n + \alpha)^{-1}$ is introduced in both Tikhonov and Ridge Regression methods, which does not allow the modified spectral function to become infinitely large. Moreover, because equation (5) is a projection analog of the differential equation (2), one comes to the idea that stabilisation can be achieved directly through modifying the Boussinesq differential equation, without use of eigenfunction expansion.

The following general method of obtaining the limited norm approximations to unbounded differential operators was proposed by Levin (1974). Apparently, Courant (1966) was taking a similar approach to unbounded operators. The method therefore may be provisionally called the "Levin-Courant approximation". The chief element of the method is the representation of the differential operator in the form of an integral, with a kernel of a generalised function. The modification consists of replacing the kernel by its smooth regularised counterpart. It must be said that there were also earlier developments, where derivatives are represented in the form of integrals. Such is the Riemann-Liouville fractional derivative, which for an arbitrary order of differentiation $\nu$ can be written as an integral: (for a very readable account on Fractional Calculus see Ross, 1975):

$$D^\nu f(x) = \frac{1}{\Gamma(\nu)} \int_0^x f(t)(x - t)^{-\nu-1} dt = f(x) * \Phi_\nu(x) \qquad (13)$$

where $\Gamma(\nu)$ is a gamma-function, and $\Phi_\nu(x)$ in the limit, defines a fractional derivative of the delta function.

$$\Phi_\nu(x) = \lim_{\mu \to -\nu} \frac{x^{\mu-1}}{\Gamma(\mu)} = \delta^{(\nu)}(x) \qquad (14)$$

Once the derivatives of delta-function have been defined, any differential operator $A^{-1}$ can be represented as an integral operator of the convolution type (Levin, 1974) (the convolution is denoted by *):

$$A^{-1}f(x) = A^{-1}(\delta * f) = A^{-1}\delta * f \tag{15}$$

The Boussinesq differential equation can thus be written as a convolution with derivatives of the delta-function $\delta(x,y)$ :

$$q = A^{-1}h = \iint_{\Omega} \left[ \frac{\partial}{\partial \xi}\left( T \frac{\partial \delta}{\partial \xi} \right) + \frac{\partial}{\partial \eta}\left( T \frac{\partial \delta}{\partial \eta} \right) \right] h[(x - \xi),(y - \eta)]\, d\xi\, d\eta \tag{16}$$

In the Levin-Courant method the delta-function in (16) is replaced by a point-spread function $\delta_\alpha(x,y)$ , defined as a twice differentiable function on finite support $|x,y| \leq \gamma$ . Thus obtained operator $A_\alpha^{-1}$ is an approximation to a true differential operator $A^{-1}$ ; the resulting recharge $q_\alpha$ , is also an approximation to $q$ .

$$\begin{aligned} q_\alpha &= A^{-1}\delta_\alpha * h = A_\alpha^{-1} * h \\ &= \iint_{\Omega} \left[ \frac{\partial}{\partial \xi}\left( T \frac{\partial \delta_\alpha}{\partial \xi} \right) + \frac{\partial}{\partial \eta}\left( T \frac{\partial \delta_\alpha}{\partial \eta} \right) \right] h[(x - \xi),(y - \eta)]\, d\xi\, d\eta \end{aligned} \tag{17}$$

As seen from (17), the differential operator in the integrand acts upon a twice-differentiable (at least) function $\delta_\alpha(x,y)$ , instead of being applied to the data $h(x,y)$ . The kernel $K_\alpha(x,y)$ given by (18) is thus square-summable; the wider the support of $\delta_\alpha(x,y)$ , controlled by parameter $\alpha$ , the smoother is $K_\alpha(x,y)$ .

$$K_\alpha(x,y) = A^{-1}\delta_\alpha(x,y) \tag{18}$$

The ill-posed problem of recharge estimation has been thus modified to a stable, well-posed problem of performing a two-dimensional convolution of the data $h(x,y)$ with a smooth, bounded function $K_\alpha(x,y)$ of (18) fulfilling the requirements of the Theorem.

## 5. DIFFERENTIAL EQUATION FORM OF THE LEVIN-COURANT METHOD

Although, in principle, there is a wide choice of point-spread functions $\delta_\alpha(x,y)$ (the two-dimensional Gaussian being the obvious candidate), one can get further insight into the problem and obtain a computationally efficient algorithm by a special representation of $\delta_\alpha(x,y)$ as a fundamental solution of a certain ancillary partial differention equation with an operator $G$ , for example as in (19).

$$G\delta_\alpha(x,y) = \left[ \delta_\alpha + \alpha \left( \frac{\partial^2 \delta_\alpha}{\partial x^2} + \frac{\partial^2 \delta_\alpha}{\partial y^2} \right) \right] = \delta(x,y) \tag{19}$$

This is the equation of the stretched membrane, supported uniformly by distributed elastic support with unit elasticity and loaded by a concentrated force $\delta(x,y)$ . The coefficient $\alpha$ controls the tension in the membrane. One can easily visualise that the shape of membrane flexure will be approaching a delta-function if the tension in the membrane tends to zero; from the equation (19) it is seen that when $\alpha = 0$ , $\delta_\alpha$ coincides with $\delta(x,y)$ .

We shall obtain now the recharge $q_\alpha(x,y)$ expressed as a solution of a stable differential equation, namely that of the described elastic membrane, by formally writing the following sequence of manipulations, using (17) and (19):

$$G\delta_\alpha = \delta \tag{20a}$$

$$\delta_\alpha = G^{-1}\delta \tag{20b}$$

$$\begin{aligned} q_\alpha &= A^{-1}\delta_\alpha * h = A^{-1}(G^{-1}\delta) * h = G^{-1}(A^{-1}\delta) * h \\ &= G^{-1}A^{-1}(\delta * h) = G^{-1}A^{-1}h \end{aligned} \tag{20c}$$

Applying formally the operator $G$ to the first and last terms in (20c) gives:

$$Gq_\alpha = GG^{-1}A^{-1}h \tag{21}$$

Returning to definitions of $G$ and $A^{-1}$ in (19) and (2), one obtains $q_\alpha$ as a solution of:

$$\left[q_\alpha + \alpha\left(\frac{\partial^2 q_\alpha}{\partial x^2} + \frac{\partial^2 q_\alpha}{\partial y^2}\right)\right] = \frac{\partial}{\partial x}\left(T\,\frac{\partial h}{\partial x}\right) + \frac{\partial}{\partial y}\left(T\,\frac{\partial h}{\partial y}\right) \tag{22}$$

The boundary conditions for $q_\alpha$ are immaterial as soon as the domain of the stretched membrane exceeds in size that for $h(x,y)$ , because $\delta_\alpha$ given by a solution to (19) decays with distance. Therefore the infinite domain for $q_\alpha(x,y)$ is suitable theoretically.

The computer solution of equation (22) can be obtained by the standard finite-element or finite difference method applied simultaneously on the same grid to both parts of (22). The recharge $q_\alpha(x,y)$ , thus obtained, is stable with respect to "noise" in $h(t)$ . Assuming $T$ constant in (22) and taking a two-dimensional Fourier Transform of (22) in the infinite domain, one obtains a frequency representation of the link between $\tilde{h}(\omega_1\omega_2)$ and $\tilde{q}_\alpha(\omega_1\omega_2)$ . ( $\omega_1$ and $\omega_2$ are frequency coordinates; the sign $\sim$ means the Fourier transformed functions):

$$\tilde{q}_\alpha(\omega_1,\omega_2) = \frac{T(\omega_1^2 + \omega_2^2)}{1 + \alpha(\omega_1^2 + \omega_2^2)} \cdot \tilde{h}(\omega_1,\omega_2) = \tilde{A}_\alpha^{-1}\tilde{h} \tag{23}$$

If one sets $\omega_1$ and $\omega_2$ tending to infinity, the asymptotic high spatial frequency relationship (equal actually to the norm of the operator $A_\alpha^{-1}$) is:

$$||A_\alpha^{-1}|| = \sup \frac{\tilde{q}_\alpha}{\tilde{h}} = \frac{T}{\alpha} \tag{24}$$

and with $\alpha \neq 0$ the solution of (22) satisfies the conditions of the Theorem.

Let us see what is the effect of the small-size numerical grid on the solution $q_\alpha$. By substituting $\bar{h}(x,y)$, given by equation (8), into (22), one obtains (with $T = \text{const}$ ):

$$|\text{Error in } q_\alpha| = \frac{\varepsilon T \dfrac{\pi^2 n^2}{\ell^2}}{1 + \dfrac{\alpha\pi^2 n^2}{\ell^2}} \tag{25}$$

As the grid size $a = \frac{\ell}{n} \to 0$, while $\alpha \neq 0$, one has the bounded error:

$$|\text{Error in } q_\alpha| = \frac{\varepsilon T \pi^2}{a^2 + \alpha\pi^2} \to \frac{\varepsilon T}{\alpha} \tag{26}$$

which is, of course, in agreement with the norm of the operator $A_\alpha^{-1}$ given by (24).

If one wants to obtain stronger suppression of "noise" in hydraulic head, the equation of the higher order than (19) can be considered. One particular choice would be to take the equation of bending of the elastic plate on the elastic foundation, the load to which is given by the same right-hand part as in equation (22):

$$\left[q_\alpha + \alpha\left(\frac{\partial^4 q_\alpha}{\partial x^4} + 2\,\frac{\partial^4 q_\alpha}{\partial x^2 \partial y^2} + \frac{\partial^4 q_\alpha}{\partial y^4}\right)\right] = \frac{\partial}{\partial x}\left(T\,\frac{\partial h}{\partial x}\right) + \frac{\partial}{\partial y}\left(T\,\frac{\partial h}{\partial y}\right) \tag{27}$$

The analog of (23) then follows from (27):

$$\tilde{q}_\alpha(\omega_1,\omega_2) = \frac{T(\omega_1^2 + \omega_2^2)}{1 + \alpha(\omega_1^4 + 2\omega_1^2\,\omega_2^2 + \omega_2^4)}\;\tilde{h}(\omega_1,\omega_2) \tag{28}$$

Representation of $q_\alpha$ by means of deflections of an elastic plate in (27) has a direct link with splines; we shall not pursue this connection any further here.

One can not fail to notice, that while the theory has been linear so far, the non-linear Boussinesq equation can be easily incorporated.

For example, a non-linear counterpart of (22) with $T = T(h)$ would be:

$$\left[q_\alpha + \alpha\left(\frac{\partial^2 q_\alpha}{\partial x^2} + \frac{\partial^2 q_\alpha}{\partial y^2}\right)\right] = \frac{\partial}{\partial x}\left[T(h)\,\frac{\partial h}{\partial x}\right] + \frac{\partial}{\partial y}\left[T(h)\,\frac{\partial h}{\partial y}\right] \tag{29}$$

Recharge $q_\alpha$ is found from solution of the linear left part of (29); the non-linear right part is used to compute the forcing term (the load) acting on the elastic membrane.

Equation (29) might be promising for reconstruction of spatial location of sources of pollutants from non-linear diffusion of moisture in soil as in Philip (1975).

## 6. CONNECTIONS WITH SPATIAL FILTERING

Returning to equation (18), one can specify the function $\delta_\alpha(x,y)$ by its Fourier transform in the form, more general than in (23) and (28), namely:

$$\tilde{\delta}_\alpha(\omega_1,\omega_2) = \frac{1}{1 + \alpha S(\omega_1,\omega_2)} \tag{30}$$

In those rare cases, when some information about the spectrum of expected recharge $F_q(\omega_1\omega_2)$ and the spectrum of noise $F_n(\omega_1\omega_2)$ in hydraulic head is available, one can form the ratio of the spectra for $S(\omega_1\omega_2)$ :

$$S(\omega_1,\omega_1) = \frac{F_n(\omega_1,\omega_2)}{F_q(\omega_1,\omega_2)} \tag{31}$$

The necessity to know the spectra hinders, however, practical use of the Wiener Filter (31).

It seems that although the differential forms (22), (27) are the particular cases of a theoretically more general convolution form (17), these differential forms are, in a sense, more practical, permitting the recharge estimation for a non-linear equation (29). The situation here seems to resemble that of the history of the Kalman Filter, where a particular differential form of a general Wiener-Hopf integral equation has wider applications than the Wiener-Hopf equation itself (Kalman, 1975).

Use of the forms (22) and (27) does not require assumptions of stationarity of the random fields for $h(x,y)$ , nor does it require knowledge of spectra as in (31).

The procedure of recharge estimation by (22) and (27) consists simply in solving numerically by any established method the differential equations, using kriging if necessary, to interpolate $h(x,y)$ from sparse measurements.

It is also hoped that (22) and (23) may find their use in Remote Sensing, where the unsophisticated and unstable analog of finite-difference approximation (7) goes under the name "Laplacian filtering" for edge enhancement (Hall, 1979).

The trade-off between accuracy of an approximate solution versus stability depends on the parameter $\alpha$ . This will be examined in a subsequent paper.

## 7. CONCLUSIONS

1. Recharge estimation is an ill-posed problem and the finite-difference type computations of derivatives as in equation (7) must be used with caution. The error estimate is given by equation (9).

2. The general stable method for computing recharge is by use of (17). Recharge is computed by taking a convolution of hydraulic head measurement with a smooth kernel.

3. From the previous method stems the differential equation form, given by (22) and (27). This form of recharge estimation does not require convolution and is applicable to the non-linear Boussinesq equation.

4. The stable methods of obtaining partial derivatives from measurements may have applications in other fields of Science, particularly in Remote Sensing for edge enhancement.

## 8. REFERENCES

[1] Aboufirassi, M, and M.A. Marino (1984). A geostatistically based approach to the identification of aquifer transmissivities in Yolo Basin, California. Math. Geology, Vol.16, No.2, pp.125-137.

[2] Allison, H.T. (1979). Inverse unstable problems and some of their applications. Math. Scientist, 4, pp.9-30.

[3] Allison, H. and Peck, A. (1985). Inverse problems for groundwater research. I.E. Aust. Hydrology and Water Resources Symposium, Sydney, pp.20-24.

[4] Courant, R. and Hilbert, D. (1966). Methods of Mathematical Physics, Vol.II, p.792, Wiley, New York.

[5] Dietrich, C.R., Anderssen, R.S. and Jakeman, A.J. (1986). Solving the inverse groundwater flow problem, in I.E. Aust. Hydrology and Water Resources Symposium, Brisbane, pp.83-86.

[6] Emsellem, Y., and G. de Marsily (1971). An automatic solution for the inverse problem, Water Res. Research, Vol.7, No.5, pp.1264-1283.

[7] Freeze, A.R. (1983). Regional groundwater analysis in surface and subsurface hydrology. In Int. Conf. on Groundwater and Man, Sydney, pp.101-119.

[ 8] Hall, E.L. (1979). Computer image processing and recognition. Acad. Press, New York, N.Y.

[ 9] Hoerl, A.E. and R.W. Kennard (1970). Ridge regression: biased estimation for nonorthogonal problems. Technometrics, Vol.12, No.1, p.55-67.

[10] Hunt, B. (1983). Mathematical analysis of groundwater Resources. Butterworths, London.

[11] Kalman, R.E. (1978). A retrospective after twenty years: from pure to applied. In Applications of Kalman Filter to Hydrology, ed. by C.L. Chiu, Pittsburgh, pp.31-53.

[12] Kisiel, C.C. and Duckstein, L. (1976). Ground-water Models, in Biswas, A.K. (ed.), Systems Approach to Water Management, McGraw-Hill, New York.

[13] Levin, G.E. (1974). Use of functional analysis methods to input force reconstruction. Vibrotechnika, Acad. of Science Lit. SSR, No.2(23), pp.161-173 (in Russian).

[14] Peck, A.T., Johnston, C.D. and Williamson, D.R. (1981). Analysis of solute distributions in deeply weathered soils. In Agricultural Water Management, 4, Elsevier, Amsterdam, pp.83-102.

[15] Philip, T. (1975). Water movement in soil, in Heat and Mass Transfer in Biosphere, Scripta Book, Washington, pp.30-45.

[16] Phillips, D.L. (1962). A technique for the numerical solution of certain integral equations of the first kind. J. Assoc. Comput. Mach., 9, pp.84-97.

[17] Ross, B. (1975). A brief history and exposition of the fundamental theory of fractional calculus. In Lecture Notes in Mathematics, 457 ed. by Dold, A. and B. Eckmann, Springer-Verlag, Berlin. pp.1-36.

[18] Sharma, M.L. (1979). Evatranspiration and stream salinity as a consequence of land use change in south-western Australia . In H.T. Morel-Seytoux (ed.), Modelling Hydrologic Processes, Water Res. Publications, Colorado, USA, pp.779-791.

[19] Tikhonov, A.N. and V.Y. Arsenin (1977). Solutions of Ill-posed Problems. John Wiley & Sons, New York.

# ESTIMATING NATURAL RECHARGE OF GROUND WATER BY MOISTURE ACCOUNTING AND CONVOLUTION

J. Willemink
Institute for Ground-water Studies
University of the Orange Free State
P.O. Box 339
Bloemfontein 9300
South Africa

ABSTRACT. A numerical model has been developed for calculating the ground-water recharge at areas without or with a thin soil cover (up to 20 cm) underlain by hard- rock formations. The system is simulated by hydrologic routing through a number of serially arranged linear reservoirs. The point rainfall is being converted, by daily moisture accounting, into effective rainfall. This, in turn, is being converted into the water-table response by the convolution operation. The unknown model parameters are calibrated, using an automatic optimization procedure based on the criterion to minimize the error differences between the recorded water-table recovery and the simulated one. The minimum required input data are daily readings of the water table in an observation well, near the topographic water divide (or just downstream an aquifuge), and daily rainfall figures from a rain gauge close to the well. The proposed model has been applied to a site in the central part of South Africa with promising results.

## 1. INTRODUCTION

The presented study is a part of a more general study on the exploitation potential of ground-water resources in the Karoo basin of South Africa.

Evidenced by neutron moisture gauging and water-level monitoring, most of the ground-water recharge in the studied area seems to occur in the elevated hilly terrains and along ridges. These areas are underlain by hard-rock formations, partly covered with shallow soils. Although the effect of rainfall on the water-table is usually quite clear under undisturbed natural conditions, a quantitative correlation is often difficult because:

- differences in effective porosity will cause the water table to rise unevenly,
- part of the rainfall may not reach the water table at all because of surface runoff and moisture deficits in the unsaturated zone,



*I. Simmers (ed.), Estimation of Natural Groundwater Recharge, 283–299.*

- lateral ground-water flow and seepage from greater depth may affect the water-table position.

Several numerical modelling techniques with which the vertical recharge component from rainfall can be determined, have been developed by others in the past to simulate the dynamic processes of moisture movement in granular textured media. However, these techniques are not easily applicable to unsaturated hard-rock formations. For instance, the soil moisture retention curve is essential for solving the differential equation governing unsaturated flow (Skaggs, 1981). One cannot imagine how to determine the moisture retention curve of, for instance, a jointed quartzite.

Ground-water level responses to natural recharge have been simulated by means of time series analysis treating ground-water levels as a random (Markov) process (Viswanathan, 1982). This method may be of interest for forecasting water-table fluctuations, but is generally not suitable for representing the water balance of the unsaturated zone.

Natural tracers as tritium and chloride may be helpful for obtaining estimates of long-term average values of recharge. However, the variability of both the rainfall amount and its tritium concentration make the input function difficult to evaluate. The chloride method must be treated with caution as accession of chloride near the soil surface because of evapotranspiration, may violate the assumption of a steady state chloride flux density throughout the unsaturated zone (Allison *et al,* 1984). Furthermore, recharge under conditions of extremely high rainfall with a long recurrence period is likely to influence the chloride concentration of ground water to a high degree, resulting in an overestimate of the mean annual recharge. Especially when the available ground-water storage is small in terms of mean annual recharge, it is a prerequisite to determine the variability of annual recharge in order to provide the right guidelines for the management of local ground-water resources.

The purpose of the proposed model is to calculate the daily effective point rainfall which causes water-table recovery during the rainy season. Sites underlain by hard-rock formations, with or without a thin soil cover (up to 20 cm), near the topographic water divide or just downstream of a low permeable intrusive dyke acting as a barrier boundary of a ground-water compartment, are of particular interest. The minimum data needed are daily water-table readings in an observation well and daily rainfall figures from a rain gauge close to the well. Additional data with respect to potential evapotranspiration and, if possible, other physical processes, is desirable.

## 2. DEVELOPMENT OF THE RECHARGE MODEL

### 2.1. Underlying concept

The operation of the hydrologic system is simulated by hydrologic routing through a number of serially arranged linear reservoirs (Fig. 1). Hydrologic routing is a generally utilized technique to problems in surface hydrology by which the hyetograph of excess rainfall is being transformed into the outflow hydrograph from a watershed (Viessman *et al,* 1977). Analogous to watershed routing, the hyetograph of seasonal rainfall can be transformed into the hydrograph of the water-table rise, superimposed on the hydrograph of the water table that would have occurred, had there been no recharge. A signal is being modified as it passes through the system, which is simulated by employing the equation of continuity with an assumed linear relationship between storage and outflow of each reservoir. The assumption of linearity regarding the unsaturated zone may be crude, as it acts in reality as a highly non-linear system.

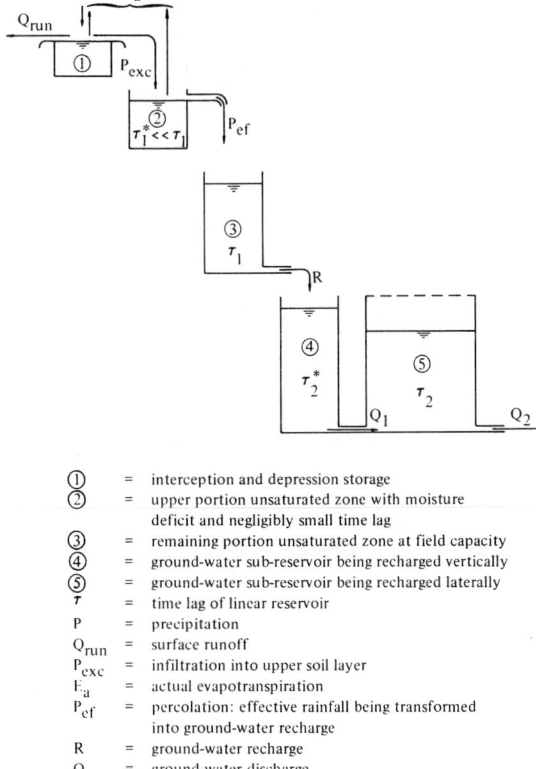


Figure 1. The conceptual approach of the recharge model.

An initial quantity of the daily rainfall is needed to replenish interception and depression storage. Replenishment of these reservoirs is assumed to take place instantaneously and any surplus of rainfall with respect to the reservoir capacity will partly become surface runoff and partly infiltrate into the upper soil layer (rainfall excess). The interception- depression storage will be depleted completely on the very same day by evaporation and possibly by infiltration into the sub-surface system should the surface detention storage be higher than the evaporative demand.

The unsaturated zone is subdivided in two portions, viz. the upper portion acting as a reservoir, with a negligibly small time lag, in which a moisture deficit with respect to field capacity due to evapotranspiration may exist. The remaining portion acts as a linear reservoir with a significant time lag, which is being considered at permanent field capacity. The thickness of the upper zone may locally, on a small scale, differ considerably if plant rooting is well-developed in fractures of the hard rock. The latter is generally filled up with clayey weathered rock material. Therefore, the upper soil layer is thought to have an equivalent depth represented by a reservoir with average moisture conditions at a certain time.

The daily rainfall excess minus daily evapotranspiration losses will replenish the upper reservoir of the unsaturated zone. The moisture excess with respect to field capacity (effective rainfall) is assumed to drain out of the upper reservoir at the end of the same day, thus escaping evapotranspiration. The effective rainfall will percolate downward through the lower reservoir of the unsaturated zone and finally reach the ground-water table with some time delay to recharge the ground-water storage.

Because ground-water recharge from rainfall is mostly restricted to terrains under a shallow soil cover, a major portion of the aquifer may be recharged by lateral ground-water discharge from adjacent sub-reservoirs, with only the latter being recharged directly from rainfall. The last reservoir is incorporated in the conceptual model as displayed in Fig. 1, just for the sake of completeness, but have a negligible influence on the recharge from rainfall.

The hydrograph of the water table reflecting the storage condition of the ground-water reservoir, results from recharge and ground-water discharge which in turn is directly proportional to the dischargeable storage. Because of the transition of moisture excess through the lower portion of the unsaturated zone into the ground-water reservoir, the recharge pattern is delayed and attenuated in comparison with the hyetograph of the effective rainfall. Consequently, the water-table recovery is the response of a subsystem consisting of two serially arranged linear reservoirs, viz. the lower portion of the unsaturated zone and the ground-water storage, to the effective rainfall characterized by a time series of block inputs with a duration of one day.

The unit hydrograph method is an important tool to be used in the transformation of a hyetograph of effective rainfall into a hydrograph which represents the difference between the water table with and without recharge. If the 1 day-unit hydrograph, the water-table response to a unit block input of effective rainfall, is known, the water-table response can be computed by the convolution operation of the unit hydrograph and the effective rainfall inputs obtained from moisture budgetting in the upper portion of the unsaturated zone.

The computational procedure, viz. moisture accounting and convolution, is repeated to reproduce the 'observed' difference in water table with and without any recharge within some range of accuracy. The term 'observed' is formally incorrect, because the water-table recession graph cannot be measured and is reconstructed from recession curves during dry seasons.

## 2.2. Mathematical structure

The infiltration and evapotranspiration processes have been described mathematically in terms of rainfall, interception- depression storage capacity, actual soil moisture storage, maximum available soil moisture and the infiltration capacity which in turn is expressed as a function of the moisture deficit with respect to field capacity. The mathematical equations in the recharge model contain parameters which bear a physical relationship to actual processes, e.g. infiltration capacity, potential evapotranspiration, soil moisture capacity and the time lag of the lower portion of the unsaturated zone.

A crucial problem was to find a suitable mathematical expression of the daily infiltration rate. In watershed hydrology the "threshold concept" is being used, which indicates that all precipitation is turned into basin retention (infiltration plus surface detention) as long as the retention capacity (potential maximum retention at a certain time) exceeds the depth of precipitation. All precipitation in excess of this limit becomes surface runoff (Kraijenhoff van de Leur, 1972). This simplified concept does not hold, however, if a daily rainfall below the daily retention capacity may induce considerable runoff.

The "curve number method" developed by the U. S. Soil Conservation Service (Kraijenhoff van de Leur, 1972) provided the basis of the relationship between the daily infiltration quantity and the daily rainfall. The mathematical model underlying the curve number method expresses the relation of the amount of rainfall in one day and the corresponding daily amount of rainfall excess that will subsequently be transformed into direct runoff as follows:

$$Q_{run} = (P - I_{max})^2 / (P - I_{max} + S) \tag{1}$$

where

$Q_{run}$ = the daily amount of surface runoff (mm)
$P$ = the daily amount of rainfall (mm)
$S = F + I_{max}$ = the retention capacity (mm)
$F$ = the infiltration capacity (mm)
$I_{max} = 0.2\ S$ = the initial losses (mm)

The initial losses are the initial quantity of interception- depression storage and initial infiltration that must be satisfied by any rainfall before runoff can occur. The retention capacity is virtually the only model parameter and is determined by antecedent moisture conditions and by the physical characteristics of the drainage basin. From algebraic analysis of

Eq. 1, it appeared that $P - Q_{run} > S$ for $P > 4.2\ S$, which is, of course, in contradiction with the concept of retention capacity. A small retention capacity could therefore imply a definite restriction to the method's applicability.

That part of the daily amount of rainfall that infiltrates into the upper soil layer during the day (here defined as rainfall excess) equals

$$P_{exc} = P - Q_{run} - I_{max} \qquad , P_{exc} > I_{max} \quad . \tag{2}$$

Substituting Eq. 2 into Eq. 1 yields

$$P_{exc} = ((F + I_{max})\ (P - I_{max}))\ /\ (P + F) \quad . \tag{3}$$

However, Eq. 3 as well holds only for $P < 4.2\ (F + I_{max})$, as the excess of rainfall exceeds the infiltration capacity at higher rainfall quantities. To overcome the limited applicability of a mathematical expression of the rainfall excess $P_{exc}$, Eq. 3 has been modified into

$$P_{exc} = F\ (P - I_{max})\ /\ (P + F) \qquad , P_{exc} > I_{max} \tag{4}$$

where $L_{\infty}\ P_{exc} = F$ and the interception- depression storage capacity $I_{max}$ is considered to be constant and therefore not to be related to the infiltration capacity. Eq. 4 may underestimate the rainfall excess, especially in case of a small infiltration capacity, but yields certainly a better approximation than the above mentioned "threshold concept".

The infiltrability (the potential maximum infiltration rate at a certain time) will normally decrease during a storm event because of splashing raindrops, swelling colloids and increasing soil moisture content. The infiltrability - time curve starts at a relatively high value and then falls rapidly during the early stages of a storm, finally levelling off and approaching a constant value. Several equations describing the time-rate process of infiltration have been developed (Skaggs, 1981). The use of these time-based models is hampered, however, as soon as the rainfall intensity drops below the infiltrability. This problem has been coped with by introducing the soil moisture storage as the dependent variable. The model, developed by Holtan (Skaggs, 1981 and Fleming, 1975) relating the infiltrability to the available storage in the surface layer, has been chosen as a basis for a mathematical expression of the infiltration capacity (the potential maximum amount of infiltration during one day) in the present model, which reads as follows:

$$F = a\ (ST_{max} - ST_{n-1})^{b} + F_{min} \tag{5}$$

where

$F$ = the infiltration capacity during the n-th day (mm)
$F_{min}$ = the lower limit of the infiltration capacity (mm)
$ST_{max}$ = the soil moisture storage at field capacity in the upper zone (mm)
$ST_{n-1}$ = the actual soil moisture storage at the end of the (n-1)-th day (mm)
a, b = model parameters

The model parameters in Eq. 5 have no physical interpretation and must be evaluated from the calibration procedure.

Evapotranspiration is the total process of moisture removal from the surface occurring in a variety of ways, i.e. from interception and depression storage, saturated or unsaturated surface soil zone and vegetation. In an attempt to simulate the evapotranspiration, two concepts have been introduced, viz. the "potential evapotranspiration" and the "evapotranspiration opportunity" (Fleming, 1975). The potential evapotranspiration may be defined as the maximum rate at which water leaves the surface at a certain time assuming an unlimited supply of available moisture. The potential evapotranspiration rate is governed by meteorologic parameters and the concept of a potential provides therefore an upper limit to the losses of moisture to the atmosphere. The second concept accounts for the condition where the moisture supply is restricted, and is here defined as the maximum quantity of moisture available for evapotranspiration during a certain day. In case the moisture supply is limited, the potential evapotranspiration will not be achieved and the quantity of moisture lost to the atmosphere will be termed the actual evapotranspiration.

In the present model the total evapotranspiration is subdivided into two components as follows:

$$E_p = E_i + E_{sp} \tag{6a}$$

and

$$E_a = E_i + E_{sa} \tag{6b}$$

where

$E_p$ = the total daily potential evapotranspiration (mm)
$E_a$ = the total daily actual evapotranspiration (mm)
$E_i$ = the daily evaporation from interception- depression storage (mm)
$E_{sp}$ = the daily potential evapotranspiration from soil moisture storage (mm)
$E_{sa}$ = the daily actual evapotranspiration from soil moisture storage (mm)

During rainfall, the potential evapotranspiration may reduce considerably below the average daily potential rate. The daily potential evapotranspiration is thought to decrease asymptotically as the daily amount of rainfall increases.

$$E_p = (E_{p,max} + E_{p,min}\,P) \,/\, (P + 1) \tag{7}$$

The interception- depression storage will lose water at the full potential rate as water is freely available until this source is exhausted. If the evaporative demand is not met by losses from the interception-depression storage, then the remaining potential is applied to the moisture storage in the upper unsaturated zone. Evapotranspiration opportunity is assumed to control the evapotranspiration from this zone. Boughton's linear relationship between the actual evapotranspiration from subsurface moisture storage and moisture deficit (Fleming, 1975), has been adopted in the present model. The actual evapotranspiration rate varies from zero at the wilting point to the maximum value at field capacity. For moisture levels greater than the wilting point level, the actual evapotranspiration equals

$$E_{sa} = E_{sp}\ (ST_{n-1} / ST_{max}) = (E_p - I)\ (ST_{n-1} / ST_{max}) \qquad , I < E_p$$

or

$$E_{sa} = 0 \qquad , I \geq E_p \qquad (8)$$

where

$E_{sa}$ = the actual evapotranspiration from the upper unsaturated zone storage during the n-th day (mm)
$E_{sp}$ = the residual potential evapotranspiration applied to the upper unsaturated zone storage during the n-th day (mm)
$I$ = the actual interception- depression storage level during the n-th day (mm)
$ST_{n-1}$ = the actual moisture level in the upper unsaturated zone at the end of the (n-1)-th day (mm)
$ST_{max}$ = the moisture level at field capacity (mm)

To satisfy the concept of evapotranspiration opportunity, the daily amount of evapotranspiration from the soil moisture storage cannot exceed the current storage quantity. By including the constraint

$$E_{sp,max} = E_{p,max} \leq ST_{max} \qquad (9)$$

in the model, violation of this concept will be prevented.

Soil water budgetting of the upper portion of the unsaturated zone is based on two somewhat crude assumptions, viz.

- excess of soil moisture ($P_{ef}$) drains downward instantaneously at the end of the day,
- interception- depression storage depletes completely and instantaneously at the end of the day.

The soil moisture level at the end of the n-th day is computed as follows:

$$ST^*_n = ST_{n-1} + P_{exc} - E_{sa} \qquad , I < E_p$$

or

$$ST^*_n = ST_{n-1} + P_{exc} + I - E_p \qquad , I \geq E_p \qquad (10a)$$

$$P_{ef} = 0 \qquad , ST^*_n \leq ST_{max}$$

or

$$P_{ef} = ST^*_n - ST_{max} \qquad , ST^*_n > ST_{max} \qquad (10b)$$

$$ST_n = ST^*_n \qquad , ST^*_n < ST_{max}$$

or

$$ST_n = ST_{max} \qquad , ST^*_n \geq ST_{max} \qquad (10c)$$

So far, the computational procedure, with respect to the daily effective rainfall based on soil moisture accounting, has been elucidated. To employ the convolution operation, a mathematical formula of the 1-day unit hydrograph of the water-table recovery should be established.

The linear impulse (instantaneous input with unity depth) response of a system consisting of two serially arranged linear reservoirs can be derived by convolution of the instantaneous unit hydrograph of the second reservoir and the input constituted by the instantaneous unit hydrograph of the first reservoir (Sugawara and Maruyama as quoted by Kraijenhoff van de Leur, 1972). Because of the assumption that linear response is not influenced by previous inputs (Viessman, 1977), the unit hydrograph is in fact the hydrograph of the additional discharge rate, superimposed on the hydrograph that would have been shown if no impulse input did occur. The mathematical expression of the instantaneous hydrograph reads as follows:

$$u_{\Delta Q}(0,t) = [\ \exp(-t/k_1) - \exp(-t/k_2)\ ] \ / \ (k_1 - k_2) \tag{11}$$

where

$u_{\Delta Q}(0,t)$ = the additional discharge rate from the second reservoir effectuated by the impulse input (L/T)
$k_1$ = the time lag of the first reservoir (T)
$k_2$ = the time lag of the second reservoir (T)

From linear systems analysis is known that the 1-day unit hydrograph of the outflow rate resulting from a block input with a duration of one day and unity depth can be derived from the instantaneous unit hydrograph as follows:

$$u_{\Delta Q}(1,t) = \int_{t-1}^{t} u_{\Delta Q}(0,\Omega)\, d\Omega \qquad , \ t > 1 \text{ day} \tag{12}$$

In a linear reservoir the outflow is directly proportional to storage which is reflected by the water table in case of a phreatic aquifer, according to

$$Q_{grw} = \mu\, H \ / \ k_2 \tag{13a}$$

and in marginal terms

$$\Delta Q_{grw} = \mu\, \Delta H \ / \ k_2 \tag{13b}$$

where

$Q_{grw}$ = the ground-water discharge (mm/day)
$\Delta Q_{grw}$ = the additional ground-water discharge caused by an effective rainfall pulse (mm/day)
$\mu$ = the effective porosity of the aquifer (-)
$H$ = the water table elevation with respect to the aquifer base (mm)
$\Delta H$ = the difference in water table with and without an effective rainfall pulse event (mm)
$k_2$ = the time lag of the ground-water reservoir (day)

Substituting Eq. 11 and Eq. 13b into Eq. 12 yields the following expression of the 1-day unit hydrograph of the the water-table recovery

$$u_{\Delta H}(1,t) = k_1 k_2 [(\exp(1/k_1) - 1) \exp(-t/k_1)] / [\mu (k_1 - k_2)] - $$
$$- k^2_2 [(\exp(1/k_2) - 1) \exp(-t/k_2)] / [\mu (k_1 - k_2)], \quad t > 1 \text{ day} \qquad (14)$$

where $k_1$ and $k_2$ are the time lags of the lower portion of the unsaturated zone and the ground-water reservoir, respectively. The time lag of the ground-water reservoir can de derived from the recession limb of the water-table hydrograph, as will be shown further on. The difference between the water-table hydrograph with and without recharge during a certain period (e.g. the rainy season) at a certain time can be calculated by convolution of the 1-day unit hydrograph and the preceding daily effective rainfall inputs. The convolution procedure, which is nothing else than the application of the principle of superposition can be employed by the matrix method (Viessman et al, 1977) as follows:

$$\begin{bmatrix} \Delta H_1 \\ \Delta H_2 \\ \cdot \\ \cdot \\ \cdot \\ \cdot \\ \Delta H_{n-1} \\ \Delta H_n \end{bmatrix} = \begin{bmatrix} P_{ef,1} & 0 & \cdot & \cdot & \cdot & \cdot & \cdot & 0 \\ P_{ef,2} & P_{ef,1} & 0 & \cdot & \cdot & \cdot & \cdot & 0 \\ \cdot & \cdot & \cdot & \cdot & \cdot & \cdot & \cdot & \cdot \\ \cdot & \cdot & \cdot & \cdot & \cdot & \cdot & \cdot & \cdot \\ \cdot & \cdot & \cdot & \cdot & \cdot & \cdot & \cdot & \cdot \\ \cdot & \cdot & \cdot & \cdot & \cdot & \cdot & \cdot & \cdot \\ P_{ef,n-1} & P_{ef,n-2} & \cdot & \cdot & \cdot & P_{ef,2} & P_{ef,1} & 0 \\ P_{ef,n} & P_{ef,n-1} & \cdot & \cdot & \cdot & P_{ef,3} & P_{ef,2} & P_{ef,1} \end{bmatrix} \times \begin{bmatrix} u_1 \\ u_2 \\ \cdot \\ \cdot \\ \cdot \\ \cdot \\ u_{n-1} \\ u_n \end{bmatrix}$$

where

$\Delta H_n$ = $H_{act,n} - H_{rec,n}$

$H_{act,n}$ = the actual water table with respect to a datum plane at the end of the n-th day (mm)

$H_{rec,n}$ = the hypothetical water table effectuated by long term ground-water storage recession at the end of the n-th day (mm)

$P_{ef,n}$ = the amount of effective point rainfall during the n-th day (mm)

$u_n$ = the unit hydrograph ordinate at the end of the n-th day (mm)

One must bear in mind that the time base of a unit hydrograph of the water-table recovery can extend over a period of several years, and that the duration of a rainy season, for example, is therefore in general considerably smaller. Consequently, just a limited portion of the unit hydrograph will be used during the convolution operation.

Moisture accounting and convolution are being repeated during the model calibration, which involves manipulating the model to reproduce the water-table recovery within some range of accuracy. The objective of the fitting or calibration procedure is to achieve an optimal set of values of process parameters such as infiltration and soil moisture capacity which cannot readily be assessed by measurement, by minimizing the error function

$$Z = \Sigma_n \, (\Delta H_{obs} - \Delta H_{calc})^2 \tag{15}$$

where

$Z$ = the objective function to be minimized
$\Delta H_{obs}$ = the observed water table recovery
$\Delta H_{calc}$ = the simulated water table recovery
$n$ = the number of records at daily time intervals

An automatic parameter optimization technique based on the Rosenbrock or Nelder-Mead method (Clarke, 1973), can be used for this purpose. It is necessary to have a measure of the goodness of fit of the model outcome to the observed data, once the model parameters have been estimated. The coefficient of determination is a useful criterion for the goodness of fit and reads

$$R^2 = 1 - \Sigma_n \, (\Delta H_{obs} - \Delta H_{calc})^2 \, / \, \Sigma_n \, (\Delta H_{obs} - \overline{\Delta H_{obs}})^2 \, , \tag{16}$$

so that the nearer to unity the value of $R^2$, the better the correlation between the observed and simulated data.

In order to derive the "observed" data from the measured water-table hydrograph, a mathematical expression of the water-table recession curve should be established. Substituting the flow equation

$$Q_{grw}(t) = \mu \, (H(t) - H_{base}) \, / \, k_2 \tag{17}$$

into the continuity equation

$$Q_{grw}(t) = - \mu \, d(H(t) - H_{base}) \, / \, dt \tag{18}$$

results in a differential equation which has as solution satisfying the initial condition $H(t) = H_0$ at $t=0$

$$H(t) = H_{base} + (H_0 - H_{base}) \exp(-t/k_2) \tag{19}$$

where

$H(t)$ = the water-table elevation during recession at time t (m + mean sea level)
$H_{base}$ = the elevation of the aquifer base (m + mean sea level)
$H_0$ = the water-table elevation at the starting time of the recession period (m + mean sea level)
$k_2$ = the time lag of the ground-water reservoir (day)

The unknown parameters in Eq. 20 can be obtained from non-linear regression (Kuester et al, 1973). Eq. 20 can be applied to any time period just by resetting the water table elevation at a certain starting time.

## 3. APPLICATION OF THE RECHARGE MODEL

The proposed model has been applied to a site near the topographic water divide of a catchment area in the central part of South Africa. The area is underlain by fine-textured sandstones interbedded with mudstones of the Karoo Sequence. The water table ranges between 15 and 20 m below the surface. The calibration period starts at the beginning of the first rainy

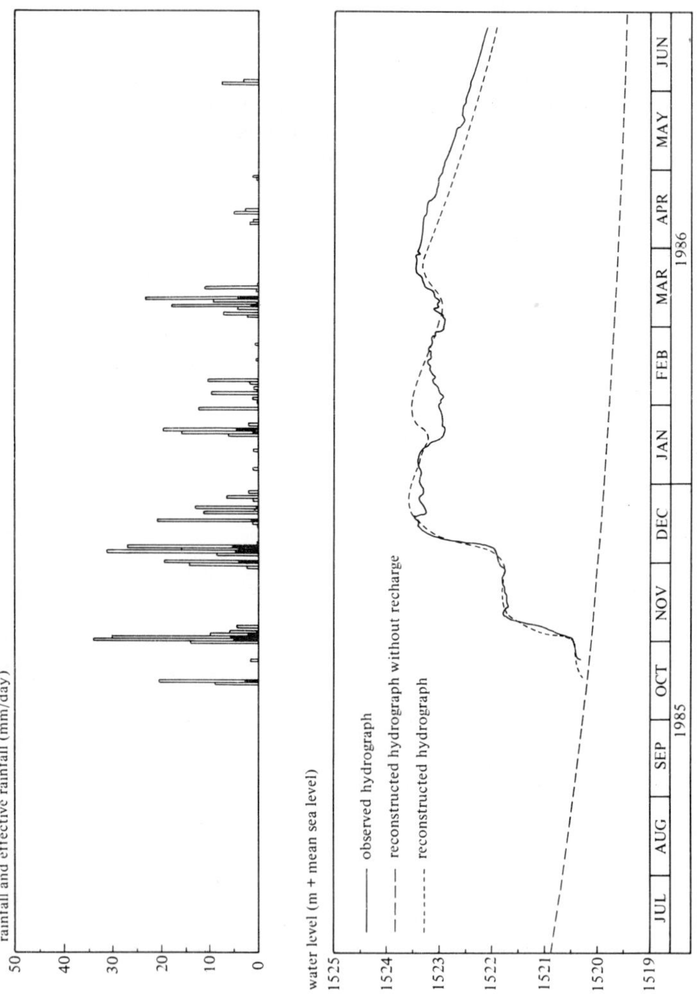


Figure 2. Point rainfall and simulated effective rainfall diagram, and observed and simulated water-table hydrograph at test site

day of the wet season preceded by a period of continuous drought, so that the initial soil moisture level in the upper portion of the unsaturated zone could be set at zero. Because rainfall is mostly occurring from late afternoon till early in the morning of the following day, the selected daily time interval runs from 8 o'clock a. m. till 8 o'clock a. m. next day, in order to prevent splitting up a single rainfall event.

The Nelder-Mead algorithm has been used for optimizing the constrained model parameters.

Fig. 2 shows graphically the results of the model calibration for the rainy season from the beginning of October 1985 till the end of March 1986. The reconstructed water-table hydrograph has been obtained by superimposing the simulated water-table recovery effectuated by ground-water recharge on the reconstructed recession graph. Although the recorded hydrograph is slightly distorted by pumping and consequently a somewhat unfortunately selected test example, the simulated hydrograph fits quite well ($R^2 = 0.96$). However, a good fit of simulated and observed outputs may be an indication, but is no proof, of the correctness of the conceptual model. Such proof can only be derived from comparing the set of computed values of the model parameters with physical information about the hydrologic system. The resulting values of the process parameters are:

$I_{max}$ = 2,27 mm (Eq. 4), $ST_{max}$ = 6.44 mm (Eq. 5 and 8), a = 0.345 (Eq. 5), b = 1.31 (Eq. 5), $F_{min}$ = 9.88 mm (Eq. 5), $E_{p,max}$ = 6.04 mm (Eq. 7), $E_{p,min}$ = 4.45 mm (Eq. 7), $k_1$ = 5.32 day, (Eq. 14), $\mu$ = 0.00754 (Eq. 14).

The interception- depression storage capacity in the Karoo can be as high as 5 mm (Roberts, 1978). Viessman showed that the depression storage capacity may reduce considerably with increasing steepness of the terrain slope (Viessman, 1977). As the slope at the test site is covered with scarce vegetation, and its steepness is more than 10%, the computed value of the interception- depression storage capacity seems quite realistic.

The maximum available soil moisture and the infiltration characteristics at the test site with a very shallow soil cover of a few centimetres and outcropping hard rock cannot be measured in the field properly. From field experiments at sites underlain by heavy clayey soils, it is known that ponding conditions will occur soon during a storm event. The potential infiltration rate drops rapidly because of crusting of the soil surface by swelling colloids.

The optimized model relationship between potential daily evapotranspiration and daily rainfall (Eq. 7) fits reasonably well with the swarm of dots representing daily Penman values derived from hourly meteorologic parameter values (Fig. 3). The relatively small difference in computed maximum and minimum potential evapotranspiration is likely because of the fact that most of the rain is falling during the late afternoon and the night, during which the evaporative demand of the atmosphere is proportionally small.

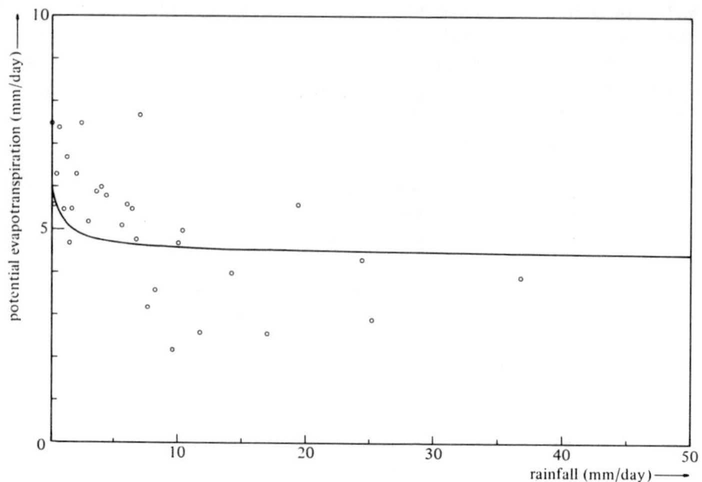


Figure 3. Daily Penman evapotranspiration (dots) and simulated daily potential evapotranspiration vs. daily rainfall.

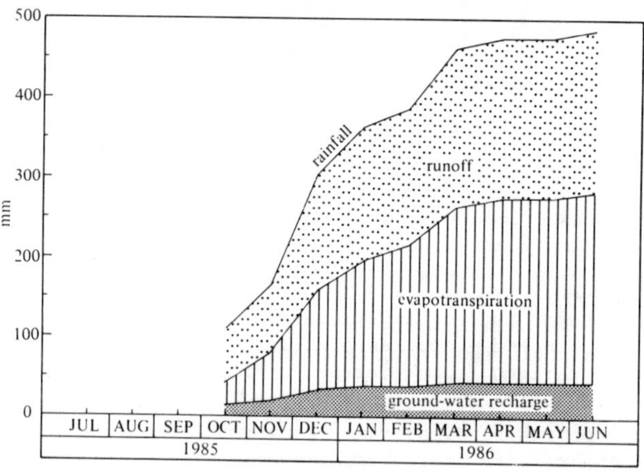


Figure 4. Cumulative monthly rainfall and its distribution over the various water balance components, resulting from the optimized recharge model.

The time lag of the lower portion of the unsaturated zone seems quite realistic, since one can see from Fig. 2 that the water table is responding to a significant rainfall event within a few days.

The effective porosity of Karoo aquifers is rather low as shown from pumping test results, and ranges between 0.5 and 1%. The optimized value amounts 0.75% and agrees well with the order of magnitude obtained from pumping tests.

The computed daily amounts of the water balance terms such as surface runoff, evapotranspiration and ground-water recharge (effective rainfall), have been added up for each calendar month of the calibration period. The cumulative monthly totals of the various water balance terms are shown in Fig. 4. The calculated totalized surface runoff is more than 40% of the total rainfall which may be possible if one realizes that the slope at the test site is covered with a thin soil layer and scarce Karoo vegetation, and its steepness is more than 10%. The total ground-water recharge during the rainy season as computed with the model, amounts to approximately 9% of the total rainfall which seems quite realistic, compared with currently accepted average recharge-rainfall ratios in the region (Bredenkamp, 1978).

In case the calibrated model could be verified and recalibrated for various rainy seasons with divergent rainfall conditions, then soil moisture accounting, using the optimized set of process parameters, could be applied to historical rainfall data for generating recharge data.

Multiple linear regression can be applied to the simulated amounts of daily effective rainfall, as shown in the rainfall diagram of Fig.2, and the preceding daily rainfall quantities. The resulting equation can be used to generate recharge data. Multiple regression analysis showed that only the amount of rainfall during the present and previous day affects the amount of effective rainfall during the present day significantly. The resulting recharge formulae reads

$$P_{ef} = 0 \qquad , \qquad 0.156\ P + 0.072\ P_{-1} < 1.165$$

or

$$P_{ef} = -1.165 + 0.156\ P + 0.072\ P_{-1} \qquad (R^2 = 0.88) \tag{20}$$

It goes without saying that Eq. 21 should be applied to the test site with the highest caution as long as the calibrated model is not yet verified.

## 4. DISCUSSION

The underlying concepts of the recharge model in regard to the physical processes of infiltration and evapotranspiration, as well as the reservoir approach, are largely derived from watershed hydrology. In this respect, the proposed model can be considered as a new approach in geohydrology for establishing a relationship between the local ground-water recharge from rainfall and the preceding rainfall with respect to areas underlain by hard-rock formations with or without thin soil cover. The application of the model may be restricted to arid and semi-arid regions characterized by clearly defined wet and dry periods, which will allow the reconstruction of a water-table recession graph during the rainy periods.

As demonstrated by the testing results, the model seems rather sound. However, except for the possible errors already emphasized previously, the model may be subject to some erroneous assumptions. Both the time lag of the lower unsaturated zone and the effective porosity of the capillary fringe are being considered as parameters that do not depend on the water-table elevation. A water-table rise implies, however, a reduction of the thickness of the unsaturated zone and consequently, a decrease of the time lag as well. The effective porosity may vary considerably with depth. The variability of both the time lag and the effective porosity depend therefore on the depth range within which the local water table is fluctuating, with respect to the site surface.

Calibrating the model for different rainy periods may result in considerably different estimates of certain model parameters. This large sampling variability of the concerning parameters indicates that large changes in the value of these parameters will have but little effect on the value of the error function. In consequence of this insensitivity to certain parameters, model calibration may yield absurd parameter values resulting in unrealistic daily water balances. Therefore, during optimization the range of possible parameter values should be constrained within which the parameter values have a physical significance. For that purpose, it is advisable to collect data with respect to potential evapotranspiration and, if possible, interception- depression storage and infiltration capacity.

The proposed model yields just point recharge estimates. It goes without saying that the model should be applied to various sites in the region, in order to get a picture of the spatial variability of the natural recharge from rainfall.

## 5. ACKNOWLEDGEMENTS

The development of the recharge model is part of a more general study on the exploitable ground-water potential of Karoo aquifers, funded by the Water Research Commission. Their financial assistance is acknowledged gratefully.

## 6. REFERENCES

ALLISON, G. B., STONE, W. J. and HUGHES, M. W. (1984) 'Recharge in karst and dune elements of a semi-arid landscape as indicated by natural isotopes and chloride'.*Journal of Hydrology* , 76, 1-25.

BREDENKAMP, D. B. (1978) *Quantitative estimation of ground-water recharge with special reference to the use of natural radioactive isotopes and hydrological simulation.* Technical Report 77, Hydrological Research Inst., Dept. of Water Affairs, Pretoria, RSA.

CLARKE, R. T. (1973) *Mathematical models in hydrology.* FAO Irrigation and drainage paper 19.

FLEMING, G. (1975) *Computer simulation techniques in hydrology.* Elsevier environmental science series.

KRAIJENHOFF VAN DE LEUR, D. A. (1972) 'Rainfall-runoff relations and computational models'.*Drainage principles and applications.* Int. Inst. for Land Reclamation Wageningen, The Netherlands. Publication 16, Vol. II.

KUESTER, J. L. and MIZE, J .H. (1973) *Optimization techniques with Fortran.* McGraw-Hill Book Company.

ROBERTS, P. J. T. (1978) *A comparison of the performance of selected conceptual models of the rainfall-runoff process in semi-arid catchments near Grahamstown.* Report 1/78, Hydrological Research Unit, Rhodes Un.,RSA.

SKAGGS, R. W. (1981) 'Infiltration'. *Hydrologic modeling of small watersheds,* 121-166, Haan, C. T., Johnson, H. P. and Brakensiek, D. L. (Eds), Am. Soc. of Agr. Engineers.

VIESSMAN, W. , KNAPP, J. W., LEWIS, G. L. and HARBAUGH, T. E. (1977) *Introduction to hydrology.* Harper & Row Publishers.

VISWANATHAN, M. N. (1982) 'The rainfall/water-table level relationship of an unconfined aquifer". *Ground Water,* Vol. **21** , No. 1, 49-56.

# NATURAL GROUND WATER RECHARGE ESTIMATION METHODOLOGIES IN INDIA

B.P.C. Sinha
Chief Hydrogeologist & Member

Santosh Kumar Sharma
Senior Hydrogeologist
Central Ground Water Board: Jamnagar House:
Mansingh Road: New Delhi-110011 (India)

INTRODUCTION. Ground water is one of the most important and widely distributed resources of the earth. Ground water development forms the bulk of irrigation development programmes in most of the states of India. The remarkable increase in food production in areas utilizing ground water irrigation bears a testimony to realisation of groundwater as an important source of irrigation.

For planned development of ground water resources it becomes essential to quantify the ground water resources of different administrative units/basins on a realistic basis. Since ground water is a dynamic and replenishable resource, its potential is generally estimated from the component of annual recharge which could be developed by means of suitable ground water structures.

## RECHARGE ESTIMATION TECHNIQUES

Rainfall is the most important source of groundwater recharge in the country. In the major part of India the maximum fluctuation of ground water levels and hence ground water recharge takes place during the south-west monsoon, but in some parts especially in the south and south-east, north-east monsoon rainfall is the dominant source of recharge.

Rainfall infiltration primarily depends upon duration and intensity of rainfall, soil moisture characteristics, topographic slopes, land use pattern, agronomic practices, weather conditions preceding, during and succeeding rainfall periods, and depth-to-water table. These conditions further impose a limit on threshold values of rainfall required to effect ground water recharge.

The natural recharge to an aquifer in a groundwater basin from precipitation is computed by various methods. Some of the methods in vogue in India are :

I. Empirical methods
II. Hydrologic budgeting methods
III. Ground water level fluctuations.

## I. EMPIRICAL METHODS



*I. Simmers (ed.), Estimation of Natural Groundwater Recharge, 301–311.*

Based on the studies undertaken by different scientists and organisations regarding correlation of ground water level fluctuation and rainfall, some empirical relationships have been derived for computation of natural recharge to ground water from rainfall. Some of these empirical relationships for different hydrogeological situations in India are described hereunder.

i) Chaturvedi Formula

Based on water level fluctuations and rainfall amounts, Chaturvedi (1943) derived an empirical relationship to arrive at the amount of rainfall that penetrates into the ground when rainfall exceeds 15 inches (38 cm) :

$$R_P = 2\ (P - 15)^{0.4}$$

with further work, the formula was modified to :

$$R_P = 1.35\ (P - 14)^{0.5}$$

Where $R_P$ = ground water recharge in inches, and P = annual rainfall in inches.

Subsequently, Sehgal (1973) developed a formula commonly called the 'Amritsar Formula' using regression analysis for certain interstream alluvial tracts in Punjab. With the same notations as above, the formula given below was found to accord with actual observations where rainfall was between 58 and 71 cm.

$$R_P = 2.5\ (P - 16)^{0.5}$$

ii) Studies in Vedavati River Basin Project, Karnataka, India by Central Ground Water Board

Under these Project studies, the continuous monitoring of ground water levels in upper and lower reaches and the central part of the basin underlain by hard rock formations was carried out to determine rainfall infiltration. The total rainfall affecting the rise in ground water levels was correlated with the increment to the ground water body, utilising a uniform factor 0.025 which represents average specified yield of the aquifers in the basin. The following relationship was derived between rainfall (R) and percolation (P) based on statistical analysis:

$$P = 0.26\ (R-23)$$

P and R are expressed in millimeters. The threshold value of rainfall resulting in significant rise of ground water level is 23 mm. The correlation co-efficient between P and R is highly significant and has a value of 0.96.

## II. HYDROLOGIC BUDGETING METHODS

In this method it is presumed that rainfall in excess of evapotranspiration losses is utilised in bringing the soil moisture to its field capacity and the rest is available for ground water recharge and runoff. The hydrological budget technique is a statement to account for water gains and losses over selected periods in an area and can be expressed as follows :

$$P + I = R + ET + \Delta SW + \Delta SM + \Delta GW$$

where, P = rainfall;
I = applied irrigation water
R = is runoff; ET is evapotranspiration;

$\Delta SW$ is change in surface water storage;
$\Delta SM$ is change in soil moisture; and
$\Delta GW$ is change in ground water storage.

Further $\Delta GW = \Delta H.yg$ in which
$\Delta H$ is change in ground water level and yg is specific yield.

The complex sum of the revised equation $P-R-\Delta SW-ET-\Delta SM$ is equal to H.yg, which is equivalent to ground water recharge.

The estimation of ground water recharge was achieved using hydrologic budget techniques and also by adopting a soil moisture budget technique in the Ghaggar Project, covering semi-arid areas of parts of Haryana and Punjab States of India under the UNDP assisted Phase III Project. These case studies are described hereunder.

i) Estimation of Recharge by Analysis of Hydrometeorological Data

The Ghaggar River Basin covers an area of some 42,200 km2 extending over parts of several states in north-western India : Hryana, Himachal Pradesh, Punjab, Rajasthan and Union Territory of Chandigarh. The Ghaggar River basin is part of the large Indo-Gangetic Quaternary basin and consists of alluvial deposits. The area is also fed by surface water irrigation through a canal network.

For water balance studies the Project area was divided into ten zones; the nine canal command areas plus the one rain-fed zone in the north-eastern part of the Project area, where there was no canal irrigation.

Water balance studies were carried out separately for each zone for each of the two crop seasons-Kharif (June to October) and Rabi (November to May) for the two years 1976-1977 and 1977-1978.

Using monthly rainfall data from the 114 existing stations and 19 project statins, total rainfall data for Kharif and Rabi seasons were derived for both years. Isohyetal maps of the project areas were prepared for each season and average rainfall converted into volume of water received for each zone.

Rainfall during the November-May period amounts to barely 20 per cent of the annual total and hence is not sufficient to raise the soil moisture to field capacity. Losses from actual evaporation plus soil moisture retention were therefore limited to rainfall in this period for all zones.

Rainfall was above normal in both 1976-1977 and 1977-1978 by about 30 and 20 per cent, respectively. 1976 and 1977 departures from normal annual rainfall were thus calculated for all water balance zones; in 1976, all zones received more rainfall than in 1977; moreover, groundwater recharge in 1976 was greater than in 1977 in all the water balance zones, as shown in table I.

TABLE I.
GROUND WATER RECHARGE AND DEPARTURE FROM NORMAL RAINFALL IN WATER BALANCE ZONES, 1976 AND 1977

| Water balance zones | Groundwater Recharge (mm) 1976 | Rainfall Departure (% above normal) 1976 | Groundwater Recharge (mm) 1977 | Rainfall Departure (% above normal) 1977 |
|---|---|---|---|---|
| 1. | 40.8 | 45.0 | 20.9 | 26.5 |
| 2. | 70.8 | 33.8 | 69.6 | 29.0 |
| 2(A). | 193.7 | 59.0 | 124.5 | 47.6 |
| 3. | 106.0 | 24.1 | 59.6 | 16.5 |
| 4. | 37.6 | 14.3 | 42.0 | 7.5 |
| 5. | 105.0 | 12.5 | 43.8 | 8.3 |
| 6, 7. | 72.9 | 29.2 | 40.7 | 17.5 |
| 8. | 114.8 | 59.0 | 90.2 | 47.6 |
| 9, 9(A). | 113.2 | 59.0 | 95.8 | 47.6 |
| 10. | 87.6 | 10.7 | 51.1 | 8.9 |

(A) by hydrometeorological method.

ii) Estimation of Recharge from Rainfall by Soil Moisture Budget Technique

The soil moisture budget technique was applied at 25 selected stations representing all water balance and climatic zones of the Project area. The findings of this technique in different climatic zones were as follows:

(a) Arid zones

(i) There is no contribution to ground water in the arid zone from the mean areal seasonal rainfall of 190 mm.

(ii) If the areal rainfall during a given year is as much as 1.5 times the mean seasonal rainfall there is no contribution to the ground water recharge.

(iii) If the seasonal areal rainfall in a given year is approximately double the mean seasonal rainfall, rainfall will contribute positively to ground water recharge.

(iv) Hence, direct recharge from rainfall in the arid zone is possible only in wet years (double or more than double the mean seasonal rainfall). In 1976 the seasonal rainfall received in the zone was 441.7 mm, which accounted for a recharge of 34.0 mm, or about 7.6 per cent. In 1977 the recharge from rainfall was nil although the recorded rainfall was about 43 per cent above the mean seasonal value.

(b) Semi-arid zone

(i) The long-term mean areal seasonal rainfall in the semi-arid zone, 312 mm, does not contribute to ground water recharge. Only in wet years, with above-normal falls, does rainfall contribute to recharge.

(ii) Rainfall contributes to ground-water recharge only in those years when seasonal rainfall is more than 10 per cent above the mean, i.e. 340 mm or higher.

(iii) In 1977, the areal seasonal rainfall was 456.6 mm, or about 116.6 mm above threshold value (340 mm), recharge was 13.8 mm, which was 12 per cent of the amount in excess of the threshold value. In 1976, the areal seasonal rainfall was 548.0 mm, or about 208 mm above the threshold value; this accounted for a recharge of 85.1 mm, which was 41 per cent of the amount in excess of threshold value.

(c) Sub-humid zone

(i) The areal seasonal rainfall in 1976 and 1977 was 35 and 46 per cent, respectively above mean areal seasonal rainfall. The recharge contribution was about 24 per cent of seasonal rainfall in each year :

(ii) Every year in which rainfall is equal to or greater than the mean seasonal rainfall, ground water recharge takes place.

(d) Humid zone

(i) The humid zone contributes significantly to runoff in the order of 50 per cent of total rainfall received.

(ii) In 1976 and 1977 the areal seasonal rainfall was about 24 and 18 per cent, respectively above mean seasonal rainfall, contributing 16 and 15 per cent respectively to ground water recharge. In this zone, ground water recharge occurs almost every year. Recharge from rainfall is about 15 per cent in a normal rainfall year; if rainfall is more than 25 per cent above mean seasonal rainfall in a year, the percentage contribution to recharge would also be greater than 15 per cent.

### (iii) Ground water recharge according to the soil-moisture balance method in Sikar Luni and Bikaner Basins of Rajasthan, under UNDP Phase-II Project

A project entitled "Ground Water Surveys in Rajasthan and Gujarat" was taken up by the Central Ground Water Board with UNDP assistance. The Rajasthan Project area is located in the semi-arid to arid western part of the State of Rajasthan and covers about 42000 km2. The area is divided into three ground water basins-the Sikdar, Bikaner and Luni basins. The area is underlain by aeolian deposits, semi-consolidated and consolidated formations. The ground water occurs under phreatic as well as confined conditions.

Soil moisture balance computations were made in semi-arid zones with an areal extent of 14,464 km2 in the Sikar basin and 26,398 km2 in the Bikaner and Luni basins of Rajasthan to obtain ground water replenishment.

In the absence of any notable surface runoff in the Rajasthan project area, it is assumed that the computed total runoff, or "water surplus" represents ground water recharge.

Considering a root zone approximately 2 metres deep, the maximum available soil moisture, or soil moisture storage capacity, is estimated at 100 mm for the eastern part of the project area.

North of the project area, lying in a surface water irrigation project (Rajasthan Canal Project) the available soil moisture is between 9.4 and 12 per cent. Accordingly, a value of 150 mm for the maximum water-holding capacity in the Bikaner area has been used.

The estimated annual water surplus, for soil moistures of 150 or 100 mm, has been computed for 11 stations for the years 1971, 1972 and 1973 to show the areal variation, as depicted in table 2. As expected, the areal variation is very large. This is explained by the very local nature of thunderstorms in this area. In 1973 for example, most of the annual water surplus at Nagaur (27°12', 73°45'), about 280 mm, was carried by a single storm in August. At the same time the annual water surplus at Didwana (27°24', 74°34') which was not hit by any cloudburst was probably zero.

During the three project years 1971-1973 the average annual water surplus was about 40 mm on an average for the Sikar basin, assuming 100 mm available soil moisture; that is, somewhat higher than the long-term mean of about 33 mm (average of water surplus for Didwana and Jhunjhunu). For the western part of the project area, if 150 mm available soil moisture is assumed, water surplus for the three years is about 32 mm.

In fact, these computations of recharge using a soil moisture budget technique are valid only when the moisture surplus percolates quickly to the water table. If infiltration capacity is lower than rainfall intensity, part of the water remains on the soil and evaporates directly. If the percolation rate in the subsurface is slowed by compact materials or clay beds, evaporation can also take place during the dry season through roots of scattered trees which penetrate deeply into the soil.

It is, therefore, quite essential to monitor the ground water levels

TABLE 2.

ESTIMATED WATER SURPLUS FOR 11 STATIONS DURING THE PERIOD 1971-1973 (IN MILLIMETRES)

| Area | Station | 1971 | | 1972 | | 1973 | |
|---|---|---|---|---|---|---|---|
| | | Annual Rain-fall | Water Sur-plus | Annual Rain-fall | Water Sur-plus | Annual Rain-fall | Water Sur-plus |
| Sikar | Churu | 298 | 0 | 191 | 0 | 304 | 11 |
| Basin & | Didwana | 402 | 4 | 160 | 0 | 334 | 0 |
| Luni | Jhunjhunu | 535 | 15 | 234 | 10 | 379 | 5 |
| Basin | Parbatsar | 857 | 391 | 239 | 20 | 453 | 68 |
| 1/ | Ratangarh | 438 | 36 | 146 | 0 | 406 | 26 |
| | Sikar | 575 | 12 | 160 | 22 | 447 | 49 |
| | Sujangarh | 472 | 89 | 159 | 0 | 540 | 112 |
| | Average | 510 | 78 | 184 | 7.5 | 409 | 39 |
| Bikaner | Bikaner | 211 | 0 | 207 | 0 | 209 | 0 |
| Basin | Didwana | 402 | 0 | 160 | 0 | 334 | 0 |
| 2/ | Dungargarh | 231 | 0 | 247 | 0 | 344 | 8 |
| | Nagaur | 467 | 98 | 331 | 0 | 731 | 277 |
| | Average | 328 | 24 | 236 | 0 | 404 | 71 |

1/ Soil moisture storage capacity 100 mm
2/ Soil moisture storage capacity 200 mm

to observe the effect of moisture surplus on a ground water regime, and as such the soil moisture budget technique needs to be checked with actual ground water responses.

## III. GROUND WATER LEVEL FLUCTUATION OR CHANGE IN GROUND WATER STORAGE METHOD

In this method use is made of the data on seasonal variation in ground water level, rainfall, applied irrigation water etc. This method is generally not applicable to confined aquifers. The rise and fall of water table over an area is a measure of change in ground water storage which is computed as a product of specific yield, the average rise in level and the area over which the change occurs. The recharge or change in storage can be expressed by the following relationship:

$R = \Delta h . Sy$

Where R = Ground water recharge

$\Delta h$ = Change in water level due to rainfall

Sy = Specific yield of the formation in the zone of fluctuation.

The rainfall recharge method utilising ground water level fluctuation and specific yield is widely used in India. Ground water level fluctuation is a true reflection of the input to and out put from the ground water regime and is used in computations of ground wter resources all over India.

Ground water levels are being monitored by different organisations in India, varying from two to four times a year based on the rainfall pattern. The frequency of measurement is greater in areas where the intensity of ground water development is high or where some specific problems as regards quality of ground water etc. are encountered. Estimation of specific yield of the different types of aquifers is quite essential for precision in computing recharge.

Estimation of ground water recharge and the extent of its utilisation in various parts of the country for speedy ground water development has been based on various empirical methods. During the last decade, significant steps were taken when extensive and intensive scientific studies through multi-disciplinary ground water projects were undertaken by the Central Ground Water Board (CGWB) and State Ground Water Organisations to evaluate the ground water resources available for development. These studies yielded valuable data from the varied hydrogeological situations in India and concerted efforts were made to rationalise recharge estimation methodology all over the country. In 1982 the Government of India formed a "Ground Water Estimation Committee" (GEC) with members drawn from various organisations engaged in hydrogeological studies and ground water development. This committee, after reviewing the data collected by Central and State agencies, research organisations, universities etc. recommended (1984) the norms for ground water resource estimation.

The Committee recommended that ground water recharge estimations should be based on the <u>ground water fluctuation method</u>. However, in areas where ground water level monitoring is not adequate in space and time, the norm of rainfall infiltration may be adopted. The rainfall infiltration contributing to ground water recharge has evolved from studies undertaken in various ground water projects in India. The recommended norms for recharge from rainfall under various hydrogeological situations are listed in Table 3.

The normal rainfall figures are taken from the Indian Meteorological Department, the main agency for collection and presentation of rainfall data. The ranges of rainfall infiltration factor are recommended as a guideline only and need to be adapted according to prevailing hydrogeological conditions.

(i) Norms for estimation of Ground Water Recharge by the Water Level Fluctuation method

Based on the analysis of well hydrographs, the water level fluctuation and specific yield approach is recommended for recharge estimation.

a) Ground Water Level Fluctuations

A well hydrograph generally follows a definite trend, like a stream

TABLE 3

RAINFALL INFILTRATION FACTOR IN DIFFERENT HYDROGEOLOGICAL SITUATIONS

| | Hydrogeological Situation | Rainfall infiltration factor (percent of normal rainfall) |
|---|---|---|
| 1. | Alluvial areas | |
| | a. Sandy areas | 20 - 25 |
| | b. Areas with higher clay content | 10 - 20 |
| 2. | Semi consolidated sandstones | |
| | Friable and highly porous | 10 - 15 |
| 3. | Hard rock areas | |
| | a) Granitic terrain | |
| | i) Weathered and fractured | 10 - 15 |
| | ii) Unweathered | 5 - 10 |
| | b) Basaltic terrain | |
| | i) Vescicular and jointed basalt | 10 - 15 |
| | ii) Weathered basalt | 4 - 10 |
| | c) Phyllites, limestones, sandstones, quartzites shales etc. | 3 - 10 |

hydrograph, with a peak followed by a recession limb. The recession limb in a post-recharge period is characterised by two distinct slopes-one steep (from August to October/November) and the other more gentle (from October/November to June). The steeper limb signifies a quick dissipation of a major part of the recharge during the later part of the recharge period itself. This recession is very sluggish in alluvial areas compared with hard rock areas, in which a substantial recession occurs within one or one and half months after the peak water level is reached. Due to less demand and adequate soil moisture in the later half of the recharge period, and under prevailing agricultural practices in India, the rapidly receding limb of the hydrograph is not considered for computation of utilisable recharge. The utilisable recharge estimate is based on pre-recharge (pre-monsoon: April-May) to post-recharge (Post-monsoon: November) water level fluctuations for the areas receiving a South-West monsoon. For the areas receiving a North-East Monsoon, a similar approach of taking pre-monsoon (November) and post-monsoon (March) water level fluctuation can be adopted.

The network of water level monitoring stations needs to be adequate in space and time to smooth inconsistencies in observations which may arise due to varied hydrogeological and other factors.

b) Specific Yield:

The specific yield of a geological formation in the zone of water level fluctuation may be computed from pumping tests. As a guide, the following values computed from various studies are recommended (all values in percent) :

| | | |
|---|---|---|
| 1. | Sandy alluvial area | 12 - 18 |
| 2. | Valley fills | 10 - 14 |
| 3. | Silty/clayey alluvial area | 5 - 12 |
| 4. | Granites | 2 - 4 |
| 5. | Basalts | 1 - 3 |
| 6. | Laterite | 2 - 4 |
| 7. | Weathered phyllites, shales, schist and associated rocks | 1 - 3 |
| 8. | Sandstone | 1 - 8 |
| 9. | Limestone | 3 |
| 10. | Highly karstified limestone | 7 |

Normalisation of Rainfall Recharge

The water table fluctuation in an aquifer reflects the rainfall of the year of observation. The estimated rainfall recharge should then be corrected to the long term normal rainfall for the area as given by the Indian Meteorological Department.

For calculating annual recharge during a monsoon the expression indicated below may be adopted.

$$\text{Monsoon Recharge} = (\Delta S + DW - RS - Rigw - Ris) \times \frac{\text{Normal Monsoon RF}}{\text{Annual Monsoon RF}} + RS + Ris$$

where

$\Delta S$ = Change in groundwater storage volume during pre- and post-monsoon period (April/May to November), (million cubic metre or mcm).

DW = Gross ground water draft during monsoon (million cubic metre)

RS = Recharge from canal seepage during monsoon (mcm)

Rigw = Recharge from recycled water from groundwter irrigation during monsoon (mcm)

Ris = Recharge from recycled water from surface water irrigation during monsoon (mcm)

RF = Rainfall (metre).

To eliminate the effects of drought or surplus rainfall years, the recharge during a monsoon is estimated as above for a period of 3 to 5 years and an average figure is taken for long term recharge. Recharge from winter rainfall may also be estimated along the same lines.

In addition to natural ground water recharge estimation, recharge due to seepage from surface water canals, return seepage from irrigation fields, seepage from tanks and lakes, contribution from influent seepage, potential recharge in waterlogged and flood prone areas, and computation of ground water resources in confined aquifers is taken into consideration.

REFERENCES

1. Agricultural Refinance (1979) and Development Corporation. Report of the Ground Water Overex-Exploitation Committee.

2. Baweja B.K. and Karanth K.R. (1980) Ground water recharge estimation in India. Unpublished Report Central Ground Water Board.

3. Ministry of Irrigation (1984) Govt. of India. Report of the Ground Water Estimation Committee. G.W. Estimation Methodology.

4. Sinha B.P.C. (1983) 'A critical review of the ground water resources evaluation methodologies and estimates in India'. Seminar on Assessment Development and Management of Ground Water Resources (Keynote address).

5. U.N.D.P. (1976) Ground water surveys in Rajasthan. Gujarat Technical Report.

6. United Nations (1985) Ground water studies in Ghaggar River basin in Punjab, Haryana and Rajasthan.

Final Technical Report- Volume-I.

# BALSEQ - A MODEL FOR THE ESTIMATION OF WATER BALANCES, INCLUDING AQUIFER RECHARGES, REQUIRING SCARCE HYDROLOGIC DATA

J. P. Lobo Ferreira
J. Delgado Rodrigues
Laboratório Nacional de Engenharia Civil
Av. do Brasil 101
P - 1799 LISBOA CODEX - Portugal

ABSTRACT. A mathematical model for the estimation of daily sequential water balances is presented. The following hydrologic variables are updated daily by the model: precipitation, runoff, evapotranspiration, soil moisture and deep recharge of aquifers. The model was used with good results in the study of water resources of several portuguese watersheds.

## 1. INTRODUCTION

The model presented in this paper was developed within the project reported in FERREIRA, 1982, to answer some questions which arise when the estimation of deep recharge of aquifers and the evaluation of water resources are tried. It has been applied to the study of several portuguese watersheds.

In regions where river discharge records are scarce or absent the evaluation of the available water resources becomes a very difficult and unreliable task. In these circumstances, computer simulation is one of the recommended procedures.

In arid and semiarid climates, it happens that monthly potential evapotranspiration usually exceeds monthly rainfall, this making it inadequate to resort to the usual evaluation of a water balance on a monthly basis.

Daily sequential water balances may have a significant role in overcoming some of the above mentioned drawbacks.

With this model it is possible to compute the deep infiltration and, thus, to have an evaluation of the amount of water available for the deep recharge of aquifers, in regions with scarce hydrological and hydrogeological data.

## 2. DAILY SEQUENTIAL WATER BALANCE

### 2.1. General

The main reason for the failure of monthly water balance models to give



*I. Simmers (ed.), Estimation of Natural Groundwater Recharge, 313–319.*

plausible results in arid and semiarid regions is the time step they consider.

In such regions, values of potential evapotranspiration are much higher than those of rainfall. In these circunstances, values of runoff as well as deep infiltration are, almost always, zero when computed on a monthly basis. Futhermore, the computed real evapotranspiration, being similar to total rainfall, may, very often, reach abnormal amounts.

In the case of some arid and semiarid regions, rainfall is of frontal rather than orographic origin which means that strong showers may appear in short periods of time. In this case daily precipitation may be almost similar to the shower value.

## 2.2. Development of a computer program for the estimation of water balances

The model presented hereinafter (called BALSEQ) uses the following data as input:

- daily rainfall;
- monthly potential evapotranspiration ;
- runoff curve number according to the U.S. Soil Conservation Service;
- maximum amount of water available for evapotranspiration in the upper soil layer;
- soil moisture at the first day of the balance;

The balance gives a daily updating of the values of the hydrologic cycle variables and gives as output data the following:

- total values of input parameters (rainfall and potential evapotranspiration), their annual and monthly mean values and standard deviations;
- estimated results of runoff (overland flow), real evapotranspiration and deep infiltration; displayed tables give total values as well as monthly and annual mean values and standard deviations;
- soil moisture on the last day of each month included in the balance period;
- data related to overland flow (the date of the event, the total rainfall and the computed amount of overland flow).

The evaluation of the runoff is made according to the method developed by the United States Soil Conservation Service (USSCS) and proposed for Continental Portugal by DAVID et al., 1976. The method can be briefly described as follows:

a) determination of the type or types of soils that exist in the basin under study and their identification according to the soil classification of the USSCS. (Soils are classified according to the amount of runoff they will produce);

b) determination of the soil use and type of plant cover;

c) calculation of the weighted runoff curve number, NC;

d) determination of the net precipitation for runoff, $P_u$, from the total precipitation, $P_t$, by the equation:

$$P_u = \frac{25.4\ (\frac{P_t}{25.4} - \frac{200}{NC} + 2)^2}{\frac{P_t}{25.4} + \frac{800}{NC} - 8} \quad \text{(mm)} \qquad (1)$$

The model allows for initial losses of precipitation in order to permit saturation of the surface layer of the soil. For this reason, it will only consider those storms whose precipitation values exceed the minimum needed to saturate the surface layer of the soil, producing afterwards surface runoff.

The value of the total precipitation after which runoff occurs depends on the type of soil and on its surface occupation. This fact is considered in the model as a function of the runoff curve number NC by equation:

$$P_t > \frac{5080}{NC} - 50.8 \quad \text{(mm)} \qquad (2)$$

If daily precipitation is lower than that minimum value it will be totally ascribed to increase the soil moisture content of the soil water zone, being evapotranspired afterwards up to the daily potential evapotranspiration value.

If daily precipitation exceeds that minimum value, the model computes the runoff through equation (1) and uses the remainder for evapotranspiration and to increase the soil moisture content. These three parcels are computed following the order just presented.

Deep recharge of aquifers will occur whenever computed soil moisture exceeds the AGUT value. The excess of moisture will be net recharge. In this model the variable AGUT is the amount of water necessary to increase the soil moisture of the evapotranspiration zone from its lowermost value to the specific retention of the soil.

In order to estimate the AGUT variable it is necessary to have a rough estimation of the soil depth subjected to evapotranspiration.

Direct evapotranspiration from the soil decreases exponentially with depth and it usually becomes negligible from one meter downwards. The depth subjected to plant withdrawals depends on the plant cover of the soil. For crops and other small plants the maximum depth usually attained by roots does not exceeds 50 to 70 cm. For trees, the mean depth of the root zone way attain two or three meters.

## 3. APPLICATIONS OF THE MODEL

### 3.1. General

The three case-studies presented hereinafter illustrate some common situations that can arise when aquifer recharge estimation are necesary.

Porto Santo island is a semiarid region where hydrologic and hydrogeologic data are almost completely absent. Calibration of the model could only be done in a rather indirect way. However, the results computed for some characteristic regions of the island proved that their different hydrogeologic behaviour can be predicted by the model.

Faro region, is Southern Portugal, is also a semiarid region where some reasonable hydrologic and hydrogeologic information already exists. The model was run with some average input values. Runoff curve number was chosen according to the soil type and vegetation cover of the region and the AGUT value was considered the one that characterizes the average conditions of Continental Portugal. The output data was compared with recharge estimations presented in a recent hydrogeologic report for the region.

Rio Maior case-study illustrates a quite different situation. The climate is much more wet and top soil much better developed. AGUT values should certainly be higher, in spite of the absence of real references available about this point. The existence of a very good knowledge about the hydrogeologic conditions of the area made this case particularly suitable for the assessment of the importance of the top soil conditions and of the vegetation cover in the estimation of the AGUT variable and, consequently, of the aquifer recharge rates.

## 3.2. Porto Santo Island

The Portuguese island of Porto Santo belongs to the Madeira archipelago situated in the North Atlantic Ocean.

It has a mean annual precipitation of 360 mm and a mean annual potential evapotranspiration of 846 mm.

Agricultural soils are very thin and poor and vegetation cover is very scarce.

The water balance for Porto Santo using BALSEQ model gave the following results.

- mean annual runoff = 13.2 mm
- mean annual aquifer recharge = 29.7 mm

In order to get some information on the validity of the estimated results a study of the surface water balance for some reservoirs was performed.

Owing to the oversized dimensions of the dams there are no recharges. All inflow water in the reservoirs is used for irrigation or is lost by evaporation or by leakage.

The equation which describes this balance is:

$$PREC + FLOW = EVP + IRR \pm LEAK \quad (3)$$

where:

PREC - Direct rainfall in the reservoir
FLOW - Runoff calculated with BALSEQ model

EVP - Direct evaporation from the reservoir (Penmam with 5% of albedo)

IRR - Water distributed for irrigation

LEAK - Includes leakage through underground drainage from the reservoir and possible (but certainly very small) incoming water by drainage from upstream formations.

The mean value obtained by this equation for the variable LEAK, which could not be quantified in the field, was 20% of the variable FLOW. This mean result was considered acceptable, confirming the validity of the USSCS method for computing the runoff.

The comparison of the computed aquifer recharge with actual exploitation rates in several groundwater catchment areas showed that computed values are realistic. These values also confirmed the exhaustion of the groundwater reserves in several zones now being exploited. At the same time additional groundwater reserves were forecasted for other zones.

### 3.3. Faro region, Algarve

Faro is the capital city of Algarve, in the south of Portugal.

This region has a mean annual precipitation of 509 mm and a mean annual potential evapotranspiration of 835 mm.

Ground surface has gentle slopes and, in general agricultural soils are not very thick. According to soil surface conditions a value of 72 for the NC variable was considered. A value of 100 mm was attributed to the AGUT variable.

BALSEQ model gave the following results:

- mean annual runoff: 18 mm
- mean annual aquifer recharge: 160 mm

The accuracy of these results can be established by comparison with the values presented in the hydrogeological study of Algarve performed by DGRAH/UNESCO/PNUD, 1981. In this study, the region of Faro is reported below the 50 mm isoline. The reported aquifer recharge for the southern areas of the Algarve is between 100 and 200 mm. The recharge computed by BALSEQ program is, thus, within the range of the value presented in that report.

### 3.4. Rio Maior lignite basin

Rio Maior lignite basin is a small graben filled with a thick layer of white sands and a lens shaped lignite and diatomite complex.

Hydrogeological studies performed in this basin provided a very good knowledge of the hydraulic characteristics of the main aquifer - - the white sands - and a precise definition of the piezometric surface.

BALSEQ model was applied to Rio Maior region for some sets of AGUT values.

Rio Maior basin is covered with well developed agricultural soils where large trees are very abundant; in these circumstances a thick root zone is likely to occur and fairly high amounts of water available for evapotranspiration certainly exist. Therefore, it was considered that several AGUT values over 150 mm should be used to run the BALSEQ program.

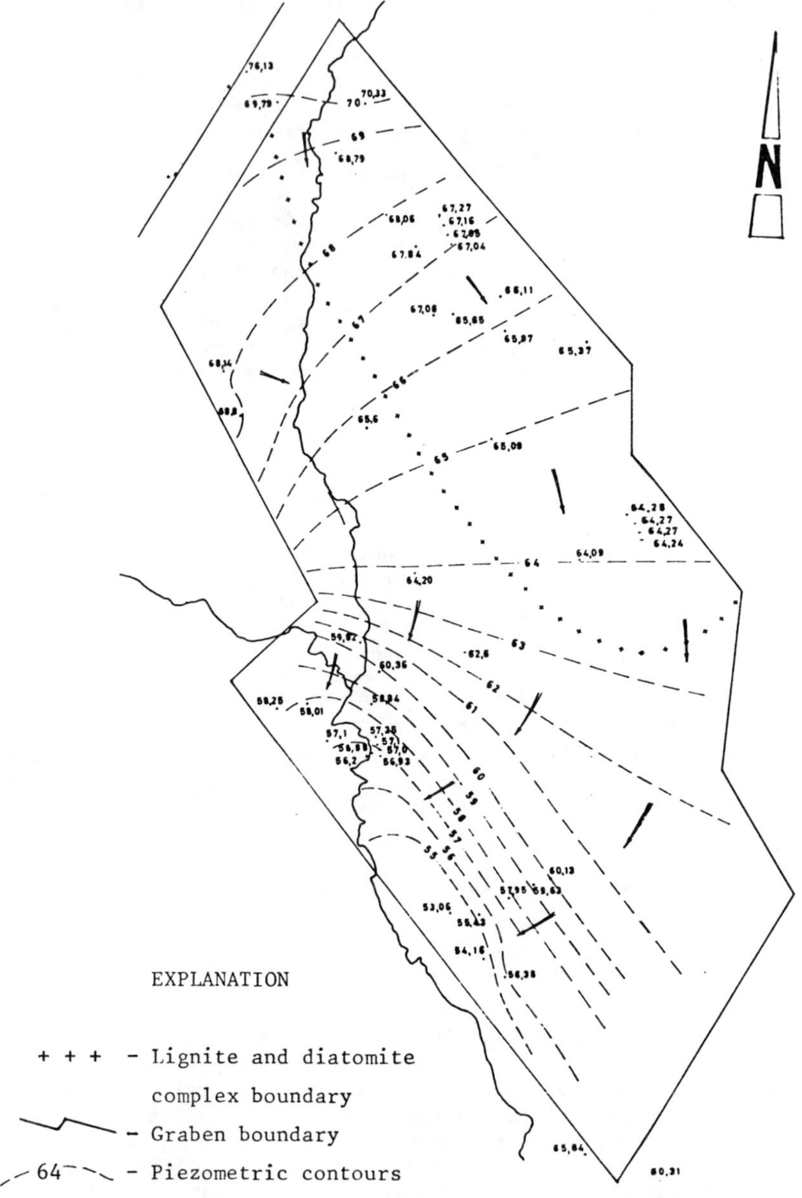


Fig. 1 - Piezometric contours of the Rio Maior aquifer (May,1980)

Taking advantage of the existence of some suitable groundwater conditions a calibration was tried by using the groundwater flow of the basin.

Fig. 1 shows the piezometric contours in May 1980. The Maior river is a natural discharge zone of the aquifer mainly in the southwestern side of the basin. In this zone, piezometric levels change very little from Winter to Summer and, thus, the situation in May 1980 represents reasonably well the average discharge conditions. Taking into account the geometry of the basin and its boundary conditions it was considered that the discharge along the 58 m piezometric contour would integrate all the basin discharge; with these assumptions a discharge of 140 l/s was estimated by this method. Assuming that the groundwater drainage basin is not very different from the surface watershed, an estimated recharge of 260 mm was computed.

The comparison between computer results and hydrogeological determinations showed that the same infiltration rate is obtained for an AGUT value of about 230 mm.

This case-study was considered to be meaningful as regards the model validation and very important for the assessment of the role played by some input parameters, for example AGUT.

## 4. CONCLUSIONS

It is believed that the BALSEQ model may be used successfully in regions with climatic conditions similar to those presented in this paper. The generalization of its use to regions with different climates should be done carefully and, whenever possible, results should be confirmed by other hydrological and hydrogeological procedures.

The results obtained with this model are however highly dependent on the values ascribed to NC and AGUT variables which have a difficult calibration procedure.

Experience gathered with the use of this model shows that very precise results should not be expected from its use. However, the existence of an acceptable range of values for runoff and aquifer recharge may be very useful in water resources evaluation, when only scarce hydrologic and hydrogeologic data are available. Computed values, even when they cannot be used directly as definite results, may be used as hydrogeologic tools in the overall assessment of recharge conditions.

## REFERENCES

DAVID et al., 1976 - Determinação de Caudais de Ponta de Cheia em Pequenas Bacias Hidrográficas. Relatório. Lisboa, Laboratório Nacional de Engenharia Civil.

DGRAH/UNESCO/PNUD, 1981 - Évaluation des Ressources en Eaux des Systèms Aquifères de L'Algarve. Lisboa, Direcçao Geral dos Recursos e Aproveitamentos Hidráulicos.

FERREIRA, J. Lobo, 1982 - Actualização do Estudo Hidrológico da Bacia Hidrográfica do Rio Maior. Relatório. Lisboa, Laboratório Nacional de Engenharia Civil.

# APPLICATIONS AND CASE STUDIES

# QUANTIFICATION OF GROUNDWATER RECHARGE IN ARID REGIONS: A PRACTICAL VIEW FOR RESOURCE DEVELOPMENT AND MANAGEMENT

S S D Foster

British Geological Survey (Hydrogeology Group)
Wallingford (Oxford) OX10 8BB, G.B.

Pan-American Health Organisation (CEPIS Laboratory)
CP 4337, Lima 100, Peru

ABSTRACT. The rate of aquifer replenishment is one of the most difficult of all factors in the evaluation of groundwater resources to measure. Estimates, by whatever method, are normally, and almost inevitably, subject to large error. Through discussion of a logical series of questions, the manner in which this factor affects practical resource development and management are defined and the limitations of various evaluation methods are discussed, with reference to examples from arid regions, especially the Botswana Kalahari and the Peruvian Atacama. When confronted by such uncertainty technical and economic analyses of sensitivity of development and management decisions to error in recharge estimates is essential. A pragmatic approach involving careful monitoring of aquifer response to heavy medium-term abstraction in a representative area and subsequent data evaluation using a distributed-parameter numerical aquifer model is advocated. This will often be the only realistic way forward given normal project constraints and data limitations under complex hydrogeological conditions and in developing nations, and will also allow time for appropriate field research into groundwater recharge mechanisms and storage controls.

## 1. WHAT IS THE PRACTICAL SIGNIFICANCE OF GROUNDWATER RECHARGE ESTIMATES?

Quantification of the current rate of natural groundwater recharge is a basic prerequisite for efficient groundwater resource management, and is particularly vital in arid regions where such resources are often the key to economic development. It will be especially critical where large and concentrated demands for groundwater supplies exist, such as many urban, industrial and mining requirements and for most, except very small-scale, agricultural irrigation. In contrast, the demands associated with rural domestic water-supply and livestock watering are generally very small and widely dispersed. Since the associated groundwater abstraction rates do not exceed the equivalent of 2 mm/a, and are often significantly less than this value, they are unlikely to overtax the storage resources of even minor aquifers in the long-term. Thus although groundwater recharge will still be of interest, specific



*I. Simmers (ed.), Estimation of Natural Groundwater Recharge, 323–338.*

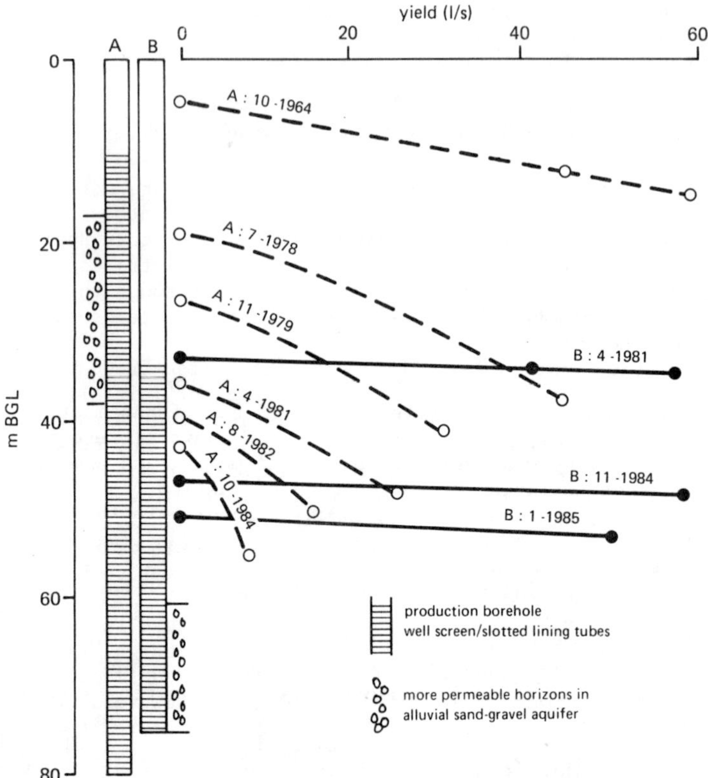


Figure 1. Production performance of pumping boreholes in the over-developed alluvial fan aquifer of Lima, Peru. Their contrasting behaviour reflects the falling groundwater level and varying depth at which more permeable horizons occur.

investigations would not normally be justified or necessary in this connection.

If uncontrolled overdevelopment of an aquifer occurs as a result of false assumptions about, or gross overestimation of, active groundwater recharge (or by default on the part of those with responsibility for allocating and controlling the use of water resources), there are often serious consequences. These vary widely with hydrogeological conditions but can include:

a) increased pumping costs, yield reductions, and even complete failure of production boreholes (Figure 1),
b) the encroachment of saline water into freshwater aquifers in some coastal and inland basin situations,
c) land subsidence consequent upon settlement of underconsolidated lacustrine, deltaic or estuarine sedimentary aquifers.

Nevertheless, aquifer overdevelopment is not necessarily bad practice.

It can form sound resource evaluation or management strategy if positively planned and carefully controlled, with the technical and economic side-effects having been realistically evaluated.

## 2. HOW SHOULD GROUNDWATER RECHARGE IDEALLY BE SPECIFIED?

A simple categorisation of natural aquifer recharge in the more arid regions is given in Figure 2, with a basic subdivision into:

a) diffuse recharge from rainfall, excess to soil moisture deficits and short-term vegetation requirements, which infiltrates directly,

b) localised recharge resulting from infiltration through the beds of perennial or (more typically) ephemereal surface watercourses and other processes,

although in practice, many intermediate conditions also frequently occur.

For most practical purposes it will be sufficient initially to estimate recharge rates within two of the scale increments indicated on Figure 2, but even this may sometimes prove difficult. Any more precision can await the analysis of aquifer response to significant medium-term abstraction. The actual frequency of recharge events is also important. Major, but very infrequent, recharge is a totally different proposition to more regular, if smaller, replenishment, since in the former case the negative side-effects of overdevelopment may have already occurred prior to the next recharge event.

It should be borne in mind that estimates of diffuse recharge are likely to be more reliable than those of localised recharge, except where the latter is sufficiently regular to produce a quasi-steadystate aquifer flow pattern which can be readily analysed. However, shallow, apparently steady-state, hydraulic gradients may also be relics of palaeorecharge features from which natural recession of groundwater levels continues to occur to the present time.

## 3. HOW FAR CAN THESE IDEALISED REQUIREMENTS BE FULFILLED IN PRACTICE?

The main techniques that can be employed specifically to estimate current groundwater recharge rates in arid regions may be divided (Table 1) into those for which some of the required data are often available or can be readily collected and those for which more specialised and expensive facilities are needed. The applicability and potential accuracy of any given method depends largely on two semi-independent facets of the ambient conditions:

a) the superficial geological environment, which determines the spatial variability of the recharge process and the extent of development of surface runoff,

b) the vegetation system and whether native or agricultural, with or without irrigation.

For a detailed description of these techniques the reader is referred to the state-of-the science papers in this volume; only their principal limitations in relation to practical objectives will be discussed here.

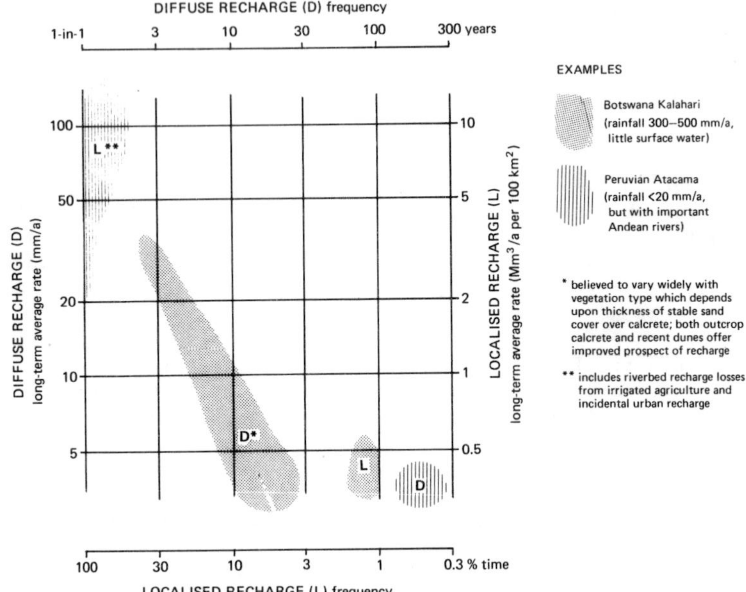


Figure 2. Categorisation of aquifer recharge for the practical evaluation of groundwater resource development and management options.

## 3.1 Hydrometeorological Data Processing

Analysis of historic rainfall data is the method most frequently applied to assess the likelihood and magnitude of diffuse groundwater recharge. It commonly suffers from a number of limitations such as the inadequate length of historic meteorological records and their lack of areal representativity due to low rainfall station density and highly localised rainfall patterns in arid regions. However, commonly the most important deficiencies of the approach are:

a) the lack of reliable data in respect of daily or hourly rainfall intensity during major rainfall episodes and complementary information on soil infiltration capacity to permit estimation of the proportion of excess rainfall which actually infiltrates and that which forms runoff,

b) uncertainty about the evaporation regime of soils and vegetation in the arid zone, and the magnitude of accumulated soil moisture deficit at the end of drought, since infiltration estimates will be highly sensitive to variation in these factors (Figure 3).

The errors inherent in the soil water balance approach can be significantly reduced in situations where the prediction of infiltration can be corroborated and calibrated by unequivocal fluctuations of a shallow groundwater table or where the main sources of infiltration is over-irrigation of agricultural crops.

| TECHNIQUE | APPLIC-ABILITY | TYPICAL ORDER OF COST* | NEED FOR SPECIALIST FACILITIES |
|---|---|---|---|
| (A) Conventional Methods | | | |
| 1. Hydrometeorological Data Processing (soil water balance) | D(L)$^{Ø}$ | c*** | |
| 2. Hydrological Data Interpretation | | | |
| - water-table fluctuations | D(L) | c*** | |
| - differential stream or canal flow | L | c*** | |
| 3. Chemical & Isotopic Analyses from Saturated Zone | D+L | c-b** | + |
| (B) Modern Techniques | | | |
| 4. Chemical & Isotopic Profiling of Unsaturated Zone | D$^{Ø\#}$ | b-a** | ++ |
| 5. Soil Physics Measurements | D$^{Ø}$ | a | ++ |

D/L diffuse/localised recharge distribution
Ø only suitable for relatively uniform soil profiles
# inappropriate for irrigated agricultural areas
* for an aquifer area of about 1000 km$^2$
** isotopic analyses increase cost substantially
*** excluding construction and operation of basic data collection network

a = >US$50,000; b = US$10-50,000; c = US$<10,000

Table 1. Principal techniques used for groundwater recharge estimation in arid regions.

### 3.2 Hydrological Data Analysis

Another commonly applied method is the interpretation of natural water-table fluctuations in terms of an aquifer recharge input. The method, when applied in isolation, is not reliable unless accurate values of aquifer unconfined storage coefficient are available from a reliable independent method, which is rarely the case. It can be grossly misleading if the fluctuations are confused with those due to pumping, barometric or other causes, the latter of which are common in observation boreholes drilled and screened in permeable, but self-confined, sections of thick aquifers.

Differential flow gauging along the length of surface watercourses is a direct approach to the evaluation of linear sources of groundwater recharge. However, measurements are often subject to large errors because of the practical difficulties associated with the extreme flow variations and the intermittent nature of most surface watercourses in arid regions, and the associated difficulty of locating and maintaining suitable sections for gauging.

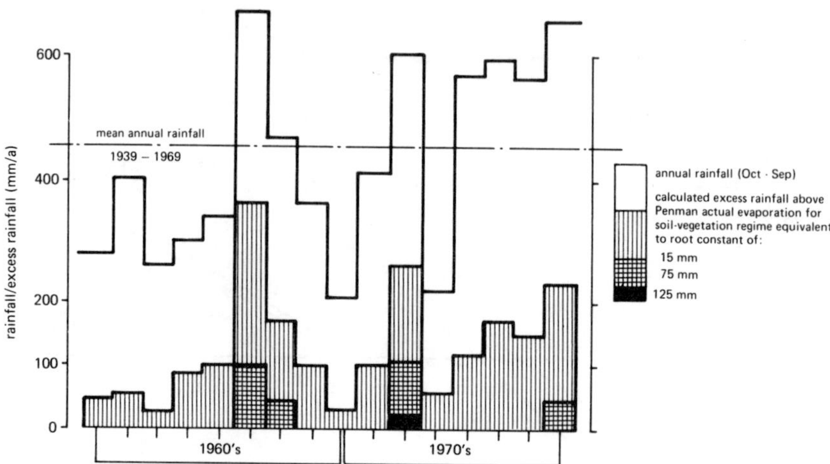


Figure 3. Summary of analysis of daily hydrometeorological data for a site in the Botswana Kalahari (Kweneng District) during 1961-77. Infiltration rates and frequencies exhibit marked sensitivity to variation in soil moisture-vegetation regime which is in practice laterally variable and difficult to quantify but groundwater table fluctuations suggest that widespread recharge was limited to a single rainfall episode in the 1971-72 wet season.

### 3.3 Chemical & Isotopic Analyses from Saturated Zone

This approach is commonly used to provide supporting evidence for the presence or absence of active aquifer replenishment, and is very useful to corroborate or to refute the hydrogeologist's conceptual model of aquifer behaviour (Figure 4). However, it is rarely possible to obtain quantitative estimates of present recharge rates by these methods.

Moreover, the ability to interpret such data often reduces markedly if non-specialist groundwater sampling methods are used, and data can be also misleading in a general sense because of such factors as:

a) low salinity groundwater is equally likely to have originated from historic recharge during past pluvial episodes as to be modern recharge,

b) the absence of thermonuclear tritium ($^{3}H$) and/or modern carbon ($^{14}C$) in such samples may not necessarily indicate an absence of modern recharge because it could variously be attributed to sampling the wrong aquifer depth interval, unsaturated zone time-lag for modern recharge, dissolution of fossil carbonate from the aquifer matrix, and isotopic exchange with older water in micropores by aqueous diffusion,

c) the stable isotopes ($^{2}H$-$^{18}O$) can give useful information about the evaporative history of groundwater but quantitative interpretation in relation to modern recharge requires a detailed body of

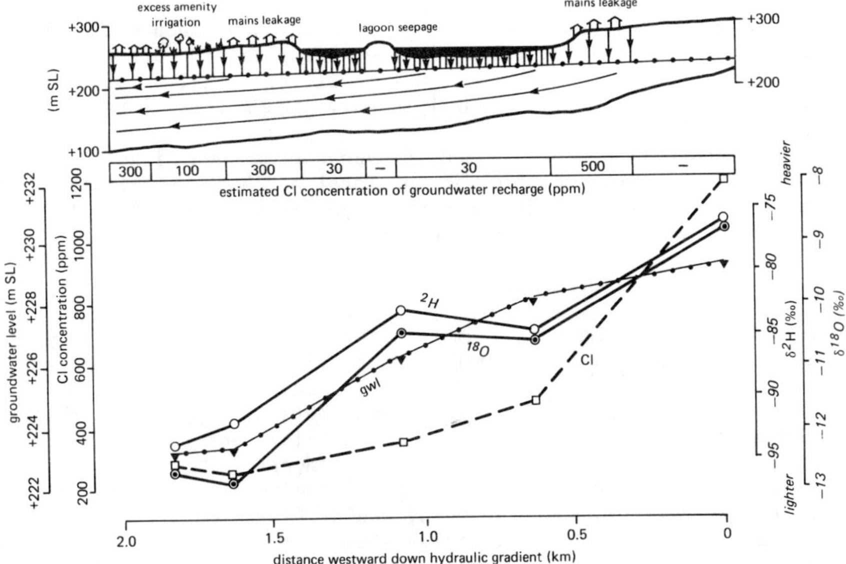


Figure 4. Chemical and isotopic data from the saturated zone of an embayment of the Lima alluvial aquifer, Peru, corroborating the concept of incidental aquifer recharge from human activity at the land surface. The area concerned is one of minimal and very infrequent rainfall and no natural surface watercourses.

complementary isotopic analyses for current rainfall, soil drainage and riverflow to permit unequivocal interpretation.

### 3.4 Chemical & Isotopic Profiling of Unsaturated Zone

Profiling the concentration of chloride (Cl), tritium ($^3$H) and deuterium ($^2$H) in the pore water of the unsaturated zone or superficial deposits overlying aquifers can yield valuable data to aid evaluation of the recent history of diffuse groundwater recharge (Figure 5), but the method is costly, especially when isotopic determinations are included. Its use is more justified in the most arid areas with native vegetation and uniform, unconsolidated soil profiles when it should provide more reliable indications of deep infiltration than soil water balance calculations.

However, obtaining adequate volumes of uncontaminated samples to well below the depth of deeper rooted vegetation (10-20 m or more) can pose formidable practical problems in the low moisture content profiles characteristic of arid regions. The method also has a number of scientific limitations including:

a) reliable historic data on the chemical and isotopic composition of rainfall (and other deposition) needed to obtain satisfactory mass balance interpretation are rarely available, and reliance has to be put upon generalised and extrapolated data for peak and/or average concentrations in rainfall,

b) the chloride mass balance at many sites may have been radically altered as a result of the agricultural application of irrigation water or unknown quantities of KCl fertilisers, animal slurries and manures, or of chloride removal in harvested salt-resistant crops, for example,

c) the method will be invalidated if bypass flow down fissures and root casts, or lateral flow at ephemeral perched water-tables occurs, which will be frequent phenomena in many superficial hydrogeological environments in arid regions.

### 3.5 Soil Physics Measurements

This method involves the regular monitoring, over a period of 1-2 years or more, of the variations in soil water suction and moisture content in a nest of tensiometers and neutron moisture access tubes to depths of up to 10 m depending on site conditions. Such data are capable of analysis to yield soil water flux, evaporation and infiltration rates usually through application of the zero flux plane concept. They are potentially a positive method of advancing knowledge of the soil moisture-vegetation regime in arid regions, which in turn, would allow a more confident interpretation of groundwater recharge from historic hydrometeorological records.

However, it must be recognised that such methods require specialist field equipment and personnel, are very time consuming, and highly site-specific. Thus, it is only realistic to apply them in the groundwater resources context for extensive areas of uniform, deep, soil profile. Moreover, they will be much more suited to the study of shallow-rooted, non-irrigated agricultural crops than to native vegetation, for example. When interpretation requires the experimental determination of the unsaturated hydraulic conductivity-moisture content relationship at low field moisture contents, this will present significant technical problems also.

## 4. HOW CAN UNCERTAINTY ABOUT AQUIFER RECHARGE BE RECONCILED WITH PRESSURE TO REACH DEVELOPMENT AND MANAGEMENT DECISIONS?

Groundwater recharge estimates in arid regions will nearly always be subject to considerable uncertainty and large error. This will be evident from:

a) the numerous limitations of each of the techniques described,

b) the wide spatial variability characteristic of rainfall and runoff events,

c) widespread lack of lateral uniformity in soil profiles and hydrogeological conditions,

d) frequent inadequacies in the hydrogeological database, especially in developing nations.

Thus it will normally not be possible to estimate the average diffuse recharge rate, for example, to within one scale increment in

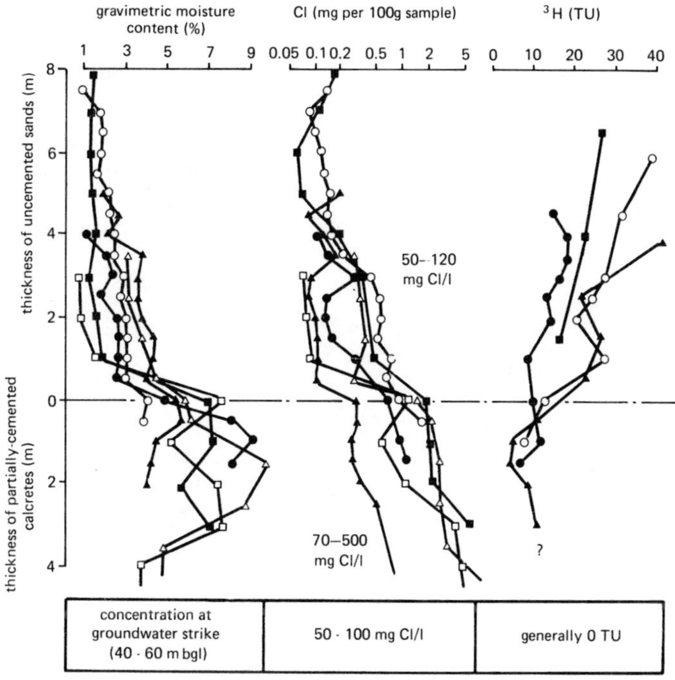


Figure 5. Moisture content, pore water chloride and tritium profiles for the Botswana Kalahari sand-cover (October-December 1977). These relate to the same area as Figure 3 and show strongly evaporative features which suggest that, despite an average rainfall of some 450 mm/a, little deep infiltration has occurred during the preceeding 10 years (and probably over a much longer period) as a result of deep-rooted vegetation down to, and beyond, the base of the sand profiles.

Figure 2. Moreover the necessary database improvement or application of new techniques to achieve more reliable recharge estimates will be more time consuming that those required to improve all other factors affecting the groundwater resources evaluation, especially when precise results are needed. It will thus not be possible because of logistical or time-table constraints on project development or on management decisions.

When confronted by such uncertainty, it is strongly advisable for project design or management strategy to be sufficiently flexible as not to require radical change in the event of initial predictions proving subject to considerable error, due to wrong assumptions about recharge rates and other hydrogeological factors. Flexibility in wellfield design, for example, might involve such features as installing borehole pumps capable of operating over a wide range of drawdowns, conservative well-spacing to allow increase in the number of production boreholes by drilling at immediate sites, over-sizing key sections of wellfield internal pipeline to allow for redistribution of abstraction (Figure 6). In terms of groundwater management the required flexibility could be

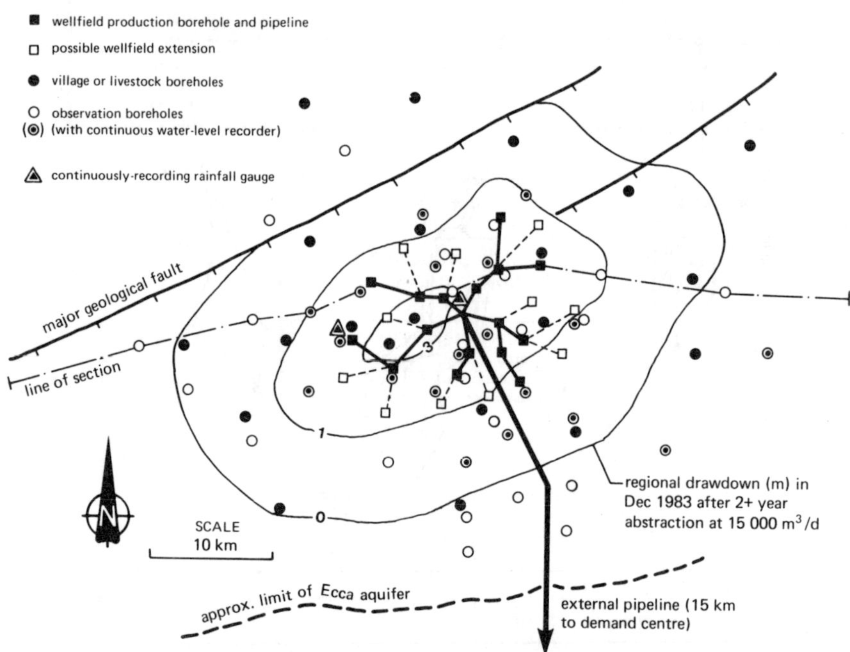


Figure 6. The Jwaneng northern wellfield in the Ecca sandstone aquifer of the Botswana Kalahari illustrating flexible design and comprehensive monitoring network. This was rapidly designed and implemented to provide a water-supply of some 15000 $m^3$/d for at least 7 years in an area of complex hydrogeology and questionable aquifer recharge.

achieved by such policies as having the potential to install reticulation mains to allow supply from an alternative groundwater source, or by earmarking a surface water source for possible conjunctive use.

Project design and policy formulation to achieve the required level of flexibility to accommodate initial uncertainty about groundwater recharge estimates is greatly aided by use of a distributed-parameter aquifer model to analyse critically the sensitivity of its output (in terms of aquifer response to groundwater abstraction) to errors in key parameters such as groundwater recharge rates, including implications of no active replenishment if this remains a possibility.

If sufficient groundwater level data, from a well-designed observation borehole network and corresponding to a period of significant medium-term abstraction, were already available, together with sound information on aquifer properties, hydraulic boundaries and discharge, then such a model could be used to determine historical groundwater recharge.

While computerised numerical prediction modelling has long been used to aid development and management of groundwater resources in arid

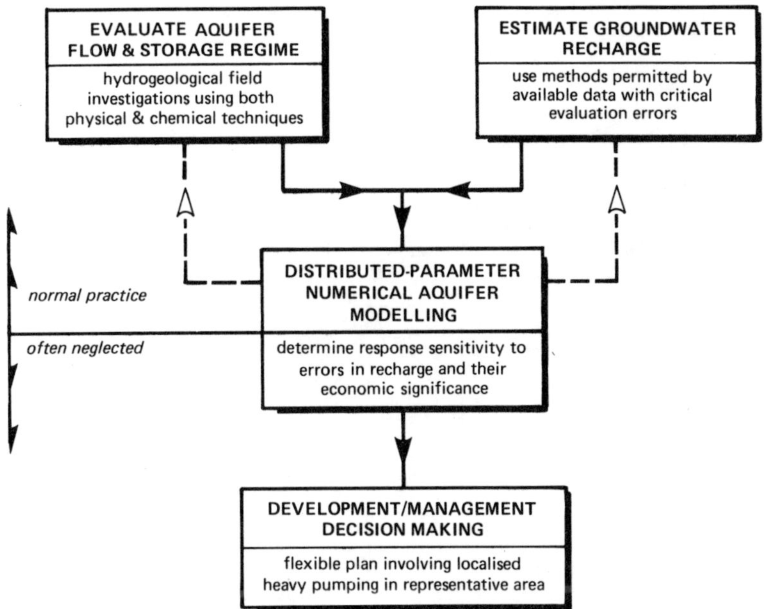


Figure 7. Organisational scheme for preliminary evaluation of groundwater resource options, including sensitivity to error in estimation of recharge rates, and for evolution of a flexible development and management plan.

regions, the specific and critical use of models in this mode is less commonplace, although it is the logical extension of conventional groundwater balance studies. In essence the technique involves varying aquifer recharge rates and distributions in the model so as to achieve calibration with historic groundwater level data (Figure 8). It is very important that the sensitivity of such a calibration to variation in recharge estimates be fully tested.

A limitation of the method, other than the requirement for a substantial body of related data, is the fact that the calibration achieved may not be unique and it may prove impossible to distinguish variations in recharge rates from those of other parameters, such as unconfined storage coefficient, for example, which also can be difficult to determine accurately by an independent method.

If insufficient hydrogeological data were initially available to adopt this technique, projects can normally be structured (Figure 9) to allow collection of the type of aquifer response data needed to calibrate the numerical model, and to allow time for the implementation of parallel investigation programmes to evaluate further aquifer recharge mechanisms and rates, storage properties and boundaries. As the calibration of the model is refined, so the accuracy of recharge estimates will improve (Figure 10).

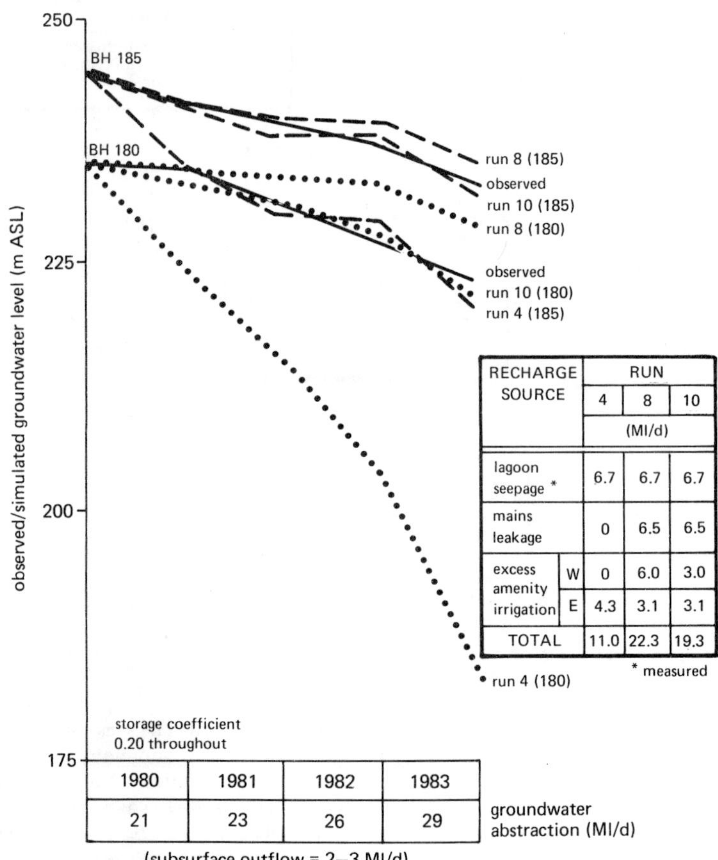


| RECHARGE SOURCE | | RUN 4 (Ml/d) | RUN 8 (Ml/d) | RUN 10 (Ml/d) |
|---|---|---|---|---|
| lagoon seepage * | | 6.7 | 6.7 | 6.7 |
| mains leakage | | 0 | 6.5 | 6.5 |
| excess amenity irrigation | W | 0 | 6.0 | 3.0 |
| | E | 4.3 | 3.1 | 3.1 |
| TOTAL | | 11.0 | 22.3 | 19.3 |

| 1980 | 1981 | 1982 | 1983 |
|---|---|---|---|
| 21 | 23 | 26 | 29 |

Figure 8. Evaluation of aquifer recharge by calibration of a numerical simulation model with observed groundwater levels. This relates to the same area as Figure 4; run 10 is considered most realistic but the calibration is not unique since a lower unconfined storage coefficient would indicate higher recharge rates.

An essential element of this integrated operational approach to groundwater recharge evaluation is that sufficient effort goes into monitoring aquifer response (e.g. Figure 6) to ensure that adequate data are collected, since short-term economies in this respect are likely to prove counterproductive in the longer run. In many instances significant groundwater abstraction will already exist although the aquifer response has not been monitored in sufficient detail or over a long enough period, but the same approach can be developed. In areas of complex hydrogeology this approach will be the only practicable way to improve the reliability

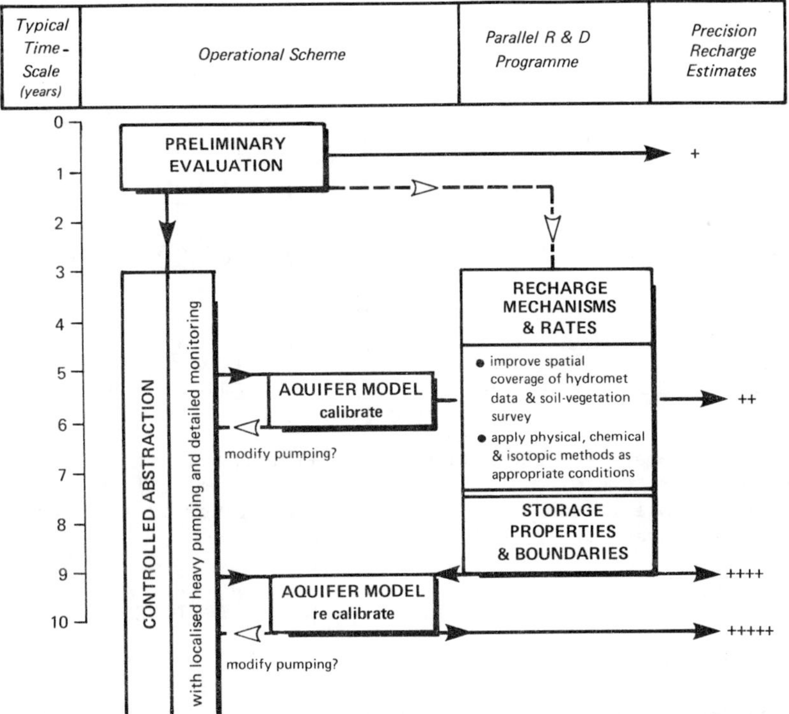


Figure 9. Outline scheme for the integrated operational approach to improving estimates of aquifer recharge rates and storage parameters.

of groundwater recharge estimates, and in many less complex situations it will often still be the most cost-effective way.

Because the exploitable storage of most aquifers is very many times greater than their average annual recharge this operational approach to improving groundwater recharge estimates is widely applicable, but there are circumstances in which the scope for this approach will be more limited. These include situations (Figure 11) where:

a) the minimum viable first stage water demand is very large relative to exploitable aquifer storage, because of the need to justify major financial investment in a lengthy external pipeline to a remote water demand centre or of the design of an industrial plant for which the water-supply is required,

b) the profitability of proposed groundwater use is highly sensitive to energy costs, such as might be the case in some agricultural irrigation schemes,

c) there is significant risk of saline encroachment in an aquifer as a consequence of medium-term overdevelopment, with salinity being an important constraint in the use of the groundwater supply.

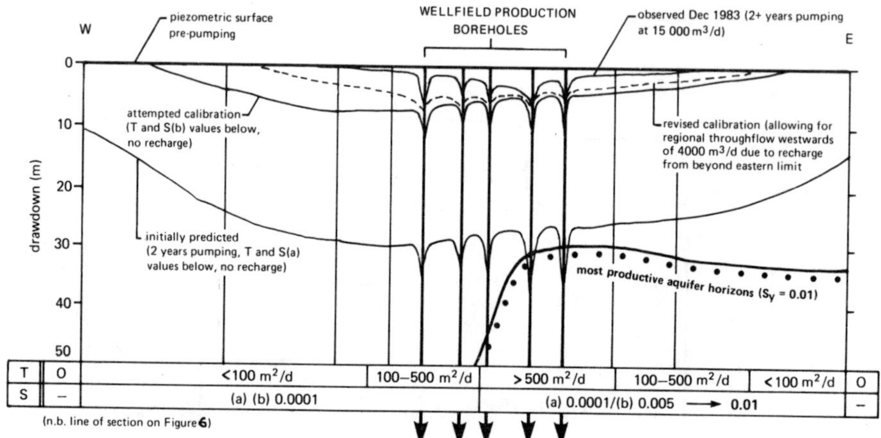


Figure 10. Schematic cross-section of the Jwaneng northern wellfield, Botswana Kalahari, showing initial predictions of the aquifer numerical model and the calibration based on operational monitoring experience with revised storage and recharge parameters.

## 5. ARE THERE ANY OTHER FACTORS THAT NEED TO BE CONSIDERED?

### 5.1 Groundwater Quality

In the author's view it is equally important to consider, and to estimate, the quality of active natural groundwater recharge in arid regions as well as its quantity. This can greatly affect the acceptability or usefulness of the associated groundwater resources, and it is often erroneously regarded as constant or essentially stable.

In regions that are currently subject to more arid climate than in their recent geological history, the salinity of any active groundwater recharge would, as a result of salt accumulation in the soil and unsaturated zone (Figure 5), be much greater than that of the groundwater stored in the aquifer, and recharge can lead to progressive quality deterioration. The reverse situation is equally common where small volumes of current groundwater recharge can be the only source of fresh water in an otherwise saline aquifer.

### 6.2 Incidental Modifications to Recharge Regime

It is also very important to recognise that human activity, in terms of agricultural development and urbanisation, can lead incidentally to radical changes in the groundwater recharge regime (both quantity and quality) in arid regions.

Amongst agricultural activities in arid regions the major modifications are likely to be associated with irrigation schemes,

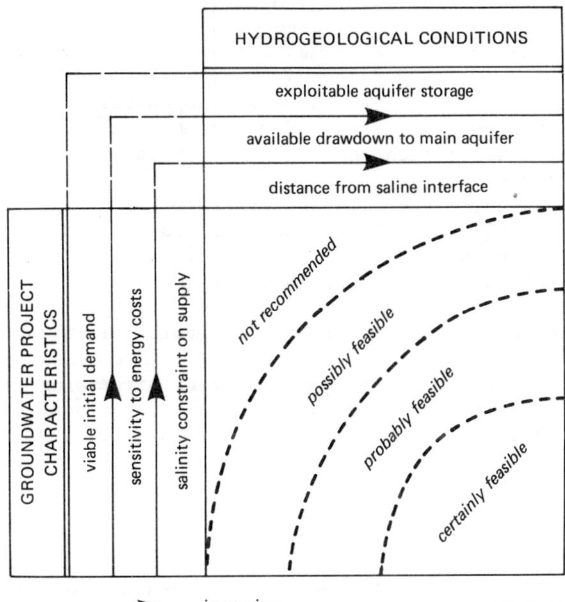


Figure 11. Feasibility of adopting an integrated operational approach to the evaluation of aquifer recharge rates and storage parameters.

although deforestation and soil erosion may also have significant effects on groundwater recharge. In the case of irrigation schemes, distribution canals normally have large seepage losses, sometimes to deep groundwater, and at field level rates of excess application will depend on irrigation methods and scheduling, and on subsoil characteristics, but can be as high as 5 mm/d (or 1000+ mm/a), which can recharge groundwater if subsoils are permeable. The quality of this recharge will depend upon the source of irrigation water, the irrigation efficiency and soil management but it can include increasing salinity load or significant concentrations of nitrate or certain pesticides.

Urbanisation may also have major impact on aquifer recharge through such processes as infiltration from unsewered sanitation, leakage from water mains, soakaway drainage of paved and roofed areas, overirrigation of gardens and amenity areas. This can lead to greatly increased rates of groundwater recharge (Figure 4, 8) and sometimes to increasing groundwater salinity or pollution by nitrates and synthetic organic compounds.

## 7. CONCLUDING SUMMARY

a) Where the hydrogeological database and project constraints permit, the application of various independent physical and chemical methods is likely to be the best way to improve knowledge of aquifer recharge mechanisms and rates. The selection of techniques depends mainly on the hydrogeological environment, vegetation system, and on whether diffuse or localised recharge is likely to predominate.
b) The critical use of aquifer numerical distributed-parameter models will normally be the most powerful and comprehensive method of improving recharge estimates. Such models also allow appraisal of the sensitivity of groundwater development options and management decisions to errors in the estimation of aquifer recharge rates and storage parameters, which is an essential procedure.
c) In many instances, especially in developing nations, there are inadequate data, particularly on the aquifer response to groundwater abstraction, to reach adequate model calibration. However, in most cases, if a flexible approach to project design or management strategy is adopted, the necessary data can be collected within the operation of normal groundwater production provided a suitable observation borehole network is installed.
d) It is important to consider the quality of present groundwater recharge, as well as its quantity, and the radical modifications to the groundwater recharge regime which often occur in arid regions consequent upon agricultural development or urbanisation.

## 8. ACKNOWLEDGEMENTS

This paper is based on personal experience gained during consulting on groundwater development and management problems in the arid regions of developing nations, a substantial proportion of which was funded by the British Overseas Development Administration. The author benefitted from valuable discussion, on the field data contained in the paper, with Moremi Sekwale (GSD - Botswana) and Carlos Valenzuela (SEDAPAL - Peru). No references to related published work are given in the text because this volume contains state-of-the-science papers on the various techniques mentioned and a comprehensive bibliography on the subject concerned.

# GROUNDWATER RECHARGE STUDIES IN SEMI-ARID BOTSWANA-A REVIEW

J.J. de Vries
Institute of Earth Sciences, Free University Amsterdam
P.O. Box 7161
1007 MC Amsterdam
The Netherlands

M. von Hoyer
Department of Geological Survey
Private Bag 14
Lobatse
Botswana

ABSTRACT. Some 80 % of the surface of Botswana is mantled with thick layers of Late-Cretaceous to Recent sandy deposits of the Kalahari semi-desert. Sedimentary and crystalline rocks of pre-Cretaceous age occur at or near the surface in the remaining eastern part of the country. Groundwater basins have been encountered in the bedrock below the Kalahari deposits all over the area, but no unanimity has been reached on the question of the existence of any present-day recharge. Replenishment of the aquifers in the outcrop areas of solid rocks is not disputed, but no reliable figures on the percolation rate could be established as yet.

The various investigations and methods applied in recharge studies in Botswana - including mass balance methods and isotope studies - will be reviewed.

## INTRODUCTION

The Republic of Botswana is a landlocked country, situated on the South African Plateau at an elevation of about 1000 m.

It covers an area of 560,000 km$^2$ and has a population of about one million. Over 80 % of the country is part of the Kalahari desert, the largest continuous sandbody in the world. The remaining, eastern part of Botswana is dominated by outcrops of the Precambrian African Shield. The majority of the people live in the eastern areas (Fig. 1).

Botswana has a semi-arid climate with erratic rainfall during the hot season, varying from 250 mm in the extreme southwest to 650 mm a year in the north-east. Open water evaporation reaches about 2000 mm a year. Perennial water is restricted to the far north of the country where the Okavango and the Chobe Rivers bring water from catchments in Angola and Zambia.



*I. Simmers (ed.), Estimation of Natural Groundwater Recharge, 339–347.*

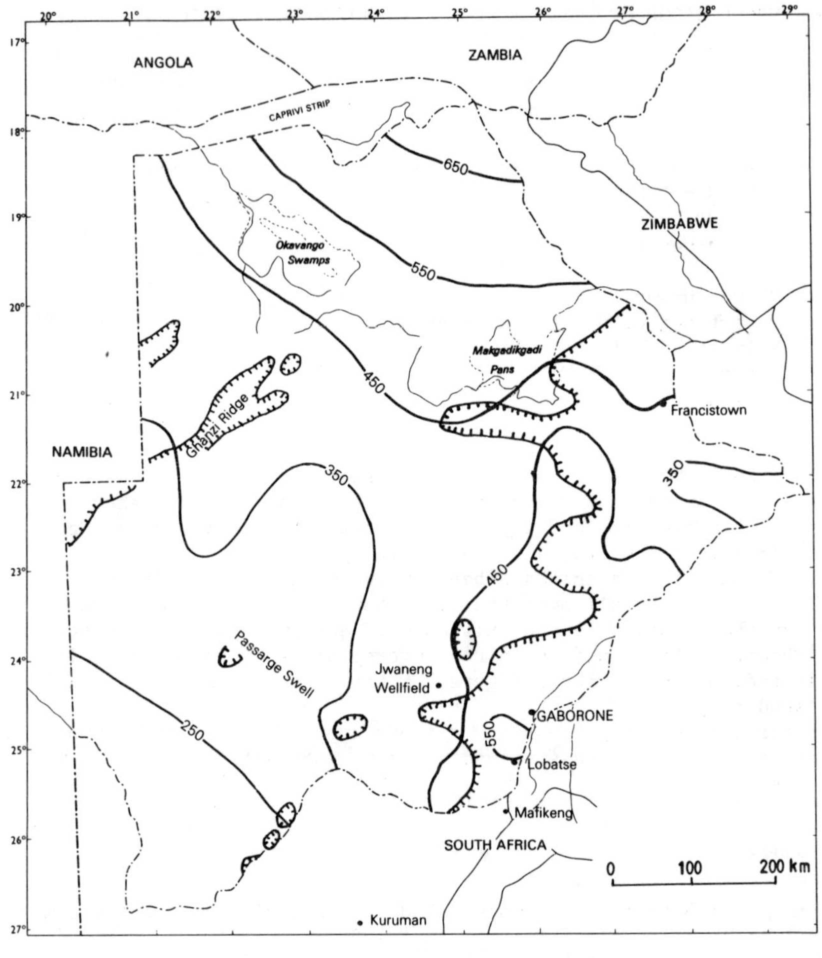


Fig. 1 Botswana, features and locations refered to in text

Successful boreholes have been drilled all over Botswana in various geological formations for the water supply of small communities and cattle, where the demand does not exceed a few ten $m^3$ per day. Systematic exploration for high yielding aquifers to satisfy the need of the major villages, towns and mining industry, commenced in the last decade and has revealed several economically exploitable groundwater basins with a capacity ranging from a few thousand up to 15,000 $m^3$ per day.

It has been estimated (VIAK, 1984) that the total economically recoverable groundwater potential in Eastern Botswana is of the order of 200,000 $m^3$ per day, which will meet about 30 % of the water demand (excluding irrigation) in the next 25 years; the rest has to be obtained from surface water with the help of dams. However, no reliable information on groundwater replenishment is available, thus it is not known if the planned abstraction would mean a depletion of the groundwater resources in the long run.

The present paper reviews the groundwater recharge research to date.

## THE KALAHARI BASIN

The Kalahari Beds comprise eolian, fluvial and lacustrine sandy sediments from Late-Cretaceous to Recent, with an average thickness of over 50 m. Interspersed are concretionary layers of calcrete and silcrete, derived from chemical activity of percolating water. These chemical deposits cover larger areas, especially in the south of the Kalahari and can reach a thickness of over 30 m. Relatively well sorted median grained sands prevail in the upper Kalahari Beds.

The Botswana Kalahari is rather densely vegetated and classifies as a bush and tree savanna with alternating grassland. It is normally characterized as a desert because of the virtual absence of surface water and even springs. Real desert conditions in the sense of sparsely vegetated and unstable moving sands prevail only in the extreme southwest.

Groundwater has been encountered in a large number of boreholes in the Karoo (Paleozoic-Mesozoic) and the Precambrian bedrock underneath the Kalahari Beds, but the Kalahari sands do not form extensive aquifers at present. Only locally perched water bodies are found, related to depressions and duricrust horizons.

### Soil moisture retention studies

The possibility of groundwater recharge by infiltration of rainwater through the Kalahari Beds under the present climatic conditions is a matter of continued debate, dating back to the beginning of this century. A.W. Rogers concluded from observations in the Kalahari of the Northern Cape Province (Republic of South Africa) as early as 1907 : "The sand of the sandveld allows rainwater to penetrate the ground rapidly, and yet is not coarse enough to let the moderate to small quantity received run through to bedrock or less pervious layer; the water is thus held in the sand by capillarity". (quoted by Van Straten, 1955).

Experiments on soil moisture retention, carried out in the laboratory of the Geological Survey of Botswana by O.J. van Straten, led to the conclusion that Kalahari sand with a thickness of over 6 m will impede active rainfall infiltration because of seasonal moisture retention and subsequent evaporation and evapotranspiration. He observed that it requires 80 mm of rainfall surplus to bring 6 m of sand to field capacity (Van Straten, 1955; Boocock and Van Straten, 1962).

## Isotope studies

C.H.M. Jennings, who carried out hydrological studies all over Botswana for many years, supported Van Straten's ideas in view of the occurrence of an overall diffuse recharge, but found convincing evidence for local groundwater accumulation under favourable conditions such as depressions of pans and old valleys with a permeable surface. Duricrust beds related to such collectors of episodic run-off often form cracks and solution cavities through which rainwater can rapidly percolate beyond the reach of evaporation. Settlements in the Kalahari are concentrated around this type of feature as well as near locations where the bedrock is at sub- or outcrop.

Isotope analyses on water samples taken from a large number of boreholes and wells in the Kalahari gave indeed evidence for recent rain recharge, notably in the northern part with an annual precipitation of over 400 mm (Jennings, 1974; Mazor et al., 1974; Mazor et al., 1977). It must be emphasized however, that sampling was rather selective by making use of existing wells which are normally found in the above mentioned favourable infiltration sites with relatively shallow groundwater.

Isotope studies on soil moisture down to a depth of 5 m by B.Th. Verhagen and Jennings (Jennings, 1974) and subsequently by Foster et al. (1982) up to a depth of 8 m, revealed the occurrence of high tritium levels below 5 m, suggesting that present rainfall penetrated at least to that depth. The latter authors however, from a combined analysis of the tritium profile with the vertical distribution of soil moisture and chloride, noticed that neither the distribution nor the total amounts of tritium reflect the sequence of post-1957 rainfall through a piston-like movement. They explain the observed profiles by a complex up- and downward movement in both the liquid and the vapour phase, including a possible tritium enrichment by fractionation. Their conclusion is that it would be imprudent to assume active recharge through Kalahari sand cover of more than about 4 m, in areas with less than 400 mm of rainfall (see also Mazor, 1982).

## Hydraulic approach

De Vries (1983) considered a possible replenishment of the sub-Kalahari aquifers in the Central Kalahari in the light of the eventuate hydraulic gradient and hydraulic depth. He argues that because of the large distance (~ 500 km) between the regional divide (the Ganzhi ridge and the Passarge swell) and the discharge base (the Makgadikgadi salt pans), an overall annual recharge of the order of 1 mm or more would have resulted in a shallow groundwater table and the emergence of springs and seepage zones, upstream of the discharge area.

The present groundwater depth and the hydraulic gradient can be explained as either in equilibrium with a present day recharge of less than 1 mm a year or as a residual from a head decay since the last major pluvial period, which according to Grey and Cooke (1977) ended some 12,000 years back.

Regional subsurface recharge at the eastern edge of the Kalahari from the Precambrian and Karoo outcrop areas however is possible. For instance, large scale abstractions from a sub-Kalahari Karoo aquifer at about 50 km from the bedrock outcrops takes place for the Jwaneng diamond mine. Although 5 million $m^3$ a year has been pumped for the last seven years now, no significant drawdown has been generated as yet.

### Discussion

It can be concluded that there is no convincing evidence for an active present day diffuse recharge through the Kalahari sands. Local and episodic deep percolation is quite likely, but its contribution to the regional aquifers seems minor.

It is evident that the tree-grass savanna with its root systems at different levels is reponsible for the depletion of all or almost all soil moisture surplus. Eagleson and Segarra (1985) showed that especially the tree-grass community in their competition for water and energy must be able to develop a stable equilibrium between rainfall and evapotranspiration.

The ability of some trees to abstract water even from deep aquifers was illustrated by Jennings (1974) who recovered living tree roots from a depth of 68 m from an unused borehole in the Central Kalahari. Since the water table was at 141 m, it is quite possible that the tree (probably boscia albitrunca) used the borehole to reach the water table.

## EASTERN BOTSWANA

The Precambrian rocks of Eastern Botswana consist of an Archean gneiss complex and supracrustal sediments and volcanites of the Transvaal and Waterberg Supergroups. Paleozoic-Mesozoic sediments of the Karoo Supergroup are locally outcropping at the edge of the Kalahari and in a few synclines in the Precambrian area. The Precambrian sediments are formed by sandstones, quartzites, shales, mudstones and dolomites. The Karoo deposits consist predominantly of sandstones and siltstones, partly covered by basalt.

Aquifers have mainly developed in fracture zones within synclinal structures. Primary permeability only occurs in Karoo sandstones thus forming aquifers of larger extent. Groundwater is also encountered in fractured and weathered zones in crystalline rocks, but such aquifers are normally of limited size.

Most of Eastern Botswana is rather flat with isolated inselbergs and ridges of resistant, mostly arenitic rocks which rise up to 400 m above the flat planation surfaces. The valleys of the ephemeral rivers are often ill-defined and weathering is shallow. A dense bush and tree vegetation covers the area, that is classified as a tree savanna.

The main aquifer systems have developed in arenitic and dolomitic rocks. These groundwater basins often correspond to surface catchments because of a geological and structural control on the geomorphology. Thus the fractured and permeable zones of the catchment are normally characterized by depressions and therefore often covered with rubble and colluvial soils.

The main water bearing fractures are generally encountered to a depth of 150 m whereas the static water level is at a depth of less than 40 m. The groundwater surplus under natural conditions is discharged by subsurface flow and leakage to lower areas where it evaporates or contributes to run-off. In closed basins the water balance is regulated by the feed-back mechanism that is formed by the coupling of water table depth and evaporation.

## Water balance studies

There is no doubt about present day recharge of the aquifers in this area, but little quantitative information is available. Deep percolation occurs from more or less unchanneled run-off, and by infiltration from ephemeral water courses. We will refer here to the first type of process; the latter is predominantly related to percolation into the alluvial sand bed of present or former valleys.

Early investigations were performed by Jennings in the 1950's and 1960's and reported in Jennings (1974). Measurements were carried out in several small (~ 10 $km^2$) groundwater basins in dolomite, quartzite and sandstone near Lobatse, exploited for the public and industrial water supply of that small town. An analysis of hydrographs and pumping figures, including the zero-water level change method, brought this author to an estimated recharge of 4 % of the annual rainfall. This means a replenishment of 22 mm for this area with an average annual precipitation of 550 mm.

Field observations made clear that recharge is concentrated in fan like colluvial deposits at the foot of the slopes, where small episodic rivulets discharge the run-off, collected from the slopes. Jennings' 4 % recharge therefore refers only to the contributing sloping areas, which cover about 50 % of the catchments. Accordingly, the recharge expressed as a water depth over the whole catchment thus amounts to 2 %, which equals an average water depth of 11 mm.

Local measurements with a gauging weir in the main rivulet gave a run-off of the sloping area of 4 % during the 1970-1971 season with 579 mm rainfall, and 10 % run-off during the next season with 648 mm. A subsequent percolation assessment with the aid of tritium in a 5 m deep profile half-way the slope and the valley axis, resulted in a figure of 7 %.

Jennings' results match rather well with observations in a similar climatic and geological environment in the Northern Cape Province in the Republic of South Africa. A water balance study was made by P.J. Smit for the Ghaap Plateau near Kuruman, a 1140 $km^2$ dolomitic catchment. Direct water surplus measurements could be performed here because of the prevailing concentrated discharge through a few large springs. Smit (1978) reports a recharge of 2.5 % of the annual rainfall for the relatively dry period 1962-1970, with an average precipitation of 346 mm. The average rainfall for the period 1940-1970 was 445 mm, thus the long term recharge would equal a water depth of 11 mm over the whole area.

Closer to Botswana was a recharge study of a dolomitic aquifer near Mafeking with analyses of hydrographs and thirteen tritium profiles, by Bredenkamp et al. (1974). They arrive at a recharge of 5.4 % of the annual rainfall in this area with an average precipitation of 560 mm a year.

## Chloride balance approach

A chloride balance approach was recently applied to a water balance study of a groundwater basin formed along a fracture zone in dolomite, shale and quartzite in the same Lobatse area. This Nywane Basin covers an area of 10 $km^2$, of which 2.5 $km^2$ forms the aquifer proper under a flat valley bottom, whereas the remainder of the catchment consists of rather impervious pediments and a few small inselbergs. The valley is filled with decomposed cherty dolomite, shale and quartzite rubble, and colluvial and alluvial soil locally to a depth of 30 m.

The total wet and dry chloride deposition has been measured since June 1983 at the Geological Survey Department in Lobatse. A deposition of 1821 mg Cl per $m^2$ was inferred from analyses of rain gauge samples for the period June 1983 - June 1986. The deposition increased from 425 mg in 1983/1984 to 678 mg in 1984/1985 and subsequently to 718 mg in 1985/1986. The average annual rainfall during that dry period was 321 mm. If we assume the annual chloride deposition to be more or less constant, then we arrive at an average figure of 604 mg per $m^2$ per year with an average rainfall of 550 mm. This combination will produce a flux of moisture through the soil surface with an average chloride content of 1.1 mg/l. The average chloride concentration of the groundwater in the aquifer was determined as 11.4 mg, suggesting a recharge of the order of 605/11.4 = 53 mm a year or 9.6 % of the annual rainfall.

This recharge figure refers to the total contributing area. If we assume, in analogy with Jennings, the contributing area to be restricted to the sloping pediments, then we arrive at a recharge of the aquifer proper of 100 mm.

A straightforward application of the chloride balance method to the aquifers surveyed by Jennings is not possible because of pollution in that area. Jennings' chemical data however, suggest that chloride concentrations between 5 and 14 mg prevail in non-contaminated groundwater. So there is evidence that the chloride balance method in these areas too would lead to higher recharge figures than the water balance method.

It is possible that the chloride deposition during the dry 1983-

1986 period is not representative for the long term precipitation, or the raingauge produced too high figures by acting as a trap for re-cycled salt particles, transported by dust storms. This process can be expected notably active in a dry period. Regular rinsing of the rain gauge with distilled water during the dry winter months however, produced a dry precipitate of 20 % of the total deposition, which does not seem to be exceptional.

## Discussion

From the still incomplete observations in the Lobatse area it can be concluded that recharge varies widely with respect to place and time, and that one should be careful in applying these figures to other areas in Eastern Botswana. The geomorphological situation, the nature of the covering layers and the type of vegetation have to be taken into account.

Favourable sites for deep percolation are constituted by areas where the aquifer rocks are in direct contact with the surface through exposed fissures and fractures, or through a cover of permeable rock fragments, allowing rapid infiltration into the bedrock beyond the reach of the roots of trees and shrubs. A maximum recharge under prevailing geological conditions may be expected in case of an equilibrium between the infiltration capacity of the recharge area and the run-off production on the contributing slopes of the catchments.

## REFERENCES


Boocock, C. and O.J. van Straten, 1962. "Notes on the geology and hydrogeology of the Central Kalahari region. Bechuanaland Protectorate". Trans. Geol. Soc. S. Africa. 65 : 125-171.

Bredenkamp, D.B., J.M. Schutte and G.J. Du Toit, 1974. "Recharge of a dolomitic aquifer as determined from tritium profiles". Int. Atomic Energy Agency (IAEA), Vienna, pp. 73-95.

De Vries, J.J., 1983. "Holocene depletion and active recharge of the Kalahari groundwaters - A review and an indicative model". J. Hydrology, 70 : 221-232.

Eagleson, P.S. and R.I. Segarra, 1985. "Water-limited equilibrium of savanna vegetation systems". Water Resources Research, 21 : 1483-1493.

Foster, S.S.D., A.H. Bath, J.L. Farr and W.L. Lewis, 1982. "The likelihood of active groundwater recharge in the Botswana Kalahari". J. Hydrology, 55 : 113-136.

Grey, D.R.C. and H.J. Cooke, 1977. "Some problems in the Quaternary evolution of the landforms in northern Botswana". Catena, 4 : 123-133.

Jennings, C.M.H., 1974. _The hydrogeology of Botswana_. PhD-thesis, University of Natal, Pietermaritzburg.

Mazor, E., 1982. "Rain-recharge in the Kalahari - A note on some approaches to the problem". J. Hydrology, 55 : 137-144.

Mazor, E., B.Th. Verhagen, J.P.F. Sellschop, N.S. Robins and L.G. Hutton, 1974. "Kalahari groundwaters : their hydrogen, carbon and oxygen isotopes". Int. Atomic Energy Agency (IAEA), Vienna, 203-225.

Mazor, E., B.Th. Verhagen, J.P.F. Sellschop, H.T. Jones, N.S. Robins, L.G. Hutton and C.M.H. Jennings, 1977. "Northern Kalahari groundwaters : hydrologic, isotopic and chemical studies at Orapa, Botswana". J. Hydrology, 34 : 203-234.

Smit, P.J., 1978. "Groundwater recharge in the dolomite of the Ghaap Plateau near Kuruman in the Northern Cape, Rep. of South Africa". Water SA, 4 : 81-92.

Van Straten, O.J., 1955. "The geology and groundwater of the Ghanzi Cattle Route". Annual Report Geological Survey Bechuanaland Protectorate, 1955, 28-39.

VIAK, 1984. _Eastern Botswana regional water study_. Ministry of Mineral Resources and Water Affairs, Republic of Botswana.

# RAINFALL - RUNOFF - RECHARGE RELATIONSHIPS IN THE BASEMENT ROCKS OF ZIMBABWE

J.Houston
Hydrotechnica
Pengwern Court
High Street
Shrewsbury, SY1 1SR
U.K.

ABSTRACT. The resources of Basement aquifers in Victoria Province, Zimbabwe are largely dependent on rainfall recharge but no direct evidence for recharge in the form of borehole hydrographs is available. Estimates of recharge have been made by three methods: river baseflow analysis; hydrochemical analysis of groundwaters and simulation modelling. All three methods produce consistent results, suggesting that recharge amounts to between 2-5% of annual rainfall. Under such circumstances, which are widespread in Africa, it is essential to make resource estimates based on many years of data, otherwise over or under estimates are likely because of the considerable annual variation in rainfall.

## 1. INTRODUCTION

As a result of the recent drought in central Africa an accelerated programme of drought relief was commissioned by the Government of Zimbabwe and the European Economic Community in Victoria Province, Zimbabwe (see figure 1). Over an area of 22000 km$^2$, 282 boreholes were sited, drilled and equiped with handpumps. The area is one of variable relief on the edge of the Zimbabwe plateau and extending down toward the river Limpopo valley. The area is underlain by Basement granites and gneisses. The principal aquifer is composite composed of weathered regolith of low permeability, high storage, overlying fissured bedrock of high permeability and low storage.

No borehole hydrographs are available to indicate that recharge to Basement aquifers is taking place in the project area. However, surface water hydrographs show considerable baseflow from aquifers in every year for which records are available. This is strong evidence for recharge. Hydrochemical evidence also indicates the probability of active recharge. Furthermore, by analogy with other areas in central Africa (Foster et al, 1982; Houston, 1982) it would appear likely that precipitation recharge will occur whenever the annual rainfall exceeds a



*I. Simmers (ed.), Estimation of Natural Groundwater Recharge, 349–365.*

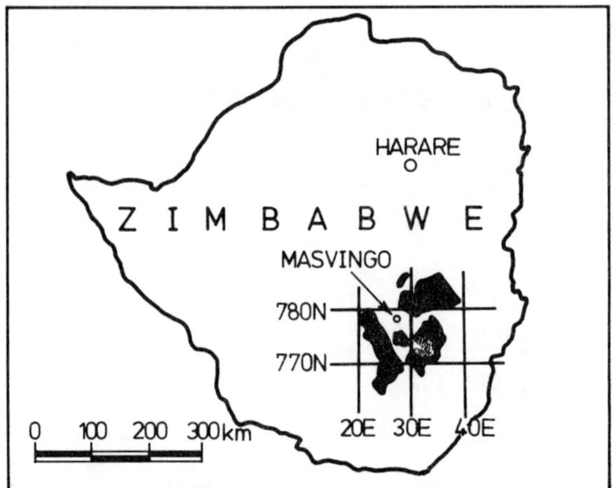


**Figure 1 : Project location map**

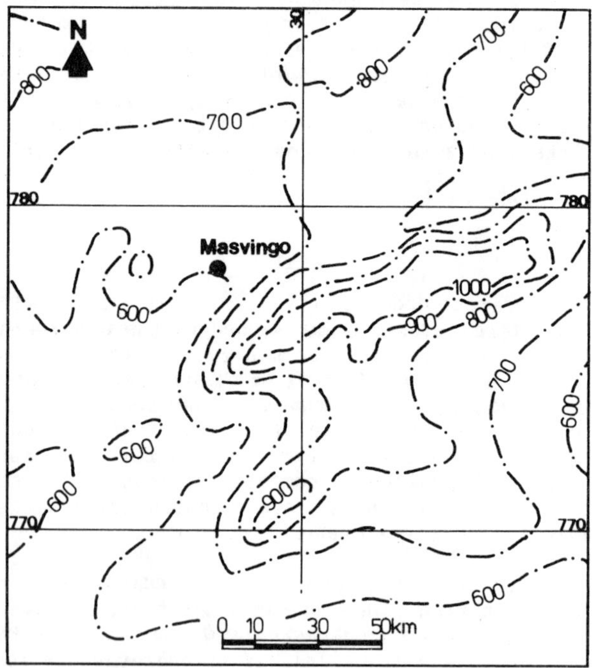


—·— Contour of mean and annual rainfall in millimetres
(Department of Meteorological Services 1968)

**Figure 2: Mean annual rainfall for Victoria Province**

threshold value. Lateral recharge from other aquifers or surface water is also a possibility, but is likely to be on a localised scale and is not considered further here.

## 2. HYDROMETEOROLOGY

Within the project area there are 22 rainfall stations with daily data in one case dating back as far as 1900. Only three stations have class A pans for the measurement of evaporation and records only go back to 1964. No systematic data is available for the calculation of evapotranspiration by an energy balance method such as Penman's so that it has been necessary to convert pan evaporation to evapotranspiration using monthly conversion factors determined empirically elsewhere in Africa (Aune, 1970; Heederick et al, 1984; Pike, 1971; Riou, 1984).

The map of mean annual rainfall produced by the Department of Meteorological Services in Zimbabwe is summarised in figure 2. It is clear that rainfall is orographically controlled and dependent upon the south-east Trade winds bringing moist air up the Limpopo valley from the Indian Ocean during summer. Summer lasts from Ocober to March when the Inter-Tropical Convergence zone moves south over northern Zimbabwe. Winter, the dry season lasts from April to September.

Changes in diurnal and seasonal rainfall are shown in figure 3 for Masvingo over a thirty year period. The rainy season usually starts in the second half of October and lasts for five months to the end of March. The most reliable months for rainfall are November, December and January. The daily intensity is also greatest during these months and greatest during the afternoon. This means that the bulk of the rainfall is concentrated in a relatively short time span, enhancing the potential for both runoff and recharge to aquifers but limiting it to short periods of the year.

Long term variations in rainfall are also of importance to recharge estimates. The mean seasonal rainfall at Masvingo over 84 years since 1900 is 623 mm but there is variation between 40% and 220% of this figure. The mean annual rainfall for the project area as a whole, found by integrating the areas on the rainfall contour map is 728 mm.

The variation in annual rainfall is shown in figure 4. The periods 1905-25, 1951-55 and 1970-80 were wetter than normal whilst the periods 1925-35, 1945-50 and 1965-70 were drier than normal. The intensity of any dry period can be determined from the steepness of the descent of the graph in figure 4b. The recent drought intensity has never been matched since records began but was almost matched in 1949-51. However the antecedent rainfall was much less in both 1949-51 and 1968-73 than in 1981-84. Assuming that recharge to Basement aquifers is largely precipitation dependent, this suggests that although groundwater levels may be falling sharply now they may not be as low as they were around 1950 and 1970.

Over the entire 84 year period there is no evidence for a trend toward drier conditions. However there is some evidence to suggest poorly developed periodicities or rainfall cycles. The evidence comes from the autocorrelation function and power spectrum of the annual rainfall series (see figure 5). The most significant periods are of 23

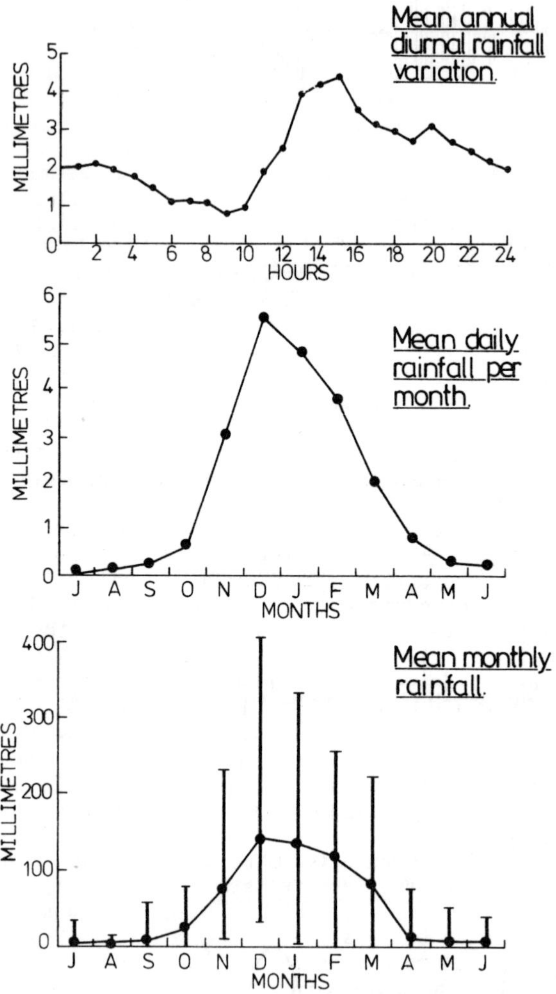


**Figure 3 : Diurnal and seasonal changes in rainfall at Masvingo for the years 1941 to 1971**

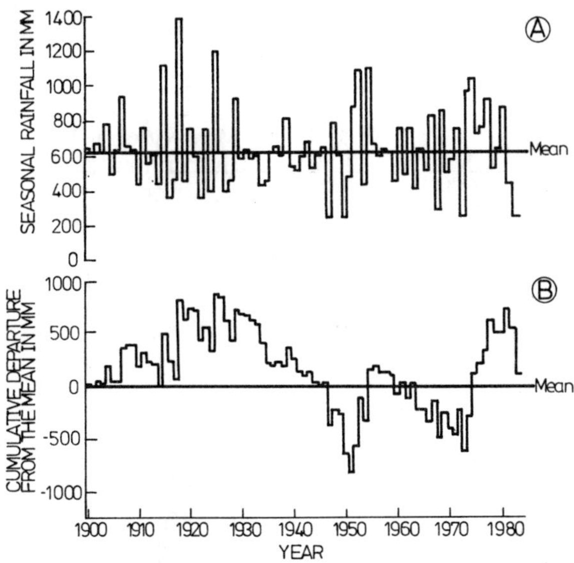


**Figure 4 : Annual rainfall time series for Masvingo**

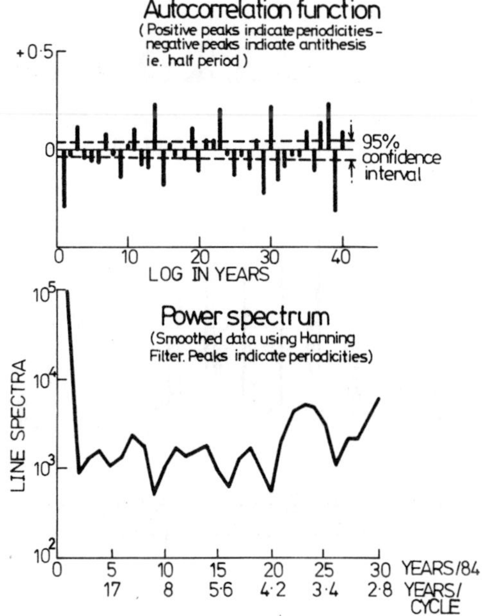


**Figure 5 : Time series analysis of annual [seasonal] rainfall at Masvingo for the years 1900 to 1983**

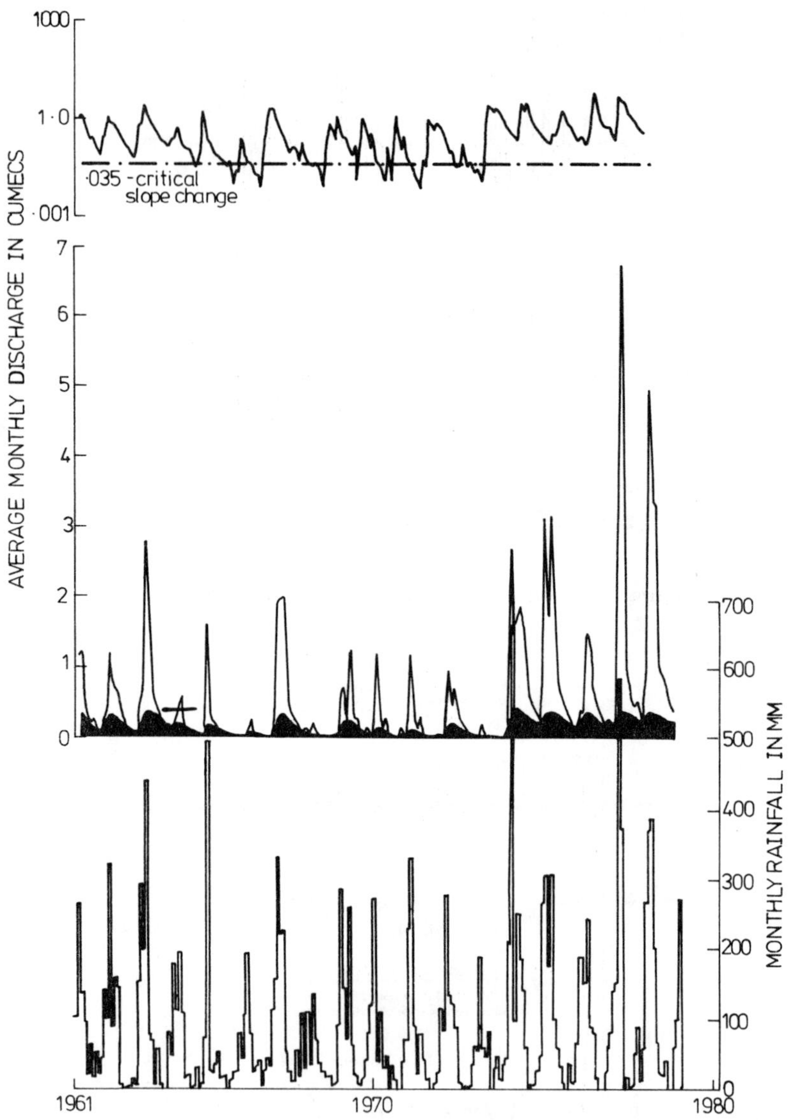


**Figure 7: The relationship between rainfall, run off and baseflow for the Mzero catchment**

years (with a harmonic at 11.6 years) and a shorter cycle of 14 years (with harmonics close to 7 and 3.5 years). Both of these periodicities are commonly found elsewhere in Africa (Houston, 1982; Lamb, 1982) and India (Vines, 1986). A reconstruction of the principal periodicities suggests that low rainfalls may continue to be a feature through the 1980's with a return to generally wetter conditions by the mid 1990's.

Mean monthly values of pan evaporation and potential evapotranspiration are shown in figure 6 for a ten year period. Evapotranspiration reaches a peak during the rainy season but the potential for excess rainfall contributing to runoff or recharge is clearly seen, despite the fact that on an annual basis evapotranspiration amounts to an average 1246 mm.

## 3. BASEFLOW ANALYSIS

Within Victoria Province four gauging stations are available with average daily flow data for periods varying from 8 to 17 years. Catchment areas vary from 100 to 5390 $km^2$, all largely on granitic and gneissic rocks.

The baseflow component of the river hydrographs shows up clearly on both arithmetic and semi-logarithmic plots (see figure 7). Baseflow discharge represents water which has percolated through the soil and vadose zones to the groundwater table but which subsequently discharges to rivers by means of gravity drainage. The simplest water balance for a catchment is given by

$$\text{discharge} = \text{recharge} + \text{change in storage}$$

Thus over periods of equivalent storage, recharge to groundwater is given by the baseflow component of the hydrograph. Each component of the hydrograph can be approximated by a relationship of the form

$$Qo = Qt \quad K^{t}$$

as shown by Barnes (1939), where Qo and Qt are the discharges at the beginning of the measurement period and after time t respectively and K is the recession constant. In the case of baseflow, K is a function of the aquifer transmissivity, storage and catchment geometry.

All hydrographs examined exhibited a critical change of slope at extreme low flows, with a marked increase in the rate of recession. The critical change in slope represents an increase in the rate of depletion of active storage in the catchment. Since it always occurs at the same discharge level for each catchment it also occurs at the same groundwater head distribution each time. It is considered likely that at this critical point an upper zone becomes depleted and a lower zone becomes active with different aquifer characteristics. This lower zone would have higher permeabilities and the critical level must thus represent the change from regolith aquifer discharge to underlying fissured bedrock aquifer discharge. It is also important to notice that baseflow continues throughout the dry season in all catchments in five out of six years, pointing to the importance of the regolith as a

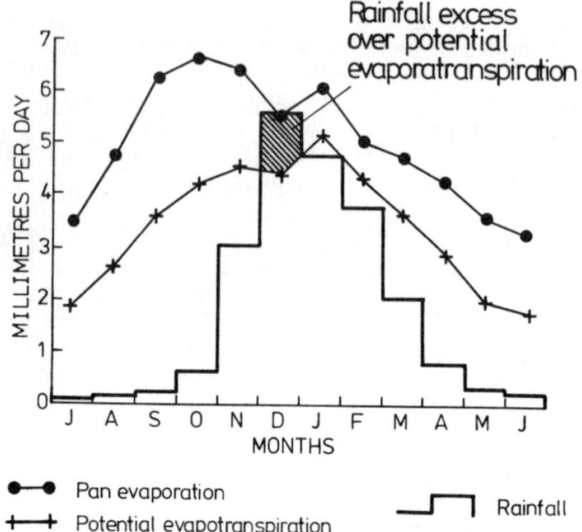


**Figure 6 : Mean monthly values of evapotranspiration and rainfall at Masvingo for the years 1974 to 1983**

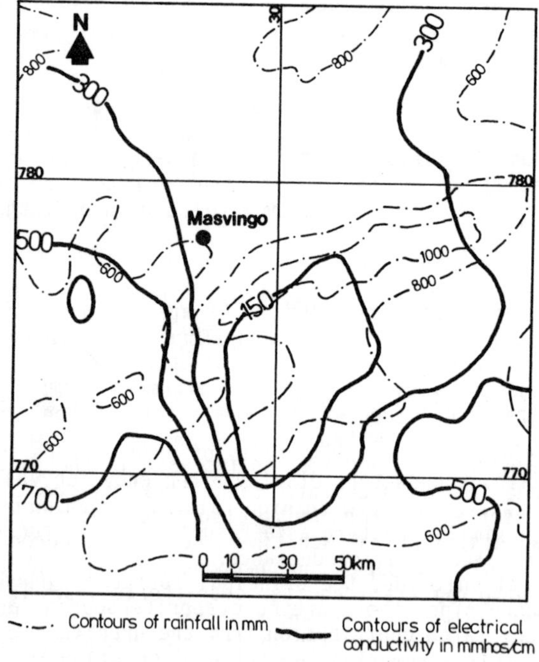


**Figure 8 : The electrical conductivity of regolith groundwaters in comparison with mean annual rainfall**

storage unit.

A regression analysis between baseflow and rainfall was carried out for each catchment. The slope of the regression line represents baseflow as a proportion of rainfall and the intercept of the regression line with the rainfall axis at the point where baseflow is zero gives the minimum critical rainfall for baseflow to have occured (see table I). Baseflow and by implication groundwater recharge averages 5% of rainfall. The Mzero catchment is different in that it has a high rainfall and is located at high altitude on the edge of the African surface, and the Lundi catchment has low rainfall and occurs at lower elevations on the post African surfaces. The minimum rainfall needed to create recharge is around 400 mm per annum for stations to the south and west of Masvingo where regolith thickness and the depth to water are greater. In the Chiredzi catchment to the north east of Masvingo the regolith aquifer is generally much thinner and water tables higher.

TABLE I Catchment characteristics for four stations

| | Chiredzi | Mzero | Musokwesi | Lundi |
|---|---|---|---|---|
| Catchment area $km^2$ | 1029 | 100 | 246 | 5390 |
| Mean rainfall mm | 818 | 1055 | 1126 | 797 |
| Mean baseflow mm | 31 | 51 | 32 | 6 |
| Baseflow as % rainfall | 4.7 | 8.5 | 4.3 | 1.9 |
| Minimum rainfall for recharge mm | 157 | 452 | 379 | 463 |
| Recession constant | .990 | .994 | .991 | .996 |
| Period of record | 1966-77 | 1961-78 | 1969-78 | 1970-78 |

## 4. HYDROCHEMICAL ANALYSIS

The variation in electrical conductivity of groundwaters is shown in figure 8. Regolith groundwaters display a tendency for values less than 150 mmhos/cm to be associated with high rainfall and runoff areas. All major ions show a similar pattern as a result of their general linear relationship with electrical conductance.

In the case of chloride, the principal source in groundwaters is from the atmosphere since there are no evaporite sources and there is unlikely to be any large contribution from the granite and gneiss host rocks. Based on this assumption the ratio of chloride in rainfall to that in groundwater is proportional to recharge thus

$$\text{recharge mm} = \text{rainfall mm} \times \frac{\text{rainfall Cl mg/l}}{\text{groundwater Cl mg/l}}$$

The relationship between chloride and rainfall is shown in figure 9. There is some suggestion here that groundwaters from areas of gneiss tend to have a higher chloride content. For each rock type a significant relationship exists between rainfall and chloride content strongly suggesting that recharge is a function of rainfall.

The chloride content of rainfall in Zimbabwe is not known but by analogy with similar areas might be expected to be about 0.5 mg $l^{-1}$ (Foster et al, 1982). Using this value and rainfall values for the area, estimates of recharge can be made (see table II). The results show that for granite aquifers recharge is likely to vary from 2-3% rainfall. For gneiss aquifers recharge rates of only 0.5-1% of rainfall are calculated. This may suggest another source of chloride in gneiss aquifers.

TABLE II Mean chloride content and recharge as a percentage of rainfall for different geological and rainfall environments

| Rainfall | Granite | | Gneiss | |
|---|---|---|---|---|
| mm | Cl mg/l | recharge% | Cl mg/l | recharge % |
| 500 | 28 | 1.8 | 104 | .5 |
| 728 | 15 | 3.3 | 59 | .9 |

## 5. RECHARGE - RUNOFF SIMULATION

In order to assess recharge more accurately a simulation model was used which requires rainfall and evapotranspiration data as input and produces estimates of recharge and runoff as output. It is essentially a routing model for flow through the soil and unsaturated (vadose) zones to the water table and to runoff. At its heart is the calculation of soil moisture deficits by the methods of Penman (1949) and Grindley (1967) which in their simplest form are given by

$$SMD_{t+1} = SMD_t + E_t - RF_t \qquad \text{for } SMD_t > 0$$

$$SMD_{t+1} = E_t - RF_t \qquad \text{for } SMD_t < 0$$

where $SMD_t$ is the soil moisture deficit at time t, E is the evapotranspiration component and RF is rainfall. The model requires the estimation of a root constant which operates a negative feedback mechanism by limiting evapotranspiration when soil moisture deficits are

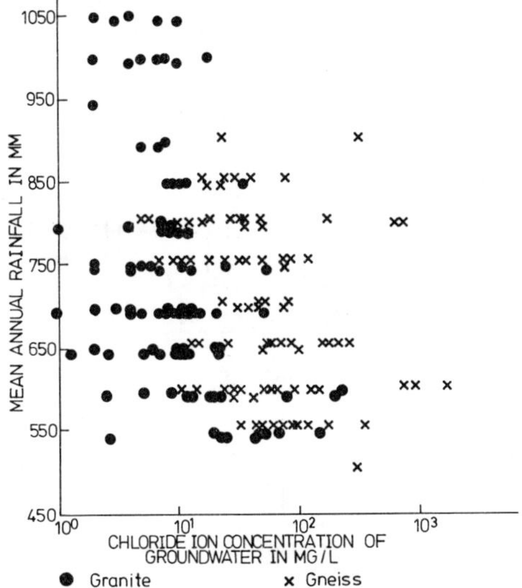


**Figure 9 : The relationship between chloride ion concentration in groundwater and rainfall at the same site· rainfall interpolated from isopleth map**

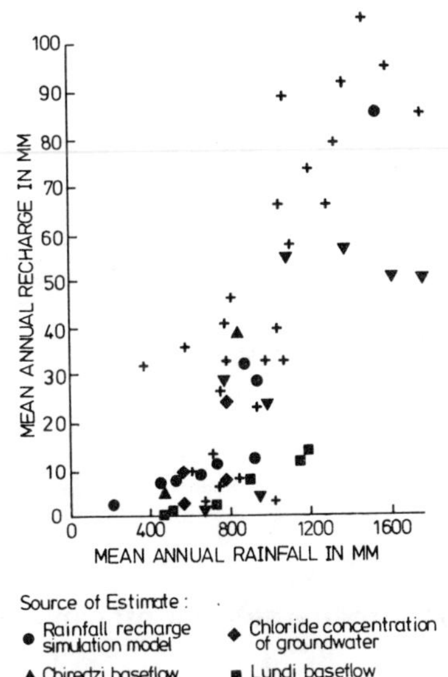


**Figure 12 : Relationship between recharge and rainfall**

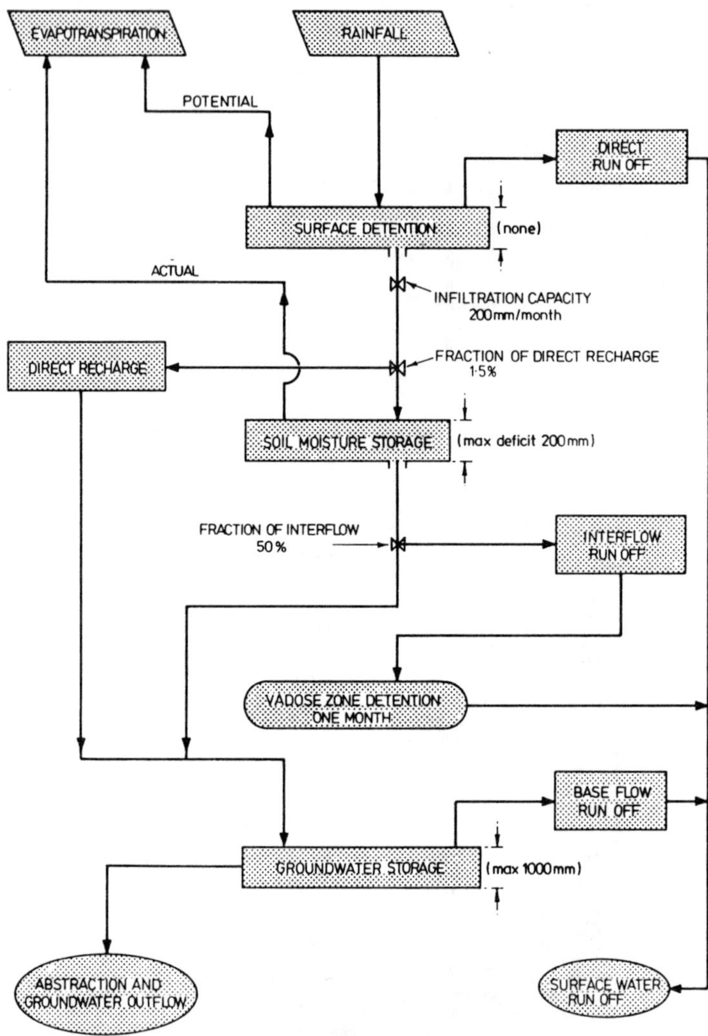


**Figure 10 : Flow diagram for recharge - run off simulation model**

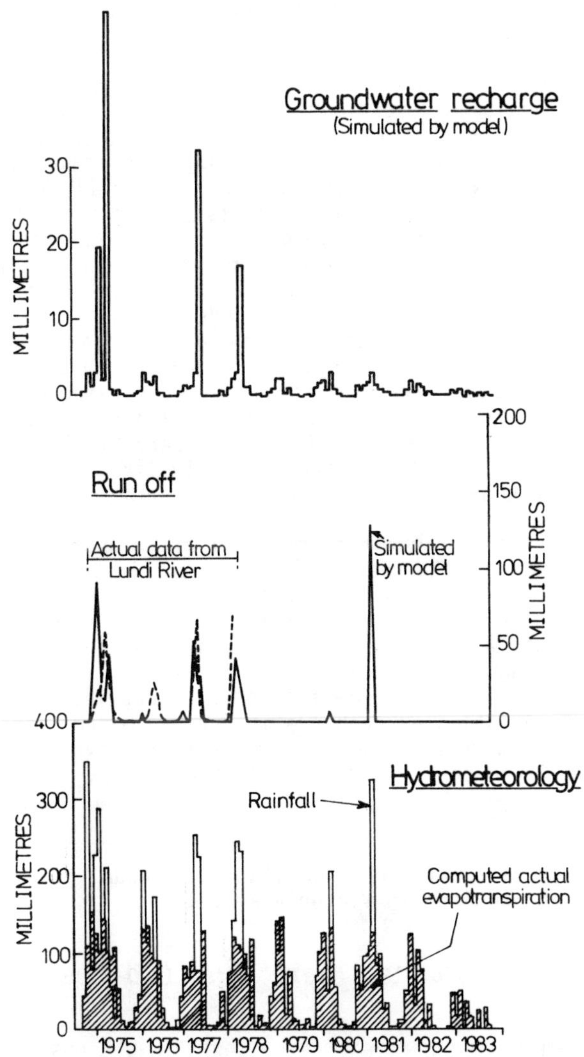


**Figure 11 : Results of recharge - run off simulation model**

high. Any rainfall in excess of evapotranspiration and soil moisture requirements is routed downwards to the vadose zone subject to the infiltration capacity where a fraction becomes interflow runoff and the remainder recharges groundwater.

A great variety of such models have been produced for temperate climates where the results have been found to be generally good (Calder et al, 1983) but tend to underestimate recharge when used on a monthly timestep basis (Howard and Lloyd, 1979). Houston (1982) has shown that such models are equally applicable to dry tropical climates providing reasonable assumptions are made. Furthermore, the error in using monthly time steps as opposed to daily is less than other potential errors, for instance that involved in determining evaporation.

Using monthly rainfall and evapotranspiration data obtained as previously described, a simulation model was developed and calibrated on the period 1974 to 1983, a period when the average rainfall was 727 mm, close to the overall mean for Victoria Province. A flow diagram for the simulation model is shown in figure 10 for run number 11. Runs 1 to 10 were used to test the sensitivity of the model by varying such factors as the infiltration capacity, the fractions of direct recharge and interflow, the maximum volumes held in surface detention and soil moisture storage, and the root constant (finally fixed at 100 mm). All values employed are considered to be realistic in the circumstances operating in Victoria Province.

The output was checked against the Lundi river hydrograph as shown in figure 11. In general the results are good confirming the validity of the model. However, in detail the actual data from the Lundi river suggests that there is slightly more interflow and baseflow than the model predicts. This would have the effect of reducing recharge to groundwater marginally. The results of the model (given in table III) suggest that some errors may be present in the first year of simulation. Since runoff is underestimated recharge is overestimated. After the first year the results settle down and mean recharge over an eight year period amounts to 22 mm equivalent to 2.5% of rainfall.

TABLE III Results of runoff-recharge simulation model for an idealised Basement aquifer in Victoria Province

| Year | 1975 | 1976 | 1977 | 1978 | 1979 | 1980 | 1981 | 1982 | 1983 |
|---|---|---|---|---|---|---|---|---|---|
| Rainfall mm | 1511 | 720 | 763 | 919 | 521 | 634 | 905 | 441 | 211 |
| Recharge mm | 82 | 11 | 39 | 27 | 8 | 9 | 12 | 7 | 3 |
| Recharge as % of rainfall | 5.4 | 1.5 | 5.1 | 2.9 | 1.5 | 1.4 | 1.3 | 1.6 | 1.4 |

## 6. RECHARGE TO BASEMENT AQUIFERS

It is possible to compare the results of recharge estimated by three completely independent methods. In order to do this the results of each estimate have been plotted as recharge aginst rainfall in figure 12. From this it can be seen that the results compare very well. Over much of Victoria Province recharge amounts to 2-5% of the rainfall on a long term basis. On a Province-wide basis with mean annual rainfall of 728 mm, recharge amounts to 12 mm/annum equivalent to 12000 $m^3$/annum/$km^2$.

In detail however, there is considerable variation in recharge resulting largely from the uneven distribution of rainfall both in space and time. On a spatial basis, those areas above 700 mm rainfall are more likely to receive significant recharge than those below 700 mm. Also, below 400 mm it is questionable whether recharge ever occurs. On a temporal basis recharge tends to be correlated with rainfall such that recharge is limited to short periods within the year and a few years with heavy rainfall are more important in producing recharge than many years of average rainfall.

If a conservative estimate of recharge is assumed to be 2% of rainfall in an area of relatively low rainfall which, for Victoria Province would be about 500 mm/annum, then it can be seen that recharge amounts to 10 mm/annum. For a rural development scheme the average demand of a hand pump is 6500 $m^3$/annum assuming 0.5 l/s over a 10 hour pumping day, 360 day year. If it is assumed that the yield is totally relient on precipitation recharge then an area of 0.65 $km^2$ is required for its sustenance at the recharge rate of 10 mm/annum. This is equivalent to a radius of 450 m around the borehole. Based on equilibrium conditions and average aquifer characteristics for Basement rocks, the cone of depression due to a hundpump would extend to at least 1000 m around the borehole and therefore recharge is not likely to be a limiting factor in resource development except in special circumstances.

## 7. DISCUSSION

It is quite clear that in any assessment of recharge in similar environments it will never be adequate to assess recharge based on a short data length. Since the necessary data for a full recharge evaluation is rarely available it is also important to use a variety of techniques for cross reference and to investigate more closely that aspect of the hydrological cycle, namely rainfall, which is more frequently better documented and can lead to a greater insight into potential recharge patterns.

Reliable estimates of recharge to aquifers in central Africa are not common, but it is worthwhile briefly considering two other estimates. In an area on the edge of the Kalahari in Botswana, Foster et al (1982) suggest that precipitation recharge is probably unlikely when the mean annual rainfall falls below 450 mm and the sand cover is greater than 4 m. In a much wetter area (mean annual rainfall 937 mm) of Zambia, Houston (1982) showed that recharge was not only dependent upon rainfall but also upon the vegetation type. Thus in areas of natural forest where evapotranspiration was high, recharge amounted to

80 mm or 9% of rainfall, whereas those areas of cleared ground with crops produced 281mm of recharge or 30% of rainfall. In both areas of Zambia infiltration rates were extremely high and runoff negligable.

The conclusion drawn from this is that for large areas of Africa, recharge is dependent on rainfall both spatially and temporaly. It is tentatively considered that rainfall below 400 mm is unlikely to produce recharge and above this value the amount of recharge will depend upon factors such as soil and vegetation type.

## 8. ACKNOWLEDGEMENTS

I would like to thank the Ministry of Water Resources and Development and Energy and the European Economic Community for allowing the results of this project to be presented here.

## 9. REFERENCES

Aune, B., 1970 'Tables for computing potential evaporation from the Penman equation.' Zambia Met. Dept.

Barnes, B.S., 1939 'The structure of discharge-recession curves.' Am. Geophys. Union Trans. **20**:721-725.

Calder, I.R., Harding, R.J. and Rosier, P.T.W., 1983 'An objective assessment of soil moisture deficit models.' J. Hydrol. **60**:329-344.

Foster, S.S.D., Bath, A.H., Farr, J.L. and Lewis, W.J., 1982 'The likelihood of active groundwater recharge in the Botswana Kalahari.' J. Hydrol. **55**:113-136.

Grindley, J., 1967 'The estimation of soil moisture deficits.' Met. Mag. **96**:97-108.

Heederick, J.P., Gathuru, N., Majanga, F.I. and Van Dongen, P.G., 1984 'Water resources assessment study in the Kiambu District, Kenya.' IAHS, Proceedings of the Harare symposium: 95-110.

Houston, J.F.T., 1982 'Rainfall and recharge to a dolomite aquifer at Kabwe, Zambia.' J. Hydrol. **59**:173-187.

Howard, K.W.F. and Lloyd, J.W., 1979 'The sensitivity of parameters in the Penman evaporation equations and direct recharge balance.' J. Hydrol. **41**:329-344.

Lamb, H.H., 1982 'Climate, history and the modern world.'
Methuen.

Penman, H.L., 1949 'The dependence of transpiration on weather and soil conditions.'
J. Soil Sci. **1**:74-89.

Pike, J.G., 1971 'Rainfall and evaporation in Botswana.'
UNDP/FAO Botswana Tech. Doc. 1.

Riou, C., 1984 'Experimental study of potential evapotranspiration in central Africa.'
J. Hydrol. **72**:275-288.

Vines, R.G., 1986 'Rainfall patterns in India.'
J. Climat. **6**:135-148.

# RECHARGE CHARACTERISTICS OF AQUIFERS OF JEDDAH-MAKKAH TAIF REGION

Y. Basmaci and M. Al-Kabir
King Abdulaziz University
Civil Engineering Department
P.O.Box 9027, Jeddah-21413
Jeddah, Saudi Arabia

ABSTRACT. Recharge characteristics of the aquifers over the Jeddah-Makkah-Taif region of the Western Saudi Arabia were studied through environmental isotope techniques. Space and time variation of oxygen-18, deuterium and tritium in rainfall and in groundwater were analyzed. Variation of oxygen-18 with respect to the altitude in rain and groundwater are expressed as $\delta O^{18} = -4h + b$ and $\delta O^{18} = -1.4h - 0.70$ respectively. The intercept of the regression equation for rain samples varies in a range of -2 to 8 ‰ $O^{18}$ indicating seasonal changes and multitude of moisture sources in the area. The recharge area is the mountainous zone. Recharge is either direct or from the floods moving to the downstream reaches.

## 1. INTRODUCTION

Jeddah-Makkah-Taif area (Figure 1) is the most crowded and industrialized part of the Kingdom of Saudi Arabia. Groundwater constitutes the major water resource. Estimation of recharge and identification of the recharge areas of the aquifers are attempted through the use of environmental isotopes.

The region is a part of the Arabian Shield which is an extensive occurrence of the Precambrian to early Paleozoic igneous and metamorphic rocks, partly overlain by younger Tertiary to Quaternary basalts and Cretaceous to Tertiary sedimentary rocks. Developed upon the shield are the sequences of Quaternary alluvial filled valleys which contain the main groundwater sources of the Western Saudi Arabia. The valleys are structurally controlled. The scarp mountains which are running parallel to the Red Sea rise abruptly from the coastal plain and reach around 4000 m asl. East of the mountains is a peneplain, named as the Sedimentary Basin. The valleys are deep and steep in the upstream reaches. They gradually join with the coastal plain and the



*I. Simmers (ed.), Estimation of Natural Groundwater Recharge, 367–375.*

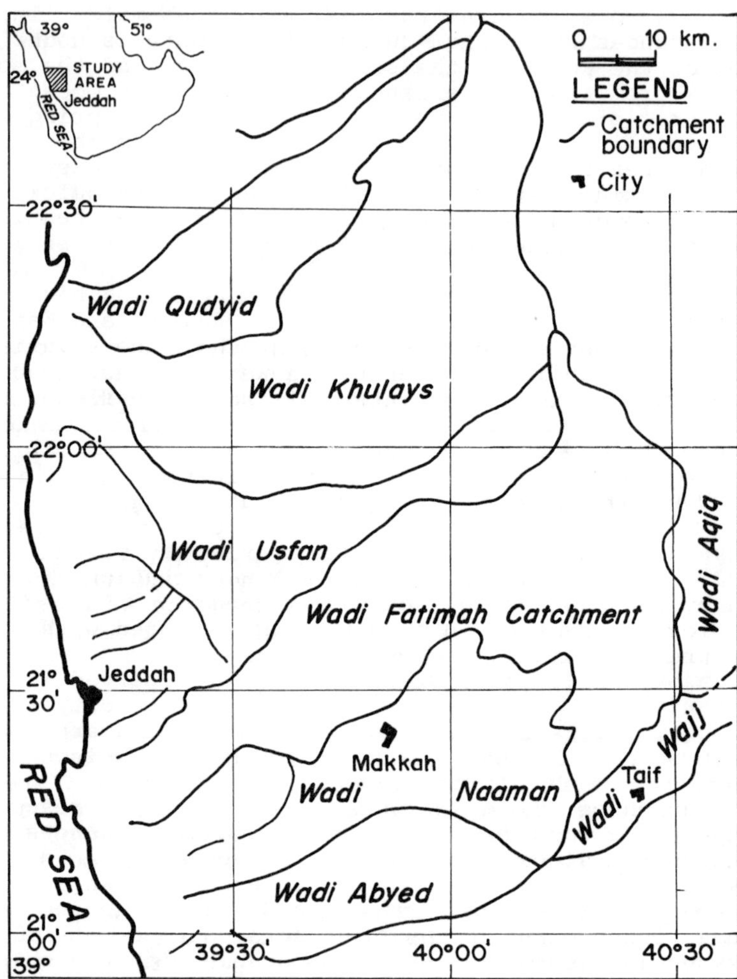


Fig. 1. Location map of the study area.

sedimentary basin.

Major portion of the area has an arid climate, characterised by scant rainfall and high summer temperature. The mountaneous zone, such as the Taif area, has mild climate. Rainfall ranges 290 mm to 430 mm. The climate is under the control of the Red Sea, the scarp mountains, and the moisture flow from the Mediterrenean and the Indian Ocean. In general precipitation is during the winter and spring seasons, developing from the moisture sources from the Mediterrenean area and during summer from the south east monsoon which is partially channelled along the Red Sea before its divergence towards the land mass by differential heating. Summer rainfall is at comparable magnitudes to those in winter and spring.

Flash floods are observed after rains having high intensities, especially in wadis Fatima, Marwani and Wajj. In other areas, most floods are intercepted by the alluvial fans at the foothills. In general runoff does not reach to the Red Sea as a result of interception of the floods by the alluvium. Downstream of the valleys the area is covered with sand dunes.

## 2. ISOTOPE COMPOSITION OF WATER SOURCES

The oxygen-18, deuterium and tritium data are given in table I. Seasonal variations in the precipitation are displayed by the changes in isotopic composition. Summer precipitation is enriched in oxygen-18 as compared with the winter precipitation.

High oxygen-18 concentrations are not observed in groundwater samples, in contrary to precipitation. Aridity and multitude of moisture sources reaching the area are considered to be responsible for high variations in isotopic composition.

Variation of oxygen-18 in precipitation with respect to altitude has a considerable scatter. Seasonal changes are significant. A slope of 4‰/km for oxygen-18 gives a series of oxygen-18-altitude curves with intercepts ranging from -2 to 8‰ $\delta O^{18}$ probably referring to different sources of moisture fluxes entering into the area. Such a rate of change for oxygen-18 with altitude is comparable with the observed precipitation over the world, such as in Central Italy of -2.82 to -3.44 ‰/km (Zuppi, et al., 1974), -2.6 ‰ /km in Nicaragua (Payne and Yurtsever, 1974), -3.0 ‰ /km on Mt Klimanjaro and the Alps (Moser and Sticher, 1970).

Isotope data from the groundwater samples are shown on Figure 2. Most data points occupy positions close to the global meteoric water line. Almost all the samples are related to the recent meteoric water, although their hydrologic history differs areally.

Table I

| Sample No. | Location | Oxygen-18 ‰SMOW | Deuterium ‰SMOW | Tritium Tu | Elevation m,as | EC micro mho | Date |
|---|---|---|---|---|---|---|---|
| Rh4 | Naaman | -1.5 | -3.5 | | 1200 | 520 | |
| Taf | Fatima | -2.4 | -5.9 | | 1700 | 810 | |
| Y4 | Naaman | -1.2 | -1.3 | | 560 | 3050 | |
| Na53 | Naaman | -0.8 | -3.6 | | 260 | 4600 | |
| F14 | Fatima | -0.9 | -6.5 | | 199 | 1200 | |
| JU1 | Jurana | -0.5 | -2.5 | 0 | -- | 700 | |
| U1 | Urana | -0.7 | -2.9 | | -- | 2600 | |
| WJ5 | Wajj | -2.3 | -6.6 | | 1700 | 3650 | |
| Na14 | Naaman | -1.3 | -2.9 | | 342 | 1600 | |
| Na12 | Naaman | -1.8 | -4.4 | | 417 | 1200 | |
| F16 | Fatimah | -2.0 | -6.0 | | 204 | 3100 | |
| F31 | Fatimah | -1.5 | -6.7 | 0 | 114 | 6150 | |
| F23 | Fatimah | -0.6 | -4.8 | | 150 | 12350 | |
| F4 | Fatimah | -1.6 | -7.1 | | 438 | 1600 | |
| Ya1 | Naaman | -2.4 | -9.6 | | 1230 | 1100 | |
| 1 | Abdiyah | 0.1 | 8.8 | 28 | | 2230 | |
| 2 | Abdiyah | -0.3 | 2.9 | 20 | | 2570 | |
| 3 | Abdiyah | -1.0 | -9.2 | 36 | | 1500 | |
| | Naaman | 2.5 | 10.5 | 15 | | 2230 | |
| | Naaman | 3.68 | 18.2 | 18 | | 1200 | |
| | Naaman | -0.2 | -2.7 | 25 | | 1042 | |
| | Naaman | -0.9 | -5.2 | 24 | | 1329 | |
| | Naaman | -1.2 | -11.2 | 38 | | 800 | |
| | Naaman | -0.6 | -3.9 | 22 | | 1242 | |
| Rain | Makkah | -2.2 | -11.2 | 62 | 417 | | 1982 |
| Rain | Jeddah | -3.1 | | 19 | | | 1970 |
| Rain | Jeddah | 0.9 | | 13.6 | | | 1970 |
| Rain | Jeddah | -1.7 | | 15 | | | 1971 |
| Rain | Taif | 1.5 | | | 1690 | | 5/1967 |
| Rain | Taif | -3.2 | | | 1690 | | 6/1967 |
| Rain | Hada | -3.4 | | | 1940 | | 5/1967 |
| Rain | Hada | -0.1 | | | 1940 | | 5/1967 |
| Rain | Hada | 1.6 | | | 1940 | | 10/1967 |
| Rain | Hada | -0.7 | | | 1940 | | 10/1967 |
| Rain | Hada | -2.3 | | | 1940 | | 10/1967 |
| Rain | Khulays | -3.0 | | | 200 | | 11/1967 |
| Rain | Hada | -3.4 | | | 1940 | | 11/1967 |
| Rain | Hada | -3.1 | | | 1940 | | 11/1967 |

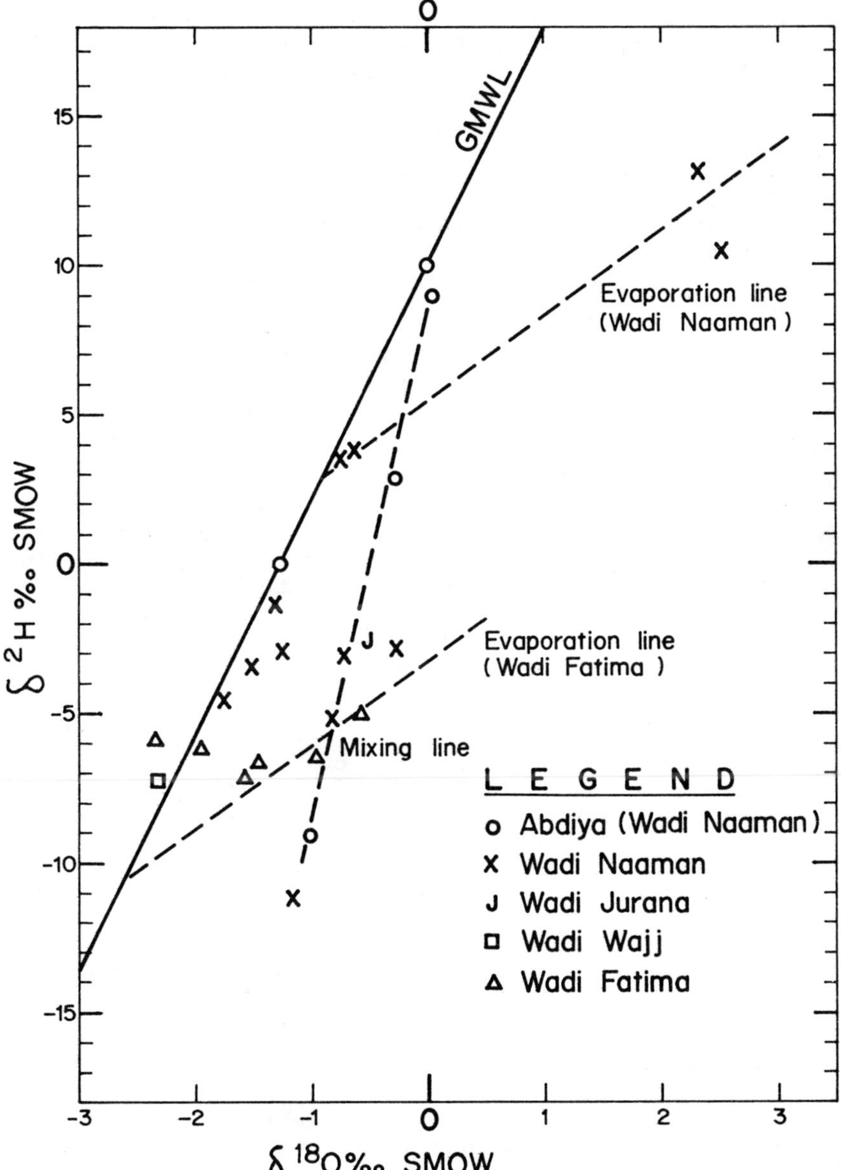


Fig. 2. The range of $^{18}O$ and deuterium values for several locations on the JMT area

Samples from the upper and middle reaches of the valleys of Naaman, Fatimah, Wajj and Khulays plot around the global meteoric water line, GMWL. Samples from middle and downstream reaches plot on the right and below the GMWL, indicating evaporation before recharge or irrigation return flow. Agricultural activities are generally concentrated over the downstream reaches. Floods are practically the only source of natural recharge,as observed from the groundwater level measurements. Samples F16 and F14 which are located about 400 m from each other accross the valley of Wadi Fatimah, have electrical conductivity and oxygen-18 contents of 3100 and 1200μmho, and -1.95 and - 0.9 ‰ respectively. The sample F16 is from a well near to the edge of the valley, and plots near the GMWL while the other indicates evaporation.

The structural features in the area are shown to be hydrogeologically active in certain localities. The sample Jul is from a catchment over the coastal plain. The well is penetrating bedrock and located nearby a fault zone which extends north east, towards Wadi Fatimah [Al Kabir, 1985]. Electrical conductivity and the oxygen-18 content of the sample are 700 μmho and -0.5 ‰ $\delta O^{18}$, respectively. The area has a low precipitation. The samples Ab2, Ab3, Ab4, Ab5 and Ab6, from the downstream reaches of Wadi Naaman plot on a mixing line, possibly groundwater through the alluvial deposits and the bedrock. Tritium data indicates relatively younger water than the upstream reaches. The examples from Jurana and Naaman are considered to be consequences of flow through fracture systems in the area.

Samples from downstream reaches of the valleys are devoid of tritium. In an investigation over the northern parts of the Red Sea coastal plain (Bosch, et al.,1980), groundwater bodies with ages older than several thousand years have been identified. The front between the old and new groundwater differs in each valley depending on the morphological characteristics of the catchment. Such a boundary is usually observed from the areal distribution of agricultural practices.

Tritium content of samples from the upstream reaches of the valleys are usually high. Turn over periods of less than 10 years are predicted for all the aquifers.

## 3. RECHARGE AREAS

An estimate of the effect of variations in altitude on the isotopic composition of groundwater over the study area may be made by use of Figure 3. The oxygen-18 contents of groundwater and rain samples are shown as a function of altitude. An attempt to correlate oxygen-18 concentrations of samples having high tritium contents and nil evaporation effect before the recharge with altitude results in

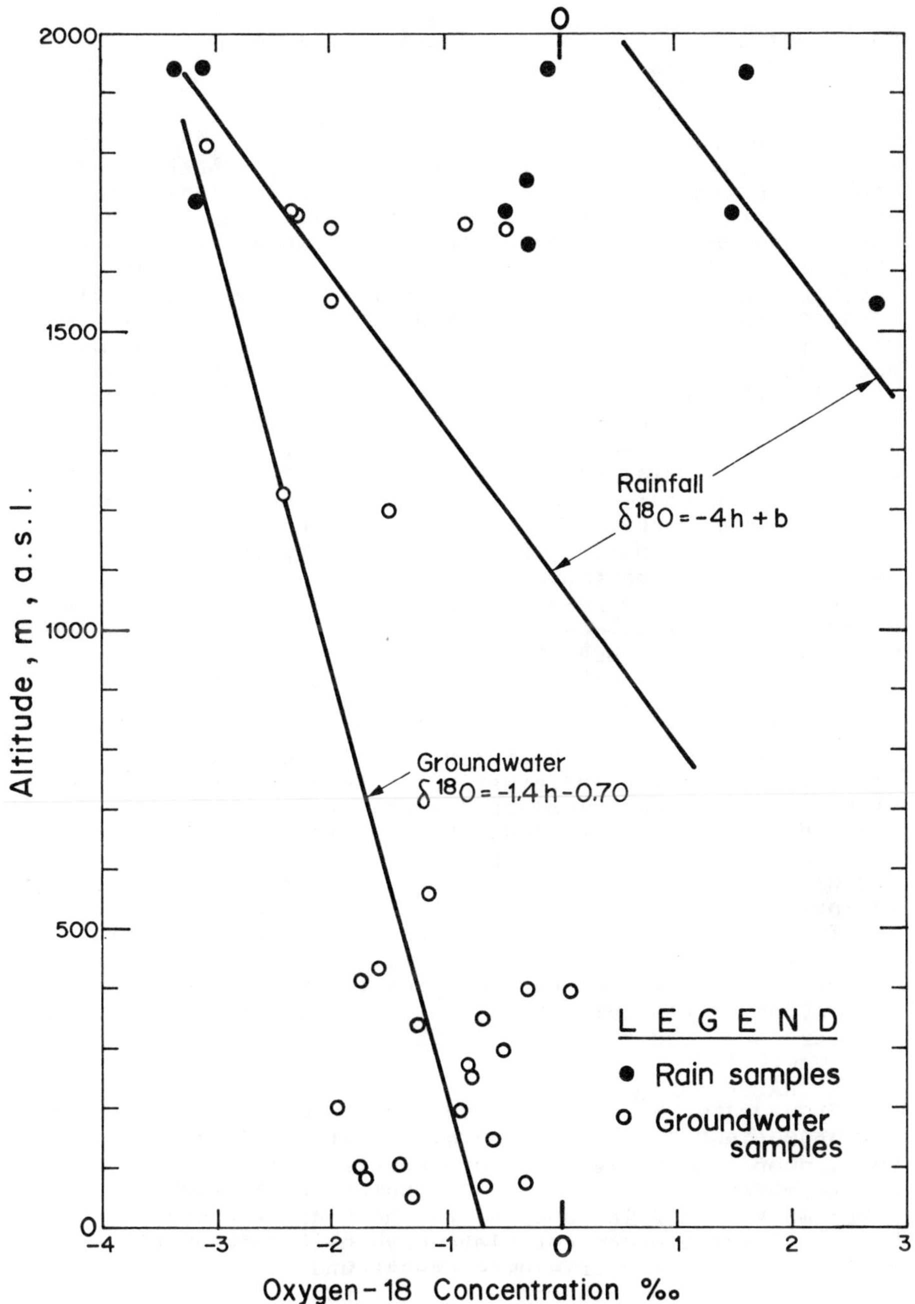


Fig. 3. Long term oxygen -18 - altitude curve .

equation:

$$\delta O^{18} = -1.4\ h - 0.70$$

where h is the altitude in km. The slope of -1.4‰/km is well below that observed for precipitation, -4‰/km. The groundwater samples from Wadis of Fatimah, Khulays, Urana, and downstream reaches of Wadi Naaman plot below the altitude - oxygen - 18 curves of rainfall samples. Samples from Wadi Wajj and upstream of Wadi Naaman plot within the envelope of altitude - oxygen-18 curves for rainfall. The main cause of the differences in the altitude dependence for precipitation and groundwater consists of percolation of higher altitude groundwater and surface water to lower levels, thus reducing the heavy isotope content of lower level groundwaters to values well below those of local precipitation. It is difficult to quantify the share of groundwater and surface water in this process. Downstream reaches of the Wadis of Fatima and Khulays are effectively recharged by the floods from the upstream reaches due to the fact that rains over the coastal plain are observed not to produce any change in groundwater conditions.

The samples from upstream reaches of Wadi Naaman, such as Yal, Rh4, N3, etc. may be the result of deep percolation of groundwater through the fracture systems. Groundwater in these parts of the catchment are found in the bedrock. Alluvium is thin, and the hydraulic gradients are steep.

The above equation, rather than describing the dependence of oxygen-18 on the altitude of recharge, correlates the isotopic composition of a parcel of groundwater within an effective altitude of discharge. Data points which are from the Wadis of Fatima and Khulays and downstream of Naaman are likely to contain comparatively higher proportions of higher altitude precipitation. The effective recharge area is above 800 m asℓ for all the valleys. This is above 1500m asℓ in Wadi Wajj and some tributaries of Wadi Naaman. The ratio of recharge area to the overall catchment is above 60 percent in valleys over the escarpment whereas it is around 40 percent in valleys draining to the Red Sea.

## 4. CONCLUSION

Groundwater aquifers of the Jeddah-Makkah-Taif area are recharged from the scarp mountains over an elevation of 800m. Mode of recharge may be directly from the rain or from floods moving towards the downstream reaches. High variations in the isotopic composition of the rains are not observed in the groundwater samples which indicates that some of the precipitation produces recharge.

## 6. REFERENCES

Al-Kabir, M., 1985. 'Recharge Characteristics of Groundwater Aquifers in Jeddah-Makkah-Taif Area', MS Thesis, King Abdulaziz Univ., Jeddah, Saudi Arabia.

Bosch, B., Oustriere, P., and Rochon, J., 1980.'Interpretation of Chemical and Isotopic Analysis of the Groundwater of the Yanbo Quadrangle'. BRGM open file report, JED-OR-19, Ministry of Petroleum and Minerals, Saudi Arabia.

Moser, H., and Stichler, W., 1970. 'Deuterium measurements on snow samples from the Alps'. Isotope Hydrology, pp 43-57. IAEA, Vienna.

Payne, B.R., and Yurtsever, Y., 1974. 'Environmental isotopes as a hydrogeological tool in Nicaragua. In Isotope Techniques in Groundwater Hydrology, pp. 193-202. IAEA, Vienna.

Zuppi, G.M., Fontes, J.Ch., and Letolle, R., 1974. Isotopes du milieu et circulations d'eaux sulfurees dans le Latium. In Isotope Techniques in Groundwater Hydrology, pp. 341-361. IAEA, Vienna.

# GROUNDWATER RECHARGE AND SUBSURFACE FLOW IN THE COMODORO RIVADAVIA AREA, CHUBUT PROVINCE, ARGENTINA. ISOTOPIC AND HYDROCHEMICAL STUDY

M. Levin, H.O. Panarello and M.C. Albero
Instituto de Geocronología y Geología Isotópica (INGEIS)
Pabellón INGEIS - Ciudad Universitaria
1428 Buenos Aires
Argentina

E. Castrillo, M. Grizinik and A. Amoroso
Universidad Nacional de la Patagonia
9000 Comodoro Rivadavia
Argentina

ABSTRACT. Groundwater in the Comodoro Rivadavia area was studied using oxygen-18, deuterium, tritium and hydrochemical evolution.

Stable isotopes in the multilayered system of the "pampa" table - lands define mainly an origin in locally melted snow with isotope composition significantly different from rain water over the lower eastern zones. Recharge has been defined as direct and autoctonous. Tritium contents suggest rather small recharge in comparison with the large volume stored in the reservoir. In the neighbouring hill zone, where discharge occurs, springs and wells show isotopic evidence of local flow. In the interhill subunit, regional flow shows salinisation and isotopic enrichment as a consequence of transit through marine facies and evaporation. In colluvial sediments groundwater exhibits higher tritium concentration, lower salinity and an isotope enrichment, due to direct infiltration of local surface water. Aloctonous to local recharge ratios can be estimated by the isotopic composition of the mixture.

In the beach ridges, regional flow and artificial recharge of imported waters, carried by an aqueduct, was isotopically evident.

## 1. INTRODUCTION

The area under study is roughly bounded by latitudes 45°50' S - 46°00' S and longitudes 67°30' W - 68°00' W and comprises the City of Comodoro Rivadavia and its sourroundings. This area is of economic importance due to its hydrocarbon production and to a highly developed oil industry. Population is about 110.000 and increases continuosly.

Water for human and industrial needs is supplied through a pipe - line coming from Muster Lake.



*I. Simmers (ed.), Estimation of Natural Groundwater Recharge, 377–393.*

The aqueduct becomes out of service several times a year.

Although at present there is no complete information about groundwater resources, researchers of the Universidad Nacional de la Patagonia claim that groundwater exploitation can be a first order contribution as an alternative supply. They have carried out hydrogeological studies since 1981, which led to definition of an important multilayered development of aquifer and a conceptual hydrogeological model (Castrillo et al., 1984).

The scope of this study is the application of environmental isotopes ($^{18}O$, $^{2}H$ and $^{3}H$) and the hydrochemical evolution to improve the knowledge about the origin of the recharge, to determine the relative importance of precipitation on the "pampas" highlands and to define whether there is some local surface water catchment during subsurface flow. Moreover, isotopes are used to verify the proposed circulation patterns concerned with the dynamics of groundwater.

### 1.1. Hydrometeorology

Although the studied region cannot be considered as an homogeneous area, the lack of available information lead us to consider only the meteorological data (1961-1970) of the coastal Comodoro Rivadavia station run by the Servicio Meteorológico Nacional. It is assumed that these data could be representative for the meteorological conditions of the whole area.

According to the Centro Nacional Patagónico rain map, rainfall over Pampa del Castillo plateau is about 250 mm and decreases to the east and to the west. Unfortunately, there are no suitable meteorological records over the plateau, but it is well known that it is frequently snow covered during the winter season.

Rain distribution is irregular during the year, with about 35 % in winter, 17 % in spring, 20 % in summer and 28 % in autumm.

The yearly average temperature is 12.8°C, January being the warmest month (19°C) and June the coldest (6.5°C).

The region is windy, with an yearly average value of 32 km/h and small variations all the year round.

The real evapotranspiration was computed according to the Turc formula as 171 mm for the area and the potential evapotranspiration (according Thornthwaite) is 720 mm. Thus, the obtained values are not reliable mainly due to the scarce information and the lack of direct field measurements. As a consequence of these uncertain data it is very difficult to asses a hydrological balance. Therefore the infiltration is unfortunately unknown.

### 1.2. Hydrogeological setting

Some details concerning groundwater geological reservoirs, with special reference to middle and upper Tertiary and Quaternary formations, are available.

In a simple brief hydrogeological view three underground sections have been recognized. The first one, the "Presarmiento", starts with the Preantracolitic basement, which has been reached by several wells. Granite and metamorphics without any information about hydrologic characteristics are present. The Bahía Laura Group follows, outcropping to the northeast of the area and going deep to the south. It is formed by ryolites and ignimbrites with alternating pyroclastic facies. Upwards there follows a thick clastic pack which belongs to Las Heras and Chubut groups, the first with some calcareous intercalations. Some levels are water and oil bearing. Generally, they can be defined like aquifers, with aquitard or aquifugue intercalations. The Salamanca Formation has aquitard and aquifer members, and it is placed between the sedimentary deposits described above and the Rio Chico Formation, which is an aquifer. "Sarmiento" section is represented by the upper part of Río Chico Formation and the whole Sarmiento Formation. Together they behave as an aquitard.

The "Postsarmiento" at the top of the sequence is the principal aquifer system of the region. It is integrated by pyroclastic and epiclastic sediments with some intercalated clay lenses. The following Formations are present: Patagonia, Santa Cruz, Rodados Patagónicos and more recent sediments. The reservoir capacity decreases from upper levels to the 50 or 60 meters of Patagonia Formation, where an aquitard 30 m thick is identified. Last and lower, about 20 m of sandy coarse material forms an aquifer. It has a large extension and its beds also stretch northeast towards the neighbouring province of Santa Cruz. Structural maps (drawn on information from well logs) show a relatively flat embossment, dipping to the southeast and losing slope gently in the same direction. Neither neotertiary tectonic evidences nor coincidence with actual morphology was found.

### 1.3. Groundwater

According to the hydrogeology setting already described, and taking into consideration water bearing units, hydrogeologists have recognized three zones. However, only the multilayered aquifer is here considered. As already defined, the system is developed in the "Postsarmiento" hydrostratigraphic unit.

The span of the multilayered system is from the unsaturated zone down to about 600 m. Phreatic water is also included.

Athough by hydrolithological and morphological position this aquifer is dominant in the area under study, it should be possible to separate two subunits of the whole. It then means that, according to the existence of anysotropy, they can be considered as subaquifers, named intermontane and littoral subunits.

Due to the lack of a sufficient density of observation points, only a schematic isophreatic map has been made. A system of elliptic morphology on the Pampa del Castillo and two big groundwater basins (to the east and west) were distinguished by hydrodynamic behavior interpretation. They become radial when affected by the "cañadones" presence. Thus, a general coincidence with the topographic relief exists.

The mean regional hydraulic gradient, 0.02, is relatively high. It increases following the direction of the flowlines, probably due to a decrease in the transmissivity coefficient.

Old inadequate wells allowed only few hydraulic surveys. The obtained values must be considered estimates. According to the Theis recovery method (without observation well), mean transmissivity for a group of five wells was 30.3 $m^3$/d.m and for a group of nine wells was 0.9 $m^3$/d.m. As a concequence permeability coefficients are extremely low.

## 2. SAMPLING AND MEASUREMENTS

A rainfall measuring station was errected in Comodoro Rivadavia City. Although snow fall collection was programmed at Pampa del Castillo, there was no snow in 1985.

Wells and springs have been sampled in the hydrological units already defined. This sampling was carried out in four zones, each with their own hydromorphologic characteristics. These zones were defined by means of satellite images and air photos by researchers of the Universidad de Comodoro Rivadavia. They will be described in Part 3., Results and Discussion.

Ocean waters and the Musters Lake-Comodoro Rivadavia aqueduct, in the Rada Tilly neighbourhood, were sampled in order to determine whether they have any influence on the aquifer system. Samples are listed in Table I and their locations are shown in the map of Fig. 1. The marked cross section is shown in Fig. 2.

Chemical composition is presented in Table II. These analyses were performed at the Universidad Nacional de la Patagonia laboratories.

Stable isotopes and tritium compositions of the water samples are also presented in Table I.

The isotopic analyses have been made at the Instituto de Geocronología y Geología Isotópica (INGEIS-CONICET). Isotope ratios $^{18}O/^{16}O$ and $^2H/H$ were determined with a McKinney type mass spectrometer, Micromass 602-D, with a double collector. $^{18}O$ concentrations were defined according to Panarello and Parica (1984) techniques. $^2H$ concentrations were determined following Coleman <u>et al</u>. (1982). Results presented include isotopic deviation (δ‰) relative to the standard V-SMOW (Craig, 1961, and Gonfiantini, 1978). The expression is:

$$\delta = 1000 \frac{R_M - R_E}{R_E} ‰$$

where

$\delta = \delta^{18}O$ or $\delta^2H$

$R_M = {}^2H/H$ or $^{18}O/^{16}O$ in the sample

$R_E$ = the same ratios in the standard

Analytical reproducibility is $\pm$ 0.2 ‰ for $^{18}O$ and $\pm$ 1.0 ‰ for $^2H$.

Tritium levels have been found on samples previously enriched according to Albero (1980), by liquid scintillation counting. The enrichment factor is about 10. Tritium concentrations are expressed in TU,

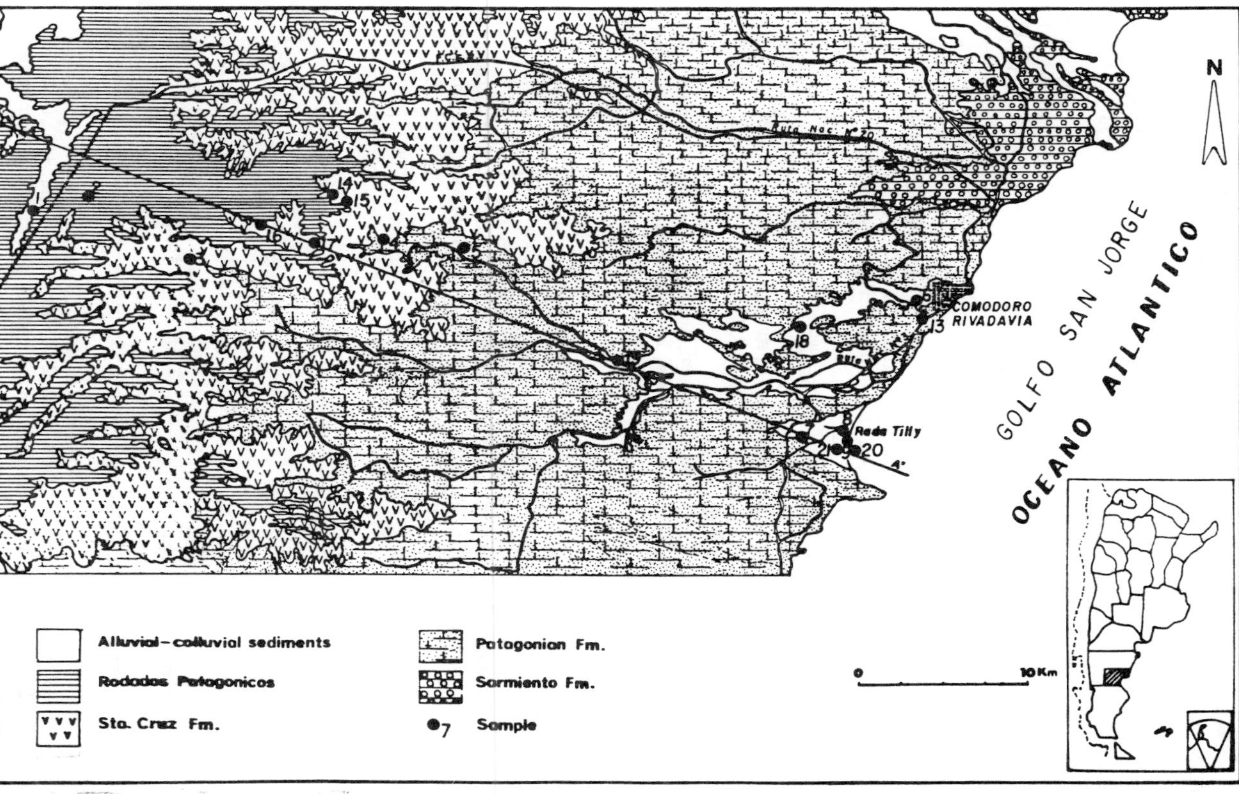


Fig. 1. Geological map and sampling points.

Tabla I. Isotopic composition of waters from Comodoro Rivadavia and surrounding area.

| SAMPLE | LOCATION | SAMPLE TYPE AND FORMATION | $\delta^{18}O$ (‰) | $\delta^{2}H$ (‰) | d (‰) | $^{3}H$ (T.U.) |
|---|---|---|---|---|---|---|
| 1 | PC | Well in Santa Cruz Fm. | -12.8 | -99.1 | 3 | 0.0 ± 0.6 |
| 2 | PC | Well in the upper multilayered aquifer | -12.2 | -99.5 | 2 | 0.0 ± 0.6 |
| 3 | CET | Spring Santa Cruz Fm. | -12.3 | -96.1 | 2 | 0.0 ± 0.6 |
| 4 | ET | Well in Santa Cruz Fm. | -11.9 | -99.0 | 4 | 0.3 ± 0.6 |
| 5 | ET | Spring Santa Cruz Fm. | -11.4 | -92.3 | -1 | 0.0 ± 0.6 |
| 6 | CR | Well in colluvial seds. | -6.9 | -69.9 | -15 | 0.6 ± 0.6 |
| 7 | RT | Well in Patagonia Fm. | -9.4 | -78.7 | -4 | 0.3 ± 0.6 |
| 8 | RT | Well in beach ridge | -5.0 | -52.9 | -13 | 7.4 ± 0.7 |
| 9 | RT | Well in beach ridge | -6.0 | -61.1 | -13 | 7.1 ± 0.7 |
| 10 | RT | Aqueduct | -5.7 | -52.1 | -6 | 6.4 ± 0.7 |
| 13 | CR | Well in Patagonia Fm. | -8.0 | -67.4 | -3 | 2.9 ± 0.9 |
| 14 | ET | Well in Patagonia Fm. | -13.4 | -105.3 | 2 | - |
| 15 | ET | Well in Patagonia Fm. | -12.8 | -103.4 | -1 | - |
| 16 | ET | Well in Patagonia Fm. | -10.6 | -94.0 | -9 | - |
| 17 | ET | Well in Patagonia Fm. | -13.3 | -111.1 | -5 | - |
| 18 | ET & CR | Well in colluvial seds. | -8.3 | -73.5 | -7 | 3.5 ± 0.6 |
| 19 | ET & CR | Well in colluvial seds. | -8.7 | -84.2 | -15 | 3.2 ± 0.6 |
| 20 | RT | Sea water (Gulf) | -2.1 | -2.4 | 14 | 1.5 ± 0.6 |
| 21 | RT | Well in beach ridge | -7.7 | -68.3 | -7 | 6.1 ± 0.7 |

PC : Pampa del Castillo
CET: Cañadón El Tordillo
ET : El Trébol
CR : Comodoro Rivadavia
RT : Rada Tilly

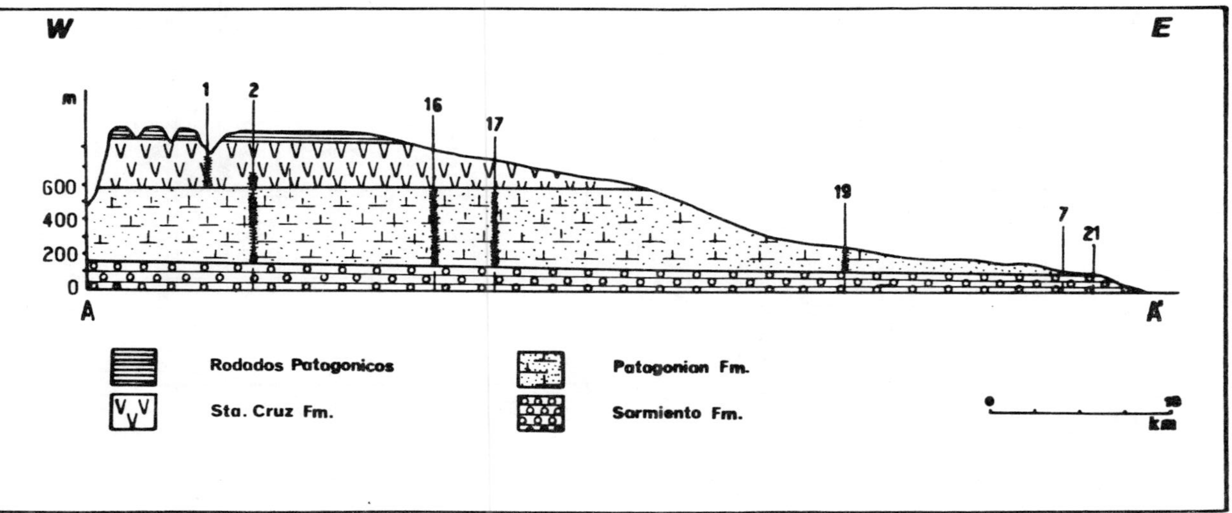


Fig. 2. E-W Geological Cross Section and representative sampled wells.

Table II. Chemical composition in mg/L of groundwaters and surface waters from Comodoro Rivadavia and surrounding area.

| Sample N° | $Ca^{2+}$ | $Mg^{2+}$ | $Na^{+}$ | $Cl^{-}$ | $SO_4^{2-}$ | $HCO_3^{-}$ | $CO_3^{2-}$ | TDS |
|---|---|---|---|---|---|---|---|---|
| 1 | 41 | 9 | ? | 31 | 10 | 87 | 4 | 245 |
| 2 | 14 | 8 | 123 | 14 | 22 | 126 | 0 | 163 |
| 3 | 12 | 3 | 49 | 21 | 23 | 141 | 0 | 156 |
| 4 | 20 | 47 | 32 | 50 | 80 | 175 | 10 | 271 |
| 5 | 12 | 16 | 91 | 53 | 84 | 156 | 0 | 216 |
| 6 | 534 | 579 | 14.000 | 7.100 | 22.000 | 380 | 38 | 30.900 |
| 7 | 481 | 217 | 7.000 | 3.400 | 11.000 | 224 | 10 | 15.600 |
| 8 | 49 | 66 | 219 | 106 | 172 | 545 | 48 | 1.000 |
| 9 | 168 | 19 | 184 | 184 | 288 | 331 | 38 | 1.510 |
| 10 | 33 | 5 | 53 | 21 | 24 | 195 | 0 | 180 |
| 13 | 534 | 496 | 14.000 | 6.700 | 23.000 | 165 | 20 | 31.800 |
| 14 | 0.2 | 0 | 86 | 28 | 38 | 117 | 7 | 170 |
| 15 | 0.2 | 0 | 102 | 32 | 64 | 122 | 5 | 344 |
| 16 | 37 | 0 | 28 | 21 | 8 | 136 | 0 | 190 |
| 17 | 2 | 0 | 86 | 35 | 32 | 107 | 10 | 286 |
| 18 | 177 | 5 | 1.500 | 440 | 2.400 | 506 | 24 | 4.000 |
| 19 | 201 | 12 | 507 | 312 | 1.000 | 185 | 5 | 2.510 |
| 20 | 419 | 277 | 14.000 | 19.600 | 4.200 | 29 | 57 | 39.000 |
| 21 | 58 | 21 | 216 | 191 | 176 | 292 | 0 | 896 |

defined as:

$$1\ TU = \frac{1\ ^3H}{10^{18}\ ^1H}$$

## 3. RESULTS AND DISCUSSION

Stable-isotope concentrations are plotted in a conventional $\delta^2H$ versus $\delta^{18}O$ diagram (Fig. 3) in which the "world meteoric water line" following the general correlation $\delta^2H = 8\ \delta^{18}O + 10$ ‰ is also shown

Contents of main anions and cations in wells, springs, Musters Lake, Comodoro Rivadavia aqueduct and sea waters are represented in the classic Schoeller diagram, Fig. 4, and the hydrochemical evolution is shown in triangular Piper form by Fig. 5.

### 3.1. Rainfall

Rain samples were collected during only three months in 1985 at the Comodoro Rivadavia station. For this reason a local meteoric line was not established and Puerto Madryn was chosen as a reference station for rain isotopic evolution. This is also a littoral station located in the same province of Chubut, 200 km north, integrated with the national INGEIS network since 1981. The weighted average for this station during 1983 was -7.7 ‰ for $\delta^{18}O$ and 7 TU for tritium, with seasonal variations showing enriched isotopic values and high tritium contents during the summer. For this reason. and taking into account that summer rain was not collected at Comodoro Rivadavia, average values found (Table III) $\delta^{18}O$: -9.2 ‰ and 3.5 TU have been considered as reliable for the area.

Table III. 1985 rain isotopic values at Comodoro Rivadavia

| sample N° | month | mm | $\delta^{18}O$ (‰) | $\delta^2H$ (‰) | $^3H$ (TU) |
|---|---|---|---|---|---|
| 11 | June | 47 | $-7.0 \pm 0.2$ | $-45 \pm 1.0$ | $3.2 \pm 0.6$ |
| 12 | July | 8.1 | $-12.6 \pm 0.2$ | $-118 \pm 1.0$ | $6.5 \pm 0.7$ |
| 22 | October | 32.3 | $-11.7 \pm 0.2$ | $-91.3 \pm 1.0$ | $3.2 \pm 0.7$ |

### 3.2. Groundwater

#### 3.2.1. Plateau-type relief mountainous zone (first level of terraces)

This zone has been defined as part of an old erosion and deposition level with a gentle slope to the east. Fluvial erosion caused minor dislocated units configurated as terrace levels, integrated by wide plateaus, locally known as "pampas" which are covered by the Rodados Patagonicos.

"Pampa del Castillo" in particular is recognized as a watershed

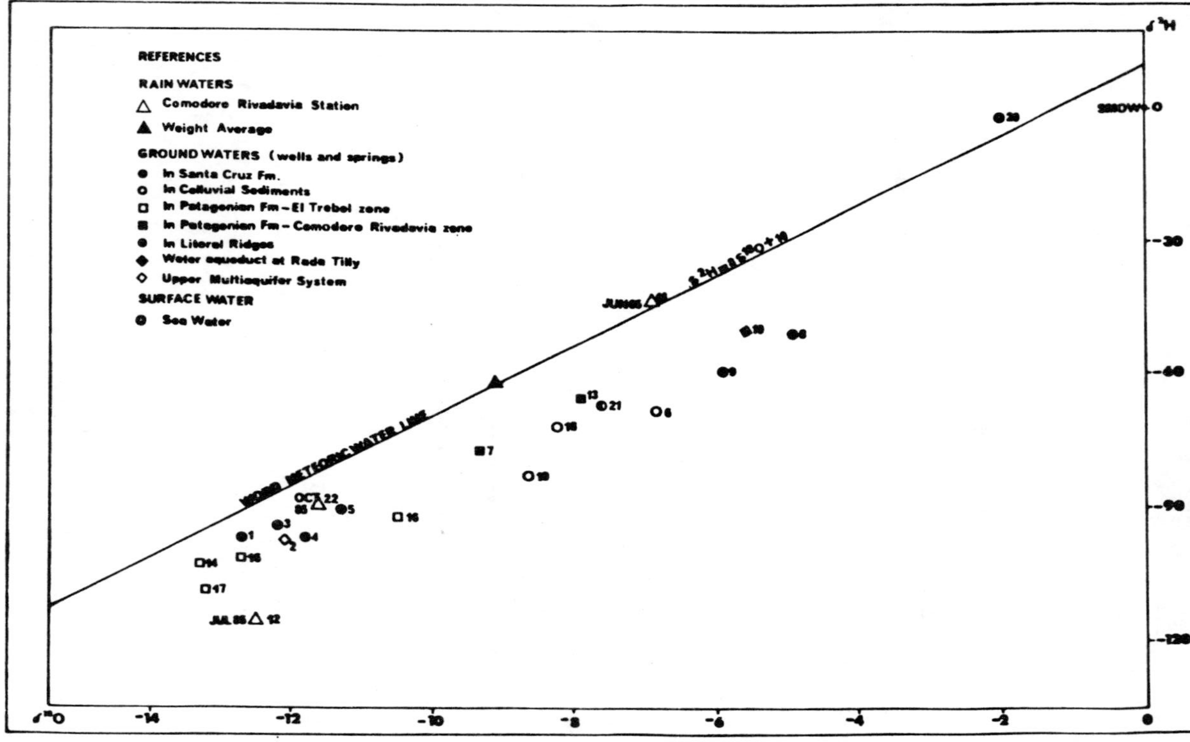


Fig. 3. $\delta^{18}O$ versus $\delta^2H$ diagram.

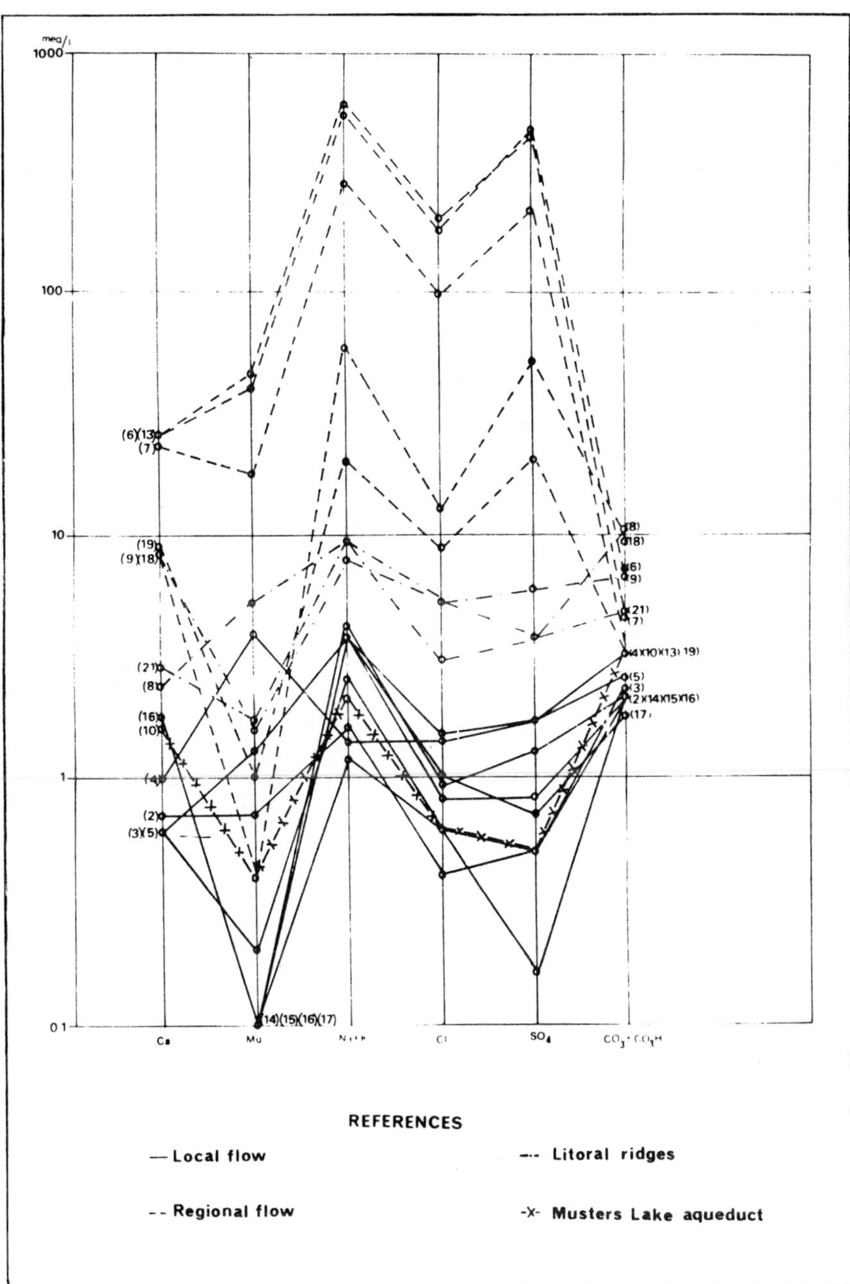


Fig. 4. Chemical composition of waters.

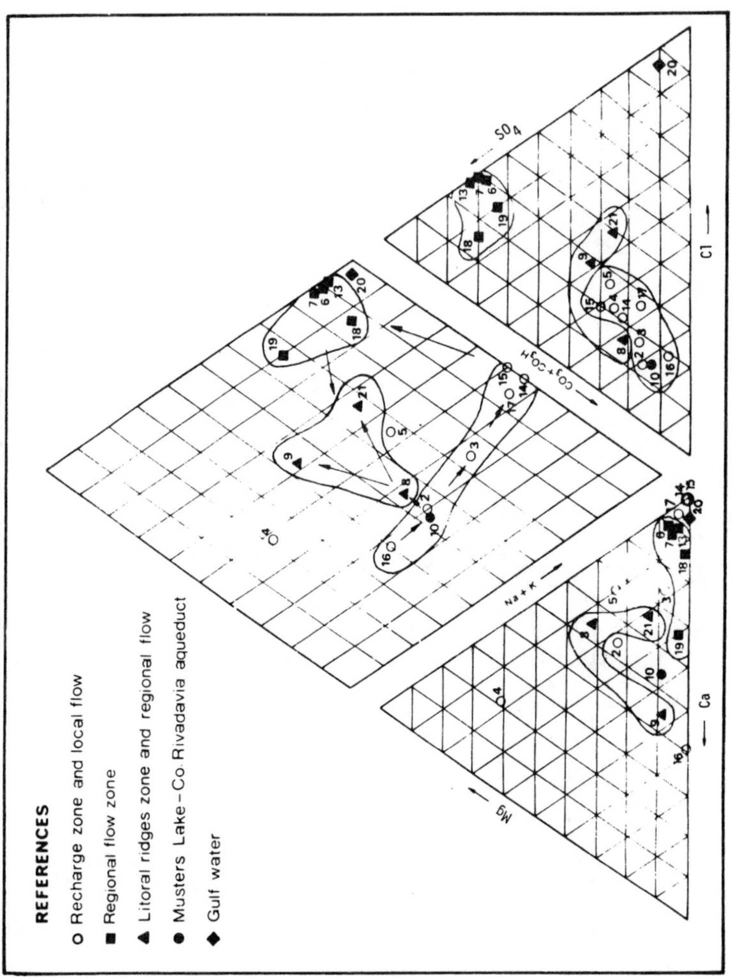


Fig. 5. Scheme of groundwater hydrochemical evolution.

with eastern and western drainage. Southward it bounds with the lower zone of the Gran Bajo Oriental. The maximum altitude is 750 m a.s.l., decreasing to the north and southeast. Drainage is poor and with centripetal character.

In this zone, representative wells in Santa Cruz Formation (N°1), in the upper multilayered aquifer (N°2) and spring of Santa Cruz Formation (N°3) were sampled.

The stable isotope composition of these samples ($\delta^{18}O$ from -12.2 ‰ to -12.8 ‰; $\delta^2H$ between -96.1 ‰ and -99.5 ‰), agrees with the typical values of snowfall at this latitude, according to the well known fact that in Pampa del Castillo snow is quite common and can be the dominant form of precipitation. On the other hand, it was expected that rain in Pampa del Castillo would be depleted of stable isotopes due to the effect of altitude. The analyzed samples exhibit a slight isotopic enrichment by evaporation, showing little shift from the meteoric water line and deuterium excess (d) lower than 10 ‰.

Tritium content of these samples is very close to dead water, indicating slow infiltration rate in the recharge pathway, or more than thirty years of transit time until their discharge as springs.

Chemically, these waters are defined as sodium-bicarbonate type, with homogeneous composition and total dissolved solids ranging from 150 to 250 mg/L.

### 3.2.2. Hilly zone

This zone is limited to the west by the foothill of the previously described unit and displays a broken morphology especially towards the eastern border, with hills of abrupt to moderate slopes, with plain tops, cut by ephemeral and intermittent rivers. Flashfloods flow to the principal collectors ("cañadones") or to lagoons where water can evaporate or infiltrate into the terrace zone. Springs are present, especially in the eastern border of Pampa del Castillo. Generally their outlets are in the intermediate beds of Santa Cruz and Patagonia Formations. Spring waters generally flow along the surface or subsurface until infiltration through the detrital intermontane deposits occurs.

Wells in Santa Cruz (N°4) and Patagonia (Nos. 14, 15, 16 and 17) Formations, and also a spring (N°5) at Santa Cruz Formation have been sampled.

The stable isotope compositions ($\delta^{18}O$ between -10.6 ‰ and -13.4 ‰, $\delta^2H$ between -94.0 ‰ and -105 ‰) of these samples are similar to those of the zone already described. Clearly, these waters have the same meteoric origin.

Tritium levels in the zone were below the detection limit in agreement with the longer pathway to the Patagonian reservoir.

Hydrochemical data show the same type of waters found in the first zone, with an increase in salinity due to the local circulation of groundwaters. Thus, the obtained values represent a slow evolution in the recharge-discharge process. A particular case is sample N°4 which is supposed to be indicative of base exchange, with loss of sodium and an increment of Mg or Ca concentration.

### 3.2.3. Intermontane zone

At both sides of the mountain plateau relief and at the northern border of the Gran Bajo Oriental, coalescent piedmont levels have developed. They show gentle slopes and are covered with unconsolidated psephite. Drainage in the area is linear and ephemeral and vanishes when slopes are broken. Drainage density is smaller in this area than in that of the Hilly zone. Brine water lagoons and evaporite concentrations occur in distal places.

Different representative wells have been sampled: one reaching the Patagonia Fromation (N°13), another that partially penetrates the Patagonia Formation (about 30 to 40 m depth) (N°7) and three in colluvial sediments (Nos. 6, 18 and 19) resting on the Patagonia Formation. This formation in the zone has changed to a well defined marine facies.

The plotted $\delta^{18}O$ vs. $\delta^{2}H$ values in the diagram of Fig. 2 are a group significantly different from those involving the samples of the first two zones. The reason is that they are more enriched ($\delta^{18}O$ from -6.9 ‰ to -9.4 ‰; $\delta^{2}H$ between -69.9 ‰ and -78.7 ‰). Moreover, sample values are located on an evaporation line (d:between -4 and -15) following an evolutionary projection of the isotopic values of wells and springs of the two zones already described. This claimed evolution can be explained in two possible ways: groundwaters flowing near surface or infiltration of meteoric local waters evaporated at the surface. Field evidence shows that surface runoff or temporary stagnant waters frequently occur and consequently evaporation can occur. Thus, the values found represent mixing of waters. According with the above reasons, sample Nos. 7, 13, 18 and 19 show tritium levels (between 2.9 TU and 3.5 TU) which indicate, with no doubt, an infiltration of local surface meteoric water. These $^{3}H$ levels, lower than the average for local rains, would indicate a dilution with dead waters flowing through Patagonia Formation. Some of these mixed waters, enriched in stable isotopes, can disminish their tritium content due to the time elapsed in the pathway between the recharge area and the sampling point. For instance, sample N°6 constitutes the extreme case for Patagonian isotopically disturbed waters discharging in colluvial sediments. The delay in the discharge zone is due to the presence of pelitic facies.

From a hydrochemical point of view, waters change to sodium sulfate and increase their salinity (up to 31 800 mg/L). This chemical typification agrees with the underground flow through Patagonia Formation marine facies with low permeability. A chemical dilution due to local recharge is evident in some wells; samples 18 and 19 (3960 and 2150 mg/L) are good examples of this dilution. Chemical and isotopic data show good correlation.

### 2.3.4. Beach Ridge Deposit Zone

The last sea transgression-regression processes in quiet beaches, protected in the terminals by more resistent and topographically high rocks, has produced subparallel coastal beach deposits. Some of these accumulations are beheaded by erosion and partially covered by eolian deposits. In the zone there is virtually no drainage system.

In this zone, wells in beach ridges of Rada Tilly Locality (Nos. 8, 9 and 21) and the aqueduct Musters lake-Comodoro Rivadavia have been sampled.

The $\delta^{18}O$ and $\delta^{2}H$ values have clearly defined a group, based on results obtained from 2 wells for domestic use ($\delta^{18}O$: -5.0 ‰ and -6.0 ‰; $\delta^{2}H$: -52.9 ‰ and -61.1 ‰) and the representative sample of lake water collected from the aqueduct ($\delta^{18}O$: -5.7 ‰ and $\delta^{2}H$: -52.1 ‰). On the other hand, it is also seen in Fig. 2 that these isotopic values are different from those of the other sampled zones. An exception to the rule for Rada Tilly is represented by the value of sample N°21. It's stable isotope composition ($\delta^{18}O$: -7.7 ‰ and $\delta^{2}H$: -68.3 ‰) is similar to that of the intermontane group. This situation would reflect the partial incorporation of Patagonian groundwater flowing towards the beach ridges.

Tritium levels in sample Nos. 8 and 9 (7.1 TU and 7.4 TU) resemble the expected average values in rain for this zone. Musters Lake tritium content (6.4 $\pm$ 0.7 TU) is almost similar, suggesting that Patagonia Formation flow dilution is negligible.

Hydrochemical data by themselves would not be able to define groundwater evolution in this environment. The ion contents tend to reflect what might be the circulation of the aqueduct type water, sample N°10, through sediments with dissolved marine salts.

## 4. CONCLUSIONS

Taking into account that the isotopic studies were programmed on the basis of an existent conceptual hydrogeological model, the interpretation allows proposal of an evolution model in the studied zones as follows:

a) The stable isotope ($^{18}O$ and $^{2}H$) composition of groundwater in Pampa del Castillo plateau reflects its origin in the snowfalls, similar to that of wells and springs towards El Trebol and El Tordillo zones (local discharge zones) and defines the origin and the recharge area as direct and autoctonus, as well as the existence of a local flow. Some moderate stable isotope enrichment is evident in groundwater samples as a consequence of: 1. a possible evaporation of snow accumulation under windy conditions; 2. stable isotope enrichment in Santa Cruz Formation water produced at surface in slow infiltration processes through calcareous cemented Rodados Patagonicos; 3. partial recharge of conjunctive water bodies. In any case, since no tritium was found in the studied wells, an underground path is assumed for the poorly enriched waters in the recharge route. From a quantitative point of view, the recharge must be rather negligible in comparison with the large volumes stored in the multilayered aquifer, and thus infiltration water is diluted by tritium dead waters.

The low groundwater salinity, as well as being characteristically sodium bicarbonated, also supports the idea that the upper plateau zone is the recharge area.

b) Groundwater isotope concentrations in the terrace zone define an underground flow quite different from that verified for wells and springs in Santa Cruz Formation reservoir. Here waters of Patagonia Formation have percolated through the upper multilayered aquifer foot or near it, producing a regional flow with enriched stable isotope composition tending toward the discharge. The enrichment in some cases has been produced because somewere along the water pathway there is a direct infiltration of local surface waters. This new source of enriched evaporated water is characterized by a meteoric origin of rains heavier in $^{18}O$ and $^{2}H$ than those of Pampa del Castillo. In some of these wells in the colluvial sediments that rest on Patagonia Formation, valuable tritium levels have been found, indicating local and recent recharge. However should these wells only point to local isotope rain values, $\delta^{18}O$ and $\delta^{2}H$ would be more positive than the true observed values. Thus, it is possible to explain the observed isotope composition by a regional flow (Patagonia Formation recharge zone type waters), interrupted and discharged in that kind of colluvial sediments. Of course, regional flow makes the local recharge waters more depleted of stable isotopes and dilutes the original tritium concentration. A very important fact is that the verified isotope mixing concentration helps to reach, rather well, a rate of mixing or type of recharge component, being about 50 % for each of the studied cases. A very interesting case on this point was found when Patagonia Formation waters in their flow near the surface, enriched their stable isotope composition after their discharge in colluvial sediments.

Regional flow through marine facies was also recognized by groundwater chemical composition. Waters are sodium-sulfate type and with a high salinity. Local recharge has also been defined by the saline dilution of some groundwaters.

c) Isotopic values allow definition of a regional flow pattern in coastal ridges. These sedimentary accumulations should behave as a temporary hydraulic barrier. Thus, regional flow discharge is temporarily interrupted in its pathway towards the Atlantic Ocean.

A particular case has been found in the wells on the coastal-ridges at Rada Tilly. According to the isotopic and chemical data, underground water evolution there has no connection with regional flow. Stable isotopes $^{18}O$ and $^{2}H$ in these wells are similar to those of the aqueduct Muster lake-Comodoro Rivadavia water. Tritium also shows a direct correlation between contained water and wells in the beach ridges. Moreover, in the zone there are no conditions for a significant local recharge. Buildings make water infiltration practically impossible where samples were taken. Furthermore, after a rain, surface release runoff to the sea is seen in the field. In summary , isotopic data resemble indirect recharge by irrigation (lake waters become more enriched by spray evaporation with sprinkler systems) and by sewage waters flowing into exploitation wells.

Unfortunately, recycled waters in the zone make it difficult to explain origin and evolution by hydrochemical evaluation.

ACKNOWLEDGEMENTS

We thank the ex-Rector of the Universidad Nacional de la Patagonia (UNP), Prof. Ing. A. López Guidi, for encouraging this study and partial financial support. We also thank Prof. Dr. E. Linares, Director of the Instituto de Geocronología y Geología Isotópica (INGEIS) for his collaboration. Special thanks to Prof. Dr. B. Dougharty for his help with the English text. We are grateful for the analytical service of the staff of the stable isotopes and tritium laboratories of INGEIS; R.G. Giovanelli, C. Ospital, E. Llambías, A. Bresba, M.C. Vera and P. Galán. Thanks are due to the staff of the chemical laboratory of the UNP for chemical analyses.

This work is Contribution N°102 of the Instituto de Geocronología y Geología Isotópica.

REFERENCES

ALBERO, M.C., 1980.'Medición de la Concentración de Tritio en Aguas en Niveles Naturales'.INGEIS. Reporte interno.

CASTRILLO, E., M. GRIZINIK and A. AMOROSO, 1984.'Contribución al Conocimiento Geohidrológico de los Alrededores de Comodoro Rivadavia, Chubut-Argentina'.To be published in Actas del IX Congreso Geológico Argentino.

COLEMAN, M.L., T.J. SHEPERD, J.J. DURHAM, J.E. ROUSE and F.R. MOORE, 1982.'A Rapid and Precise Technique for Reduction of Water with Zinc for Hydrogen Isotope Analysis'.Anal. Chem. **54**: 993-995.

CRAIG, H., 1961.'Standards for Reporting Concentrations of Deuterium and Oxygen-18 in Natural Waters'.Science **133**: 1833-1834.

GONFIANTINI, R., 1978.'Standards for Stable Isotope Measurements in Natural Compounds'.Nature **271**: 534-536.

PANARELLO, H.O. and C.A. PARICA, 1984.'Determinación de Oxígeno-18 en Aguas. Primeros Valores en Aguas de Lluvia de Buenos Aires'.Rev. Asoc. Geol. Argentina, **XXXIX (1-2)**: 3-11.

# GROUNDWATER RECHARGE OVER WESTERN SAUDI ARABIA

Y. Basmaci and J.A.A. Hussein
King Abdulaziz University
Civil Engineering Department
P.O.Box 9027
Jeddah 21413
Saudi Arabia

ABSTRACT. Recharge characteristics of the groundwater aquifers over the western Saudi Arabia are investigated. Average yield of each aquifer is proportional to that part of the catchment in the semiarid zone. Recharge is directly from the rain over the outcropping rocks and the basalt plateau and results from the floods in the alluvial aquifers. Pediment zones intercept much of the runoff.

## 1. INTRODUCTION

Saudi Arabia, in general, is an arid country. The most water short area is the Western Saudi Arabia which contains four of the seven large cities of the Kingdom. Renewable water resources, practically all groundwater, are estimated as $225 \times 10^6 m^3$ per year (Ministry of Planning 1980). Overall demand exceeds this capacity. The deficit is met through desalinisation. All the available conventional water supplies are planned to be diverted from the irrigation sector to meet the domestic and municipal water demand outside the major cities.

## 2. GEOMORPHOLOGY OF WESTERN SAUDI ARABIA

Western Saudi Arabia is a part of the Arabian Shield which is an extensive occurance of Precambrian crystalline and metamorphic rocks. Geomorphology of the region is strongly influenced by the rift structure of the Red Sea in Tertiary, during which the Red Sea coastal plain has been lowered by about 3000 meters below the shield and the shield tilted eastward (Figure 1). The boundary has formed the escarpment, a huge shear step of over 1000m. Therefore, the mountains are referred to as scarp mountains. They are developed on folded and faulted Precambrian rocks. The valleys and ridges are structurally controlled. In the south western part of the peninsula the mountains rise abruptly from the coastal plain and reach 3760m elevation. The escarpment forms the water divide of the valleys which drain to the Red Sea and to the sedimentary basin. East of the escarpment forms part of an old land



*I. Simmers (ed.), Estimation of Natural Groundwater Recharge, 395–403.*

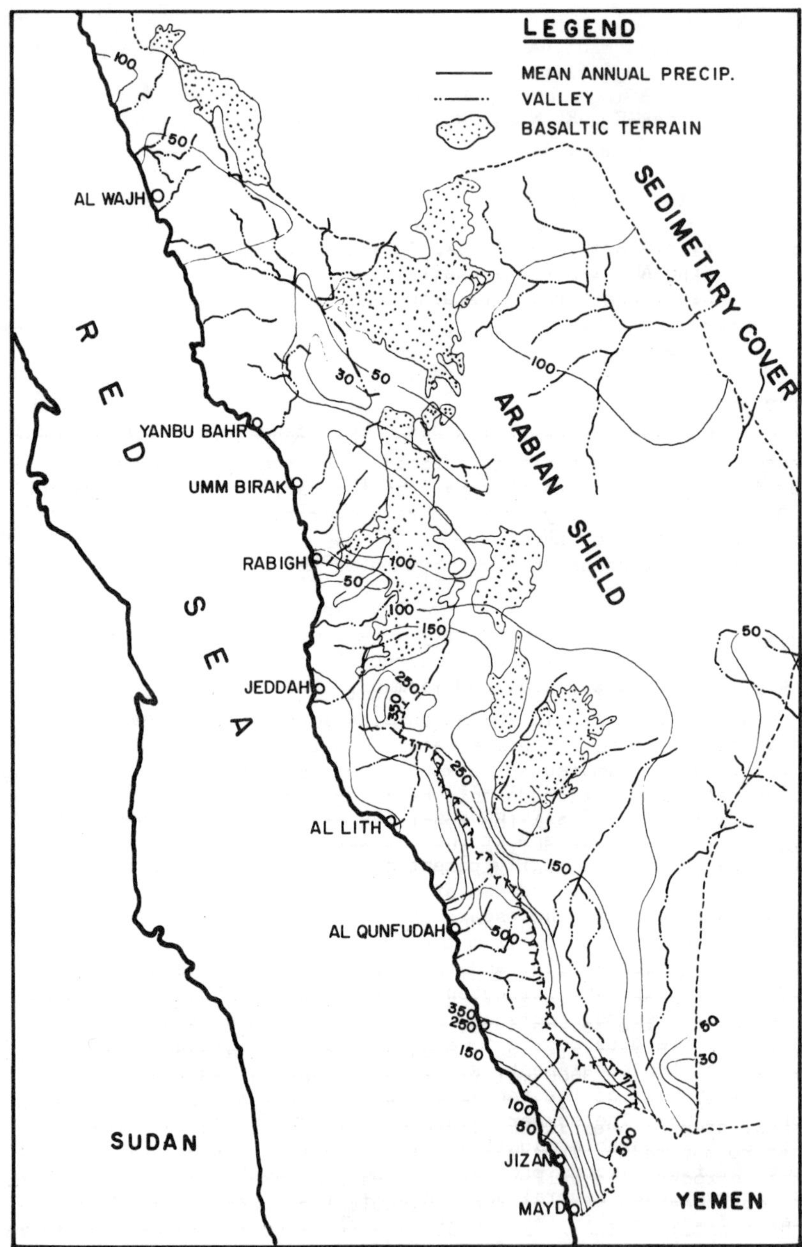


Fig. 1 – Morphology and rainfall distribution over the western Saudi Arabia.

surface which had been peneplained and then uplifted. The upper reaches of valleys such as Wadi Wajj, Wadi Liyyah, etc., have the appearance of recently eroded valleys. They are relatively deep, very steepsided and have very winding courses. Further downstream the relief decreases progressively. The escarpment disappears towards the north of a line which connects Makkah and Taif cities. Wadi Naaman is over such a transition zone. Elevation of mountains decrease. Basalt lavas of Harrat Rahat form extensive cover spreading out eastwards and westwards. Their form is controlled by the prebasaltic drainage network which in general consisted of broad shallow valleys over peneplained Tertiary topography. Khulays and Usfan basins rise on the Harrat Rahat plateau at elevations around 1300m, and run from a divide formed by a broad, poorly marked ridge. Wadi Hawrah, a tributary of Wadi Fatimah originates on the basalt plateau.

Alluvial deposits in the valleys start from a sharp contact of the sharply rising escarpment. They are found in the form large fans which are limited to the uppermost reaches. These fans form a large band in the south and narrow gradually towards the north. Alluvium is uniform and highly permeable. Spring flow from the crystalline rocks disappears in these deposits in short distances just after the bedrock contact. Apparently, large volumes of runoff are transformed into recharge. Deep groundwater levels provide an immense groundwater storage zone. Rapid percolation minimises losses due to evaporation.

Basalt flows and the buried valleys form good groundwater aquifers. The city of Madina and its vicinity depend on the groundwater from the Harat Rahat. Recharge rates are observed to be comparatively higher than crystalline rocks (Adam and Basmaci, 1983).

## 3. CLIMATE

The climate of Western Saudi Arabia, in general, is arid. All the area is in net moisture deficit. However, the elevated areas, above 1000m a.s.l., such as the escarpment, have relatively low temperatures. Rainfall is above 200m. Daily potential evaporation rates are above 7mm at lower elevations and 4mm at higher altitudes. Climate of mountaneous zone along the escarpment is therefore classified to be semiarid. Precipitation is from the cyclones over the Red Sea, the Mediterrenean moisture flux and the monsoons. The Red Sea is the generation area for the infrequent storms that move northward bringing rare and scattered downpours and flashfloods to the desert.

Any moisture flux moving towards the east is influenced by the topography of the escarpment. Rains are in spring, fall and winter. Approximately half of the annual rainfall is received, on the average, in April and May. The summer rainfall is from the monsoon system, sometimes at comparable magnitudes with those in spring. Runoff is in the form of flash floods produced by high intensity and short duration rains. It continues several days in some of the catchments and gradually disappears in the desert. Annual flood volumes fluctuate within two to three orders of magnitude.

## 4. RECHARGE

Groundwater recharge is a complex function of meteorologic, gemorphologic and hydrologic characteristics of a given region. The function differs from humid to arid climates significantly. Storms which recharge aquifers in an arid region would not be effective in a humid region. Thin vegetal cover, outcropping rocks, fractures, rains with high intensity in short durations favor rapid percolation of rainfall and flood development in certain parts of the catchments. All the recharge process is confined in a short time period by reducing the impact of high rates of evaporation. Depression and interception losses are small and lost through evaporation. Moisture flow through fine grain sediments does not reach to groundwater table because of high rates of exfiltration. The threshold value of annual precipitation for producing recharge over the sand dunes of the eastern Saudi Arabia was estimated to be 50mm (Dincer, et al, 1974). For finer sand this limit has to be increased in inverse proportion to the grain size. Ghurm and Basmaci (1983) showed that recharge to rainfall ratio could be as high as 41 percent in the upstream reaches of valleys. Rains at high intensities are capable of producing recharge. Such rains are observed a few times in a year. Al Kabir (1985) notes that deuterium excess of groundwater samples prove recharge from high intensity summer precipitation as well.

Direct recharge of rainfall through fractures is a local phenomena and not significant. Groundwater in vast areas of out-cropping rocks in the Western Saudi Arabia is affected least by precipitation. There is a significant difference in water quality in wells penetrating bedrock and alluvium in close vicinity. Some fracture systems and lineaments do transfer groundwater at large distances (Al Kabir, 1985). However direct recharge over basalt terrains is a reality as observed from the groundwater quality differences. Recharge from floods is not possible because of immature drainage network.

The alluvial aquifers are recharged by floods. The upstream reaches with steep slopes and barren outcropping rocks generate runoff even at low rainfall intensities. However, floods are observed in only certain valleys. Downstream reaches of most of the catchments which drain to the Red Sea are dry. The floods running over steep slopes disappear in alluvial fans. Environmental isotope studies over the coastal area of the Red Sea indicate groundwater ages exceeding several thousand years (Bosch, et al, 1980).

Some catchments, such as Wadi Fatimah, generate floods which flow long distances. Highly permeable alluvial deposits in the flood plains provide ideal conditions for groundwater recharge.

Irrigation in the Kingdom of Saudi Arabia depends on groundwater abstractions. Surface runoff has only marginal importance. Therefore, size of cultivated area in a catchment is proportional to groundwater recharge in a long period provided that grounwater would be utilized up to the margins of safe yield for maximizing the agricultural output. This was the case in the Western Saudi Arabia until 1970. The period following 1970 has witnessed ever increased groundwater exploitation in all valleys. Deep wells were drilled. Galleries and dug wells were dried up.

Table I

| Name of Catchment | Symbol | Recharge Area(Km$^2$) | Cultivated Area(Km$^2$) | Daily yield, (m$^3$) |
|---|---|---|---|---|
| Baysh | B | 4400 | 360 | |
| Itwad | I | 1300 | 84 | |
| Haly | H | 4080 | 206 | |
| Yiba | Y | 2320 | 130 | |
| Qanunah | Qa | 1200 | 81 | |
| Ahsibah | Ah | 660 | 25 | |
| Lith | AL | 1540 | 48 | |
| Naaman | N | 180 | 6 | 13500(1) Sogreah,1980 |
| Fatimah | F | 2300 | 18 | 10900(2) Italconsult,1967 |
| Yamaniah | Ym | 620 | | 14250(3) Bazuhair, 1981 |
| Usfan | Us | 200 | 9 | 25600(2) Italconsult,1967 |
| Khulays | Kh | 2380 | 18 | 97700(2) Italconsult,1967 |
| Rabigh | R | 1230 | 17 | |
| Qudaid | Q | 210 | 12 | 20600(4) Al-Hajeri, 1975 |
| Wajj | W | 300 | | 11600(2) Italconsult,1967 |
| Liyyah | Liy | 273 | | 16000(2) Italconsult,1967 |

(1) Sogreah, 1980
(2) Italconsult, 1967
(3) Bazuhair, 1981
(4) Al Hageri, 1975

Daily yields, cultivated areas and recharge areas in several valleys are listed on Table I. Cultivated areas and catchment areas are estimated from aerial photographs and the geographical maps(USGS, 1972) which are based on observations the years preceeding 1970. Recharge areas, however, are approximated on the basis of isohyetal maps. The relationship between recharge area and cultivated area of several catchments is shown on Figure 2. The catchments to the south and to the north of Taif plot in the figure in two groups, with the following regression equations:

Southern catchments; $A = 1.36 \times 10^{-2} \; RA^{1.2}$

Northern catchments; $A = 5.00 \; RA^{0.17}$

with A cultivated area and RA recharge area in $Km^2$. The southern catchments have steep relief. Rainfall increases proportionally with the elevation. Floods are intercepted through the alluvial band along the foothill of the escarpment. The northern catchments, such as Wadi Fatimah, Khulays, Qudaid and Usfan have relatively low relief. Annual precipitation decreases towards the north. Therefore the cultivated area is comparatively small. Major portion of Wadi Fatimah is covered by crystalline rocks. Floods are regular. The other catchments, Qudaid, Usfan, Khulays and Rabigh are over the basalt plateau.

Recharge area is compared with daily yield in Figure 3. The aquifers in the wadis of Fatimah, Khulays, Usfan and Liyyah are known to be overused. The first three of these have large aquifer volumes. Groundwater abstractions far exceed the natural recharge. Groundwater levels of these aquifers were falling continuously in the last decade. The aquifers in the wadis of Naaman and Wajj are fully developed. Abstractions are in equilibrium with recharge because of the small aquifer volumes. Daily yield, Dy in $m^3$, is correlated with the recharge area, RA, as

$$Dy = 30RA$$

It is not possible to extrapolate the recharge volumes over the Western Saudi Arabia on the basis of these relationships. The data is not enough to cover the whole region. Recharge process over the basalt plateau is not understood well. Therefore the results are qualitative.

## 5. CONCLUSION

Recharge of aquifers over the Western Saudi Arabia is influenced by the geomorphology and basalt topography. The southern catchments have high yields because of high precipitation over the escarpment. Floods are efficiently transferred into recharge through the alluvial fan over the foothill. The northern catchments benefit from direct recharge over the basalt plateau. Therefore, the recharge of northern catchments are comparable with the south although precipitation is relatively low.

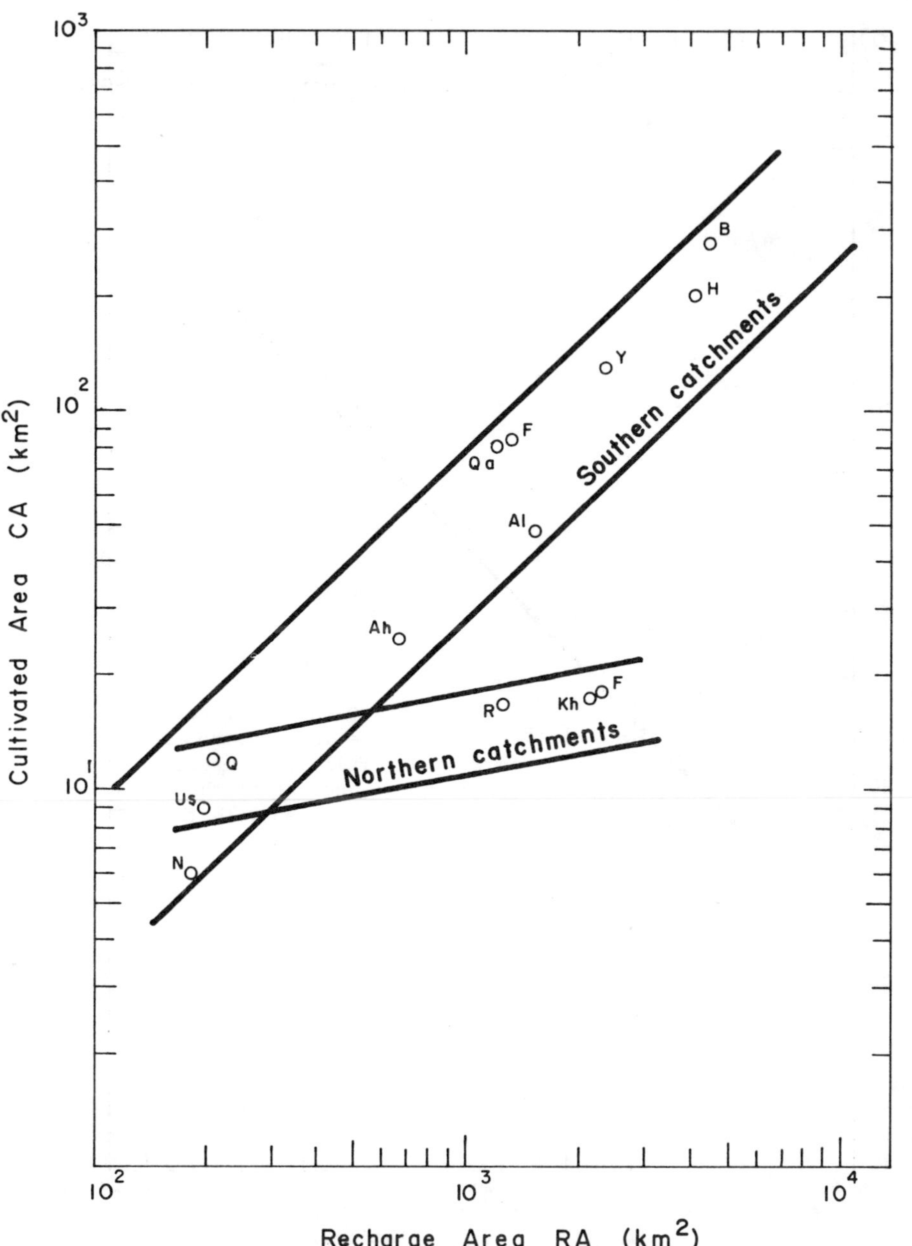


Fig. 2 - Distrbution of agricultural density over the western Saudi Arabia.

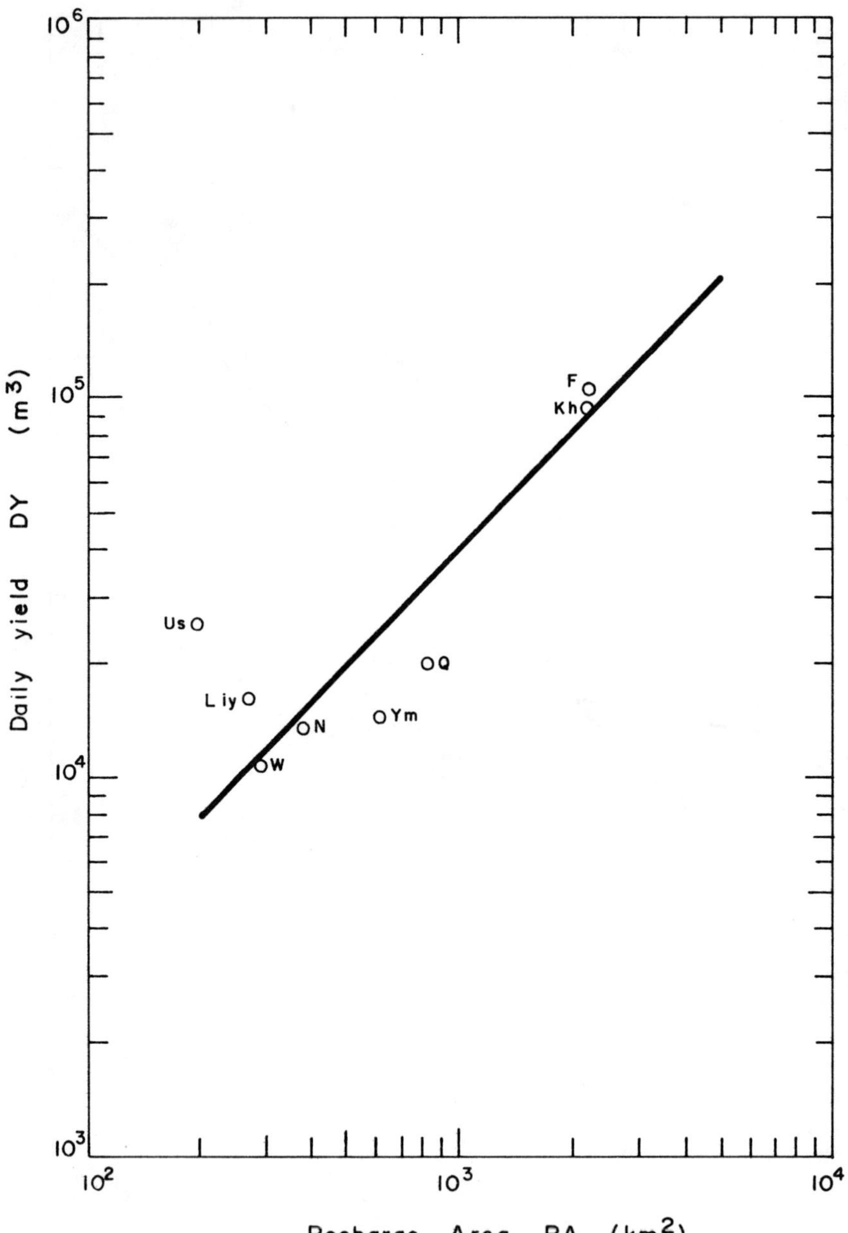


Fig. 3- Yield of aquifers over the western Saudi Arabia.

## 6. REFERENCES


Adam, E.G.E., and Basmaci, Y., 1983, 'Groundwater potential in the south of Medina Al-Munawwarh' in Seminar on Water Resources in the Kingdom of Saudi Arabia, 17-20 April, King Saud University, Riyadh, Saudi Arabia.

Al-Hajeri, F.Y., 1975. Groundwater Studies of Wadi Qudaid, MS Thesis King Abdulaziz University, Jeddah, Saudi Arabia.

Al-Kabir, M., 1985. Recharge Characteristics of Aquifers over the Western Saudi Arabia. MS Thesis, King Abdulaziz University, Jeddah, Saudi Arabia.

Bazuhair, A.G., 1981. Hydrogeology of Wadi Yamaniah. MS Thesis, King Abdulaziz University, Jeddah, Saudi Arabia.

Bosch, B., Oustriere, P., and Rochon, J., 1980. 'Interpretation of Chemical and Isotopic Analyses of the Groundwater of the Yanbu Quadrangle' BRGM open file report, JED-OR-19, Ministry of Petroleum and Minerals, Saudi Arabia.

Dincer, T., Al-Mugrin, A., and Zimmermann, U., 1974. Study of infiltration and recharge through the sand dunes in arid zones with special reference to the stable isotopes and thermonuclear tritium. Jour. of Hydrology, 236 pp. 79-109.

Ghurm, A., and Basmaci, Y., 1983. Hydrogeology and Wadi Wajj, in Seminar on Water Resources in the Kingdom of Saudi Arabia, 17-20 April, King Saud Univer., Riyadh, Saudi Arabia.

Italconsult, 1967. 'Final Report on Water Supply Survey for JMT Area' Ministry of Agriculture and Water, Riyadh, Saudi Arabia.

Ministry of Planning, 1980. Third Five Year Development Plan, Riyadh, Saudi Arabia.

Sogreah, 1980. 'Wadi Naaman Underground Dam' Ministry of Municipal and Rural Affairs, Riyadh, Saudi Arabia.

USGS, 1972. 'Topographic Map of the Arabian Peninsula' Ministry of Petroleum and Minerals, Riyadh, Saudi Arabia.

# NATURAL RECHARGE OF KARST AQUIFERS IN WESTERN TAURUS REGION (SOUTHWESTERN TURKEY)

Gültekin Günay
Hydrogeological Eng. Dept.
Hacettepe University, Eng. Faculty
Beytepe 06532, Ankara-TURKEY

ABSTRACT. In the Western Taurus region that extends between the Lake District and the Mediterranean Sea, South-western Turkey, some of the largest karst aquifers in the world are to be found. These are, from east to west, namely, the karst aquifers of the Manavgat river basin, the Köprüçay river basin, the Aksu river basin and karst aquifers of the travertines of the Antalya plateau and the Kırkgöz Springs.

Computation of recharge and discharge of the above listed aquifers has been made during the intensive and detailed hydrogeological investigation of the relevant basins. The calculations revealed that a significant contribution, as natural recharge of the aquifers, comes from the adjacent basins.

This paper is to discuss the studies on the natural recharge from the adjacent basins and their results, in detail.

## 1. INTRODUCTION

About one third of Turkey is underlain by carbonate rocks most of which are highly karstified. Intensive karstification is present in the south-central Taurus karstic region where a great water resources potential exists and where most of the huge hydrotechnical constructions are to be found which are of great importance from the standpoint of the economic development of the country (Fig. 1).

The karst area in south-central Turkey has many complex karst water resources problems in need of solution. Only a limited number of such problems are those related to the estimation of natural recharge and underground storage capacity.

The study area is subdivided into 7 subbasins in order to simplify the water balance computations (Fig. 2).

This paper deals primarily with the deficit water because the interconnections between the karstic subbasins are quite complex to be determined.

Based on the water balance computations natural recharge for each subbasin is estimated and thus hydrogeological behavior of the whole system is explained. Various methods are used in water balance computa-



*I. Simmers (ed.), Estimation of Natural Groundwater Recharge, 405–422.*

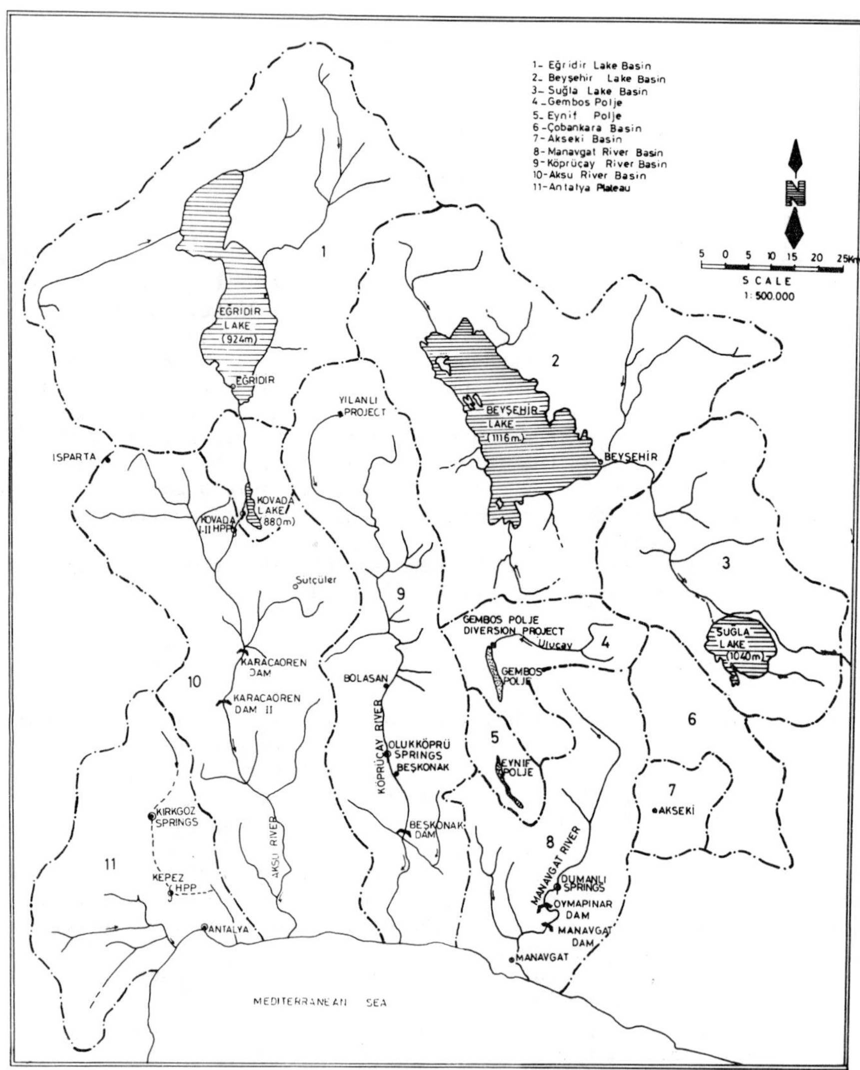


Fig. 1 Hydrotechnical projects constructed and under-construction at the study area .

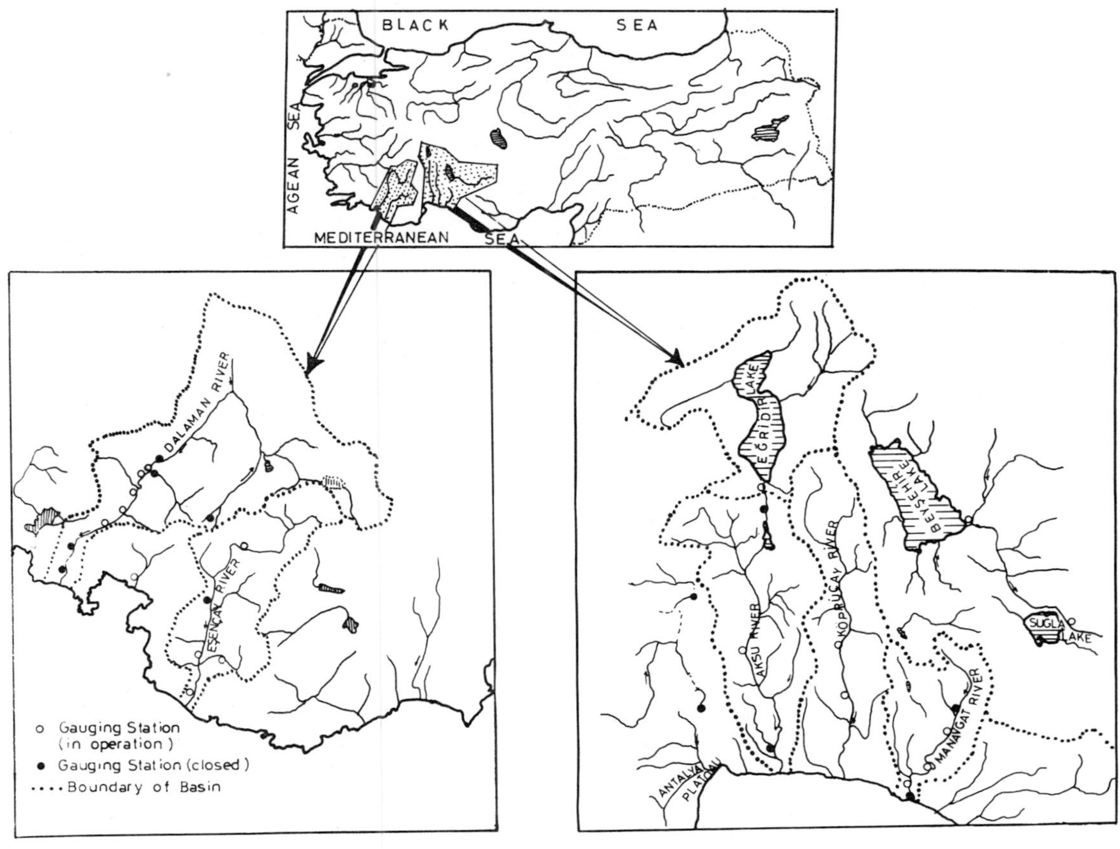


Fig. 2 Subbasins in the study area.

tions each of which reflects a different objective of the computations. Since this paper deals with the natural recharge but not with the methods of water balance computations, only a brief explanation is given about the methods used in here, in computations. The accuracy of the results of the computations depends on the precision of the measurement of several terms in the water balance which is questioned in order to achieve a more correct interpretation and more reliable conclusions.

The study area is subdivided into the Manavgat river basin, the Beyşehir lake basin, the Köprüçay river basin, the Antalya travertine plateau, the Aksu river basin, the Dalaman river basin and the Eşençay river basin.

The geology of the study area is very complex. The oldest rocks are of the Paleozoic era. Vast areas are covered by limestone of the Cretaceous era and the Miocene age.

The Antalya basin is a Tertiary and Quaternary sedimentation basin with travertine of several hundred meters thick. In the Köprüçay valley, the Miocene formations were of primary concern; these are of relatively impermeable Beşkonak formation and the permeable and karstified conglomerate formation. Most of water recharge, however, comes from the high plains and mountain flanks composed of Mesozoic limestone.

Groundwater storage and the routes of subsurface water movement are dominated by structural lithologic features. Fissures, fractures and faults are of utmost importance for the occurrence of aquifers and their springs. Investigation and identification of their occurrence and distribution are of paramount interest in studying the water budgets and consequently the development of water resources potential of this area.

## 2. AN OUTLOOK TO HYDROGEOLOGY OF THE STUDY AREA

The stratigraphy of the region contains various units from the Cambrian to the recent age. Carbonate units are mainly Devonian,Permian,Triassic, Jurassic, Cretaceous and Tertiary.

The structure is quite disordered. A lot of upthrusts, carbonate and non-carbonate units reach the thickness of thousands of meters.Among Paleozoic metamorphic rocks there are recrystallized limestones, marbles and calcschist layers. Between carbonate layers, schists, Triassic, Liassic, Cretaceous and Tertiary clastics, sandstone, shale etc. generally form impervious and non-karstic barriers.

The region is the most important and largest karst region in Turkey. The most common features of karst morphology are dolines, uvalas, poljes, sinkholes, springs, underground rivers, caves,submarine springs.

Hydrogeological maps of the subbasins are attached to the paper (Figs. 3, 4, 5, 6, 7).

## 3. ESTIMATION OF NATURAL RECHARGE

The natural recharge for each subarea is estimated using the water budget computation results.

## 3.1. Manavgat River Basin

The Manavgat river basin has been the subject of extensive and detailed geologic and hydrogeologic studies during the last fifteen years mostly because of the construction of the Oymapınar Dam and reservoir (Fig. 3).

It is concluded from previous studies that the water drains into the Manavgat river not only from its surface watershed but also from the closed surface basins to the north, west and especially east of the Manavgat river. This watershed can be found by computing the water budget of the basin. The surface watershed area of the Manavgat river at the Homa gauging station is about 928.4 $km^2$. Monthly and annual discharge are too high for such a small drainage area, so a considerable subterranean contribution must be coming from the adjacent areas which is called natural recharge.

The average precipitation is 1487 mm. The drainage area is 1221 $km^2$. In this case, the flow is 4.7 billion cubic meter/year and this figure is as large as 2.6 times the precipitation. If the drainage area is considered -the area computed at Homa gauging station is 928.4 $km^2$- then the flow becomes 3.4 times larger than the precipitation.

## 3.2. Beyşehir Lake Basin

The Beyşehir lake area covers the north side of the study area(Fig. 4). It performs an important water regulation role. Historically the lake is drained both by a natural channel and valley east toward the Konya-Çumra region and by water leaking underground along its northwestern side. Water enters the lake both by the surface and the subsurface flows.

The water budget of Beyşehir lake is influenced by well developed karst features that inflow and outflow.

The present Beyşehir lake fluctuates between a level of 1,121.00 and 1,125.00 meters. The lake's average annual outflow was 340 million cubic meter in the period 1933-1981. Tha lake's area is 745 $km^2$ at 1125.5 and 400 $km^2$ at 1117.00.

The water budget of the lake, when determined by correlation and the water balance equation shows a certain amount of variation over the years.

Leakage and evaporation loses are higher during high lake levels than during the low lake levels.

Leakage losses are larger during the wet period and influence the surface and subsurface inflow and outflow only when the water level exceeds approximately 1123.00 meter. This suggests that leakage occurs through the sinkholes existing on the western and northwestern side of the lake and there is no way for water to escape through the bottom of the lake.

Computations were made for every year on a monthly basis over a period of 23 years between 1960 and 1982. The results have revealed that the leakage starts in November -the beginning of the wet period- and stops (if the level goes down to 1123 meters) in the end of May or in June -the end of the wet period. During the dry period, contribution from non-gauged springs with up to 20 l/s discharging through the Anamas

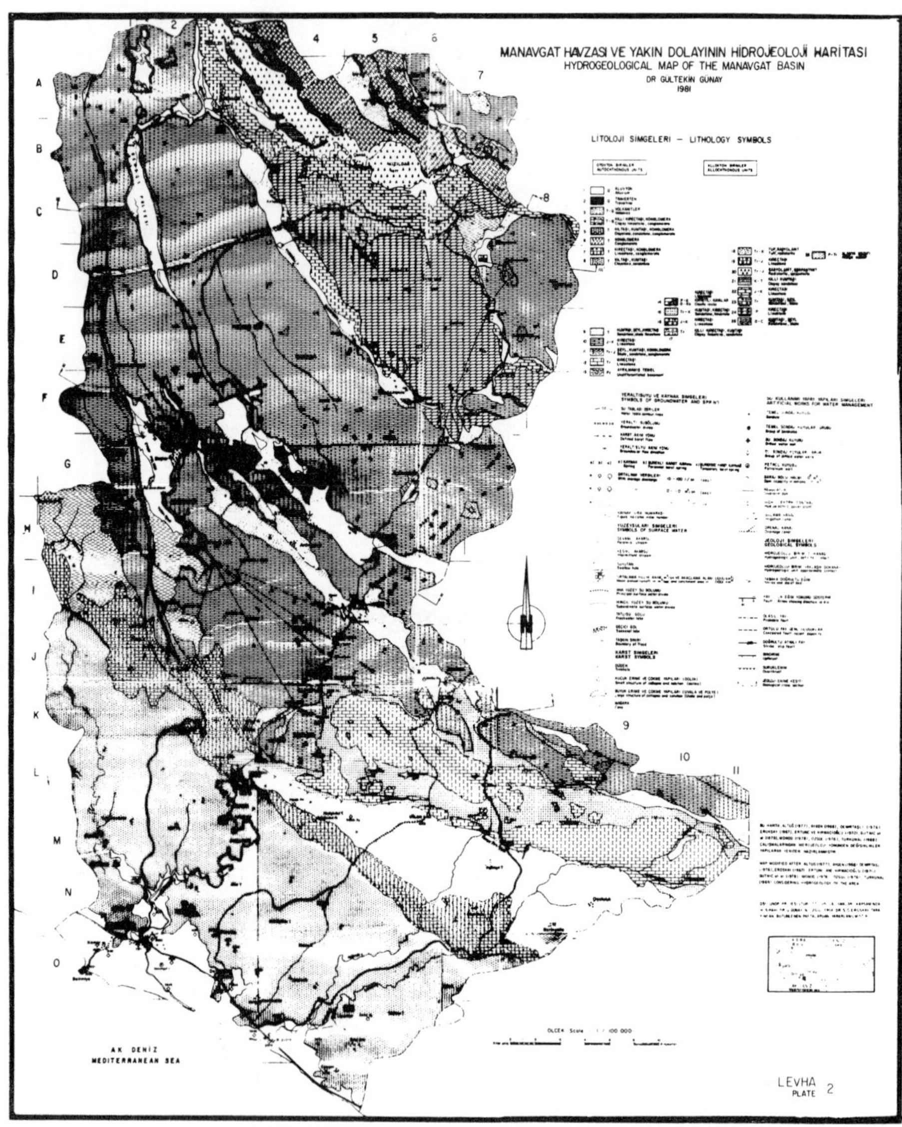


Fig. 3 Hydrogeological Map of Manavgat river basin and its vicinity (after, Günay, 1981).

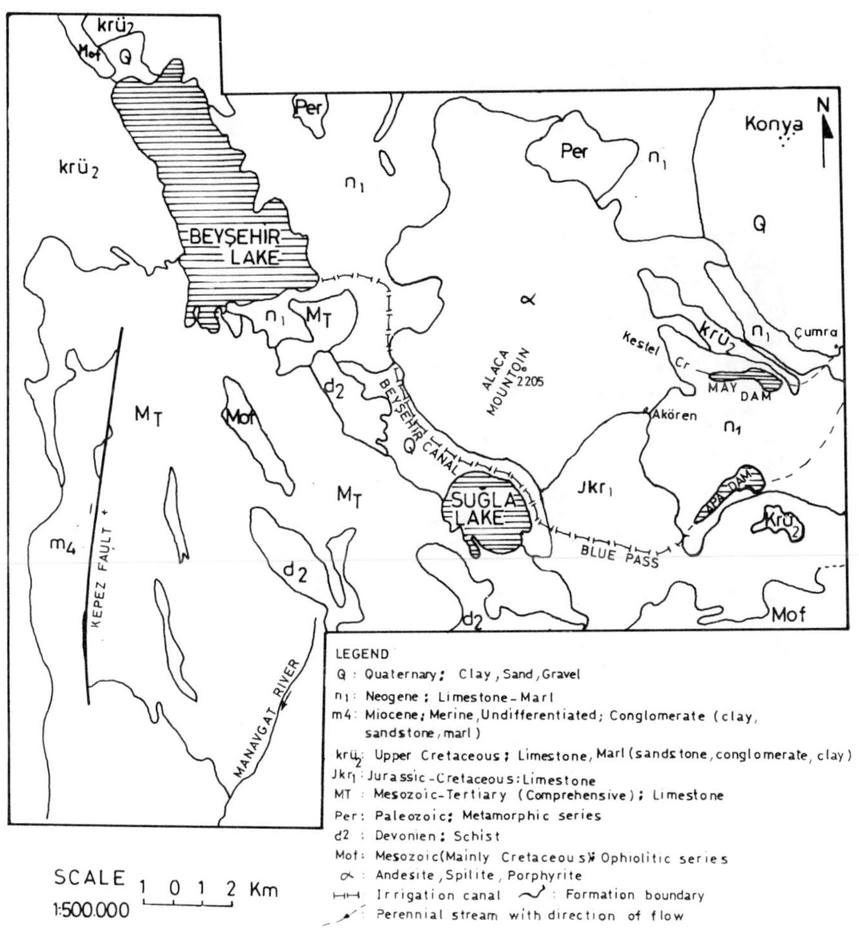


Fig. 4 Hydrogeological Map of Beyşehir-Suğla Lake basin and its vicinity.

Mountains located to the west of the lake was found to be significant. The leakage from the lake reached a maximum of 28.7 $m^3/s$ in March 1975 when the level of the lake reached 1125.00 meters.

Average leakage from the lake, also taking into account the contributions from the non-gauged springs of the Anamas Mountains, has been found to be 5 $m^3/s$. The water balance computations for Beyşehir lake have been based on correlation analysis and mass transport equations.

## 3.3. Köprüçay River Basin

The Köprüçay river basin is in the central part of the study area(Fig.5). It's watershed has an area of about 1942.4 $km^2$ at the most downstream gauging station. The average measured discharge some 20 km from the sea is 117 $m^3/s$. It represents an average runoff of 1,560 mm over the surface drainage area. Evidently, there is an additional runoff in the river flow that cannot be accounted for by surface catchment and precipitation. The greatest contribution to the Köprüçay river is from the Olukköprü springs emerging along the 2 km long narrow gorge. The minimum discharge of the system of the Olukköprü springs during the dry season was 35 $m^3/s$.

Clearly, the surface watershed does not coincide with the subsurface water contributing area.

According to water budget computations, one third of the Köprücay river streamflow comes from outside the drainage area which is the natural recharge.

The computations were made for the period between 1963 and 1971. This limitation was due to the lack of streamflow records in Bulasan streamflow gauging station after 1971.

The general characteristics of the two stations are given in Table 3.1.

Table 3.1. General Characteristics of Köprüçay Basin

| Station Name | Elevation (m) | Drainage Area ($km^2$) | Total Flow (million) ($m^3$) | Average Flow ($m^3/s$) | Flow over Drainage Basin (mm/year) | Average Prec. (mm) |
|---|---|---|---|---|---|---|
| Beşkonak | 117 | 1942.4 | 2932 | 93 | 1509 | 1700 |
| Bulasan | 370 | 1538.4 | 968 | 31 | 629 | - |

Drainage area of the Beşkonak gauging station is 404 $km^2$ larger than the area at Bulasan station. The ratio is 26%. On the other hand, the ratio of the flow recorded at the stations is 203%. Taking the effective precipitation as 60% of the precipitation, the annual flow may be computed as

$$1700 \times 0.6 = 1020 \text{ mm.}$$

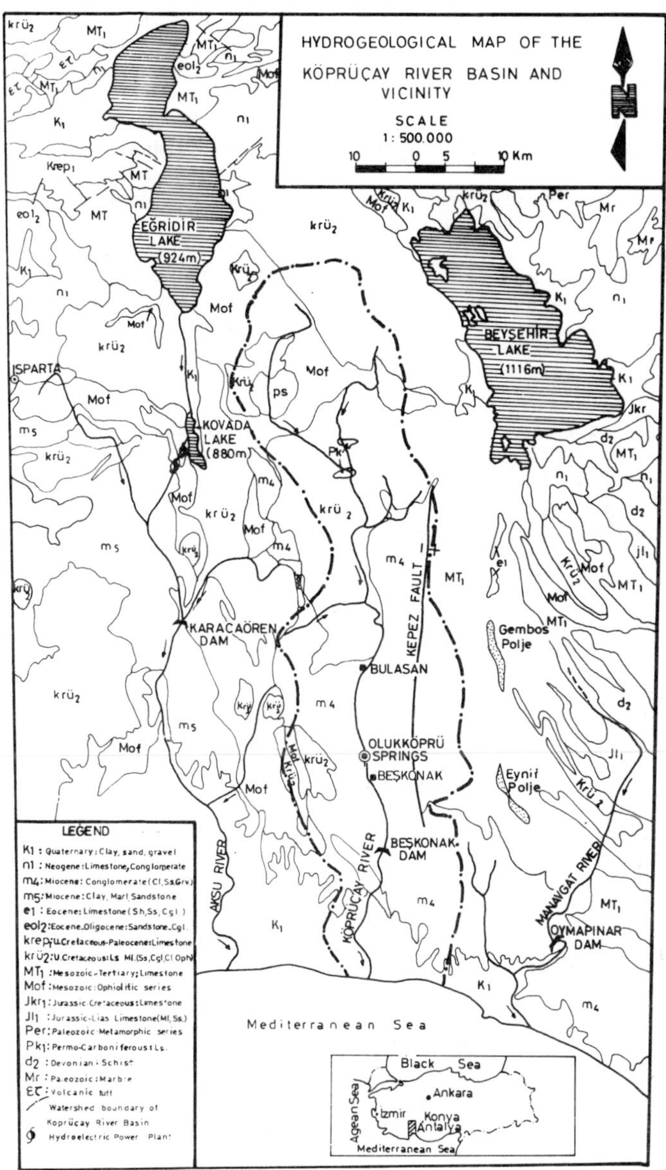


Fig. 5 Hydrogeological Map of Köprüçay river basin and its vicinity.

In contrast, the recorded flow in the region is

$$1964 \times 10^6/404 \times 10^6 = 4861 \text{ mm.}$$

It seems that it is impossible to collect so much water from the small area between Beşkonak and Bulasan which is only 404 $km^2$. These figures suggest a significant natural recharge from the adjacent basins to the Köprüçay river basin.

For the whole basin together with the two drainage basins at Beşkonak and Bulasan gauging stations, the water budget computations give the following results.

The annual average flow at Beşkonak station is 1932 x $10^6$ $m^3$,the drainage area is 1942.4 $km^2$, the average precipitation 1700 mm. and the flow coefficient is 60%, thus the total flow at Beşkonak might be

$$1.7 \times 0.6 \times 1942.4 \times 10^6 \text{ m}^2 = 1981 \times 10^6 \text{ m}^3/\text{year.}$$

But the recorded total flow at Beşkonak is 2931 x $10^6$ $m^3$/year.

There is a difference of about 1000 x $10^6$ $m^3$/year which corresponds to one third of the total flow.

The excess must be coming from the Eğridir lake or the springs emerging from the Anamas Mountains.

## 3.4. Aksu River Basin

The Aksu river basin is situated between Kovada lake and Antalya bay. It has a watershed area of 2,852 $km^2$. The natural drainage of Kovada lake joins the Aksu river on the river's northern end. The studies showed no large karst springs in the valley as is the case with the adjacent valleys of the Köprüçay and Manavgat rivers, though medium and smaller size karst springs are present.

According to Sipahi (1976), the discharge rate depends almost totally on the precipitation. However, water budget computations have revealed that there is natural recharge from the adjacent basins also in this basin.

The average precipitation is 906.42 mm. This figure varies in the northern part (868.64 mm) and in southern part of the basin (1113.5 mm).

The drainage area of the upper part is 1595 $km^2$. Therefore, the total precipitation equals

$$0.869 \times 1595 \times 10^6 = 1386 \times 10^6 \text{ m}^3/\text{year}$$

30% of this figure is assumed to be loss by evapotranspiration, so that the net input becomes

$$1386 \times 10^6 - 416 \times 10^6 = 970 \times 10^6 \text{ m}^3/\text{year}$$

On the other hand, the average flow recorded at Kargı station is 593.9 x $10^6$ $m^3$/year. So an excess of 376 x $10^6$ $m^3$/year comes from the adjacent basins as natural recharge. According to the average precipitation of 868.64 mm., this excess water must be coming from an area of 430 $km^2$.

Taking the whole basin as a unit, the drainage area is 2067 $km^2$. So the total precipitation is

$$0.906 \times 2067 \times 10^6 = 1872 \times 10^6 \; m^3/year$$

Subtracting the losses by evapotranspiration (30% of the figure found above);

$1872 \times 10^6 - 562 \times 10^6 = 1310 \times 10^6 \; m^3/year$, whereas the average flow recorded at the most downstream gauging station Boztepe is $945 \times 10^6$ $m^3/year$. Thus, the excess water is computed to be

$$1310 \times 10^6 - 945 \times 10^6 = 365 \times 10^6 \; m^3/year$$

coming from the adjacent basins, most probably recharging partially the Köprüçay river and Kırkgöz springs.

## 3.5. The Antalya Travertine Plateau

This subarea includes about 800 $km^2$. However, the subarea reflects a larger surface because it represents the catchment area of the large Kırkgözler springs. This surface covers the Upper Antalya Plateau,the Döşemealtı plain of an average altitude of 250-300 m., the lower plateau of the Varsak and Düden plain at an elevation between 50 and 150 m., and the watershed of the Kırkgözler springs. The most important karst features are the Kırkgözler springs, which discharge between 16 and 60 $m^3/s$, the swallow hole and spring system of Bıyıklı swallow hole - Varsak Düdenbaşı spring. The latter system is developed in the karstified travertine of the Quaternary age. Travertine is the dominant geologic unit in the plateau area of this subarea.

For the travertine plateau the total recharge is $500.5 \times 10^6 \; m^3/y$. Whereas the total discharge is $830.5 \times x0^6 \; m^3/year$. On the other hand, the total discharge of the Kırkgöz springs is $497.5 \times 10^6 \; m^3$ whereas the total recharge is $827.5 \times 10^6 \; m^3/year$. Consequently, it can be stated that excess water of $330 \times 10^6 \; m^3/year$ comes from the Mesozoic limestone which forms the reservoir of the Kırkgöz springs and feeds the travertine plateau.

## 3.6. The Dalaman River Basin

The Dalaman river basin is situated to the west of the Antalya travertine plateau and has an area of 5380 $km^2$ (Fig. 6). The general characteristics of the basin with its subbasins are given in Table 3.2. The water balance computations are based simply on the mass transport equation.

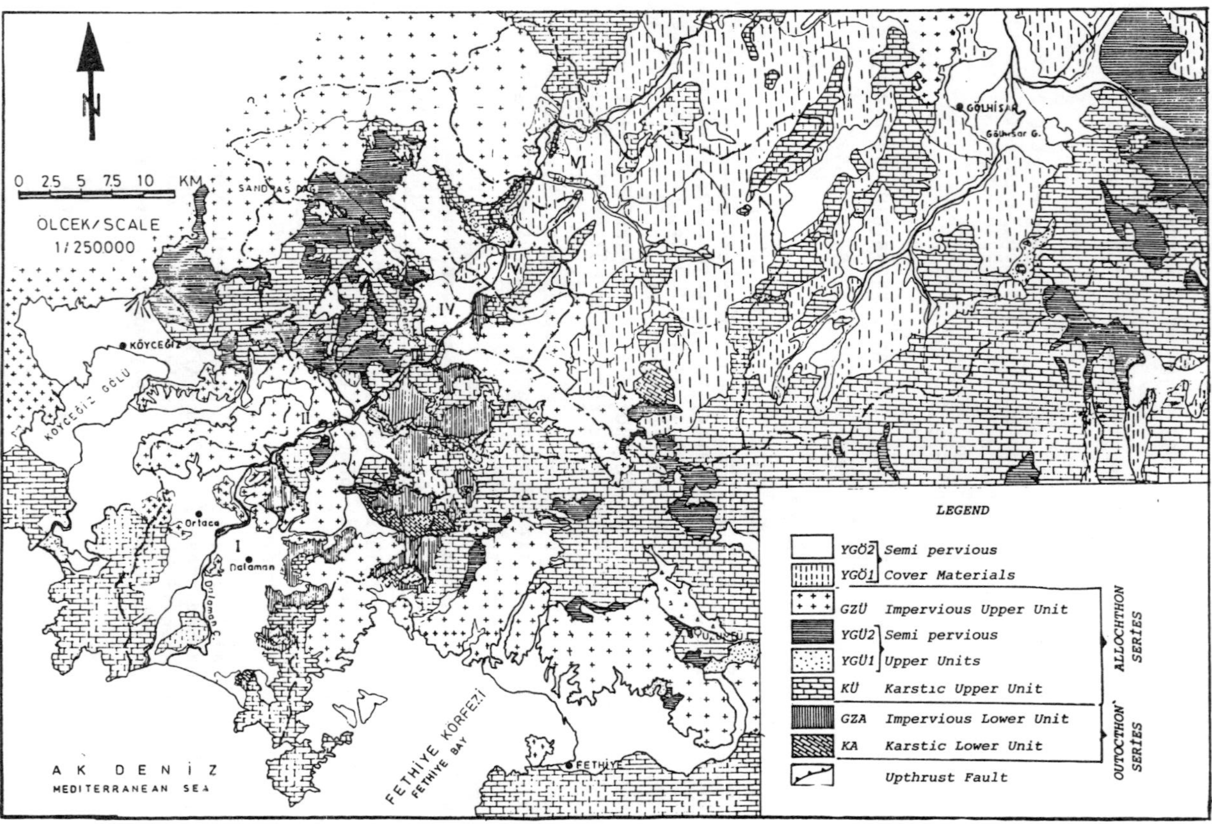


Fig. 6 Hydrogeological Map of Lower Dalaman river basin and its vicinity (after Yeşertener, 1986).

Table 3.2. The results of water budget by the way of simple precipitation-runoff-evapotranspiration

| Basin Name | No. | Drainage Area | Prec. | Runoff | Evap. | Dif. | % Dif. |
|---|---|---|---|---|---|---|---|
| Dalaman Plain | I | 420.7 | 484.2 | 686.3 | 290.5 | -492.6 | -101.7 |
| Dalaman Dam | II | 44.3 | 57.6 | 64.3 | 34.6 | - 41.3 | - 71.6 |
| Akköprü Dam | III | 377.4 | 473.8 | 168.1 | 284.3 | 21.4 | 4.5 |
| Meşebükü FGS | IV | 112.5 | 143 | 105.7 | 85.8 | - 48.5 | - 33.9 |
| Narlı Dam | V | 265.1 | 296.1 | 391.8 | 177.6 | -273.4 | - 92.3 |
| Sandàlcık Dam | VI | 631.1 | 709.5 | 413.5 | 425.7 | -129.7 | - 18.3 |
| L.Dalaman Basin | | 1851.1 | 2131 | 1765 | 1278.6 | -912.6 | - 42.8 |
| Dalaman Basin | | 5380 | 4340 | 2293 | 2604.4 | -557 | - 12.8 |

Evap. = Prec. x 0.60

All units are $10^6$ $m^3$/year.

The results have shown that an amount of excess water of $912.6 \times 10^6$ $m^3$/year contributes to the large karstic aquifer in the region which equals 43% of the amount provided by precipitation; 30% of this excess water comes from the Upper Dalaman basin while the other 13% comes from other subareas. According to the computed figures (Table 3.3), the catchment area must be 2644.3 $km^2$ which is as much as 43% of the drainage area.

Table 3.3. The calculation results of real drainage area for the study area

| Basin Name | No. | Observed Prec. | Calculated Prec. | Calculated Area | Real Area | Added Area |
|---|---|---|---|---|---|---|
| Dalaman Plain | I | 484.2 | 976.8 | 751.4 | 420.7 | 330.7 |
| Dalaman Dam | II | 57.6 | 98.9 | 76.1 | 44.3 | 31.7 |
| Akköprü Dam | III | 473.8 | 452.4 | 360.4 | 377.4 | -17.1 |
| Meşebükü Dam | IV | 143 | 191.5 | 150.7 | 112.5 | 38.2 |
| Narlı Dam | V | 296.1 | 569.5 | 509.9 | 265.1 | 244.8 |
| Sandalcık Dam | VI | 709.5 | 839.2 | 746.4 | 631.1 | 115.3 |
| Lower Dalaman | VII | 2131 | 3043.6 | 2644.3 | 1851.1 | 793.2 |
| Dalaman Basin | | 4340 | 4897 | 6072.9 | 5380.1 | 692.9 |

### 3.7. The Upper Eşençay Basin

The Eşençay river basin is situated to the west of the Dalaman basin (Fig. 7).

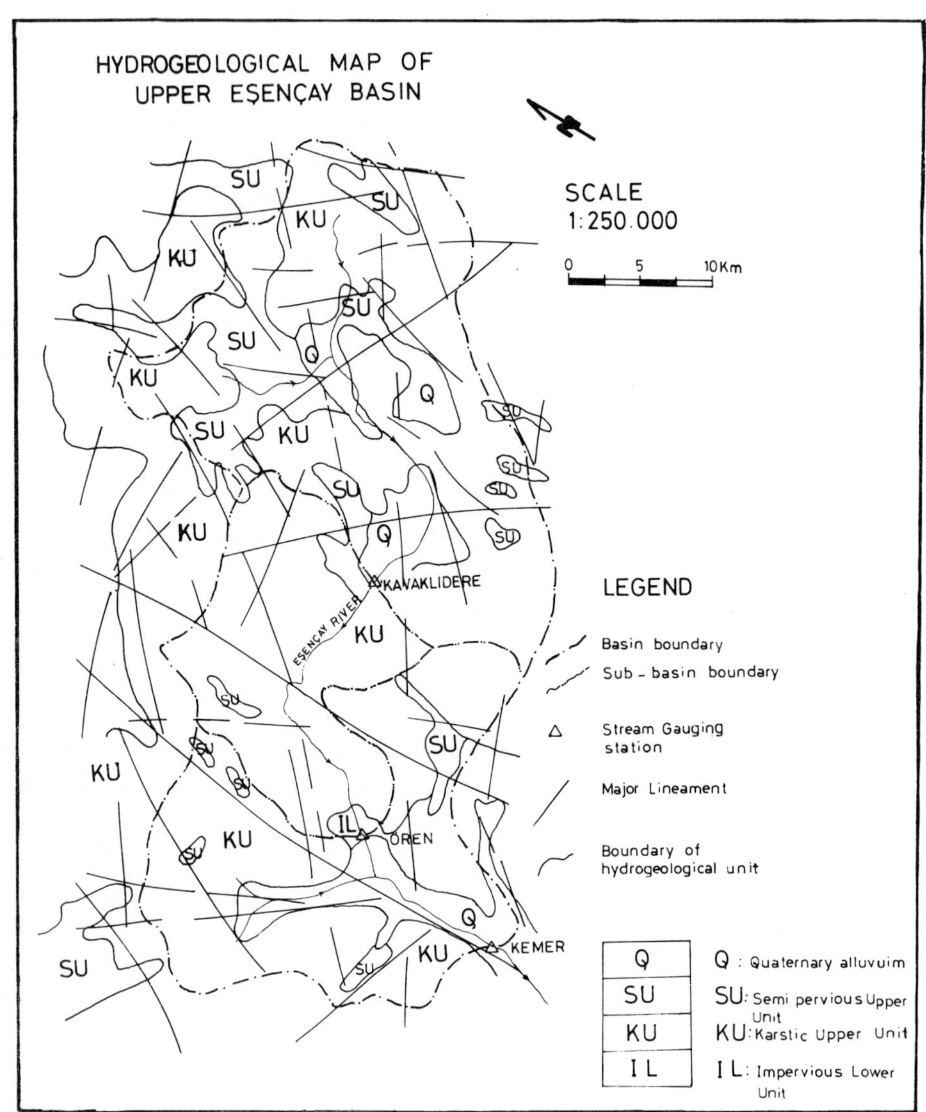


Fig. 7 Hydrogeological Map of Eşençay river basin and its vicinity (after Bayarı, 1986).

Based on the simple water balance equation, the water budget was computed for this basin. The general hydrogeological characteristics of the basin are given in Table 3.4.

Table 3.4. The results of the water budget

| Hydrologic Parameter(1) | Basin Kavaklıdere | Ören | Kemer | Ören-Kavaklıdere(2) |
|---|---|---|---|---|
| Flow | 391.2 | 975.5 | 889.2 | 566.3 |
| Precipitation | 759.1 | 835.8 | 934.8 | 2171.1 |
| Evaporation(3) | 455.5 | 501.5 | 560.9 | 601.7 |
| Flow Deficit | -87.6 | -632.2 | -515.3 | -1769.9 |

Explanation:
(1) All units are mm/year
(2) It covers the area between the Ören and Kavaklıdere Flow Gauging stations
(3) Evapotranspiration = Prec. x 0.60

The results revealed that 566.87 x $10^6$ $m^3$/year of this excess water comes from the adjacent basins as natural recharge. The total precipitation is 411.25 x $10^6$ $m^3$/year and the total discharge is 978.12 x $10^6$ $m^3$.

## 4. ACCURACY OF COMPUTATIONS AND SOURCES OF ERRORS

The water balance is a stochastic relation between the water balance components which are random variables in time and space with usually unknown probability distribution. The independent input variable precipitation is transformed in the hydrological system into the dependent output variables evapotranspiration and runoff (streamflow in karst) and into change in storage in the system. The inputs of a water balance model are precipitation, temperature, evaporation, potential evapotranspiration or other similar quantities. The output is mostly runoff. In general, precipitation is the largest quantity in a water balance equation. But it is still not known with sufficient accuracy how much precipitation actually reaches the ground. No reliable water balance computation is possible with insufficient knowledge of the spatial rainfall patterns. The main reason for errors in the areal mean precipitation is the high spatial temporal variability of precipitation and the resulting complicated statistical structure of precipitation data. The computation of areal precipitation from point precipitation values constitutes one of the most crucial problems in hydrogeology. In water balance computations presented in this paper areal precipitation is computed by the Polygon (Thiessen) method which is the most suitable method for the study area. However, the individual or sequential rainfall events can not be determined with sufficient accuracy. Furthermore there is a precipitation record only for a certain period and all computations are based on these records. This is the main and the most important source of error. Secondly, determination of site evapotranspiration is another important source of error. In general, evapotranspiration provides the second largest quantity in a hydrogeological water balance. In this paper, evapotranspiration values determined by the concept of complementary relationship between areal and potential evapotranspiration for water balance computation have been used. Turc and Thornthwaite formulae are employed in determining the real evapo-transpiration. The validity of these methods must be checked since some assumptions such as moisture content of the soils which is known as reserve water is considered to be 100 mm (Thornthwaite). Surface runoff depends on the temperature and is obtained from Turc graph, which may not be suitable for the study area. This, in fact, may influence the results of the computations and must be considered in the interpretations.

## 5. CONCLUSIONS

The water balance computations showed significant contribution by natural recharge from adjacent basins. In order to reach a higher degree of accuracy, the sources of errors must be analysed before interpretating. The natural recharge in the southwestern Taurus Mountains region reaches up to 1000 million cubic meter per year. Construction of hydrogeotechnical projects must be planned taking into account both the natural recharge and the area that provides this excess water. It is concluded that computations of water balance and consequently the natural recharge must reach a higher accuracy and new approaches and methods with well defined parameters.

## ACKNOWLEDGEMENTS

This study was performed as a part of a joint project with the United Nations Development Program (TUR/81/004). Detailed information is available in the Technical Progress Reports published by the Karst Research Center.

The author acknowledges the assistance of Mr. M. Ekmekçi and Mr.C. Denizman, and for reviewing the English text of the paper to Prof. C. Kuruç.

REFERENCES

BAYARI, C.S., 1986, Yukarı Eşençay (Fethiye KD) Havzasının Karst Hidrojeolojisi İncelemesi; Yüksek Mühendislik Tezi, Hacettepe Üniversitesi, Jeoloji Mühendisliği Bölümü, Ankara. (Unpublished)

DYCK, S., 1983, 'Overview on the Present Status of the Concepts of Water Balance Models', Proceedings of the Hamburg Workshop, IAHS publication No. 148, 3-20 pp.

DSİ, 1985, Antalya-Kırkgöz Kaynakları ve Traverten Platosu, DSİ yayını, Ankara.

EKMEKÇİ, M., 1986, Beyşehir Gölü Havzası ile Komşu Havza Akımları Arasındaki İlişkilerin Araştırılması; Yüksek Mühendislik Tezi, Hacettepe Üniversitesi, Jeoloji Mühendisliği Bölümü, Ankara.(Unpublished)

EROSKAY, O., GÜNAY, G., 1979, Tecto-Genetic Classification and Hydrogeological Properties of the Karst Regions in Turkey. International Seminar on Karst Hydrogeology-Proceedings, DSI - UNDP Project (TUR/77/015) publ., DSI Ankara, Turkey, pp. 385.

GÜNAY, G., 1981, Manavgat Havzası ve Dolayının Karst Hidrojeolojisi İncelemesi, Hacettepe Üniversitesi,Mühendislik Fakültesi Yerbilimleri Enstitüsü (Assert. Thesis).

GÜNAY, G., JOHNSON, A.I., (Editors) 1986, Karst Water Resources, International Symposium on Karst Water Resources-Proceedings, IAHS publ. no. 161, Ankara, Turkey, pp. 654.

GÜNAY, G., KARANJAC, J., (Editors) 1978, Karst Hydrogeology Symposium Proceedings, Oymapınar. DSI-UNDP Project Technical Publ. 27, DSI, Ankara, pp. 293.

GÜNAY, G., SİPAHİ, H., 1979, 'Eğridir-Beyşehir Gölleri ile Akdeniz Arasındaki Alanda Yapılan Karst Hidrojeolojisi Çalışmaları', Mühendislik Jeolojisi Simpozyumu, TJK yayını, 49-54.

GÜNAY, G., YAYAN, T.Y., 1979, Antalya Kırkgöz Karst Kaynaklarının Hidrojeolojisi (Hydrogeology of the Antalya Kırkgöz Karst Springs): DSI-UNDP Projesi (TUR/77/015) yayını, 32, 118 s.

KARANJAC, J., GÜNAY, G., and ALTUĞ, A., 1977, Kovada Lake Preliminary Water Balance, DSI-UNDP Project (TUR/77/014) no. 8.

KARANJAC, J., GÜNAY, G., 1980, 'Development of Karst Water Resources in Turkey with Emphasis on Groundwater', Natural Resources Forum 4, 61-73.

ÖZİŞ, Ü., KELOĞLU, N., 1976, 'Some Features of Mathematical Analysis of Karst Runoff',Karst Hydrology and Water Resources, Vol. I, Water Resources Publications, pp. 221-235.

ÖZİŞ, Ü., HARMANCIOĞLU, N., 1980, 'An Outlook to Karst Hydrology Studies in Turkey, Karst Hydrogeology-Proceedings, International Seminar on Karst Hydrogeology, Oymapınar, Antalya, Turkey, pp.

SİPAHİ, H., 1985, 'Karst Hydrogeology of the Aksu River Basin; Proceedings of International Symposium on Karst Water Resources, IAHS Publication no: 161, Ankara.

VAN DER BEKEN, A., HERMANN, A., 1983, 'New Approaches in Water Balance Computations', Proceedings, XVIIIth General Assembly of the International Union of Geodesy and Geophysics, Hamburg-Germany, IAHS publication No: 148, 167 pp.

YEŞERTENER, C., 1986, Aşağı Dalaman (Fethiye KB) Havzasının Karst Hidrojeolojisi İncelemesi; Yüksek Mühendislik Tezi, Hacettepe Üniversitesi, Jeoloji Mühendisliği Bölümü, Ankara. (Unpublished)

# ESTIMATION OF RECHARGE OF SAND AQUIFER OF THE ISLAND OF MANNAR SRI LANKA

D.C.H. Senarath
Department of Civil Engineering
University of Moratuwa
Moratuwa
Sri Lanka

ABSTRACT. An estimation of recharge to an aquifer can be carried out on the basis of water balance in which the top surface of the catchment and the soil moisture zone are taken into account. The data required are daily precipitation and potential evapotranspiration. The water balance can be verified on the basis of comparison of estimated stream flow with measured stream flow. The method is illustrated by application to estimate the recharge to a sand aquifer situated in the island of Mannar in north-west Sri Lanka.

## 1. INTRODUCTION

In many third world countries, much investment has taken place in recent times in the development of surface water supplies. Although the development of surface water schemes has consumed an enormous capital, the investments may be financially justified on account of the multipurpose nature of these schemes and the variety of benefits that are usually associated with such schemes. However, there are many situations in which a surface water scheme does not provide the most appropriate solution for satisfying a water demand. In such situations, a groundwater scheme may be an appropriate alternative involving much less capital in many instances.

In order to derive the optimum benefit from a groundwater scheme, a proper resource study has to be carried out. But unlike surface water processes, much of groundwater activity takes place unseen by the human eye, in the strata that lie below the ground surface. Measurement and monitoring of groundwater flow is therefore extremely difficult. As a result it has often been necessary to make several simplifying assumptions in evaluating the resources of an aquifer (Fernando, 1973), because much of the data that is required for a systematic study has not been available. When date is available it is no longer necessary to resort to such doubtful assumptions and more reliable studies involving refined techniques can be carried out.



*I. Simmers (ed.), Estimation of Natural Groundwater Recharge, 423–434.*

In addition to the geological information which is usually available with a fair degree of accuracy on the basis of many types of geological studies, the groundwater recharge, i.e. the quantity of water that percolates to the water table from the ground surface is one of the most important factors that determines the resources of an aquifer.

Although the physical properties of an aquifer such as permeability, thickness, storage properties and boundaries are factors mainly influencing the flow of water within an aquifer, an accurate estimation of recharge often enables a fairly accurate prediction about the overall yield. A detailed analysis of flow within an aquifer can be carried out using one of many mathematical techniques that are now available (Rushton and Redshaw, 1979). In such analyses, recharge is an important parameter and therefore it must be estimated accurately.

## 2.0 ESTIMATION OF RECHARGE

### 2.1 Data Requirements

As the entire recharge is primarily derived from precipitation, a reliable record of precipitation, representative of the surface area feeding the aquifer is essential for estimating recharge. It must be noted that some confined aquifers may derive their recharge from the ground surface far remote from the area of abstraction and in such situations, the precipitation records that are used must correspond to the area that feeds the aquifer and not the area that overlies the aquifer. A surface water balance is carried out, taking into account the components such as evapotranspiration, interflow, stream run-off, base flow and groundwater recharge. There are many techniques for carrying out the surface water-balance (Rushton and Ward, 1979); (Senarath and Rushton, 1984). The essential feature of these methods is that a daily balance is evaluated starting from the precipitation for each day. The daily value of potential evaporation is estimated either from Penman's formula (Penman, 1949) using meteorological data or from daily values of pan evaporation. If the latter is used, a multiplying factor of about 0.8 has to be used, to correct for the scale effect of the pan and the heat transfer from the sides of the pan.

Thus, the data required for estimation of recharge are precipitation, potential evaporation and stream run-off together with information regarding topography, geology and vegetation. Since the factors affecting processes such as evaporation change rapidly, the time step for calculations should be preferably not longer than one day. In practice a time step of one day gives good results.

### 2.2 Soil Moisture Balance

The actual evapotranspiration corresponding to each day is estimated from potential evaporation taking into account the type of vegetation covering the area and the residual moisture content of the soil. A variety of models of the conceptual type can be devised for estimating actual evapotranspiration from potential evaporation.

In one of these models (Rushton and Ward, 1979) it is assumed that as long as the soil moisture deficit (in relation to the moisture content at field capacity) is less than a certain value known as 'root constant', evaporation takes place at the potential rate, the water required for this purpose in excess of precipitation being removed from soil moisture storage, thereby increasing the soil moisture deficit. The value of root constant is a function of the type of vegetation, those with deeper roots having higher values of root constant, assuming that the spacing between individual plants for a given type of vegetation remains the same. When the soil moisture deficit is in excess of root constant, it is assumed that the soil moisture is unable to contribute readily to evapotranspiration and therefore only a small fraction (about 10%) of potential evaporation would be removed from the soil moisture storage. This roughly corresponds to the amount of evaporation that could take place from a wet bare soil. It is also assumed that recharge of the aquifer is possible only when the soil moisture deficit is completely nullified bringing the soil to field capacity condition. Any water that infiltrates when the soil is in this condition is assumed to percolate down to the water table and this component is regarded as groundwater recharge.

Many variations depending on specific physical conditions are possible in the model that is actually devised to carry out the soil moisture balance. Any model that can be physically justified is acceptable provided that the parameters that are introduced can be evaluated either from physical observations or by calibration using observed data.

## 2.3 Water Balance Model

Using a method similar to that outlined above, to represent soil moisture balance, a model can be devised on a similar basis to represent the overall water-balance of the entire surface catchment. This would consist of components such as precipitation, surface run-off, evapotranspiration, interflow, groundwater recharge, baseflow and stream run-off.

The peripheral components of such a model would be precipitation, evapotranspiration, groundwater recharge and stream run-off (Figure 1). The distribution of flow along different routes from a given region can be defined by constant distribution coefficients whose numerical values can be established by trial and error or by optimization using suitable objective functions at the stage of calibration. The model can be calibrated by comparing calculated stream run-off with gauged stream run-off and the quantity of groundwater recharge can be estimated. If the stream run-off out of the catchment is not appreciable the soil moisture balance would give the gross recharge directly. Net recharge may be calculated from gross recharge by subtracting the outflow across the aquifer boundaries.

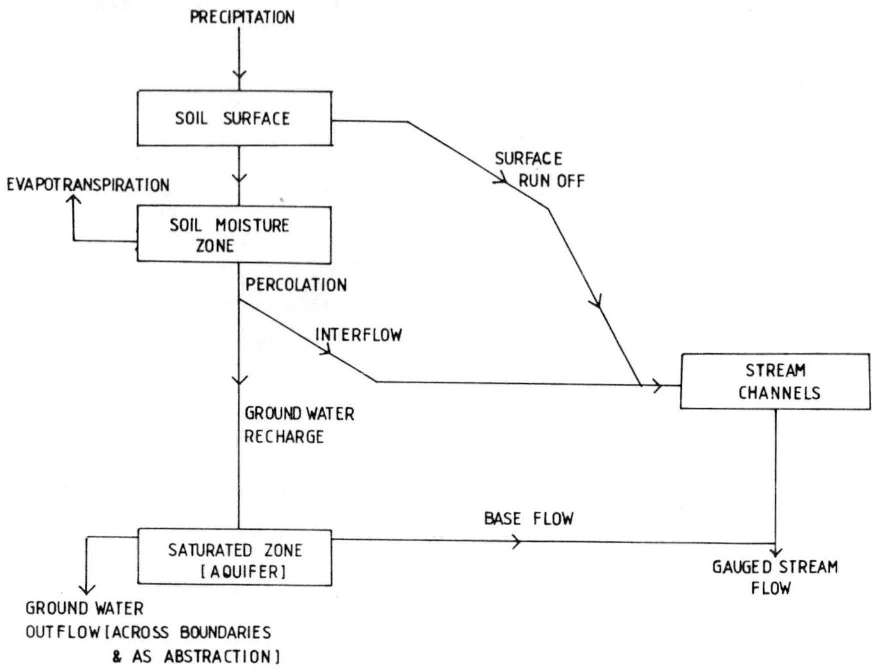


A TYPICAL FLOW PATH SYSTEM IN A CATCHMENT

FIGURE 1

## 3.0 ESTIMATION OF RECHARGE OF THE SAND AQUIFER OF MANNAR

The method outlined in the foregoing paragraphs is demonstrated by applying this method to estimate the recharge of the sand aquifer of Mannar.

### 3.1 Study Area

The island of Mannar is situated along the north-west coast of Sri Lanka and is about 260 km north of Colombo. (Figure 2). The extent of the island is approximately 128 km$^2$.

The area is generally flat and constitutes coastal plains except for the sand dune ridges rising along the axis of the elongated island. The area is devoid of any surface streams.

Geomorphologically, two contrasting formations are present namely, lagoonal deposits and windblown sand deposits. The low elevation areas consist mainly of lagoonal deposits while the higher elevations are occupied by windblown sand deposits. These quarternary deposits which together have a maximum thickness of about 20 metres are underlain by miocene limestone followed by secondary deposits of jurassic sandstone and the pre-cambrian basement which is encountered at a depth of about 250 metres (Rao, 1984). See Figure 3.

### 3.2 Climate

The Mannar island falls into the semi-arid climate area and receives a mean annual precipitation of about 975 mm. The maximum and minimum mean monthly precipitation are 240 mm corresponding to the month of November and 5 mm corresponding to the month of July respectively.

The temperature in the area ranges from a minimum of 23° C to a maximum of 35° C. The nearest pan evaporation measuring station is at Giant's Tank which is about 30 km south-east of the Mannar island. The climate at Giant's tank is similar to that at Mannar island and it is assumed that the pan evaporation values of Giant's tank are applicable to the Mannar island. The average monthly pan evaporation values for Giant's tank for the twenty year period from 1954 to 1973 range from 125 mm corresponding to the month of December to 178 mm corresponding to the month of June.

### 3.3 Hydro-geology

The hydrogeological information for Mannar has been obtained mainly from the work carried out by the National Water Supply and Drainage Board, Sri Lanka (Rao, 1984).

There is a large number of shallow dug wells spread throughout the island each of which supplies the domestic requirements of individual homes. But the overall abstraction from these wells is insignificant compared to the abstraction from deeper tube wells constructed and maintained by the National Water Supply and Drainage Board and which provide a water supply to about 20% of the population. The total abstraction from these amount to about 860 $m^3$ per day. A study was undertaken to estimate the additional abstraction that could be safely implemented to augment the present supply. This study was primarily based on an estimation of the recharge that reaches the aquifer. A detailed description of hydrogeology of the study area is given by Rao (1984).

### 3.4 Hydrological Water Balance

The hydrological water balance is a quantitative statement of the balance between the total water gains and the losses of the basin for a period of time. This may be stated as follows:

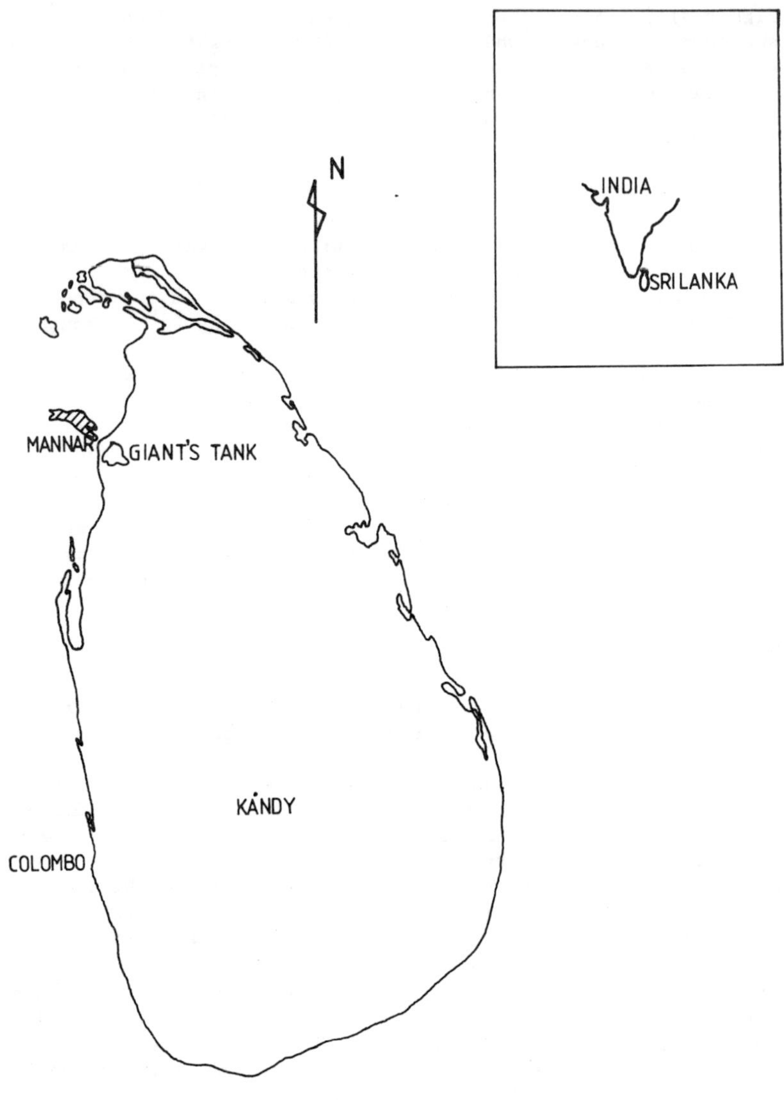


SCALE: 1:3,000,000

THE STUDY AREA

FIGURE 2

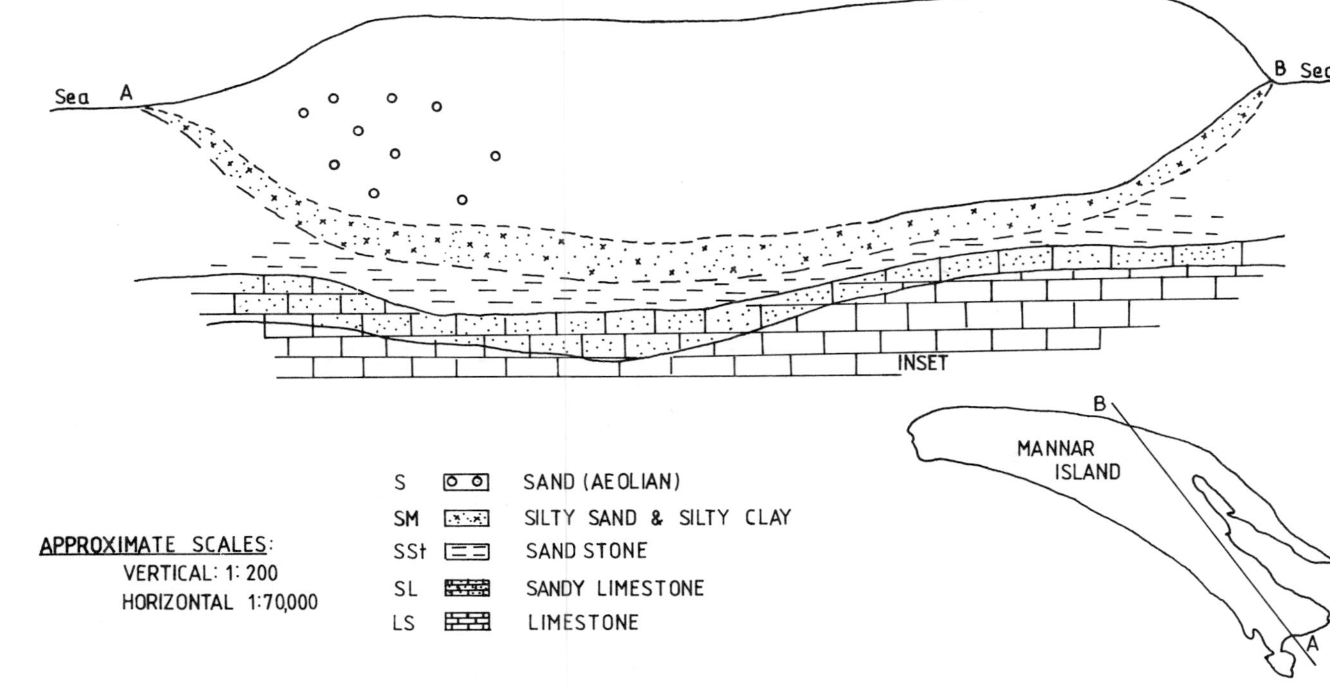


A TYPICAL GEOLOGICAL CROSS SECTION ACROSS MANNAR ISLAND

FIGURE 3

$P = E + R_s + R_g + U + dS_m + dS_g$ where,

$P$ = precipitation

$E$ = evapotranspiration

$R_s$ = direct run-off (out of the catchment)

$R_g$ = groundwater discharge

$U$ = outflow through the aquifer boundaries

$dS_m$ = change in soil moisture content

$dS_g$ = change in groundwater storage (net recharge)

The method of application of this water balance equation to estimate groundwater recharge consists of the following steps:

(a) The values of daily precipitation are obtained from records corresponding to the period of the study. In the present study the values of daily precipitation recorded at Mannar were used.

(b) The daily values of potential evaporation are obtained either from Penman's formula using the relevant meteorogical data or from pan evaporation data.

In the present study, the values of pan evaporation recorded at Giant's tank gave consistently lower values compared to those given by Penman's formula, the difference being about 14%. In order to arrive at a conservative estimate for recharge, the higher values, i.e. those given by Penman's formula were used.

(c) Direct run-off out of the catchment usually consists of stream run-off. This is expressed as a percentage of the daily precipitation, taking into account the characteristics of the catchment such as slope and permeability and the volume of stream run-off compared to the volume of precipitation reckoned over a period of a few years. However, in the present study, since the upper stratum of the catchment is highly permeable and since there are no significant surface streams, direct run-off out of the catchment is small. The component direct run-off, $R_s$,was therefore taken as a small fraction of the daily precipitation.

(d) A soil moisture balance is carried out on a daily basis using the method outlined previously. The calculations are commenced around the beginning of December when the soil moisture deficit is zero due to monsoonal rains (in the case of the dry zone of Sri Lanka). The calculations are terminated at a similar date in a subsequent year so that the net increase in soil moisture content, $dS_m$, for the period of study would be zero. In the present study, the calculations covered the 11 year period from December 1972 to November 1982.

Root Constant is a parameter that depends on the type of land use in the catchment area. Grindley (1969) suggests values of root constant for different types of vegetation in the United Kingdom ranging from 12.5 mm for rough grazing to 200 mm for woodland. In a study of the north west limestone aquifer in Vanathavilu, Sri Lanka, (Lawrence and Dharmagunawardene, 1981) the following values of root constant have been used.

| | |
|---|---|
| Annual crops such as paddy and grass land | 50 mm |
| Coconut and Mango | 150 mm |
| Forest cover | 300 mm |

Vegetation in Mannar consists mainly of randomly distributed palm trees (palmirah) and small shrub. Therefore, considering the values suggested above, the root constant for this area is probably in the region of 100 mm to 150 mm. In the present study, a series of values ranging from 50 mm to 150 mm were tried, bearing in mind that higher values give more conservative estimates for recharge.

(e) Outflow through the aquifer boundaries i.e. U,is calculated from Darcy's equation, based on estimated values of permeability and thickness of the aquifer and hydraulic gradient at the aquifer boundary, which is evaluated from data on water table elevation.

For Mannar island, the boundary length (perimeter) is 70 km, the average thickness is 7.75 metres and the water table gradient towards the boundaries is 0.0035. Based on a number of pumping test results, the average permeability was estimated as 10.2 metres per day. Corresponding to these values, the outflow through the boundaries works out as follows:

From Darcy's law,

$$\text{Discharge } Q = \text{Area (A) X Permeability (K) X } \frac{dh}{dl}$$

Therefore,

$$U = 70 \text{ X } 1000 \text{ X } 7.75 \text{ X } 10.2 \text{ X } 0.0035$$
$$= 19367 \text{ m}^3 \text{ per day}$$

As mentioned earlier, the area of Mannar island is 128 km$^2$. Therefore, the outflow of 19367 m$^3$ per day is equivalent to

$$\frac{19367 \text{ X } 10^3}{128 \text{ X } 10^6} \text{ X } 365 = 55 \text{ mm over the catchment per year.}$$

## 4.0 RESULTS

Soil moisture balance computations were carried out on a daily basis for the eleven year period from 1972 to 1982 for 30 trials using different combinations of possible values of (a) direct run-off fraction of precipitation (b) root constant and (c) the fraction of potential evaporation that is removed from the soil as evapotranspiration when the soil moisture deficit exceeds root constant. This factor is denoted as F. The mean values of recharge expressed in mm per year are shown in Table 1.

From Table 1 it is seen that the most stringent conditions for recharge are imposed in Trial 19 corresponding to the maximum values of direct run-off fraction, root constant and F and gives a mean recharge of 93 mm per year which is 10.34 per cent of mean annual precipitation. Similarly, the least stringent conditions correspond to Trial 13 which gives a mean recharge of 287 mm per year which is 31.9 per cent of mean annual precipitation.

Comparing with previous studies of a similar nature in which the recharge estimates have been verified by other methods such as by calibration against stream run-off data and by using the recharge values as inputs to flow models which are verified by comparison with recorded data, (Senarath and Rushton, 1984; Senarath, 1981) the most appropriate values of direct run-off factor, root constant and factor F for the present case are probably about 0.05, 150 and 0.10 respectively. These correspond to Trial 24 in which the mean net recharge is 53 mm per year. Making allowances for the outflow through the aquifer boundaries and abstraction (pumping) from wells, the corresponding gross recharge is 121 mm per year which is about 13.45 per cent of the mean annual precipitation.

In the present study, as there are no surface streams it was not possible to calibrate the water balance model by comparison of calculated stream run-off with gauged stream run-off. Therefore the unknown parameters had to be established on the basis of values already established for other similar catchments.

## 5.0 CONCLUSIONS

This study has shown that with reliable data on precipitation and potential evaporation, it is possible to make realistic estimates of groundwater recharge. When the recharge is correctly estimated, the quantitative resources of the aquifer can be evaluated and a scheme can be drawn up for its utilization. Further information regarding the behaviour of the aquifer would emerge during the initial stages of its utilization as more data becomes available. The estimates of recharge as obtained by the method used in this study can be confirmed by feeding these values to a flow model of the aquifer whose results can be verified by comparison with field data such as groundwater head distribution in space and time. The method used in this study also will give monthly or daily values of recharge.

| Trial Number | Direct Run-off Fraction of Precipitation | Root Constant in mm | Factor F | Mean Recharge in mm per year | |
|---|---|---|---|---|---|
| | | | | Gross | Net |
| 1 | 0 | 100 | 0.10 | 184 | 116 |
| 2 | 0.10 | 100 | 0.10 | 128 | 60 |
| 3 | 0 | 50 | 0.10 | 189 | 121 |
| 4 | 0 | 100 | 0.10 | 167 | 99 |
| 5 | 0 | 100 | 0.10 | 184 | 116 |
| 6 | 0 | 150 | 0.10 | 145 | 77 |
| 7 | 0 | 50 | 0.10 | 253 | 185 |
| 8 | 0 | 50 | 0.20 | 182 | 114 |
| 9 | 0 | 150 | 0.20 | 140 | 72 |
| 10 | 0 | 100 | 0.20 | 159 | 91 |
| 11 | 0 | 150 | 0.05 | 149 | 81 |
| 12 | 0 | 100 | 0.05 | 202 | 134 |
| 13 | 0 | 50 | 0.05 | 287 | 219 |
| 14 | 0.10 | 100 | 0.10 | 130 | 62 |
| 15 | 0.10 | 150 | 0.10 | 98 | 30 |
| 16 | 0.10 | 50 | 0.10 | 189 | 121 |
| 17 | 0.10 | 50 | 0.20 | 129 | 61 |
| 18 | 0.10 | 100 | 0.20 | 112 | 44 |
| 19 | 0.10 | 150 | 0.20 | 93 | 25 |
| 20 | 0.10 | 150 | 0.05 | 102 | 34 |
| 21 | 0.10 | 150 | 0.05 | 152 | 84 |
| 22 | 0.10 | 50 | 0.05 | 224 | 156 |
| 23 | 0.05 | 100 | 0.10 | 158 | 90 |
| 24 | 0.05 | 150 | 0.10 | 121 | 53 |
| 25 | 0.05 | 50 | 0.10 | 221 | 153 |
| 26 | 0.05 | 50 | 0.20 | 155 | 87 |
| 27 | 0.05 | 100 | 0.20 | 134 | 66 |
| 28 | 0.05 | 150 | 0.20 | 115 | 47 |
| 29 | 0.05 | 150 | 0.05 | 126 | 58 |
| 30 | 0.05 | 150 | 0.05 | 180 | 112 |

RESULTS OF TRIALS

Table 1

## 6.0 ACKNOWLEDGEMENT

The Author acknowledges with gratitude the assistance received from the National Water Supply and Drainage Board and University of Moratuwa Sri Lanka in carrying out the study described in this paper.

## 7.0 REFERENCES

1. Fernando, A.D.N. The groundwater resources of Sri Lanka, Publication of Ministry of Irrigation, Power and Highways, Sri Lanka, May 1973.

2. Grindley, J. The calculation of actual evaporation and soil moisture deficit over specified catchment areas, Hydrol, Memo. No. 38, Met. Office, Bracknell Berks., U.K., Nov., 1969.

3. Lawrence, A.R. and Dharmagunawardene, H.A. The groundwater resources of the Vanathavillu Basin, Groundwater Division, Water Resources Board, Sri Lanka, 1981.

4. Penman, H.L. The dependence of transpiration on weather and soil conditions, Joun. Soil. Sci. 1, 1949.

5. Rao, K.V.R. Groundwater investigation for community water supply in Mannar Island, Publication of National Water Supply and Drainage Board, Sri Lanka, February 1984.

6. Rushton, K.R. and Redshaw, S.C. Seepage and Groundwater Flow. Numerical Analysis by Analog and Digital Methods, John Wiley and Sons, Wiley Interscience Publication, 1979.

7. Rushton, K.R. and Ward, C. The estimation of groundwater recharge, Joun. of Hydrol. (41), 1979.

8. Senerath, D.C.H. and Rushton, K.R. A routing technique for estimating ground-water recharge, Ground Water, Vol. 22, No. 2, March-April 1984.

9. Senarath, D.C.H. Regional groundwater flow simulation using a numerical model including aquifer dewatering and recharge estimation, Ph.D. thesis, University of Birmingham, U.K., January 1981.

10. Wijegoonewardene, S.J.P. Recharge of the fresh water aquifer in Mannar Island, M. Eng. thesis, University of Moratuwa, August 1984.

# GROUNDWATER RECHARGE FROM THREE CHEAP AND INDEPENDENT METHODS IN THE SMALL WATERSHEDS OF THE RAIN FOREST BELT OF NIGERIA

Kalu O. Uma,
Dept. of Geology,
University of Nigeria,
Nsukka, Nigeria
and
Boniface C.E. Egboka,
Dept. of Geological Sciences,
Anambra State University of Technology,
Awka Campus, P.M.B. 5025,
Awka, Nigeria

ABSTRACT. This paper discusses a study to critically and comparatively evaluate groundwater recharge into the small watersheds of the Imo River Basin, Nigeria, using three independent but cheap methods. The methods include; groundwater recharge using a simplified equation that relates groundwater stage to actual recharge with the aid of the aquifer porosity; recharge from baseflow recession analysis; and recharge from a water balance method. The results show that recharge values obtained from the water balance method were consistently higher than values from the other methods and ranged from 13.15 % to 43.19 % of the annual rainfall. Values obtained from baseflow recession analysis were consistently lower and ranged from 10.86 % to 28.59 %, while the values computed from the groundwater stage were moderate and ranged from 20.67 % to 37.23 % of the annual rainfall in the watersheds. The recharge values obtained from the groundwater stage appear very reasonable. However, the method is affected by the local relief of the monitoring station and it is necessary to monitor at as many stations as are possible to increase the reliability of the results. The recharge values obtained from the different methods show good correlation with one another. In the absence of groundwater stage data, recharge may be satisfactorily estimated as the mean of the values obtained from the baseflow recession analysis and the water balance methods.

INTRODUCTION. Numerous techniques for estimating groundwater recharge have been perfected in the last decade. Some of these methods use geochemical and radiometric data while others use hydrometeorological data including seasonal fluctuations in groundwater river stages and remote-sensing. More recently, recharge has also been estimated from groundwater flow patterns using inverse modelling techniques. Many of these methods continue to be elusive to water resources planners, developers and researchers in developing countries because of lack of sufficient funds to acquire the techniques for such sophisticated methods of re-



*I. Simmers (ed.), Estimation of Natural Groundwater Recharge, 435–447.*

charge analysis. For instance, in many developing countries, there are no laboratories that can analyse radiometric properties of water samples, and computer facilities are scarce. What is needed is to improve the quality of recharge estimates based on more conventional methods involving the routine monitoring of groundwater/river stages and other hydrometeorological factors. This paper discusses the results of a study to critically and comparatively evaluate the groundwater recharge into small watersheds of the Imo River Basin, Nigeria, using three independent but cheap methods. The methods include; groundwater recharge using a simplified equation that relates groundwater stage to actual recharge with the aid of aquifer porosity; recharge from baseflow recession analysis; and recharge from a water balance specifically prepared for the areas. The watersheds studied are illustrated in Figure 1.

## PHYSIOGRAPHIC AND GEOLOGIC FRAMEWORK OF THE WATERSHEDS

The Imo River Basin is divided into three relief areas of upper, middle and lower zones. The upper zone is bounded in the northeast and northwest by prominent cuestas, the crests of which rise to about 350 m above mean sea level and to about 200 m above the surrounding plains. Between the cuestas are undulating areas comprising dissected ridges and numerous isolated hills. At the foot of some of these ridges and hills are found springs and effluent seepages. The general slope is about 0.016 southwest and southeast. The middle zone is characterised by gently undulating ridges and lowlands. The m ajor relief features are the valleys of the Imo River and its tributories, the Otamiri and Oramirukwa Rivers - lying generally 40 to 60 meters below the surrounding nearly-flat landscape. The lower reaches of the Imo River Basin can be considered as part of the River Niger Delta with the typical features of a deltaic environment.

The basin lies mostly within the rain forest belt of Nigeria. However, the northern extremity comprising about 4 % of the basin and the coastal fringe constituting about 10 %, are within the Savannah and Mangrove belts respectively. Two distinct warm seasons occur : a dry season which lasts from November to March and a rainy season which extends from April to October. The dry season is often punctuated by a few scattered rains, especially in February and March. Rainfall is brought by the northward movement of the moist Equatorial Maritime Air Mass from the Gulf of Guinea with prevailing winds from the southwest. The total annual rainfall increases from 188 mm at Okigwi in the interior to about 3723 mm at Opobo on the coast.

The bedrock of the Imo River Basin consists of a 5480 m thick sequence of sedimentary rocks which range in age from Upper Cretaceous to Recent. The Nusukka, Imo and Ameki Formations, which comprise various proportions of sandstones and shales, underlie the upper middle region of dissected ridges and isolated hills of the upper Imo River Basin. The Ajali Sandstones, Ogwashi/Asaba and Basin Formations are predominantly sandy and are characterised by gently undulating plains in the northeast and south of the basin. There formations are covered to varied

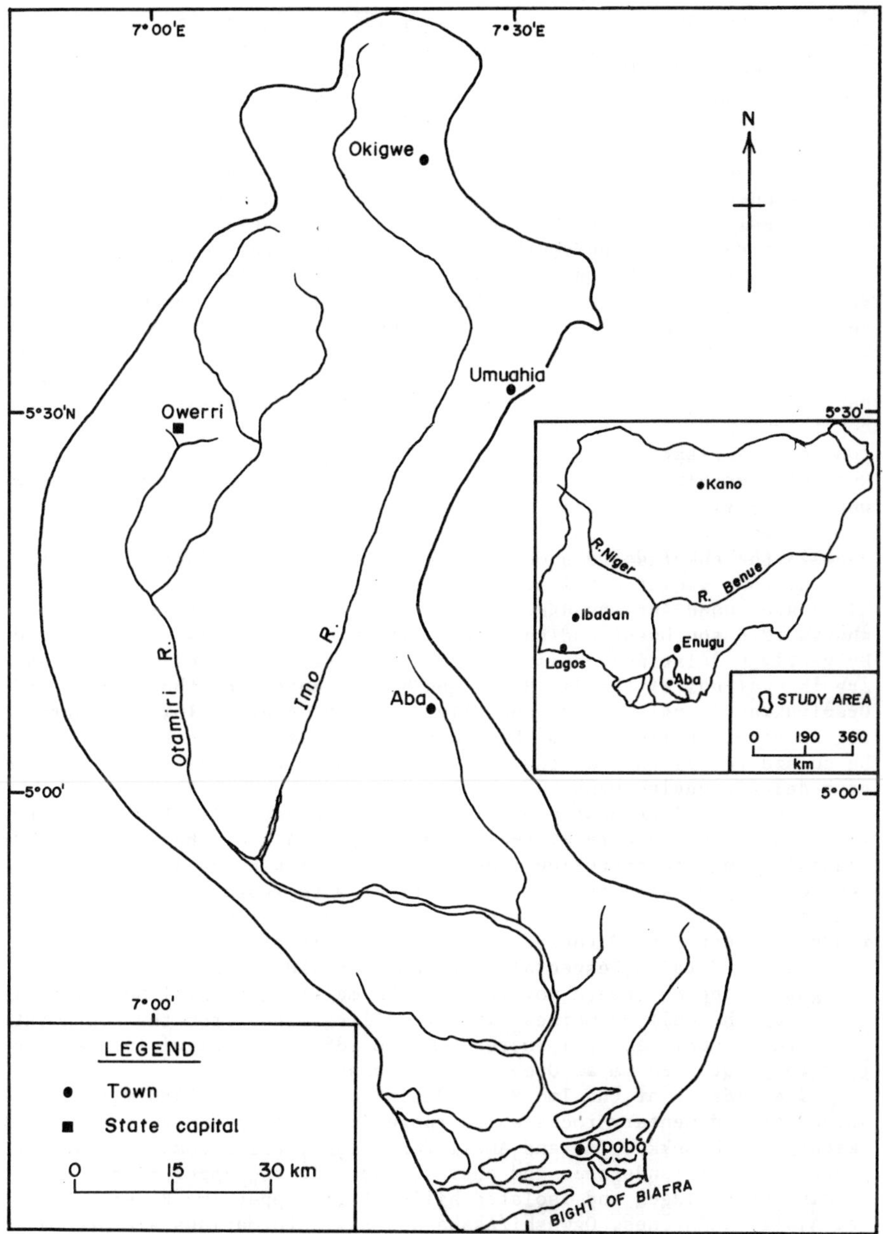


Figure 1 : Location map of the Imo River Basin showing major physiographic features; insert shows the location of the basin in Nigeria

depths by red acid sandy soils (Grove, 1951; and Uma, 1986).

## HYDROGEOLOGIC SETTING

The interlayering sandy and shaley units form extensive complex aquifer systems. Two groups of aquifer have been identified; unconfined and confined. One unconfined aquifer occurs as thin disjointed lenses which correspond with the broken ridges and isolated hills in the upper part of the basin. Thick and deep unconfined aquifers also occur at the northeastern margin and southern part of the basin overlain by thick sandy formations. The confined aquifer occurs in the upper middle part of the basin and comprises a thick sequence of interlayered sandstones, sandy clays and shales. Groundwater recharge into these aquifers is mainly through direct infiltration of rainwater. Although recharge occurs directly through outcropping areas of the unconfined aquifers, the confined aquifers are mostly recharged at the outcrop areas of their constituent sandstone units. Recharge by vertical leakage through the confining aquitards is probably insignificant, because the confining horizons contain very thick unfractured shales and clays with very low permeability.

## ANALYTICAL TECHNIQUES AND RESULTS

The amount of water recharging aquifers beneath the watersheds was evaluated by three methods. Thefirst was from direct monitoring of groundwater stage (level) in wells, and the conversion of the changes in water level to actual recharge; the second was by analysing the baseflow recession characteristics of the river systems; and the third by computing the water balance of the watersheds.

### Recharge Evaluation by Direct Monitoring of Groundwater Stage

The groundwater held in storage in any elemental volume of porous rock (aquifer) per cross-sectional area perpendicular to flow is given as :

$$R = nht \tag{1}$$

where n = porosity of the sandy aquifer

h = saturated thickness at an initial time t.

Groundwater recharge into the elemental volume is represented by an increase in the storage water and is indicated by a rise in the aquifer water level. The amount of recharge dR, after any given time $t_1$, can be calculated from the relation :

$$dR = n\,(h_{t_1} - h_{t_0}) \tag{2}$$

where $h_{t_1} - h_{t_0}$ is the increase in water level after $(t_1-t_o)$ time units. Based on equation 2, the total groundwater recharge $R_T$, during a recharge season may be expressed as

$$R_T = n \int_{t=1}^{t=m} \Delta h_t \tag{3}$$

where $t_1$ = to = time at the beginning of the recharge season
$t_m$ = time at the end of the recharge season
m = no. of groundwater level observations during the recharge season
h = saturated thickness of the aquifer
n = aquifer porosity

The volume of water that has recharged the aquifer during the entire recharge period is obtained by multiplying $R_T$ by the surface area of the aquifer.

Water level fluctuations in parts of the unconfined aquifer of the Imo River Basin are shown as an example in Fig. 2. Although these fluctuations may have in part been produced by other factors such as bank storage of nearby streams, pumping at nearby producing wells, evapotranspiration effects, etc. (Todd, 1980), the major factor is believed to be the recharge from rainfall. Field conditions and precautions taken during measurements ensured that even if these other factors were present, their effects were very minor. Figure 2 reveals that while significant rainfall starts around March and ends in November, the groundwater recharge season generally starts around late May to early July and ends in late October or early November. No porosity values have been experimentally established for the aquifers of the Imo River Basin, but a value of 0.3 (or 30 %) was used for the unconfined aquifers and is equated with that estimated experimentally for the sandy top soil (Enplan, 1975). Monthly and average annual recharge values in the watersheds were thus computed using a porosity of 30 %, groundwater stage from Fig. 2 and equation 3.

Table 1 shows that the maximum monthly recharge for 1985 in the upper Otamiri watershed ranges from 180 mm at Avu to 260 mm at Ogbaku, while minimum values vary between 10 mm and 40 mm at Avu and Ogbaku respectively. The 1985 average annual groundwater recharge in the upper Otamiri watershed was 655 mm. In the Lower Otamiri watershed, maximum monthly recharges vary between 170 and 240 mm, while minimum values ranged from 20 to 30 mm and the average annual recharge was 590 mm in 1984 and 600 in 1985. The lower Imo River watershed recorded maximum monthly recharges of 120 to 320 mm in 1984 and 180 to 290 mm in 1985. The corresponding minimum recharge ranged between zero to 20 mm in 1984 and 20 to 50 mm in 1985, while the average annual recharge varied from 547 mm in 1984 to 687 in 1985. Table one also indicates that the maximum monthly recharge occurs either in August or September while minimum values occur in November. The 1984 annual recharges were lower than the 1985 values in all watersheds monitored and this is probablye due to the difference in the volume of antecedent rainfall for those years. The average annual rainfall in the upper and lower Otamiri watersheds was 2135 mm in 1984 and 2424 mm in 1985. In the lower Imo River watershed, the corresponding values were 1934 mm in 1984 and 2224 in 1985. The variation in the monthly and annual recharge rates between the monitoring stations is probably related to the local relief, which controls the runoff component of the effective rainfall, and the natural stochastic variations in rainfall amount and intensity. This implies that the re-

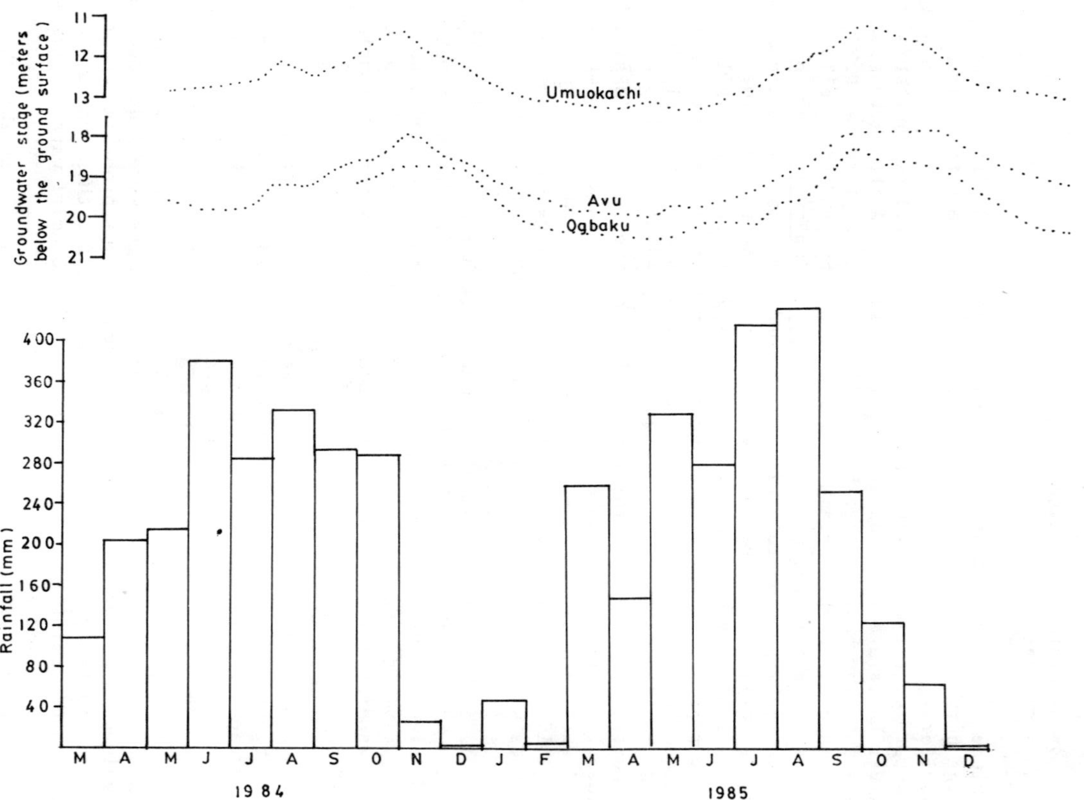


Figure 2 : Groundwater stages and average monthly rainfall in some parts of the Imo River Basin

| Watershed | Location | Year | J | F | M | A | M | J | J | A | S | O | N | D | Tot. |
|---|---|---|---|---|---|---|---|---|---|---|---|---|---|---|---|
| Upper | Avu | 1984 | - | - | - | - | - | - | 200 | 20 | 150 | 150 | 60 | - | 580 |
| Otamiri | Ogbaku | | | | | not | | recorded | | | | | | | |
| Lower | Umukire | | - | - | - | - | - | - | 140 | 140 | 150 | 230 | 20 | - | 680 |
| Otamiri | Umuapu | | | | | | | 60 | 120 | 60 | 170 | 150 | - | - | 440 |
| | Igwurita | | - | - | - | - | - | 80 | 90 | 120 | 180 | 180 | - | - | 650 |
| Lower | Ogwe | | - | - | - | - | - | - | 120 | 0 | 210 | 90 | 30 | - | 450 |
| Imo | Azumini | | - | - | - | - | - | - | 30 | 110 | 80 | 120 | 20 | - | 470 |
| River | Okeikpe | | - | - | - | - | 60 | 150 | 0 | 320 | 110 | 30 | 50 | - | 720 |
| Upper | Avu | 1985 | - | - | - | - | 60 | 80 | 130 | 180 | 160 | 10 | 10 | - | 630 |
| Otamiri | Ogbaku | | - | - | - | - | 100 | 40 | 140 | 140 | 260 | - | - | - | 680 |
| Lower | Umukire | | - | - | - | - | - | - | 120 | 200 | 140 | 80 | - | - | 540 |
| Otamiri | Umuapu | | - | - | - | - | - | 120 | 170 | 140 | 210 | - | - | - | 640 |
| | Igwurita | | - | - | - | - | 150 | 30 | 90 | 110 | 240 | 30 | 30 | - | 620 |
| Lower | Ogwe | | - | - | - | - | 80 | 140 | 60 | 60 | 290 | 80 | 20 | - | 730 |
| Imo | Azumini | | - | - | - | - | - | - | 140 | 180 | 60 | 90 | 90 | - | 560 |
| River | Okeikpe | | - | - | - | - | 100 | 50 | 160 | 80 | 290 | 90 | - | - | 770 |

\- recession begins or is in progress

Table 1 : Monthly and annual groundwater recharge (mm) in some watersheds in the Imo River Basin (based on water table data)

liability of the average recharge values computed for each watershed would depend on how closely the average relief and local rainfall of the monitored stations represent the average values for the entire watershed. Therefore the higher the number of monitoring stations in a watershed, the more reliable is the recharge estimate obtained.

Recharge from Baseflow Recession Analysis

The recession characteristics of river/stream hydrographs may be analysed using recession curves. The basic equation is (Barnes, 1939) :

$$Qr = Qo/\ 10^{t/k} \tag{4}$$

where Qo = discharge of the river at any given time
Qt = discharge after t time units
t = time interval
k = storage delay factor (or time required for the discharge to decrease by a factor of 10, or one log cycle).

The total potential discharge Qtp, which is also the total groundwater stored in a basin at the end of a recharge season (defined as the total volume of groundwater that would be discharged during an entire recession if complete depletion were to take place uninterruptedly) is calculated by integrating equation 4 from times zero to infinity. Domenico (1972) gave the final equation as

$$Qtp = Q_o K/2.30 \tag{5}$$

A part (Qf) of the Qtp is discharged from the flow system during the following recession, while the remainder (Qr) is stored in the groundwater flow system at the end of the recession. That is,

$$Qtp = Qf + Qr \tag{6}$$

The difference between the Qtp of any flow system at the beginning of recession, or end of the recharge season and the Qr left at the end of the previous year recession, is the groundwater recharge into that flow system for the season considered. For example, if the Qtp for a certain year is $Q_2$ and the remaining potential at the end of the preceeding year's recession is $Q_1$, the groundwater recharge for that year is $(Q_2-Q_1)$. Qf is calculated by integrating equation 4 over a time period equal to the period of recession and Qr is calculated from equation 6. More details of the method are given by Domenico (1972), Ralston and Williams (1979), and Hall (1978).

The $Q_o$ and K of the watersheds are given by Uma (1986). Using these values and equations 4 to 6, the groundwater recharges into the watersheds were calculated for the 1979/1980 hydrologic years. The 1978/79 and 1979/80 hydrologic years were the only periods with consistent hydrograph records throughout the Imo Basin and with this data only the groundwater recharge for 1979 could be calculated. Table 2 shows the results. Recharge into watersheds of the Upper Imo River Basin is seen to

| Watershed | Stream Gauging station | Recharge (mm) | Recharge as percent of rainfall |
|---|---|---|---|
| Upper Imo | Ndimoko | 206.37 | 10.86 |
| Upper Imo | Umuna | 224.16 | 11.80 |
| Middle Imo | Umuokpara | 259.35 | 13.30 |
| Upper Otamiri | Nekede | 424.20 | 17.39 |
| Oramirukwa | Olakwo | 146.30 | 6.65 |
| Lower Otamiri | Chokocho | 441.03 | 17.64 |
| Lower Imo | Obigbo | 743.43 | 28.59 |

Table 2 : groundwater recharge from baseflow recession analysis

range from 10.9 % of annual rainfall at Ndimoko to 13.3 % at Umuokpara. Values for the lower Imo River Basin are generally higher, with a range of 17.4 % of annual rainfall at upper Otamiri watershed at Nekede to 28.6 % for the lower Imo River watershed at Obigbo. The recharge value of 6.7 % of rainfall for the Oramirukwa watershed at Olakwo is rather low and incompatible with the other estimates. The Oramirukwa river (in the area considered) has been shown by Uma (1986) to be losing water along much of its course and its baseflow recession characteristics may not truly reflect groundwater conditions in the wateshed.

Recharge from Water Balance Analysis

A water budget for the watersheds was prepared by accounting for the various water inputs and outputs. The principal input is mainly from rainfall, the only additional possible source being inflow from adjacent watersheds. However, available hydrologic and hydrogeologic records (Balasha-Jalon, 1980; Uma, 1984; Uma and Egboka, 1986; and Uma 1986) indicate an absence of water exchange between the main Imo River Basin and adjacent basins. Based on an empirically determined evapotranspiration value for the watersheds, Balasha-Jalon (1980) gave a functional relationship between effective rainfall (total rainfall-evapotranspiration) and total rainfall of the form

$$\text{Log } Re = 2.025 \log R_t - 3.697 \tag{7}$$

where Re = effective rainfall (mm)
Rt = total rainfall (mm)

The groundwater recharge for each watershed was thus determined as the difference between the effective rainfall derived from equation 7 and surface runoff from each watershed, determined from hydrograph separation of the streamflow records. The results are given in Table 3. Equation 7 is generalised in the sense that it assumes an average evapotranspiration for all watersheds, irrespective of the varied land use and local climatic conditions.

| Watershed | Stream Gauging Location | Total Rainfall (mm) | Effective Rainfall (mm) | Runoff (mm) | Groundwater recharge (mm) | Groundwater recharge % total rainfall |
|---|---|---|---|---|---|---|
| Upper Imo at Ndimoko | Ndimoko | 1880 | 857.37 | 603.2 | 254.17 | 13.52 |
| Upper Imo at Umuna | Umuna | 1880 | 857.37 | 590.6 | 266.77 | 14.19 |
| Middle Imo at Umuokpara | Umuokpara | 2196 | 1174.37 | 785.66 | 388.71 | 17.70 |
| Upper Otamiri at Nekede | Nekede | 2135 | 1109 | 272.42 | 836.58 | 39.18 |
| Oramirakwa at Olakwo | Olakwo | 2135 | 1109 | 463.53 | 645.47 | 30.23 |
| Lower Otamiri at Chokocho | Chokocho | 2424 | 1434 | 386.96 | 1047.04 | 43.19 |
| Lower Imo at Obigbo | Obigbo | 2500 | 1527 | 520.99 | 1006.0 | 40.24 |

Table 3 : watershed recharge computed from water balance analysis

It also does not consider inter-basin groundwater exchange which should occur since the watersheds are hydraulically connected. The recharge values given in Table 3 are thus somewhat generalised. The table shows that recharge varies between 13.5 % and 17.7 % of average annual rainfall for the upper Imo River Basin watersheds and 30.2 % to 43.2 % for the lower areas.

## DISCUSSION AND CONCLUSIONS

Table 4 compares the recharge values computed from the three analytical techniques. Values obtained from the water balance method are generally high while the corresponding values from the baseflow recession analysis are relatively lower. Those obtained from the groundwater stage data fall in general between the other two methods. The high recharge values from the water balance method may be related to over generalization of the

| Watershed | Groundwater recharge from water level monitoring | Groundwater recharge from recession analysis | Groundwater recharge from water balance | Remarks |
|---|---|---|---|---|
| Upper Imo River at Ndimoko | - | 10.86 | 13.52 | slight deviation |
| Upper Imo River at Umuna | - | 11.80 | 14.19 | " |
| Upper Imo River at Umuokpara | - | 13.30 | 17.70 | " |
| Upper Otamiri at Nekede | 27.42 | 17.39 | 39.18 | Significant Variation |
| Oramirukwa at Olakwo | - | 6.65 | 30.23 | " |
| Lower Otamiri at Chokocho | 27.26 | 17.64 | 43.19 | " |
| Lower Imo River at Obigbo | 29.57 | 28.59 | 40.24 | slight deviation |

- not recorded

Table 4 : comparison of recharge values from the three methods

hydrometeorological data used. It is possible that therun off component and/or evapotranspiration effects have been underestimated. The streamflow data used in the calculation of runoff may not have accounted for all the runoff from each watershed. In addition, the water balance method does not account for groundwater exchange between adjacent watersheds, which would certainly affect the input/output relationships.

The low recharge values obtained from baseflow recession analysis, relative to the other methods, is believed to arise from alternative and unaccounted for groundwater outlets. For example, some of the recharged water is probablydischarged as deep-flow out of the watersheds and hence not recorded as baseflow. Uma (1986) has already shown that deep groundwater flow away from the watersheds is significant. Groundwater recharge based purely on the baseflow component of such watersheds is thus bound to be lower than the net natural recharge. Rechargevalues obtained from the groundwater stage data appear very reasonable, especially when compared with infiltration rates of 30 % to 38 % of annual rainfall determined experimentally for surface soils (Enplan, 1975) of the lower

Imo River Basin.

The three methods gave recharge values which are statistically quite closely related. Those obtained from the water balance and recession analysis have a correlation coefficient of 0.766, whilethose from the groundwater stage and recession analysis have a correlation of 0.924. However, the groundwater stage and water balance methods have a weak negative correlation of -0.226. It was also observed that the mean of recharge values from the water balance and the recession analysis are approximately equal to that obtained from the more reliable groundwater stage method; correlation between the values is 0.90. This observation implies that in the absence of groundwater stage data 'representative' recharge may be estimated from the average of values obtained by the other two methods.

The following conclusions may thus be drawn :

(1) annual groundwater recharge varies from 10.9 to 17.7 % of annual rainfall for watersheds in the upper region of the Imo River Basin (Umuokpara and north), and from 17.4 to 43.2 % of annual rainfall in those of the lower region;
(2) the recession analysis method for estimating recharge gives minimum values, while the water balance method gives maximum values:
(3) results from groundwater stage data give very useful estimates of the net groundwater status but it should be measured at as many locations as possible to increase the reliability of the values obtained;
(4) in the absence of groundwater stage data, recharge may be estimated satisfactorily from the mean of recharge values computed from the baseflow recession analysis and water balance techniques.

ACKNOWLEDGEMENTS

Professor E.P. Leohnert of Munster University, W. Germany, Dr. K.M. Onuoha of University of Nigeria and Dr. C.O. Okagbue also of University of Nigeria made useful contributions to the paper. The Enplan Group of Consulting Engineers funded the field work. These contributions are gratefully acknowledged.

REFERENCES

Balasha - Jalon, 1980. 'Imo state rural water supply scheme'; first planning report. Technical report prepared for the Imo State Ministry of Public Utilities.

Barnes, B.S., 1939. 'The structure of discharge recession curves'. Trans. Amer. Geophys. Union, v. 4, p. 721-725.

Domenico, P.A., 1972. Concepts and models in groundwater hydrology. McGraw-Hill, New York.

Enplan Group, 1975. Imo River Basin - pre-feasibility study. Technical report prepared for the Federal Ministry of Agriculture, Lagos, Nigeria.

Grove, A.T., 1951. 'Landuse and soil conservation in parts of Onitsha and Owerri Provinces'. Geol. Surv. Nig., Bull. no. 21, p. 25-58.

Hall, F.R., 1968. 'Baseflow recessions - a review'. Water Resour. Research, v. 4, no. 5, p. 973-983.

Ralston, D.R. and R.E. Williams, 1979. 'Groundwater flow systems in the western phosphate field in Idaho. In : W. Back and D.A. Stephenson (Guest Editors), Contemporary Hydrogeology - The George Burke Maxey Memorial volume. J. Hydrol., v. 43, p. 239-264.

Uma, K.O., 1984. Water resources potentials of Owerri and its environs. Unpubl. M.Sc. Thesis, University of Nigeria, Nsukka, Nigeria.

Uma, K.O., 1986. Analysis of transmissivity and hydraulic conductivity of sandy aquifers of the Imo River Basin, Nigeria. Unpubl. PhD Thesis, University of Nigeria, Nsukka, Nigeria.

Uma, K.O. and B.C.E. Egboka, 1986. 'Groundwater potentials of Owerri and its environs., Nig. J. Min. Geol. (in press).

# QUANTITATIVE ESTIMATION OF GROUND-WATER RECHARGE IN DOLOMITE

D.B. Bredenkamp
Directorate Geohydrology
Department of Water Affairs
Private Bag X313
Pretoria
South Africa

ABSTRACT. Quantitative estimation of annual recharge in the Bo Molopo dolomitic region (Western Transvaal) by means of the following equation, is demonstrated with a great measure of success:

$$RE(I) = A(RF(I) - B)$$

where RE(I) is the annual recharge and RF(I) the annual precipitation. B represents the threshold rainfall that is required to effect recharge and A is a lumped catchment parameter. Values for A (ranging from 0,28 to 0,35) and for B (ranging from 310 to 360) were obtained using reference recharge values that were determined

- from an interpretation of a sinkhole hydrograph
- from water balance calculations.

In checking the validity of the equation to estimate the average annual recharge $\overline{RE}$, using the average annual rainfall in different dolomitic areas, the following relationship provided excellent results (correlation coefficient of 0,989):

$$\overline{RE} = 0{,}30\ (\overline{RF} - 313)$$

where $\overline{RF}$ is the average annual rainfall.

The latter equation is regarded as the general relationship by which both annual and average recharge could be estimated for dolomitic regions in the summer rainfall areas of South Africa.

## 1. INTRODUCTION

Since 1963 the Bo Molopo dolomite region (Fig. 1) has been studied with the object of estimating ground-water recharge and to determine the exploitation potential of different dolomite compartments.



*I. Simmers (ed.), Estimation of Natural Groundwater Recharge, 449–460.*

The Grootfontein compartment, recently supplemented by the Molopo eye is the main water supply to Mafikeng/Mmabatho. Several methods were employed to estimate its ground-water recharge e.g.

- water balances
- use of natural isotopes and tritium profiles
- aquifer modelling
- rainfall-recharge equations
- chloride ratio of rainfall to that of ground-water

This paper deals mainly with the application of a simple recharge equation which provides an estimate of the average ground-water recharge and annual variability with great success as is demonstrated by comparing results with estimates derived by other methods. Its use in the simulation of the discharge of dolomitic springs is also illustrated.

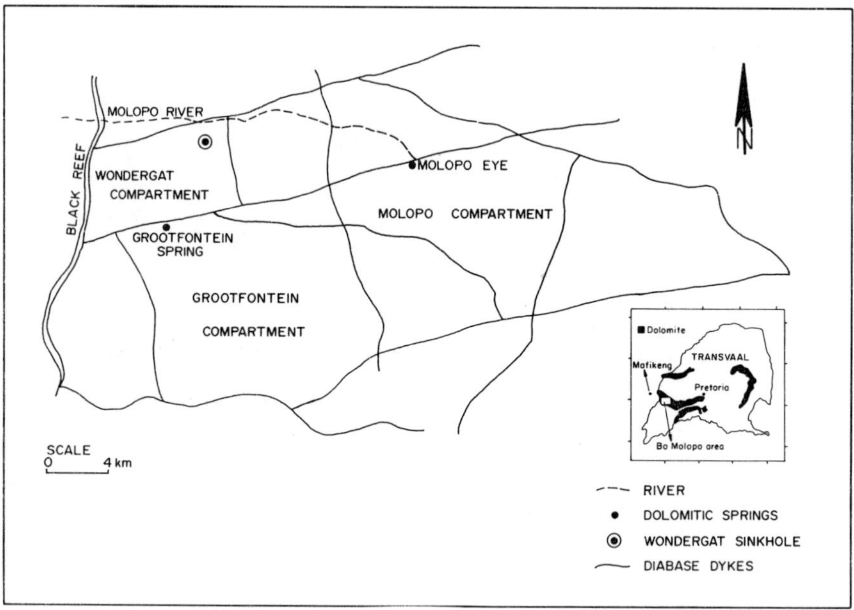


Figure 1. Map showing some of the Bo Molopo dolomite compartments.

## 2. RAINFALL RECHARGE EQUATION

### 2.1 Development

The development of a recharge equation was initially attempted using daily rainfall and climatic data (Bredenkamp 1974) but was superseded by models based on monthly input data (Schutte 1975, Bredenkamp 1978) with a substantial improvement in the simulation.

This prompted annual models to be tested yielding equally good and often better results.

The culmination of the evaluation of a number of different rainfall recharge equations was the following simple relationship:

$$RE(I) = A(RF(I) - B) - SMD \qquad (1)$$

where RE(I) denotes the recharge for year I; RF(I) is the rainfall for year I; A,B are parameters to be optimized and SMD represents accumulated soil moisture deficit with a maximum value of C.

B represents the threshold rainfall below which no recharge would take place. A is the coefficient governing the recharge quantity whilst B also determines the variability. Any negative outcome of equation 1, i.e. rainfall below B, is regarded as soil moisture deficit, but this is only allowed to accumulate to a set maximum value C. The optimisation however, showed that for most instances C = 0 implying that soil moisture deficit can be disregarded in the dolomite if annual rainfall values are used to simulate recharge.

The monthly version of equation 1 was also tested but was slightly less successful to simulate annual recharge values (c.f. Sect. 2.3).

## 2.2 Reference recharge values

The ground-water hydrograph of the Wondergat sinkhole (Fig. 2) was used to reconstruct annual values of recharge as an effective rise of the water table, according to the method proposed by Bredenkamp (1974,1978).

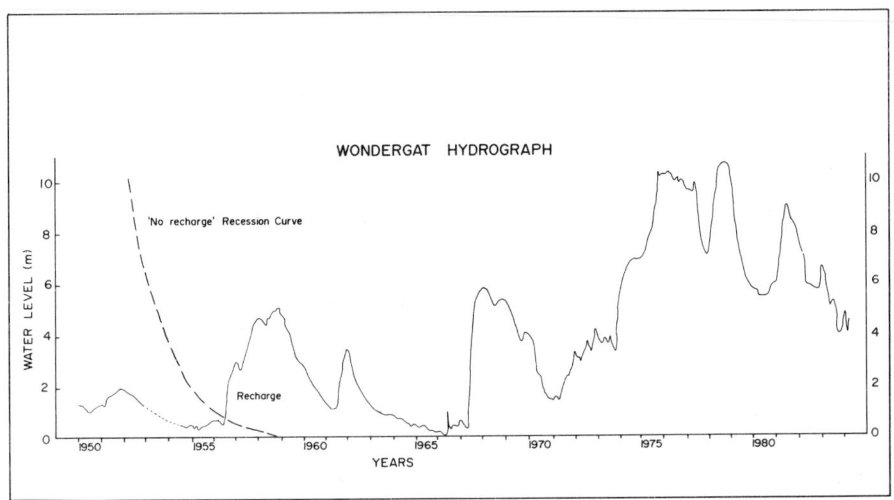


Figure 2. Hydrograph of the Wondergat sinkhole which was used to reconstruct annual recharge.

By incorporating the "no recharge" recession, the difference between the actual water level and the level to which it would have declined, had no recharge taken place, was assumed to present the recharge. The "no recharge" response of the Wondergat was composed of the 1960/1 and 1979/80 recession limbs of the hydrographs. Although monthly recharge equivalents were obtained they were combined to yield annual recharge values. This was done to overcome the rainfall recharge lag and because annual estimates of recharge are adequate for most quantitative ground-water assessments.

The Wondergat annual recharge equivalents are shown in Fig. 3. These formed the reference values to calibrate the different models and assess their ability to achieve effective simulation. The difference between the reference and simulated values were squared and summed (F value in Table I) to compare the performance of models. The correlation coefficient reflects the degree of correspondence attained.

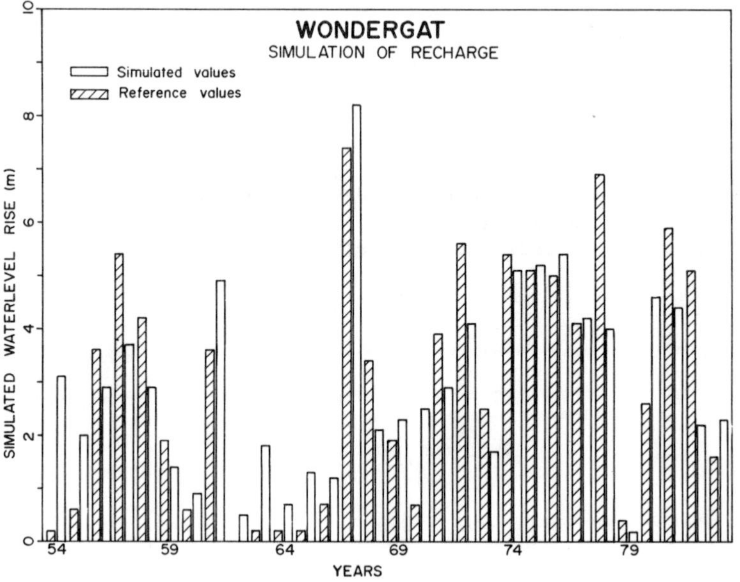


Figure 3. Recharge values derived from annual rainfall and reference recharge values reconstructed as a water level rise.

## 2.3 Optimisation

Although convergence of the F-value was sensitive to variation of B in equation 1, it was hardly affected by adjustment of A. A unique solution of A could only be achieved if either porosity of the aquifer is known or deriving the average recharge by an independent method.

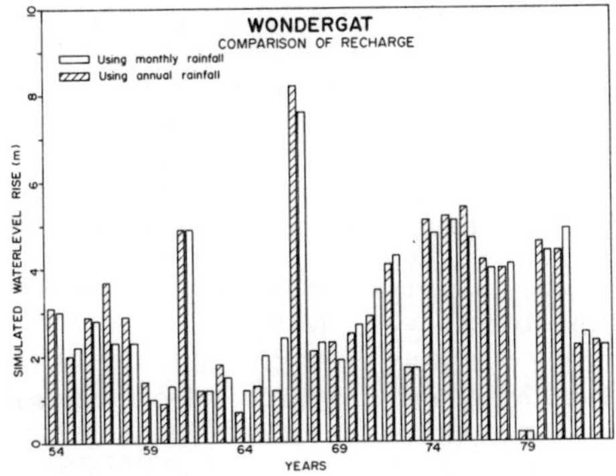


Figure 4. Comparison of recharge values obtained using annual and monthly rainfall in equation 2.

A finite element model study of the Grootfontein aquifer indicated a porosity of about 0,028. Using this porosity value A and B were found by optimisation to be 0,35 and 360 respectively (Table I). However values for A ranging from 0,32 to 0,36 and B from 340 to 360 also produced acceptable simulations.

TABLE I: CONVERGENCE OF F FOR DIFFERENT VALUES OF A, B AND C IN RECHARGE FORMULA - EQUATION 1

| A | B | C | Porosity | Simulated recharge (mm) | Residuals squared F | Correlation Coefficient |
|---|---|---|---|---|---|---|
| Annual version | | | | | | |
| 0,30 | 320,0 | 0 | 0,028 | 83,0 | 58,58 | 0,795 |
| 0,30 | 340,0 | 0 | 0,028 | 77,0 | 59,90 | 0,795 |
| 0,30 | 360,0 | 0 | 0,028 | 71,0 | 63,98 | 0,795 |
| 0,30 | 380,0 | 0 | 0,028 | 65,1 | 70,74 | 0,795 |
| 0,30 | 400,0 | 0 | 0,028 | 59,3 | 80,07 | 0,796 |
| 0,35 | 320,0 | 0 | 0,028 | 96,8 | 63,65 | 0,795 |
| 0,35 | 340,0 | 0 | 0,028 | 89,8 | 58,05 | 0,795 |
| 0,35 | 360,0 | 0 | 0,028 | 82,8 | 56,20 | 0,795 |
| 0,35 | 380,0 | 0 | 0,028 | 75,9 | 58,01 | 0,795 |
| 0,35 | 400,0 | 0 | 0,028 | 69,2 | 63,35 | 0,796 |
| 0,40 | 320,0 | 0 | 0,028 | 110,7 | 87,19 | 0,795 |
| 0,40 | 340,0 | 0 | 0,028 | 102,7 | 72,62 | 0,795 |
| 0,40 | 360,0 | 0 | 0,028 | 94,7 | 62,96 | 0,795 |
| 0,40 | 380,0 | 0 | 0,028 | 86,8 | 58,09 | 0,795 |
| 0,40 | 400,0 | 0 | 0,028 | 79,0 | 57,84 | 0,796 |
| 0,35 | 360,0 | 10 | 0,028 | 82,8 | 56,20 | 0,795 |
| 0,35 | 360,0 | 20 | 0,028 | 82,8 | 56,20 | 0,795 |
| Monthly version | | | | | | |
| 0,40 | 60,0 | 0 | 0,028 | 81,7 | 65,2 | 0,794 |

For the monthly version of equation 1, the values for A = 0,40 and B = 60 were obtained by optimisation. As is shown in Table I (F-value) the monthly model is almost as good as the annual version. The differences between simulated recharge values for monthly and annual simulations were however surprisingly small as is indicated in Fig. 4, whilst Fig. 3 shows the comparison between simulated and reference recharge values using annual rainfall as input.

## 3. MODEL VERIFICATION

Substituting the values of A and B obtained by optimisation into equation 1 gives:

$$RE(I) = 0{,}35\ (RF(I) - 360) \qquad (2)$$

Validation of this rainfall recharge relationship and its general application were sought by verifying results with recharge estimates obtained by other methods.

### 3.1 Water balance calculations

The boundaries of the Grootfontein compartment (165 $km^2$) are well delineated, and influent and effluent leakage can be assumed equal. This allows ground-water recharge to be estimated using a simple water balance over two periods (see Table II).

TABLE II: RECHARGE USING A SIMPLIFIED WATER BALANCE

| Period | Pumpage $m^3.10^6$ | Change in storage 1) $m^3.10^6$ | Rainfall | Calculated recharge |
|---|---|---|---|---|
| Sep. 80–Aug. 82 | 27,0 | 0 | 735+467 | 164 |
| Sep. 82–Aug. 83 | 13,5 | 6,26 | 480 | 48 |

1) assuming aquifer porosity 0,028

Using these recharge and rainfall values to solve for A and B indicate that A = 0,285 and B = 311.

The response of the Grootfontein aquifer has been successfully simulated by a finite element model (FEM) (Van Rensburg, 1985) and although further refinement is to be carried out the following recharge values were obtained from a water balance incorporating all grid elements (Table III).

TABLE III: COMPARISON OF RECHARGE

| Hydrological year | Rainfall mm | Recharge 1) using FEM mm | Recharge using equation 2 mm |
|---|---|---|---|
| 1980/81 | 735 | 114 | 131 |
| 1981/82 | 467 | 42,5 | 37 |
| 1982/83 | 480 | 36 | 42 |
| 1983/84 | 314 | -2,3 | -16 |

1) assuming aquifer porosity 0,028

Solving for A and B in equation 1 using the recharge values in column 3 yields A = 0,277 and B = 323, which is in close agreement with the previous estimates of these parameters.

The different sets of values obtained for A and B are not very critical as is indicated by the following comparison of recharge estimates (Table IV). There is good correspondence, except for rainfall values close to the threshold value B.

TABLE IV: COMPARISON OF RECHARGE ESTIMATES USING EQUATION 1

| Rainfall mm | Recharge (mm) (A = 0,35; B = 360) | Recharge (mm) (A = 0,285; B = 310) |
|---|---|---|
| 400 | 14 | 25,6 |
| 500 | 49 | 54 |
| 600 | 84 | 82,6 |
| 700 | 119 | 111 |

As is evident from Figure 3 there is still some degree of variability in recharge not accounted for by equation 2. This is not surprising if one considers variability of rainfall intensity, duration and spatial distribution, and the heterogeneity of the recharge area.

Attempts to obtain a better simulation by weighting A and B with climatic indices, such as the ratio of actual evaporation to average evaporation and days of rainfall, did not effect a significant improvement.

### 3.2 Reconstruction of the flow of dolomite springs

The simulation of the flow of two springs is included as another application of the recharge equation to obtain the recharge which controls the spring discharge. The method employed is that proposed to reconstruct the flow of the springs at Pretoria Fountains (Bredenkamp et al, 1985). This entails calculation of the average effective area in which recharge is responsible to sustain spring flow:

$$\bar{Q} = \overline{RE} \times \text{Area} \tag{3}$$

where $\overline{RE}$ is the average recharge and $\bar{Q}$ the average flow of the spring.

By using this calculated average area, the annual recharge values estimated from equation 2 could be converted to annual flow rates. These were distributed incorporating carry-over of flow from the preceding years in the following way:

$$Q(I) = 0{,}333(2Q(I-1) + 0{,}3QR(I) + 0{,}4QR(I-1) + 0{,}3QR(I-2))$$

where Q(I) is the flow for year I and QR(I) is the recharge converted to an annual flow rate by means of equation 3.

Fig. 5 shows the measured and simulated flow rates for the Maloney's eye which is a prominent dolomitic spring west of Krugersdorp. Although 1:1 correspondence is not obtained, fluctuations are synchronous for most of the record. The simulated flow during the high rainfall period of 1976-78 is significantly lower than the measured values, possibly indicating an addition of ground-water from adjacent compartments during years of excessive rainfall.

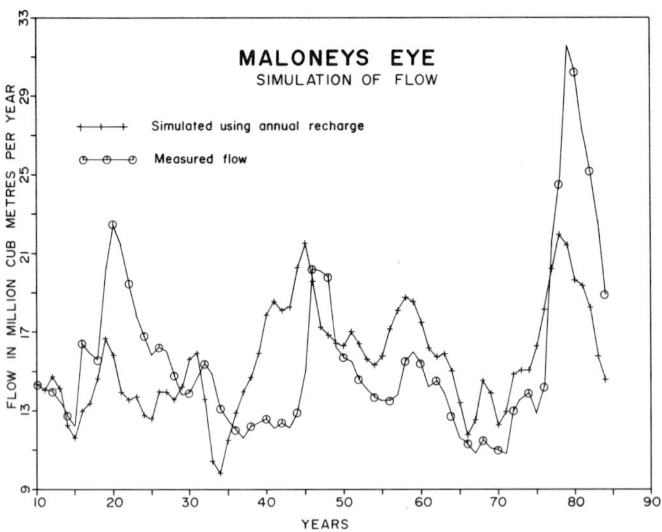


Figure 5. Comparison between the reconstructed flow of Maloney's Eye spring and the measured flow.

Reconstruction of the flow of the Molopo eye was urgently required for the design of a pipeline to convey the water to Mafikeng/Mmabatho. Existing flow measurements were unreliable and spring flow had to be inferred by correlation with the water level of the Wondergat. The other reconstruction of the spring flow using equation 2 and 3 is shown in comparison with the first values (Fig. 6). A high degree of correspondence is evident.

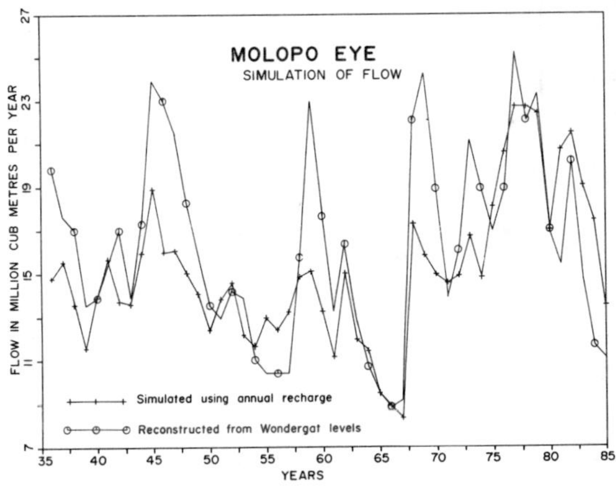


Figure 6. Comparison between the reconstructed flow of Molopo Eye spring and the values inferred by correlation with the Wondergat water level.

## 4. GENERAL RAINFALL-RECHARGE EQUATION

Quantitative estimates of the average annual recharge for different dolomite regions were used to verify the use of equation 2 in estimating average annual recharge by means of the average annual rainfall for a particular area. The comparison of these estimates indicated in Table V reflects a high degree of correspondence.

However when the recharge estimates, used as control values (column 4 in Table V) are plotted against average annual precipitation (Fig. 7) a linear relationship with slightly different values for A and B is obtained:

$$\overline{RE} = 0{,}30\ (\overline{RF} - 313) \qquad (4)$$

where $\overline{RE}$ is the average annual recharge and $\overline{RF}$ denotes the average annual rainfall.

TABLE V: COMPARISON OF RECHARGE FOR DIFFERENT AREAS IN RELATION TO ESTIMATES USING EQUATION 2

| Locality | Catchment area $km^2$ | Long-term Av. Rainfall mm | Recharge (Eq. 2) mm | Recharge using other methods (mm) | Methods used for recharge estimation (Bredenkamp et-al 1987) |
|---|---|---|---|---|---|
| 1. Zuurbekom | 130 | 672 | 109 | 104 | Ground-water balance and chemical balance |
| 2. Gemsbokfontein | 64 | 665 | 107 | 86 | Water balance over two periods |
| 3. Bank | 154 | 640 | 98 | 109 | Hill method; Average flow of spring; Inferred from spring flow |
| 4. Steenkoppies | 177 | 630 | 94,5 | 103 | Average flow of spring; Darcy equation |
| 5. Holfontein | 75 | 630 | 94,5 | 93,5 | Water balance |
| 6. Kliprivier | 470 | 700 | 119 | 134 | Based on 2 mm recharge per rainfall event |
| 7. Grootfontein | 165 | 560 | 71 | 65 | Water balance |
| 8. Grootfontein | 165 | 546 | 65 | 69,5 | Water balance finite element model |
| 9. Molopo eye | - | 590 | 77 | 87 | Reconstructed flow of spring |
| 10. Pretoria/Rietvlei | 30 | 682 | 113 | 104 | Water balance |
| 11. Sishen | - | 390 | 11 | 31 | Finite element simulation |
| 12. Sabie | 33,4 | 1 250 | 311 | 287 | Based on flow from drainage tunnel |
| 13. Pretoria Fountains | 30 | 675 | 110 | 108 | Average flow of Fountains-east springs |
| 14. Kuruman | - | 346 | 0 | 11 | Average flow of Kuruman eye |

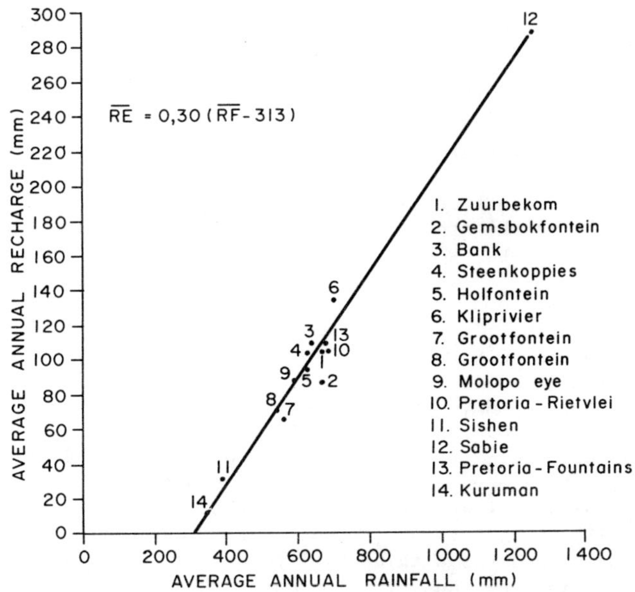


Figure 7. Comparison of average annual recharge estimates for different regions with annual precipitation.

The high correlation coefficient (r = 0,989) reflects excellent agreement between the reconstructed and reference recharge values. This provides a great deal of support for equation 4 as being the general rainfall-recharge relationship that applies to all dolomite areas in the summer rainfall regions of South Africa.

Incidently the different estimates for A and B in the Bo Molopo area (c.f. Sect. 3.1) yield an average of 0,304 for A, which is identical to the value in equation 4, and 331 for B which is only slightly higher than that obtained in equation 4.

## 5. CONCLUSIONS AND RECOMMENDATIONS

Reliable quantitative estimation of recharge in dolomite regions can be obtained using the simple equation based only on annual rainfall as input variable. Average annual recharge can be obtained by means of the average annual rainfall and annual variability of recharge by use of annual rainfall totals.

A similar equation, only using monthly rainfall as input also gives satisfactory simulation of recharge but was not as good and its advantage over the annual version is merely that monthly recharge could be inferred.

As was demonstrated the estimated annual recharge values can be used in reconstructing the flow of dolomitic springs and can also be applied to determine the size of the catchment area from the average spring discharge.

Variable recharge obtained from equation 1 can be used as input to aquifer models in order to improve the simulation of an aquifer.

Ascertaining the values of A and B in equation 1 will be pursued for areas with different climatic regimes. Application of equation 1 is also being tested for aquifers other than dolomitic systems and preliminary results are equally promising (Bredenkamp, 1987).

It seems more than likely that a relationship, similar to the one given by equation 4 could also be found for dolomitic areas in other regions of the world.

## 5. REFERENCES

BREDENKAMP, D.B. and VOGEL, J.C., 1970. 'Study of a dolomitic aquifer with Carbon-14 and Tritium'. Proc. Symp. Vienna, Isotope Hydrology 1970, IAEA, Vienna, P.349.

BREDENKAMP, D.B., SCHUTTE, J.M. and DU TOIT, G.J., 1974. 'Recharge of a dolomitic aquifer as determined from Tritium profiles'. Proc. Symp. Vienna (1974), 'Isotope Techniques in Ground-water Hydrology' IAEA, Vienna, P73.

BREDENKAMP. D.B., 1978, 'Quantitative estimation of ground-water recharge with special reference to the use of natural radioactive isotopes and hydrological simulation'. Technical report No. 77, Department of Water Affairs, South Africa.

BREDENKAMP, D.B., 1981. 'Kwantitatiewe ramings van grondwateraanvulling volgens eenvoudige hidrologiese modelle'. Trans. geol. Soc. S.A., 84 (1981), 153-160.

BREDENKAMP, D.B., FOSTER, M.B.J. and WIEGMANS, F.E. 1985. 'Dolomitic aquifer southeast of Pretoria as an emergency ground-water supply'. Proc. 17th Int. Congr. IAH, 1985, Tucson P. 761.

BREDENKAMP, D.B., 1987. 'Quantitative estimation of ground-water recharge in the Pretoria-Rietondale area'. Technical report no. 3508, Department of Water Affairs, South Africa.

BREDENKAMP, D.B. et al, 1987. 'Case studies of quantitative estimates of ground water recharge'. Technical report in preparation, Department of Water Affairs, South Africa.

KAFRI, U., et al., 1986. 'The hydrogeology of the dolomite aquifer in the Klip River-Natalspruit basin'. Tech. report 3408, Directorate Geohydrology, Department of Water Affairs, Pretoria, South Africa.

LYNCH, S.D. et al., 1984, 'Ground-water management at the Sishen mine by finite element modelling'. Proc. Int. Conf. on Ground-water Technology, Johannesburg, South Africa.

VAN RENSBURG, H.J., 1985, ''n Ondersoek na die benutting van grondwater in die Grootfontein kompartement (Wes Transvaal)'. unpublished M Sc Dissertation, UOVS, Bloemfontein, South Africa.

QUANTITATIVE ESTIMATION OF GROUND-WATER RECHARGE IN THE PRETORIA-RIETONDALE AREA

D.B. Bredenkamp
Directorate Geohydrology
Department of Water Affairs
Private Bag X313
Pretoria
South Africa

ABSTRACT. Annual values of ground water recharge expressed as an effective water-level rise were derived from an interpretation of the hydrographs of three monitoring boreholes. Independent estimates of the average annual recharge of the area were derived by means of the Darcy equation and by interpretations of natural tritium profiles in the soil overburden.

The effective porosity of the aquifer was calculated from the average annual recharge (66 mm) and the average rise of the water level derived from the hydrographs. This effective porosity (0,0093) was used to convert the annual equivalent water-level rises to a depth of precipitation. The latter were plotted against annual rainfall and indicates a linear relationship with annual rainfall in excess of a threshold value i.e.

$$RE(I) = A\ (RF(I) - B)$$

where RE(I) is the recharge and RF(I) the rainfall for year I. B is the threshold rainfall with an average value of 395 mm and the coefficient A is a lumped catchment parameter indicating the fraction of the excess annual rainfall that represents recharge. For the Rietondale area A is about 0,20 which implies that 20% of the excess rainfall constitutes recharge. The equation is similar to that obtained in the Bo Molopo dolomite area but in that case A is 0,30 and B is 313 mm.

## 1. INTRODUCTION

Quantitative estimation of recharge in the Bo Molopo area (Bredenkamp, 1986) has revealed that annual recharge could be inferred from annual rainfall by means of the following equation:

$$RE(I) = A\ (RF(I) - B) \qquad (1)$$

where RE(I) and RF(I) denote the recharge and rainfall for year I; A indicates the fraction of the rainfall in excess of the threshold value B that represents ground water recharge. The values of A and B in this case proved to be about 0,30 and 313 mm respectively.



*I. Simmers (ed.), Estimation of Natural Groundwater Recharge, 461–476.*

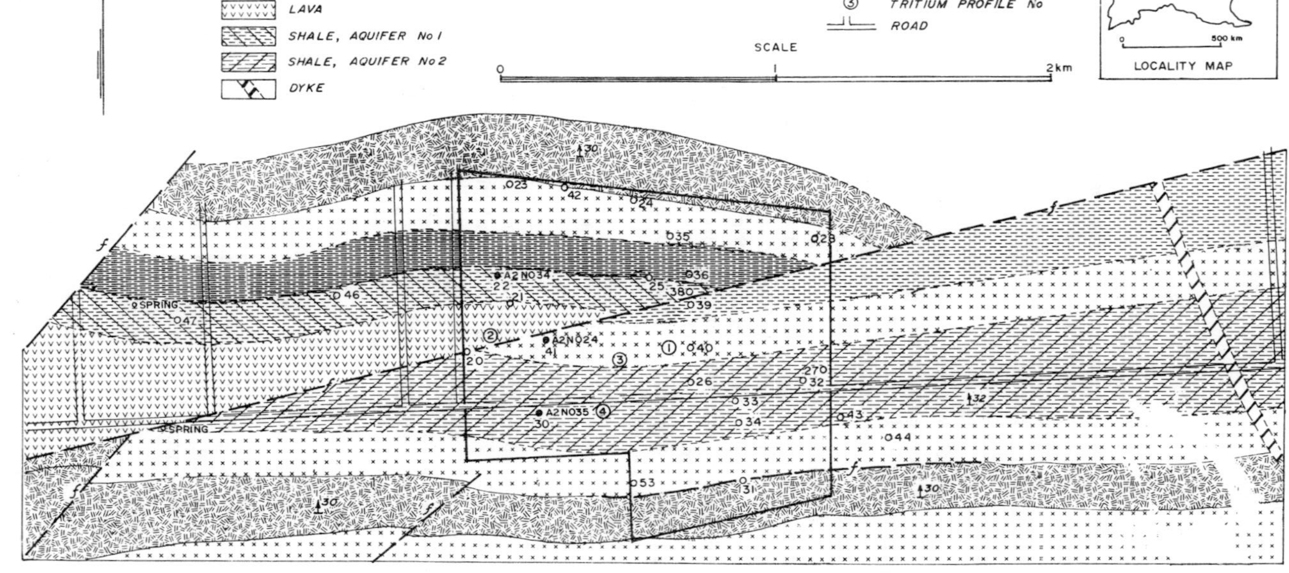


Figure 1 - Geology of the Rietondale area

In the present study a similar assessment of recharge for the Pretoria-Rietondale area was conducted.

## 2. GEOHYDROLOGY OF THE AREA

The Rietondale agricultural experimental farm shown in Fig. 1 comprises an area of about 1,5 $km^2$ and since 1955 it has also been used to conduct ground water studies. Several exploration boreholes were drilled and long-term observation boreholes were established to monitor the ground water level fluctuations.

Most of the area is covered by soil varying in thickness from about 0,6 m up to 7 m. The surface area is slightly elevated towards the middle and forms a north-south divide sloping east and west with an insignificant depression in the centre region. This however carries little water due to the flat terrain and good drainage of the soils.

The geology of the areas was interpreted from a number of exploration boreholes and is shown in Fig. 1. The geology is dominated by two prominent ridges formed by the Daspoort quartzite of the Pretoria sequence, which has been duplicated by an east-north-east to west-north-west striking fault. Although the diabase constitutes a minor aquifer the two prominent aquifers are the shale band to the north of the fault (No 1 aquifer) and that to the south bedded between two diabase sills. (No 2 aquifer) The general dip of the strata is about 32°N.

Permeable fissures have been proved by drilling to exist to depths of about 120 m and borehole 25 has a yield of about 10 ℓ/s. Ground-water is draining according to the topographical slope and surplus water is lost by effluent seepage towards the west, where springs emerge during prolonged wet periods.

Pumping tests of short duration had been carried out in 1963 on a few boreholes but the water-level drawdown did not conform to that of an isotropic aquifer. The explanation by Enslin and Bredenkamp (1963) was that the water-level response is predominantly governed by open fractures whereas the bulk of the ground water is stored in lower permeability interstices. It is therefore not possible to determine the specific yield of the aquifer by means of pumping tests.

## 3. GROUND-WATER RECHARGE

Estimation of the ground water recharge of the Rietondale area, was attempted by an analysis of hydrographs and by simulation using conceptual models based on monthly rainfall and climatic factors (Bredenkamp 1978). Following the success that had been obtained with the simulation of recharge in the Bo Molopo dolomite based only on annual rainfall, the Rietondale data was subjected to further scrutiny.

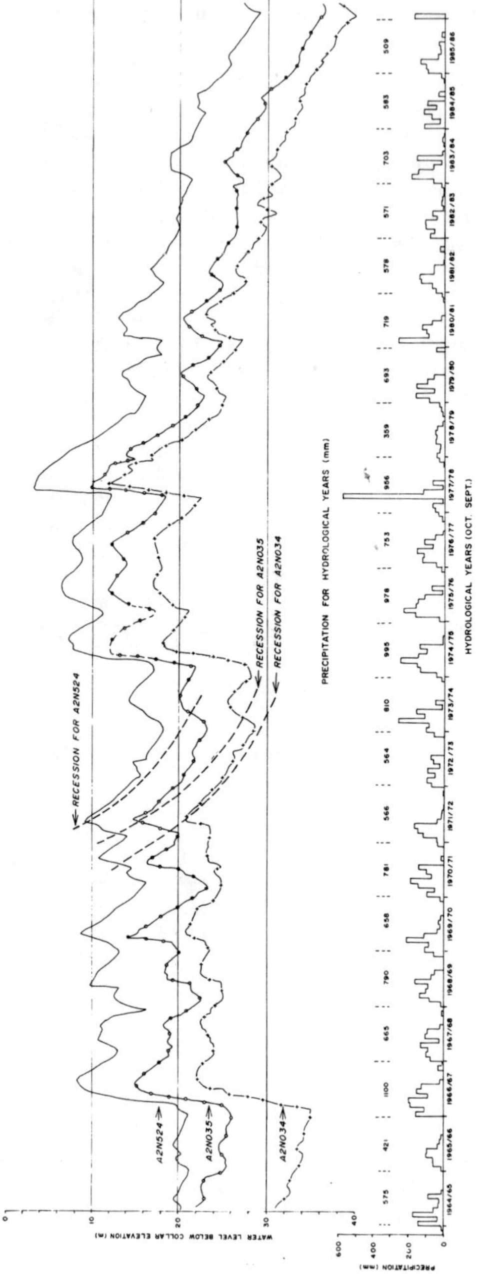


Figure 2 - Hydrographs of monitoring boreholes

### 3.1 Reconstruction of annual ground water recharge

The hydrographs of three monitoring boreholes A2N524, A2N034 and A2N035 (Fig. 2) were analysed using the method that had been devised for the Bo Molopo region (Bredenkamp 1987). By incorporation of a recession curve, representing conditions of no-recharge, the effective rise in the water level for each month could be obtained.

For each of the hydrographs the recession curve was compiled from the water level decline during periods of assumed no recharge. These recession curves are shown in Fig. 2 and they are almost identical in spite of the fact that each hydrograph represents a different aquifer.

The monthly equivalent rises in the water levels were combined to yield an effective annual water level rise corresponding to recharge for each hydrological year (October-September). These values are listed in Tables I to III.

### 3.2 Recharge estimation

The annual values of the inferred water level rise (m) can be converted to an equivalent amount of precipitation if the porosity of the aquifer is known. However as is indicated in section 2 the porosity could not be inferred from pumping tests and had to be obtained indirectly by first estimating the average annual recharge. The latter was obtained using the Darcy equation and by interpretation of natural tritium profiles (Bredenkamp 1978).

#### 3.2.1 Recharge using the Darcy equation

Considering the flow of groundwater through a cross section fo the aquifer over a period of one year.

$$RE = T.I.L.t$$

where RE is the average recharge, T the transmissivity ($m^2/d$), I the average hydraulic gradient, L the width of the flow cross section and t= 365 days. T-values obtained from pumping tests are shown in Table IV.

TABLE I: RECONSTRUCTED RECHARGE AS AN EFFECTIVE RISE (M) OF THE WATER TABLE

| HYDR. YEAR | OCTOBER | NOVEMBER | DECEMBER | JANUARY | FEBRUARY | MARCH | APRIL | MAY | JUNE | JULY | AUGUST | SEPTEMBER | ANNUAL TOTAL |
|---|---|---|---|---|---|---|---|---|---|---|---|---|---|
| 1965/66 | 0 | 0 | 0 | 0,4 | 0,55 | 0 | 0 | 0 | 0 | 0 | 0 | 0 | 0,90 |
| 1966/67 | 0 | 1,8 | 2,35 | 2,15 | 3,10 | 3,2 | 0,7 | 0,7 | 0,5 | 0,3 | 0,1 | 0 | 14,9 |
| 1967/68 | 0 | 0,6 | 0,30 | 0,90 | 0,70 | 0,9 | 0,95 | 0,7 | 0,3 | 0 | 0 | 0 | 5,35 |
| 1968/69 | 0 | 0,2 | 0,30 | 0,45 | 0,30 | 2,0 | 1,1 | 0,4 | 0,35 | 0,3 | 0,3 | 0,1 | 5,80 |
| 1969/70 | 0,8 | 0,6 | 1,9 | 1,3 | 0,4 | 0,15 | 0,3 | 0,2 | 0,25 | 0 | 0 | 0,05 | 5,95 |
| 1970/71 | 0,1 | 0,6 | 0,5 | 0,8 | 0,3 | 0,6 | 1,3 | 0,7 | 0,5 | 0,4 | 0,4 | 0,4 | 6,6 |
| 1971/72 | 0,65 | 0,6 | 1,9 | 1,7 | 0,3 | 0,05 | 0 | 0,1 | 0 | 0 | 0 | 0 | 5,2 |
| 1972/73 | 0 | 0 | 0,1 | 0 | 0 | 0,5 | 0,05 | 0 | 0,05 | 0 | 0 | 0,1 | 0,80 |
| 1973/74 | 0,5 | 1,0 | 1,2 | 0,9 | 0,6 | 0,4 | 0,05 | 0 | 0 | 0 | 0 | 0,4 | 4,45 |
| 1974/75 | 0,4 | 0,4 | 3,3 | 3,4 | 4,2 | 1,1 | 0,8 | 0,5 | 0,3 | 0,3 | 0,3 | 0,2 | 15,2 |
| 1975/76 | 0,2 | 0,2 | 2,1 | 1,0 | 1,3 | 1,3 | 1,4 | 1,0 | 0,8 | 0,4 | 0,3 | 0,7 | 10,7 |
| 1976/77 | 0,9 | 0,9 | 0,85 | 0,75 | 0,8 | 0,7 | 0,4 | 0,1 | 0,1 | 0 | 0 | 0 | 5,5 |
| 1977/78 | 0,1 | 0,1 | 0,15 | 3,6 | 4,0 | 4,4 | 0,3 | 0 | 0,1 | 1,4 | 0 | 0 | 14,1 |
| 1978/79 | 0,2 | 0 | 0 | 0 | 0 | 0 | 0 | 0 | 0 | 0,1 | 0,1 | 0,1 | 0,5 |
| 1979/80 | 0,15 | 1,0 | 0,8 | 1,2 | 1,1 | 0,4 | 0 | 0 | 0 | 0 | 0 | 0 | 4,65 |
| 1980/81* | 0 | 1,2 | 2,05 | 0,8 | 0,6 | 0,8 | 1,0 | 0,15 | 0 | 0 | 0 | 0 | 6,6 |
| 1981/82* | 0 | 0 | 0,6 | 1,0 | 0,4 | 0 | 0 | 0 | 0,05 | 0 | 0 | 0 | 2,05 |
| 1982/83* | 0 | 0,4 | 0,3 | 0 | 0 | 1,2 | 0,6 | 0 | 0 | 0 | 0 | 0 | 2,85 |
| 1983/84* | 0,2 | 0,5 | 0,7 | 0,2 | 0 | 0 | 0 | 0 | 0 | 0 | 0 | 0 | 1,6 |
| 1984/85** | | | | | | | | | | | | | |

* Results not very reliable due to pumpage.
** Results discarded due to effect of pumpage.

TABLE II: RECONSTRUCTED RECHARGE AS AN EFFECTIVE WATER LEVEL RISE (M) FROM HYDROGRAPH A2N035 - RIETONDALE

| HYDR. YEAR | OCTOBER | NOVEMBER | DECEMBER | JANUARY | FEBRUARY | MARCH | APRIL | MAY | JUNE | JULY | AUGUST | SEPTEMBER | ANNUAL TOTAL |
|---|---|---|---|---|---|---|---|---|---|---|---|---|---|
| 1965/66 | 0,05 | 0,35 | 0,5 | 0,75 | 0,6 | 0,3 | 0 | 0 | 0,3 | 0,2 | 0 | 0,2 | 3,25 |
| 1966/67 | 0,6 | 0,75 | 1,8 | 4,15 | 3,6 | 2,1 | 1,1 | 0,35 | 0,15 | 0,2 | 0,3 | 0 | 15,10 |
| 1967/68 | 0 | 0,7 | 0,25 | 1,1 | 0,6 | 0,6 | 1,25 | 1,1 | 0,2 | 0,05 | 0 | 0 | 5,85 |
| 1968/69 | 0 | 0,3 | 1,1 | 0,85 | 0,7 | 3,15 | 0,65 | 0,8 | 0,5 | 0,4 | 0,1 | 0,25 | 8,8 |
| 1969/70 | 1,3 | 1,0 | 5,7 | 0 | 0 | 0 | 0 | 0 | 0 | 0 | 0,1 | 0 | 8,1 |
| 1970/71 | 0,1 | 0 | 0,95 | 1,1 | 1,85 | 1,2 | 7,05 | 0,9 | 0 | 0 | 0 | 0,6 | 13,75 |
| 1971/72 | 6,5 | 0,6 | 2,4 | 2,4 | 1,2 | 0 | 0,1 | 0,8 | 0 | 0,35 | 0,1 | 0,05 | 7,2 |
| 1972/73 | 0 | 0,45 | 0,5 | 0,3 | 0,1 | 0,4 | 0,85 | 0,4 | 0,2 | 0,35 | 0 | 0,05 | 3,6 |
| 1973/74 | 0,4 | 0,55 | 1,0 | 1,1 | 1,9 | 0,8 | 0,9 | 0,65 | 0,4 | 0,4 | 0,4 | 0,25 | 8,75 |
| 1974/75 | 0,3 | 0,35 | 2,2 | 3,05 | 0,8 | 0,8 | 0,8 | 0,8 | 0,8 | 0,8 | 0,8 | 0,8 | 12,30 |
| 1975/76 | 0,8 | 0,8 | 1,9 | 2,3 | 2,0 | 1,6 | 1,75 | 0,7 | 0,6 | 0,5 | 0,55 | 0,4 | 13,90 |
| 1976/77 | 0,8 | 1,35 | 1,3 | 1,35 | 1,4 | 0,45 | 0 | 0 | 0 | 0 | 0,20 | 0,25 | 7,5 |
| 1977/78 | 0,5 | 0,5 | 0,5 | 1,8 | 3,75 | 3,8 | 1,7 | 0,2 | 0,6 | 0,85 | 0 | 0 | 14,20 |
| 1978/79 | 1,3 | 0,45 | 0 | 0 | 0 | 0 | 0 | 0 | 0 | 0 | 0 | 0 | 1,75 |
| 1979/80 | 0,2 | 0,8 | 1,2 | 0,9 | 1,25 | 1,2 | 0,35 | 0 | 0 | 0 | 0 | 0 | 5,90 |
| 1980/81* | 0,1 | 1,4 | 1,95 | 1,0 | 1,2 | 1,1 | 0,55 | 0 | 0 | 0 | 0 | 0 | 7,30 |
| 1981/82* | 0 | 0,3 | 0,7 | 1,0 | 0,75 | 0,05 | 0 | 0 | 0 | 0,1 | 0,1 | 0 | 3,0 |
| 1982/83* | 0 | 0,2 | 0,3 | 0,35 | 0,3 | 0,3 | 0,3 | 0,3 | 0,2 | 0,2 | 0,7 | 0 | 3,15 |
| 1983/84* | 0,3 | 1,1 | 0,8 | 0,65 | 0,8 | 0 | 0,4 | 0 | 0,1 | 0 | 0 | 0 | 4,15 |
| 1984/85** | | | | | | | | | | | | | |

* Results not very reliable due to pumpage.
** Results discarded due to effect of pumpage.

TABLE III: RECONSTRUCTED GROUND-WATER RECHARGE AS AN EFFECTIVE WATER LEVEL RISE (M) FROM HYDROGRAPH A2N524 - RIETONDALE

| HYDR. YEAR | OCTOBER | NOVEMBER | DECEMBER | JANUARY | FEBRUARY | MARCH | APRIL | MAY | JUNE | JULY | AUGUST | SEPTEMBER | ANNUAL TOTAL |
|---|---|---|---|---|---|---|---|---|---|---|---|---|---|
| 1955/56 | 0,7 | 0,9 | 1,0 | 1,9 | 2,6 | 1,1 | 1,0 | 0,7 | 0,8 | 0,6 | 0,5 | 0,8 | 12,6 |
| 1956/57 | 0,7 | 0,4 | 1,0 | 1,0 | 1,2 | 1,3 | 0,5 | 0,4 | 0,7 | 1,5 | 1,3 | 1,9 | 11,9 |
| 1957/58 | 1,8 | 1,0 | 0,7 | 0,7 | 0,2 | 0,4 | 0,7 | 0,5 | 0,4 | 0,2 | 0,1 | 0 | 6,6 |
| 1958/59 | 0,3 | 1,5 | 1,5 | 1,6 | 2,4 | 0,7 | 0,5 | 0 | 0 | 0 | 0 | 0 | 8,5 |
| 1959/60 | 0 | 0,3 | 0,6 | 0,4 | 0,2 | 0,7 | 1,4 | 0,6 | 0 | 0 | 0 | 0 | 4,2 |
| 1960/61 | 0 | 1,5 | 2,6 | 1,8 | 1,8 | 1,5 | 1,0 | 0,7 | 0,2 | 0 | 0 | 0 | 11,1 |
| 1961/62 | 0 | 0,3 | 0,3 | 1,3 | 1,0 | 0 | 0 | 0 | 0 | 0 | 0 | 0 | 2,9 |
| 1962/63 | 0,4 | 0,4 | 0,5 | 0,3 | 0 | 0,5 | 0,2 | 0,2 | 0,4 | 0,5 | 0 | 0 | 3,4 |
| 1963/64 | 0 | 0,8 | 0,9 | 2,3 | 2,3 | 0 | 0 | 0 | 0 | 0 | 0 | 0 | 6,3 |
| 1964/65 | 0 | 0 | 0 | 0,4 | 0,7 | 0,5 | 0 | 0 | 0 | 0 | 0 | 0,2 | 1,8 |
| 1965/66 | 0,6 | 1,0 | 0,5 | 0,4 | 0,8 | 0 | 0 | 0,2 | 0 | 0 | 0 | 0 | 3,5 |
| 1966/67 | 1,1 | 0,9 | 3,1 | 4,6 | 3,1 | 2,3 | 1,5 | 1,0 | 0,2 | 0 | 0 | 0 | 17,8 |
| 1967/68 | 0 | 0,5 | 2,1 | 2,2 | 0,8 | 0,6 | 0,7 | 0,7 | 0 | 0 | 0 | 0 | 7,6 |
| 1968/69 | 2,2 | 1,2 | 0,5 | 1,2 | 3,2 | 0,6 | 0,6 | 0,4 | 0 | 0 | 0,3 | 0,4 | 10,6 |
| 1969/70 | 1,5 | 1,6 | 2,6 | 1,8 | 0 | 0 | 0 | 0 | 0 | 0 | 0 | 0 | 7,5 |
| 1970/71 | 0 | 0 | 0 | 1,4 | 1,8 | 0,5 | 0,8 | 2,9 | 1,2 | 0 | 0 | 0 | 8,6 |
| 1971/72 | 0 | 1,6 | 2,5 | 1,7 | 1,8 | 0 | 0 | 0 | 0 | 0 | 0 | 0 | 7,6 |
| 1972/73 | 0 | 0 | 0,6 | 0,5 | 1,1 | 0 | 0 | 0 | 0 | 0 | 0 | 0 | 2,2 |
| 1973/74 | 0,5 | 0,7 | 1,0 | 1,4 | 1,0 | 1,4 | 0,9 | 0,3 | 0 | 0 | 0 | 0 | 7,2 |
| 1974/75 | 0 | 0,4 | 0,8 | 5,3 | 2,7 | 2,2 | 1,0 | 1,0 | 0,8 | 0 | 0 | 0 | 14,2 |
| 1975/76 | 0 | 0,7 | 1,5 | 2,4 | 2,4 | 1,6 | 1,4 | 0,9 | 0,9 | 0,5 | 0,2 | 0,3 | 12,8 |
| 1976/77 | 0,7 | 1,2 | 1,4 | 1,4 | 1,3 | 1,1 | 0,5 | 0,2 | 0,5 | 0,2 | 0 | 0,2 | 8,7 |
| 1977/78 | 0,5 | 0,4 | 0,3 | 2,0 | 8,2 | 0,6 | 0,6 | 0,4 | 0,2 | 0,1 | 0 | 0,3 | 13,6 |
| 1978/79 | 0 | 0 | 0 | 0 | 0 | 0 | 0 | 0 | 0 | 0 | 0 | 0 | 0 |
| 1979/80 | 0,6 | 0,7 | 1,3 | 1,1 | 1,0 | 0,9 | 0,5 | 0 | 0 | 0 | 0 | 0 | 6,1 |
| 1980/81* | 0,6 | 0 | 4,8 | 0,4 | 1,2 | 1,2 | 0,7 | 1,5 | 0 | 0 | 0 | 0 | 8,9 |
| 1981/82* | 0 | 0 | 0 | 1,1 | 0,6 | 0 | 0 | 0 | 0 | 0 | 0 | 0 | 1,7 |
| 1982/83* | 0 | 0,3 | 0,4 | 0,3 | 0,1 | 0,3 | 0 | 0,1 | 0 | 0 | 0 | 0 | 1,5 |
| 1983/84* | 0,3 | 1,2 | 1,7 | 0,5 | 0,4 | 0,3 | 0,2 | 0,2 | 0 | 0 | 0 | 0 | 4,6 |
| 1984/85** | | | | | | | | | | | | | |

* Results not very reliable due to pumpage.
** Results discarded due to effect of pumpage.

TABLE IV: TRANSMISSIVITY VALUES OBTAINED FROM PUMPING TESTS

| PUMPED BOREHOLE | OBSERVATION BOREHOLE | TRANSMISSIVITY ($m^2/d$) | METHOD |
|---|---|---|---|
| 25 | 36 | 21,7 | Time drawdown |
| 25 | 22, 35, 36, 37, 38 | 28 | Distance drawdown |
| 25 | 22, 35, 36 | 60,5 | Distance drawdown |
| 32 | 27 | 35,1 | Time drawdown |
| 25 | 35 | 20,0 | Time drawdown |
| 25 | 36 | 34,0 | Time drawdown |
| 25 | 37 | 38,0 | Time drawdown |
| 25 | 38 | 53,0 | Time drawdown |
| | Average | 40 | |

These are representative of aquifer 1 which has an average transmissivity of about 40 $m^2/d$, whereas according to Enslin and Bredenkamp (1963) the transmissivity of aquifer 2 is about 44,8 $m^2/d$. The water level gradient could be estimated from assumed average water level conditions and is 0,0127 for aquifer 1 and 0,0178 for aquifer 2. The width of the aquifers is 70 m and 140 m respectively which yields:

$$RE_1 = 40 \times 0{,}0127 \times 70 \times 365$$
$$= 12\ 979\ m^3/a$$

and

$$RE_2 = 44{,}8.0{,}0178.140.365$$
$$= 40\ 749\ m^3/a$$

These values correspond to an equivalent rainfall amount of 63 mm and 68 mm respectively but the estimates could vary by $\pm$ 6 mm due to uncertainties associated with I and T.

3.2.2 Recharge obtained from Tritium profiles

The average ground water recharge for the Rietondale area was derived from tritium profiles using three methods to interpret the natural tritium distribution in the soil (Bredenkamp, 1978).

The first method uses the 1962/3 and 1958 tritium spikes as a specific time reference (see Fig. 3). By assuming the propagation of soil moisture in the unsaturated zone to be effectively a piston-like displacement, the average recharge could be calculated. This method of interpretation could only be applied to profile 4 as this was the only profile to intersect the 1958 and

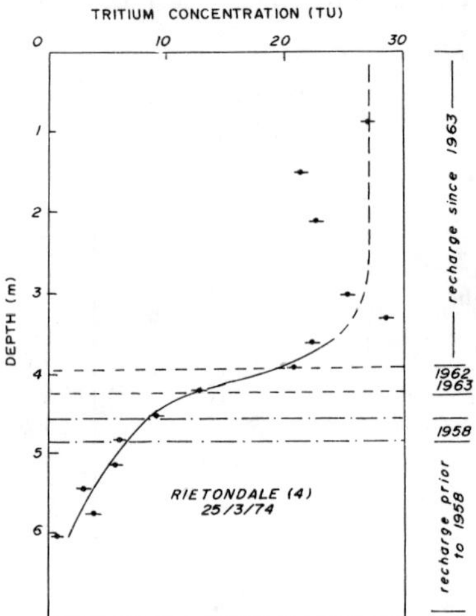


Figure 3 - Tritium profile 4 - interpretation by first method

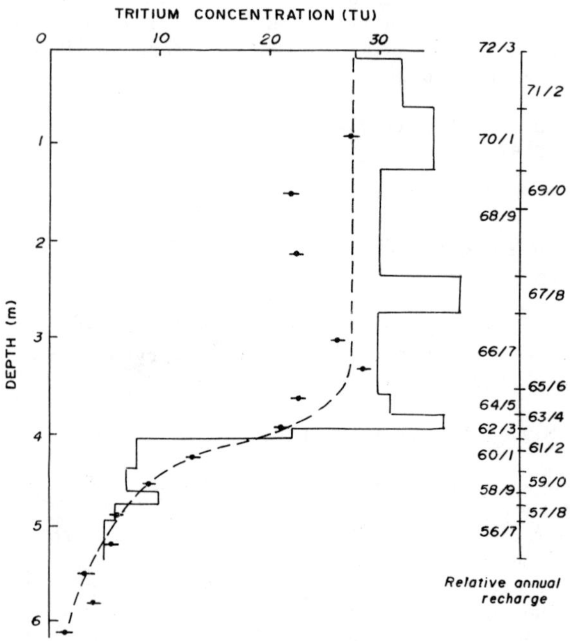


Figure 4 - Tritium profile 4 - interpretation by second method

1962/3 transition zone. An average recharge of 55-58 mm/a was obtained. (Table V)

TABLE V: AVERAGE ANNUAL GROUND-WATER RECHARGE INFERRED FROM TRITIUM PROFILES

| PROFILE NUMBER | AVERAGE ANNUAL RECHARGE (mm) | | | | | | |
|---|---|---|---|---|---|---|---|
| | METHOD (a) | | | | METHOD (b) | METHOD (c) | AVERAGE |
| | 1958 Reference | 1958* Reference | 1962/3 Reference | 1962/3* Reference | | | |
| 1 | | | 68 | | | | |
| 2 | | | 65 | | | | |
| 3 | | | 83 | | | | |
| 4 | 55,4 | 60,4 | 59,3 | 65,2 | 64 | 58 | 60,3 |

* Corrected for annual variability

(a) Based on bomb-tritium references

(b) Simulation of observed tritium profile from rainfall and relative recharge

(c) Based on total accretion of tritium in soil profile

The second method of interpretation, which again could only be applied to profile 4, attempts to reconstruct the observed tritium profile from the tritium input of rainfall and the effective annual recharge. This method yields an average recharge of 64 mm/a. (See Fig. 4 and Table V)

The third method compares the total accretion of tritium in the soil profile to that of the input precipitation - this yields an average recharge of 58 mm/a (Table V). The combined average recharge calculated from the tritium profiles using the different methods of interpretation was 60,3 mm/a.

For the other 3 profiles only the minimum recharge could be ascertained, assuming the 1962/63 reference to coincide with the deepest soil sample obtained. By combining all tritium-profile interpretations, an average annual recharge of 69 mm was obtained which is in good agreement with the estimate of 63 and 68 mm/a using the Darcy equation.

TABLE VI: ANNUAL GROUND-WATER RECHARGE INFERRED FROM HYDROGRAPHS USING THE NATURAL RECESSION REPRESENTING NO-RECHARGE

| HYDROLOGICAL YEAR OCT - SEPT | INFERRED RECHARGE FROM MONITORING BOREHOLES | | | | | | | | |
|---|---|---|---|---|---|---|---|---|---|
| | A2N034 | | A2N035 | | A2N524 | | AVERAGE VALUES | | ANNUAL |
| | EFFECTIVE wl. RISE (m) | RAINFALL EQUIVALENT (mm) | EFFECTIVE wl. RISE (m) | RAINFALL EQUIVALENT (mm) | EFFECTIVE wl. RISE (m) | RAINFALL EQUIVALENT (mm) | EFFECTIVE wl. RISE (m) | RAINFALL EQUIVAMENT (mm) | PRECIPITATION (mm) |
| 1955/56 | - | - | - | - | 12,6 | 117,2 | - | - | 832 |
| 1956/57 | - | - | - | - | 11,9 | 110,7 | - | - | 793 |
| 1957/58 | - | - | - | - | 7,2 | 67,0 | - | - | 478 |
| 1958/59 | - | - | - | - | 8,5 | 79,0 | - | - | 733 |
| 1959/60 | - | - | - | - | 4,2 | 39,0 | - | - | 501 |
| 1960/61 | - | - | - | - | 11,1 | 103,2 | - | - | 779 |
| 1961/62 | - | - | - | - | 2,9 | 27,0 | - | - | 629 |
| 1962/63 | - | - | - | - | 3,4 | 31,6 | - | - | 586 |
| 1963/64 | - | - | - | - | 6,3 | 58,6 | - | - | 686 |
| 1964/65 | - | - | - | - | 1,8 | 16,7 | - | - | 575 |
| 1965/66 | 0,9 | 8,4 | 3,25 | 30,2 | 3,5 | 32,6 | 2,55 | 23,7 | 421 |
| 1966/67 | 14,9 | 138,6 | 15,10 | 140,4 | 17,8 | 165,5 | 15,93 | 148,1 | 1 100 |
| 1967/68 | 5,35 | 49,8 | 5,85 | 54,4 | 7,6 | 70,7 | 6,27 | 58,3 | 665 |
| 1968/69 | 5,80 | 53,9 | 8,8 | 81,8 | 10,6 | 98,6 | 8,40 | 78,1 | 790 |
| 1969/70 | 5,95 | 55,3 | 8,1 | 75,3 | 7,5 | 69,7 | 7,18 | 66,8 | 658 |
| 1970/71 | 6,6 | 61,4 | 13,75 | 127,9 | 8,6 | 80,0 | 9,65 | 89,7 | 871 |
| 1971/72 | 5,2 | 48,4 | 7,2 | 67,0 | 7,6 | 70,7 | 6,60 | 61,4 | 566 |
| 1972/73 | 0,8 | 7,4 | 3,6 | 33,5 | 2,2 | 20,5 | 2,20 | 20,5 | 564 |
| 1973/74 | 4,45 | 41,4 | 8,75 | 81,4 | 7,2 | 67,0 | 6,80 | 63,2 | 810 |
| 1974/75 | 15,2 | 141,4 | 16,25 | 151,1 | 14,2 | 132,1 | 15,22 | 141,5 | 995 |
| 1975/76 | 10,7 | 99,5 | 13,10 | 121,8 | 12,8 | 119,0 | 12,2 | 113,5 | 978 |
| 1976/77 | 5,5 | 51,2 | 7,5 | 69,7 | 8,7 | 80,9 | 7,2 | 67,0 | 753 |
| 1977/78 | 14,1 | 131,1 | 14,2 | 132,1 | 13,6 | 126,5 | 13,97 | 129,9 | 956 |
| 1978/79 | 0,5 | 4,7 | 1,75 | 16,3 | 0 | 0 | 0,75 | 7,0 | 693 |
| 1979/80 | 4,65 | 43,3 | 5,9 | 54,9 | 6,1 | 56,7 | 5,55 | 51,6 | 693 |
| 1980/81 | 6,6 | 61,4 | 7,3 | 67,9 | 8,9 | 82,8 | 7,6 | 70,7 | 719 |
| 1981/82* | 2,05 | 19,1 | 3,0 | 27,9 | 1,7 | 15,8 | 2,25 | 20,9 | 578 |
| 1982/83* | 2,85 | 26,5 | 3,5 | 29,3 | 1,5 | 14,0 | 2,50 | 23,3 | 571 |
| 1983/84* | 1,6 | 14,9 | 4,15 | 38,6 | 4,6 | 42,8 | 3,45 | 32,1 | 703 |
| Average | 5,98 | 55,6 | 7,93 | 73,7 | 7,4 | 68,8 | 7,10 | 66,0 | 723,6 |

* Results not very reliable due to effect of pumpage.

### 3.2.3 Annual recharge

The average annual recharge obtained if all estimates are combined is 66 mm/a. This was used to infer the average effective porosity of the aquifer by dividing the average recharge (mm) with the average rise in the water table (m) that was effected by the recharge. Results are shown in the following table.

| MONITORING BOREHOLE | AVERAGE WATER LEVEL RISE (m) | AVERAGE EFFECTIVE POROSITY |
|---|---|---|
| A2N524 | 7,4 | 0,0089 |
| A2N034 | 5,98 | 0,011 |
| A2N035 | 7,93 | 0,0083 |
| Average | 7,10 | 0,0093 |

The average porosity was calculated using the average water level rise obtained from the different hydrograph interpetations yielding $\frac{66}{7100}$ = 0,0093.

Instead of assuming the average recharge for all three cases to be the same it is more likely that the average effective porosity is the same but that the average recharge differs. Such variation in recharge could be explained for instance by a different thickness of the overburden and variation in the soil types and their drainage characteristics. Hence by assuming the average porosity (0,0093) the average annual recharge for each of the cases could be ascertained. These results are listed in Table VI indicating a recharge of 68,8 mm for A2N524 and 55,6 mm for A2N034 and 73,7 mm in the case of A2N035. The average of these values is naturally 66 mm being the value assumed to derive the porosity value of 0,0093.

### 3.3 Relationship between rainfall and recharge

By plotting the annual recharge (mm) against annual precipitation a striking linear relationship is evident corresponding to equation 1 which was derived for dolomitic areas. The annual recharge again shows a linear relationship with rainfall in excess of an annual threshold value (See Fig. 5).

The values of A, B and the correlation coefficient applying to each of the hydrograph interpretations are given in Table VII.

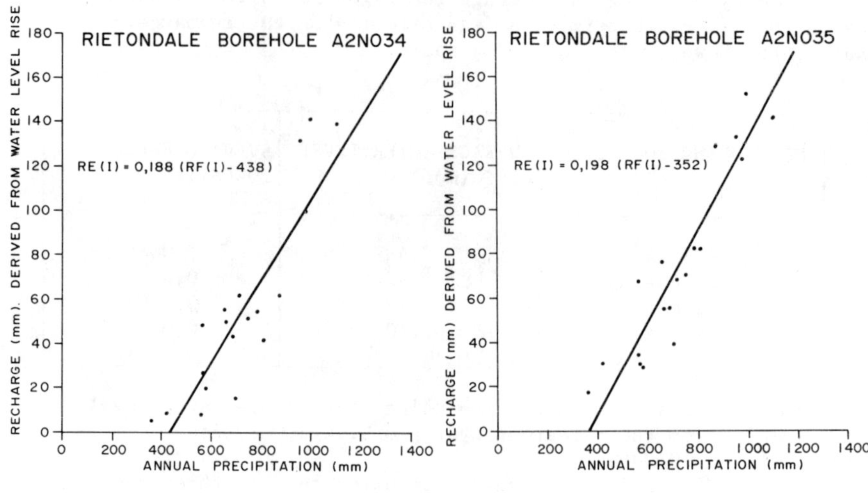


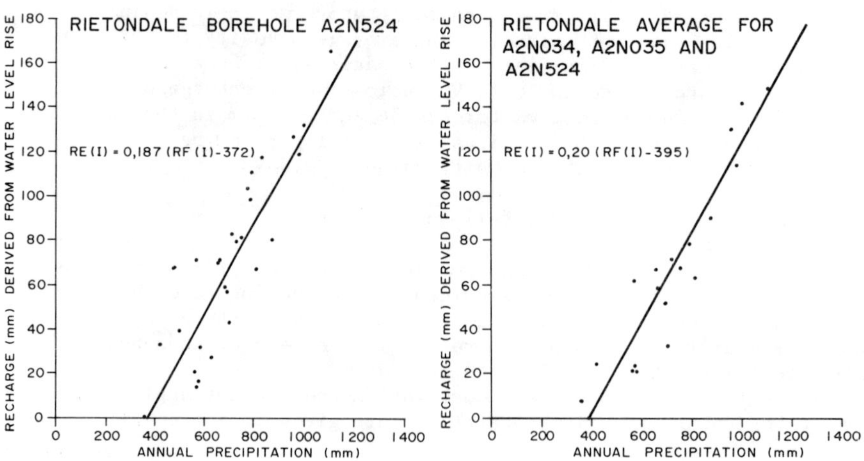


Figure 5 – Rainfall recharge relationship for monitoring boreholes

TABLE VII: VALUES OF A AND B IN RAINFALL-RECHARGE RELATIONSHIP (FIGURE 5)

| HYDROGRAPH | A-VALUE | B-VALUE (mm) | CORRELATION COEFFICIENT (r) |
|---|---|---|---|
| A2N524 | 0,207 | 372 | 0,883 |
| A2N034 | 0,188 | 438 | 0,889 |
| A2N035 | 0,198 | 353 | 0,924 |
| Average | 0,20 | 395 | 0,933 |

Assuming the average values to be representative of the rainfall-recharge relationship for conditions similar to the Pretoria-Rietondale area (equation 1) becomes:

$$RE(I) = 0,20\ (RF(I) - 395) \qquad (2)$$

This implies that 20% of the rainfall in excess of 395 mm constitutes ground water recharge.

## 4. DISCUSSION

Based on a preliminary scrutiny of the soil conditions in each of the cases presented, it appears that the differences in recharge correspond to the thickness and characteristics of the soil overburden.

In the case of aquifer 1 (A2N034) the soil is a loamy type and is more than 2 metres deep whereas for aquifer 2 (A2N035) the soil is of similar type but much shallower. This explains the lower threshold value (353 mm) and higher recharge inferred from borehole A2N035 as compared to a lower average recharge in the case of A2N034. On the other hand borehole A2N524 which is situated on the fault zone responds according to average conditions and for this reason there is good agreement between its threshold value (372 mm) and the average value of 395 mm.

An interesting result is that the value of A is apparently not much affected by the differences in the threshold values. In comparison to dolomitic regions (A=0,30) the recharge in the Rietondale area (A=0,20) is at least 10% lower for the same rainfall conditions, according to equation 2.

## 5. CONCLUSIONS

A linear relationship between annual rainfall and recharge similar to the one that has been established for dolomitic regions, appears to be valid for the Rietondale recharge situation and it is expected that the linear relationship could equally well apply to similar aquifers as is the case in dolomitic aquifers.

Further verificaton that the relationship given by equation 2 is indeed a valid one could be sought in applications such as:

- estimation of average annual recharge using different conventional methods
- aquifer simulations where the input-recharge could be varied according to the variability of the annual rainfall using equation 2.

The rainfall-recharge equation could prove a significant contribution for improved assessments of aquifer performance insofar that the recharge need no longer be an uncertain element in aquifer simulations.

## 6. REFERENCES

ENSLIN, J.F. and BREDENKAMP, D.B.: 'The value of pumping tests for the assessment of ground water supplies in secondary aquifers in South Africa' Technical report 30, 1963, Department of Water Affairs, South Africa.

BREDENKAMP, D.B.: 'Quantitative estimation of groundwater recharge with special reference to the use of natural radioactive isotopes and hydrological simulation', Technical report 77, 1978, Department of Water Affairs, South Africa.

BREDENKAMP, D.B.: 'Quantitative estimation of ground water recharge in dolomite', Paper to be presented at International Workshop on Estimation of Ground-water Recharge, March 1987, Antalya, Turkey.

# ANALYSIS OF LONG-DURATION PIEZOMETRIC RECORDS FROM BURKINA FASO USED TO DETERMINE AQUIFER RECHARGE

Dominique THIERY
Bureau de Recherches Géologiques et Minières
Département EAU
B.P. 6009
45060 ORLEANS CEDEX
FRANCE

ABSTRACT. An 8-year water-level record for an observation well in a granite environment in Ouagadougou (Burkina Faso) is analysed using a lumped-parameter hydrological model. The model computes aquifer levels from rainfall and potential evapotranspiration data, and is calibrated with observed levels. Very satisfactory calibration is achieved, although aquifer levels have been dramatically declining since 1978.

It appears that, even with small computational time-steps, a unique solution for calibration is only possible if the precise storage coefficient is known or if surface runoff data are available. In the absence of such data, multiple calibrations displaying the same agreement with observed data give different values for aquifer recharge, although relative variation is the same from year to year. When used in conjunction with a long set of **in-situ** rainfall records, however, the various sets of parameters applied result in almost-identical extension of water-level data, an important advance.

The model shows that the 1978-1985 period is typified by the lowest water levels encountered in at least 60 years, and that a return to a rainfall sequence near the long-term average would cause the level to rise, although only after a period of 7 to 10 years.

## 1. INTRODUCTION

The marked interannual variability of rainfall in the African Sahel causes still-greater variability in aquifer recharge and piezometric levels. The severe drought encountered since the 1970's has often resulted in substantial drops in water level, despite slight seasonal recovery. Long-duration observation data are only rarely available, although data sequences covering 5 to 10 years can be used to analyse the possible evolution in water level for various rainfall scenarios.

An observation well was sunk to a depth of 20 m in a granite aquifer in Ouagadougou (Burkina Faso). The well is screened from 6 to 20 m, and taps 5 m of



*I. Simmers (ed.), Estimation of Natural Groundwater Recharge, 477–489.*

granitic sand, 4 m of weathered granite, and 5 m of fresh granite. It has been monitored since 1978 by the ICHS, which has related data covering an eight-year period from 1978-1985. These data were subjected to detailed analysis using a lumped-parameter model, the aim being to extend the data sequence.

## 2. AVAILABLE DATA

### 2.1. Piezometric levels

The water levels in the ICHS observation well in Ouagadougou were manually monitored by the ICHS between 1978 and 1985 (Diluca and Muller, 1985), and a monthly sequence was deduced by interpolation between measurements. Figure 1 shows a continuous fall in the average levels, passing from a depth about 6 m below the surface in 1978 to just over 10 m in 1985 (i.e. a fall of 0,45 m/year). A rise in level of about 0,80 m every year is nonetheless recorded for several months in autumn before levels fall again.

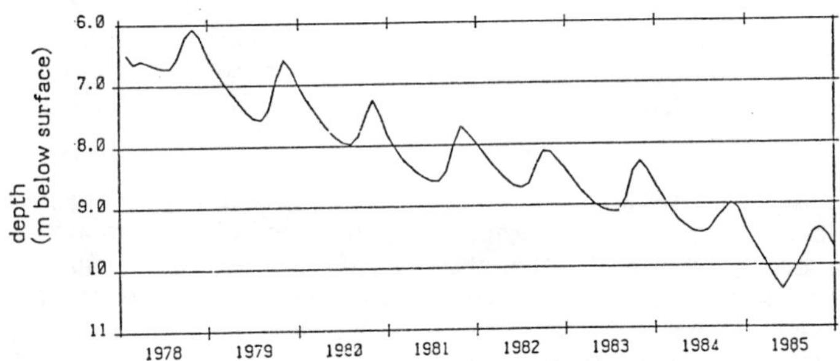


Figure 1. Evolution in water level in the ICHS observation well in Ouagadougou (Burkina Faso) from 1978 to 1985

### 2.2. Hydrodynamic characteristics

No data are available, only one very short-duration pumping test (lasting one hour) having been undertaken in May 1978 near the study area. In particular, the storage coefficient (or effective porosity) is not known, although it can be estimated at between 1 and 8 % on the basis of the rock composition. Transmissivity values deduced from this test were 3 to 7 $10^{-5}$ $m^2/s$.

### 2.3. Rainfall

Daily rainfall figures from 1959 to 1985 (27 years) were provided by ORSTOM and ICHS. Average annual rainfall for the 1959-1985 period was 825 mm, but only 690 mm from 1978 to 1985. The number of days of rainfall was 73 per annum for the 1959-1985 period, but only 60 per annum from 1978 to 1985. Monthly rainfall figures are also available for the 1929-1958 period.

## 2.4. Potential evapotranspiration (PET)

The monthly values applied for the model were calculated in Ouagadougou using Turc's formula. Only the interannual means (for an unspecified period) could be taken from an atlas (Lemoine and Prat, 1972). The total annual PET is 2,084 mm.

# 3. THE LUMPED-PARAMETER MODEL USED

The BRGM's GARDENIA model was used to calculate the balance of rainfall, potential evapotranspiration, runoff and infiltration. This lumped-parameter hydrological model makes it possible to produce a local balance in daily time-steps, and to calculate the actual evapotranspiration (ETR), runoff, infiltration, and spot water-table level. The model is described in detail by Roche and Thiery (1984), the basic principles being briefly reviewed below.

## 3.1. Operational principle

The GARDENIA model consists of three superimposed layers (fig.2). The first (RU) is characterized by its retention capacity (RUMAX) (or maximum soil moisture deficit), and represents the retention effect in the first few metres below the surface. This layer is supplied by rainfall, and is emptied by evapotranspiration. No runoff or infiltration occurs before this layer is saturated. The model takes account of effects caused by the interception in surface depressions, and schematizes the "valve effect" of unsaturated soil depending on the degree of humidity.

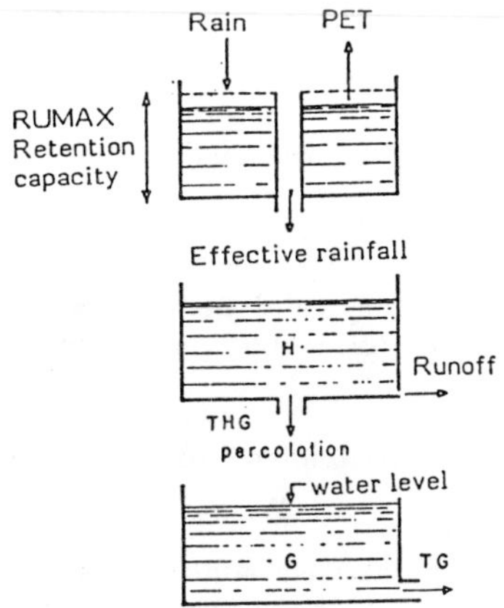


Figure 2. Principle of the GARDENIA model used to simulate piezometric levels

The second layer (H) is characterized by two parameters :
(1) a half-percolation time (THG)
(2) an equal runoff-percolation level (RUIPER)
This layer transfers water to the water table through the unsaturated zone, and controls distribution between runoff and infiltration : the higher the level in this layer (as a result of heavy rainfall), the greater the runoff proportion. When the level of the layer is the same as the equal runoff-percolation level, infiltration equals runoff ; when the level of the layer is lower infiltration is greater than runoff.

The third layer (G) is characterized only by the half-recession time, and represents exponential aquifer recession. Thiery (1985) has shown that this scheme corresponds for practical purposes to an aquifer bounded on one side by an impermeable rectilinear barrier and on the other by an imposed-level rectilinear barrier. If the observation well is positioned sufficiently far away from the imposed-level boundary, the piezometric level (NI) is deduced from the level G in layer G by the formula :

$$NI = G / EMM + NB$$

where NI is the piezometric level
G is the level in the layer
EMM is the unconfined storage coefficient or specific yield or equivalent effective porosity
NB is the base level

Four parameters are therefore to be determined :
(1) the retention capacity (RUMAX), which alone controls the value of potential evapotranspiration (PET)
(2) the half-percolation time (THG), and the equal runoff-percolation level (RUIPER) which control the runoff proportion and the delay-time between excess water in the soil and a rise in the aquifer level.
(3) the half-recession time (TG) which governs the rate of aquifer recession
and also two amplitude parameters :
(1) the base level (NB)
(2) the storage coefficient or effective porosity (EMM)

### 3.2. Principles applied for calibration of the model

In order to calculate the balance under valid conditions, sequential data (not-necessarily simultaneous) for the following parameters should ideally be available :
(1) runoff
(2) piezometric level and storage coefficient for the unconfined piezometric fluctuation surface

The parameters of the model are in this case adjusted in order best to reproduce the two sequences of data. In practice, as is the case in the example cited :
(1) data for runoff are not always available (except where runoff can be assumed to be nil or negligible)
(2) the unconfined storage coefficient is inadequately known (or sometimes a confined aquifer storage coefficient, quite unrelated to aquifer recharge).

It may however be thought, a priori, that data on the piezometric level may be used alone to determine the retention capacity (RUMAX). If the value assumed for the model is too low, the levels calculated will affect the rainfall sequences too soon and too frequently. If this value is too high, the effect of piezometric levels will occur too late and in some cases not at all.

This distribution of runoff and infiltration for effective rainfall can therefore be calculated in unique-solution form using the non-linear scheme for reservoir H, which does not give a fixed infiltration percentage but smooths the effect of heavy effective rainfall. Regulation of this smoothing of heavy effective rainfall (by means of the RUIPER parameter) will therefore regulate distribution between infiltration and runoff.

Infiltration will thus be transformed into variation in the piezometric level by the half-recession time (TG), the amplitude being inversely proportional to the storage coefficient.

This type of calibration poses no problems where piezometric levels vary rapidly in relation to sequences of isolated and non-periodic rainfall. In practice, as is dicussed in chapter 4, piezometric levels are often cushioned from the direct effects of rainfall and react only to a true "rainy season" rather than to a sequence of isolated rains. The sequence of water levels therefore describes a pseudo-sine curve in response to the pseudo-sine curve for smoothed rainfall. Calibration consists of reproducing the pseudo-sine curve for water levels, with a time-lag in relation to the rainfall curve. If the storage coefficient is inadequately known, the variations in level cannot easily be reduced to millimetres of recharge and, as will be seen, a unique solution for calibration is likely to be no longer possible.

## 4. CALIBRATION OF MODEL

It has been seen that evapotranspiration is only governed by a single parameter : the soil retention capacity (RUMAX). This parameter is therefore essential because it allows determination of the effective rainfall, or rainfall minus ETR. This effective rainfall is itself distributed between runoff and recharge by the RUIPER parameter (associated with the half-percolation time THG). The rate of recession is only governed by the half-recession time TG. This parameter has no effect either on recharge or on water balance, but makes it possible to spread recharge in time and to simulate the water level observed (when used in association with the base level and storage coefficient).

It thus emerges that two parameters are of fundamental importance for calculation of recharge :

(1) The soil retention capacity (RUMAX)
(2) The equivalent-runoff level (RUIPER).

### 4.1. Determination of soil retention capacity

Six calculations were made by fixing the soil retention capacity at 0, 10, 20, 40, 70 and 100 mm respectively. Figure 3 shows the value for effective rainfall (runoff + infiltration) obtained for the 1978-1985 period as a function of retention capacity. Figure 4 shows that the best results are obtained for a retention capacity of less than 40 mm. At capacities of 40, 70 and particularly 100 mm, the quality of calibration deteriorates, some years (1984 and 1985) displaying no rise in level. Figure 3 shows that the effective rainfall calculated is highly dependent on the retention capacity, decreasing from 482 mm/year for nil retention to 196 mm/year for a retention of 20 mm.

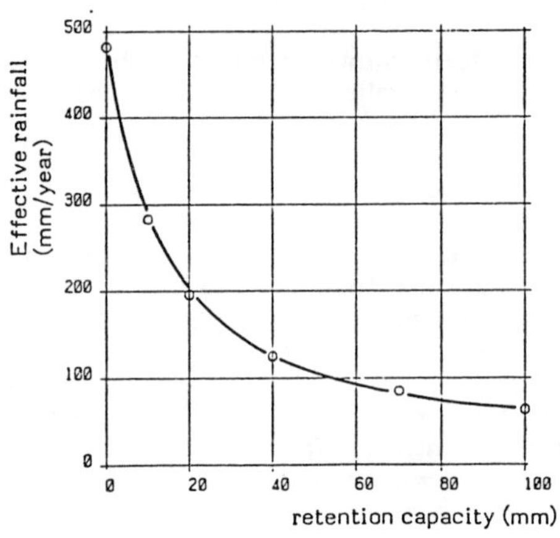


Figure 3. Relationship between effective rainfall and the retention capacity assumed for the model

This parameter is highly sensitive. When the retention capacity increases from 0 to 20 mm, the evapotranspiration increases by 286 mm/year, a factor representing 14 times the capacity. A retention capacity of 20 mm thus passes on average 14 times per annum from a saturated state to a completely-dry state. This is due both to the high potential evapotranspiration (2,084 mm/year, or nearly 6 mm/day on average), and to the irregularity of daily rainfall. The sensitivity of the retention capacity is a problem when extending these parameters to other locations, but decreases noticeably in the case of very high capacities (the evapotranspiration rises by 62 mm/year for a capacity passing from 40 to 100 mm, i.e. a factor very close to 1).

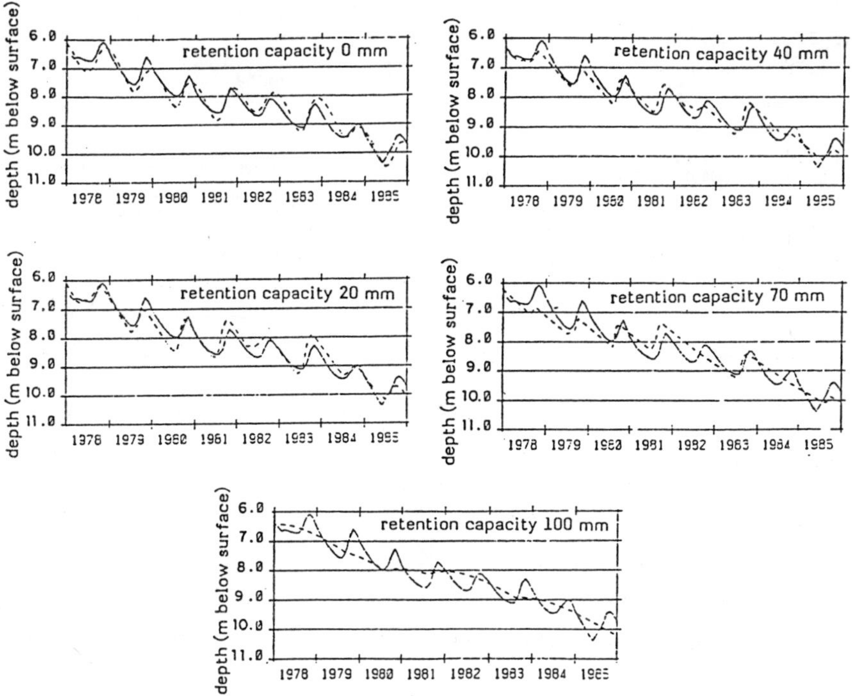


Figure 4. Simulation of water levels for a retention capacity ranging from 0 to 100 mm

## 4.2. Determination of runoff proportion

The runoff proportion is governed by the RUIPER (equal runoff-percolation level) and THG (half-percolation time) parameters. Six calculations were made for three retention-capacity values (10, 20 and 30 mm) associated with two storage coefficient values (1 % and 4 %). Figure 5 shows that the six simulations are acceptable, all typified by correlation coefficients ranging from 0,965 to 0,980. Adjust ments obtained with a storage coefficient of 1 % or of 4 % are equally satisfactory, the optimum retention capacity being 10-30 mm in both cases. The overall storage coefficient cannot therefore be determined more precisely by simple analysis of natural variations in water level.

Figure 6 shows variations in surface runoff associated with the six hypotheses used in calculation. It is very clear that runoff varies markedly from hypothesis to hypothesis, being much more sporadic for a retention capacity of 30 mm than for a retention capacity of 10 mm, and being about 50 % higher for a storage coefficient of 1 % than for a storage coefficient of 4 %.

Assuming a storage coefficient of 1 %, the average annual recharge is 23 to 45 mm/year when retention capacity is chosen between 10 and 30 mm. Assuming a storage coefficient of 4 %, the computed annual recharge would be 75 to 142 mm/year.

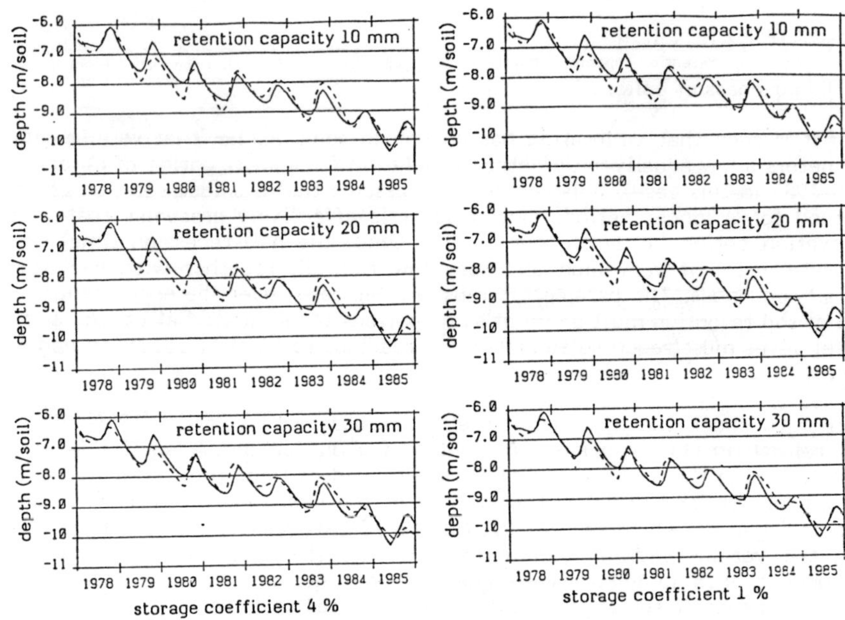


Figure 5. Simulation of piezometric level for three retention-capacity values and two storage coefficients

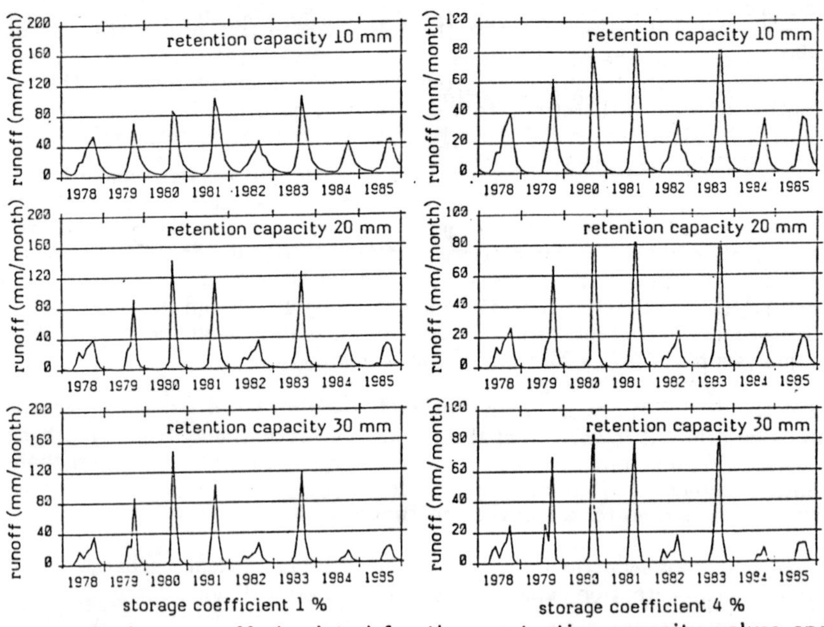


Figure 6. Surface runoff simulated for three retention-capacity values and two storage coefficients

## 5. DATE EXTENSION

### 5.1. Uniqueness of calibration

When it is seen that calibration has no unique solution, i.e. that various sets of parameters allow equally-satisfactory simulation for a short period of observation (1978-1985 in this case), it is tempting to assume that a unique solution could be obtained for a longer period. This hypothesis effectively presumes that a long observation period (at least four or five times the half-recession time) should integrate years of very high and very low rainfall, together with exceptional rainfall sequences. The parameters must in this case cover the entire range : the level of soil retention must be nil at some periods and saturated at others, and the aquifer level must reach sufficiently-low readings to identify recession and base level.

In order to check the hypothesis, the entire observation period (1959-1985) for daily rainfall in Ouagadougou was used, and evolution in the aquifer level was simulated using the various sets of parameters identified. The initial water level was assumed to be 6 m below the surface at the beginning of 1959. Results are given in Figure 7, which shows that, for the 17-year period, the levels calculated using the various sets of parameters are virtually identical. It is therefore concluded that a long observation period does not improve the possibility of establishing a single set of parameters or a single recharge value.

Detailed analysis shows that, although the algorithm used in calculation of distribution between infiltration and runoff is non-linear, low daily effective rainfall producing a larger ratio of infiltration than high daily effective rainfall, the cumulative infiltration ratio for the rainy season is approximately constant. Given the inertia of the aquifer, it is the cumulative infiltration over a period of several months which allows reproduction of observed levels. The exact distribution of such infiltration from day to day, governed by the equal runoff-percolation level of the model, thus has little effect on water levels but has a marked effect on the runoff proportion.

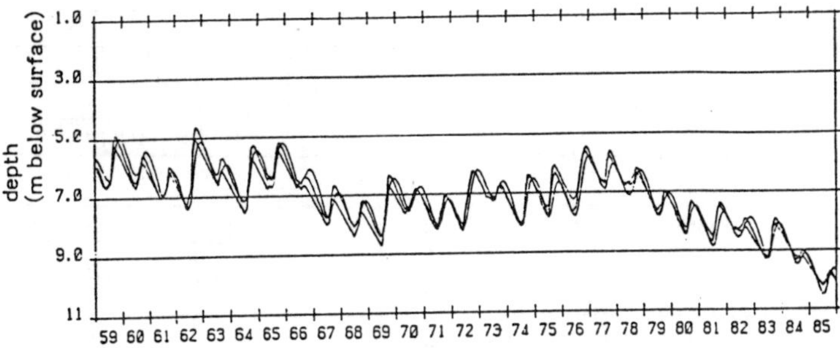


Figure 7. Simulation of the 1959-1985 period for three retention-capacity values and a storage coefficient of 1 %

For purposes of well-defined calibration, it is therefore believed that a sequence of piezometric observations is insufficient, even where the value of the storage coefficient is known. It is effectively necessary also to have access to "direct" and "delayed" runoff measurements and/or to measurements of the capillarity and water content in the unsaturated zone. This will provide all the elements in the balance, i.e. :

(1) runoff (measured)
(2) infiltration and evapotranspiration based on measurements in the unsaturated zone
(3) variation in the aquifer level (measured)

The role of the model is thus only to provide interpolation between measurement dates and to provide the most-satisfactory compromise between the three elements in the balance.

## 5.2. Reliability of model

The model is calibrated on the eight years between 1978 and 1985. Before being used to extend data, it is necessary to check the model's behaviour by extrapolation. For three retention capacities (0, 20 and 40 mm), calibration for the first four years was made, the levels for the last four years being calculated without modifying the calibration. The reverse operation was also undertaken (calibration based on the last four years being checked for the first four years). Figure 8 shows that results are very satisfactory. Calibration for the 1978-1981 period competently forecasts water levels for 1982 to 1985, although the latter are clearly lower than those for the calibration period. Calibration for the 1982-1985 period also, although less satisfactorily, allows calculation of levels for 1978 to 1981. The coefficients of adjustment for the entire period are all in excess of 0,91, reaching 0,94 for five coefficients out of six. Despite the fact that the calibration parameters do not allow unique solutions, the model can thus reliably be used to extend available data.

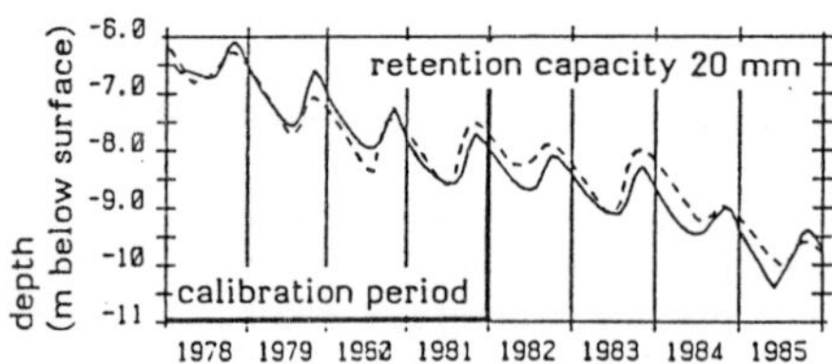


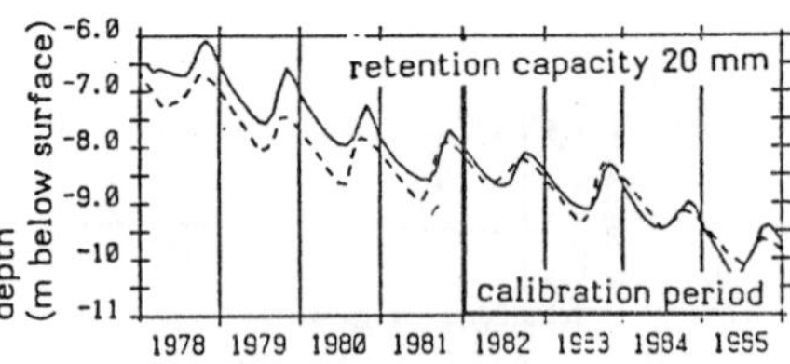


Figure 8. Control of model calibration
(a) calibration based on the 1978-1981 period
(b) calibration based on the 1982-1985 period

5.3. Simulation of return to more-abundant rainfall

Analysis of annual recharge calculated for 1959-1985 shows that, for all sets of parameters, the 1981 recharge is very close to average recharge for the period as a whole. A sequence of ten years from 1985 was therefore simulated on the basis of the 1981 rainfall in order to observe the type and rate of aquifer reaction. Figure 9, based on four hypotheses of calculation, shows a very slow rise, the aquifer taking more than ten years to reach a new equilibrium (on which seasonal fluctuation is superimposed). A single average year like 1981 will only produce a slight rise in water level after one year, and will not alone recharge the aquifer for several years, as might be assumed. This is obviously due to the inertia of the aquifer, typified by a half-recession time of more than four years. The time required to produce a rise in water levels is obviously similar to that required to produce a fall.

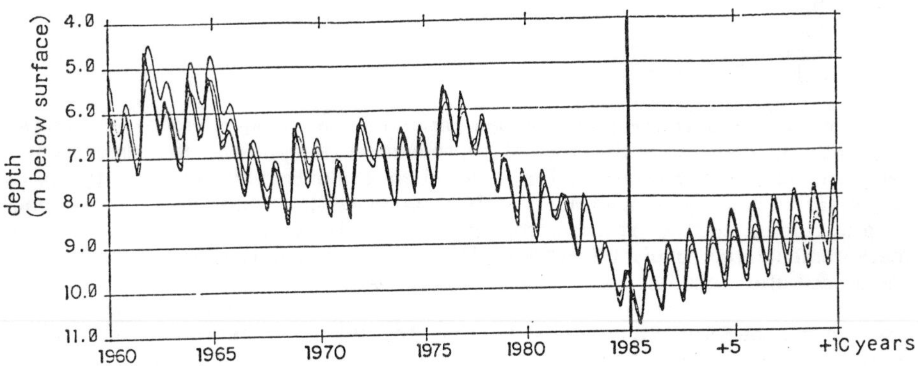


Figure 9. Simulation of a period of ten average years (identical to 1981) from 1985

In order to place the period of study (1978-1985) in its context, the entire rainfall sequence from 1929 to 1985 was used. The model was accordingly readjusted for monthly rainfall figures (the only data available for 1929-1958), and very satisfactory calibration was obtained, with a correlation coefficient of 0,96.

The entire 1929-1985 period was then simulated. In order to show the effect of a return to more-abundant rainfall, it was assumed that the 57-year sequence from 1929 to 1985 will be repeated from 1985. Figure 10 shows that levels for the 1978-1985 period are the lowest since 1929, and that, if a wetter sequence occurs after 1985, a rise in water level to a depth of 6 m below the surface will occur over a period of 7 to 10 years.

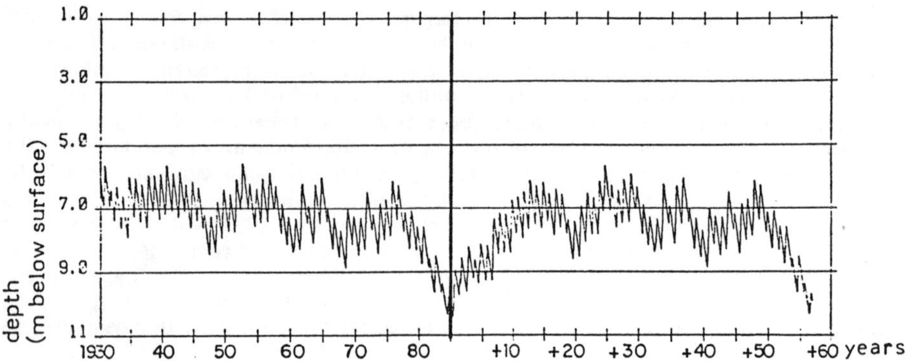


Figure 10. Simulation of the 1929-1985 period prolonged by a 57-year rainfall sequence identical to that for 1929-1985

## 6. CONCLUSIONS

It is shown that a very simple lumped-parameter hydrological model of rainfall and water levels allows the correct reproduction of a piezometric sequence in semi-arid climatic conditions. For a detailed evaluation of the recharge, it is necessary to have access, in addition to a piezometric sequence (even if a long-duration one), to the storage coefficient and to dispose of a sequence of measurements for runoff (direct and delayed) or for capillarity and humidity in the unsaturated zone.

The various calibrations made show that the water retention capacity which allows optimum simulation of the piezometric sequence is between 10 and 30 mm. Assuming that the storage coefficient is equal to 1 %, the computed average annual recharge for the period 1978-1985 is estimated at between 23 and 45 mm/year. Despite the fact that calibration has no unique solution, a single extension of aquifer levels is obtained on the basis of a long sequence of rainfall data. The model thus allows extrapolation of levels for 1929-1977 based on rainfall figures. It shows that levels for the observation period (1978-1985) are the lowest in the 57-year period of historical rainfall records, and can easily be explained by natural climatic variation.

Two scenarios were simulated for the period after 1985. The first assumes a succession of years identical to 1981 (in which recharge was close to the average for the entire period). The second assumes repetition of the 1929-1985 sequence. Both scenarios show that the aquifer, which is a priori not at this point in time in a critical state, should rise regularly and return to an average level in 7 to 10 years' time.

All calculations assume that, after recharge, the aquifer will recess by lateral flow towards a natural outlet. This outlet or flow has not been effectively identified, and it cannot be excluded that aquifer recession will occur by take-out due to evapotranspiration. Attempts to model this kind of take-out have not been successful, but the hypothesis cannot be discarded, and other algorithms taking account of a real evapotranspiration of 200-550 mm/year (depending on the hypothesis) and of a potential evapotranspiration of 2,084 mm/year should be tried.

## BIBLIOGRAPHY

**Diluca, C.,** and **Muller, W.,** 1985, Evaluation hydrogéologique des projets d'hydraulique en terrains cristallins du bouclier ouest-africain. ICHS hydrogeological series

**Lemoine, L.,** and **Prat, J.C.,** 1972, Cartes d'évapotranspiration potentielle calculée par la formule de L. Turc pour les pays membres du CIEH. Ouagadougou, January 1972

**ORSTOM,** 1977, République de Haute-Volta, Précipitations journalières de l'origine des stations à 1965. ICHS, Ouagadougou

**Roche, P.A.,** and **Thiery, D.,** 1984, Simulation globale de bassins hydrologiques. Introduction à la modélisation et description du modèle GARDENIA. BRGM report 84 SGN 337 EAU

**Thiery, D.,** 1985, Pourquoi un modèle à réservoirs permet-il de simuler correctement le tarissement d'une nappe ou d'une source ? BRGM technical record EAU 85/23, October 1985

**Turc, L.,** 1961, Evaluation des besoins en eau d'irrigation. Evapotranspiration potentielle. Ann. Agronomique 1961-12

# HUMID ZONE RECHARGE : A COMPARATIVE ANALYSIS

# HUMID AND ARID ZONE GROUNDWATER RECHARGE - A COMPARATIVE ANALYSIS

Gert Knutsson
Department of Land Improvement and Drainage
Royal Institute of Technology
S-100 44 Stockholm
Sweden

ABSTRACT. Humid climates - contrary to arid climates - are characterized by higher precipitation than evapotranspiration. This means that water balance methods for estimation of groundwater recharge are more useful in humid than in arid climates. Groundwater recharge in humid climates takes place more or less continuously by percolation in the unsaturated zone in the higher parts of the terrain and the water is discharged in the lower parts. This is contrary to the conditions in the arid climates, where the input of water is intermittent and the recharge is mainly localized to the lower parts of the terrain (river valleys, wadis) and rock outcrops. The unsaturated part of the ground with its vegetation is the crucial zone. The dominating water movement is downward in humid climates with leaching and weathering, contrary to in arid climates, where there is an upward transport and enrichment of salts. The methods for estimation of groundwater recharge based on soil water balance or soil water flow are more important in humid than in arid climates.

## 1. INTRODUCTION

Very few comparative analyses on groundwater recharge in humid and arid climates seem to have been published. L'vovich (1979) wrote a very comprehensive analysis on the regional hydrology of the world and he also took in account groundwater recharge. His figures were used by Falkenmark (1979) to compare the water balance in the coniferous forest belt of the boreal (humid) climate, the taiga, with that of the desert savannah and dry savannah respectively. Although the rainfall is higher in the dry savannah (1 000 mm) than in the taiga (700 mm) the groundwater recharge is much lower (30 mm compared to 140 mm). The main reason for this is the much higher evapotranspiration in the savannah. Balek (1983) also pointed out the great importance of evaporation losses in the arid zone of the



*I. Simmers (ed.), Estimation of Natural Groundwater Recharge, 493–504.*

Tropics on groundwater recharge as well as the transpiration losses of groundwater by root uptake and capillary rise (in one example down to 6 m) in forests of the humid zone of the Tropics. In a paper, as rapporteur of the IHP-project "Comparative hydrology", Falkenmark (1986) listed the fundamental processes to be understood. She stressed upon the soil zone with its cover of vegetation as a key zone of the hydrological cycle and briefly described the main differences in groundwater recharge and discharge in arid and humid climates (Figure 1). In the following these differences will be discussed more in detail. The basis of the discussion is groundwater recharge under natural conditions in humid climates.

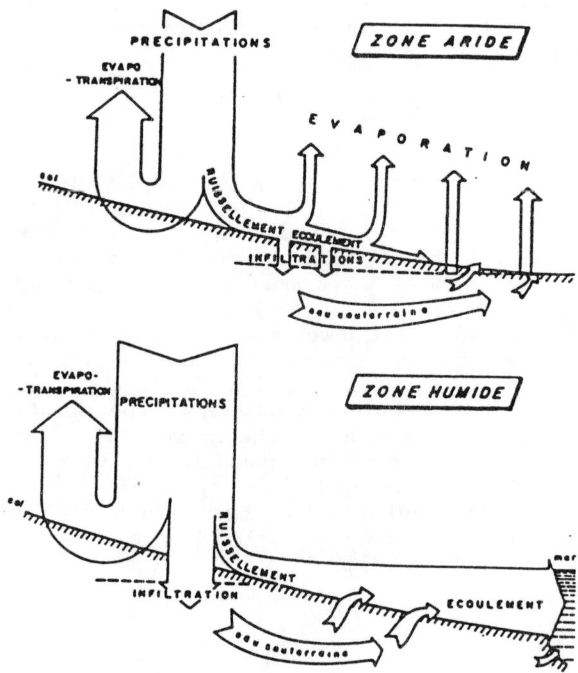


Figure 1. Groundwater recharge and discharge under humid and arid conditions (originally from Erhard-Cassegrain and Margat 1979, here from Falkenmark 1986).

## 2. BASIC CONCEPTS

### 2.1 Climate

Humid climates are characterized by higher precipitation than evapotranspiration calculated on an annual basis. This is contrary to the conditions in arid climates. The classification of humid climates is based on differences in mean annual temperature and amount as well as seasonal distribution of precipitation. It must, however, be pointed out that there is a large local difference in temperature and precipitation within a climatic region. The orographic effect plays an important role for the local and regional distribution of precipitation in all types of climate. Areas in rain-shadow only receive 20% of the rainfall on the windward side of a mountain in monsoon climate (Balek 1983), which means that semi-arid areas may be found close to a very humid region, e.g. the area east of the Western Ghats in the southernmost India.

### 2.2 Hydrology

The hydrology of humid climates is characterized by the excess of water, which means that most rivers are perennial and that surface water is stored in lakes and swamps - all contrary to the situation in arid climates. The runoff has seasonal variations, which are due to the type of climate but also to the topography and geology. The monsoon climate in India gives rise to a double maximum of flow, the subtropical dry-summer climate causes a marked decrease of the stream flow during the hot summer and a slight winter maximum, whereas the snow accumulation during the long winter of boreal and highland climates results in a winter minimum and a spring or even a summer maximum (e.g. "the mountain flood" in northern Scandinavia). Two other significant characteristics of the hard winter climate are the ice-covered lakes and the frozen ground, which sometimes affect the runoff; the frozen ground also the water movement in the unsaturated zone.

Rivers, lakes and swamps are usually discharge areas for groundwater, contrary to in arid climates. The recharge of groundwater in humid climates mostly occurs in the higher parts of the terrain but also in flat areas. Recharge from rivers is, however, of importance in some places or under certain conditions (e.g. induced infiltration by pumping).

### 2.3 Geology

The geology of humid climates is similar to that in arid climates as regards the types of bedrock and the tectonics. Some differences exist, mostly due to the geological history and how long the exogenous processes have worked in a certain climate; compare weathering depth or karst formation in the humid tropics with those in the arid climates, especially in desert areas. Unconsolidated sediments (gravel, sand, silt and clay) are found all over the world along rivers and in basins. There are great differences in stratification and thickness, in some areas thousands of meters, in others only some few meters. The latter is normally the case in those areas in temperate and boreal climates, which were glaciated during the Pleistocene. The erosion of the land-ice was very hard. A polished bedrock surface is overlain by a thin, compacted deposit from the land-ice itself: till or moraine. Glaciofluvial, lacustrine and marine sediments were deposited, when the land-ice melted away; the most striking feature is the elongated ridges of stone, gravel and sand, called eskers.

Vegetation, soiltype and land-use are directly or indirectly related to climate as well as geology.

## 3. GROUNDWATER RECHARGE ESTIMATION CONCEPTS

The above-mentioned main features in climate, hydrology and geology of the humid and arid climates influence the groundwater recharge and the possibilities for estimation of the recharge in the following ways:

### 3.1 Precipitation

Amount, type and intensity of precipitation varies between different types of humid climates and the areal and seasonal variations of precipitation - as well as of evapotranspiration - are considerable owing to topography (rain-shadow), vegetation and type of climate. The variability of annual precipitation is, however, less than 20% "departure from normal", which is much lower than in semi-arid and arid climates (Biswas 1984). Consequently, the need of data series of long duration as well as of dense data-networks is not so crucial as in arid climates. A general feature of humid climates is the excess of precipitation; though seasonally water deficit as well as water accumulation (snow) occur (Figure 2). This means that water balance methods for estimation of groundwater recharge seem to be more useful in humid than in arid climates.

### 3.2 Vegetation

Dense vegetation cover in humid climates reduces the amount of water available for recharge by interception and transpiration. But the slow movement of water through the vegetation prevents erosion of the surface of the ground, reduces overland flow and facilitates the infiltration. In fact, afforestation of an area will increase groundwater recharge. Seasonal variations of recharge due to variations in vegetation cover (crops etc) must be considered. The vegetation of arid climates is sparse but reduces the recharge of groundwater by transpiration due to the high suction capacity and the root length of many plants adapted to drought (Balek 1983).

### 3.3 Overland flow

Overland flow or surface runoff is not as important as in arid climates but is a striking feature during heavy rains in the humid tropics and the subtropics as well as on windward high slopes in the temperate oceanic climate. In temperate and boreal climates, overland flow

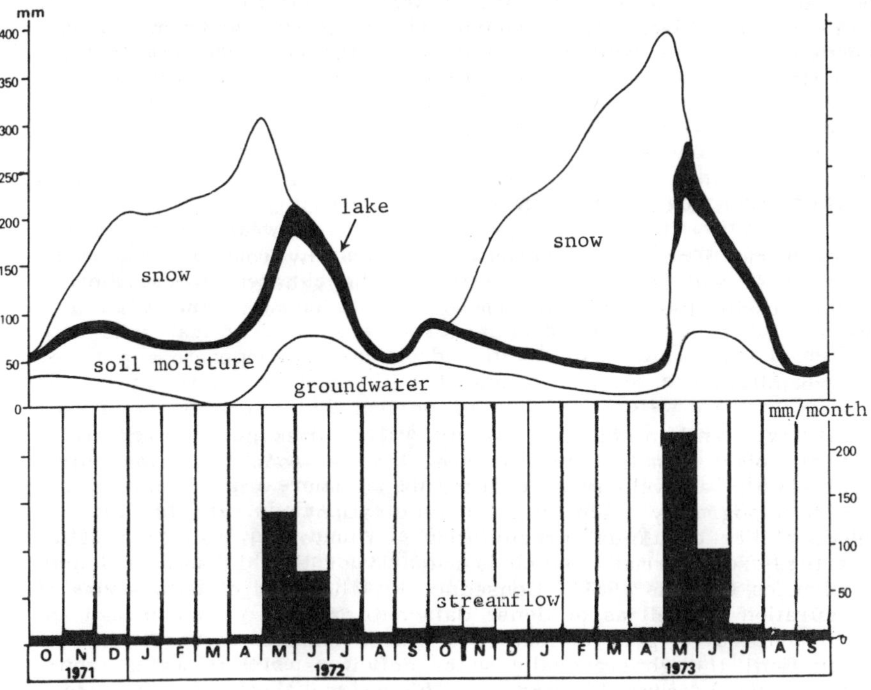


Figure 2. Water storage and runoff during Oct. 1971 - Sep. 1973 in the Lappträsket area in southern Sweden. The storages are shown related to zero storages taken as the minima during the ten years period of observations. The absolute values are only revealed for the snow storage, which is zero during the summers (from Grip and Rodhe 1985).

otherwise is a rare event in recharge areas due to less intensive rainfall, well-developed vegetation and sufficient infiltration capacity of most soils. If overland flow occurs, it does so only over short distances or during special events as intense snowmelt and frozen ground. The normal situation is that overland flow only takes place in the discharge areas, which mostly occupy a small part of a catchment.

## 3.4 Soil water

Water movements in the unsaturated zone of the ground is complicated. Downward movements due to infiltration and percolation and upward movements due to evapotranspiration and capillary rise in combination with ground frost result in a dominating downward movement (as piston flow or sometimes in certain flow paths) in the recharge areas of the humid climates. The downward movement in an unsaturated porous medium is very slow which has been determinated by following tritium-peaks (Andersson and Sevel 1974). The vertical movement is many times slower than the horizontal movement in the saturated zone of the same medium. But by piston flow there is a quick response between an infiltration event and a rise of the groundwater level. Rapid direct flow may occur through macropores of soils and open fractures of bedrock.

The downward water movement gives rise to weathering, leaching and downward transport of chemical components, some of them to the groundwater. This is contrary to in the arid climates, where there is an upward transport and enrichment of salts on the surface of the ground (salt crust). The leaching processes form typical soils as laterites in the humid tropics and podsols in the temperate and boreal climates. The processes also change the physical properties of the soil as its structure, water retention properties and hydraulic conductivity, which affects the infiltration of water. Biological and/or frost processes contribute to a higher hydraulic conductivity of the upper part (0.5-1.0 m) of the soil in temperate and boreal climates (Figure 3) as well as cracking during dry periods. In the humid tropics, weathering processes produce clay minerals, which give the soil a lower hydraulic conductivity. Differences in hydraulic conductivity by these reasons as well as by stratification and structural heterogenities may give rise to lateral flow, especially on slopes.

The dominance of a downward water-movement in the unsaturated zone of humid climates gives importance to the methods for groundwater recharge estimation based on soil water balance or soil water flow.

The consequences are that detailed data of the chemical and physical properties of the soil are necessary when using techniques and models dealing with the unsaturated zone. In order to get these data, methods such as lysimeters and introduced tracers (Knutsson 1968) can be used.

### 3.5 Frozen ground

Frozen ground in cold climates affects the groundwater recharge in several ways. The change of the soil structure is already mentioned. The infiltration capacity is decreased and all precipitation is accumulated as snow during several months (6-8 months in northern Scandinavia). However, infiltration sometimes may occur in frozen ground (Kane and Stein 1983, Engelmark, 1984). During the frost penetration there is a strong and rapid capillary movement of water in the unsaturated zone and from the groundwater table - if it is rather close to the ground surface - to the freezing front, especially in fine sand, silt and sandy-silty till. The groundwater level is successively lowered during this phase but is rapidly raised when the melting starts. The long interruption of recharge (and biological and chemical processes) as well as the internal up- and downward - sometimes also lateral - movement of water in the unsaturated zone must be considered when the techniques for estimation of groundwater recharge are applied.

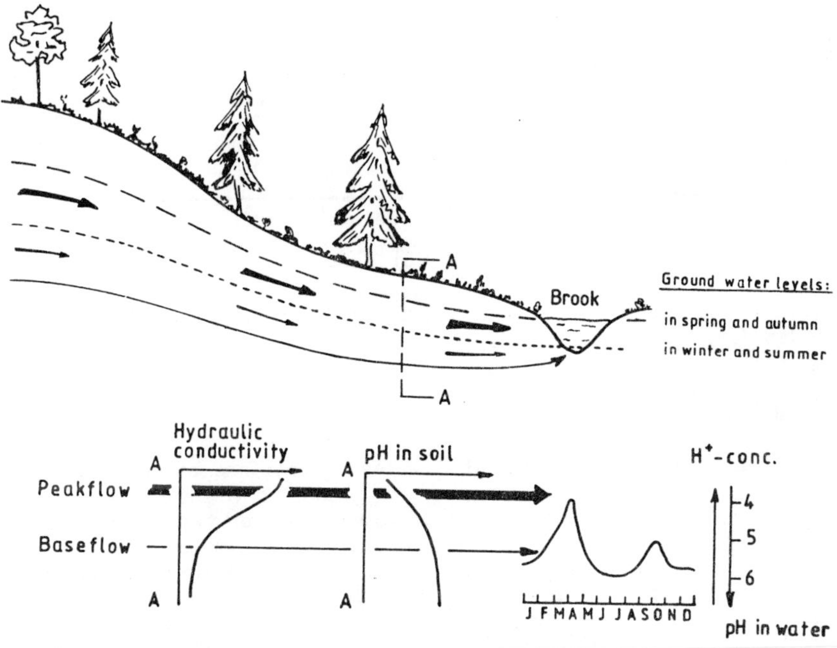


Figure 3. Hydrogeology in a slope in till during different seasons (from Jacks et al., 1984).

## 3.6 Groundwater table

Groundwater tables mostly lie comparatively close to the ground surface and follow the topography rather well. This is due to the abundant groundwater recharge in humid climates. Coupled to a well-developed vegetation this results in a closer interaction than in arid climates between the unsaturated zone including the root zone and the groundwater zone, sometimes reflected as a diurnal fluctuation of the groundwater level (Figure 4, Johansson 1986). Root-water uptake from great depths is not so important as in semi-arid and arid climates.

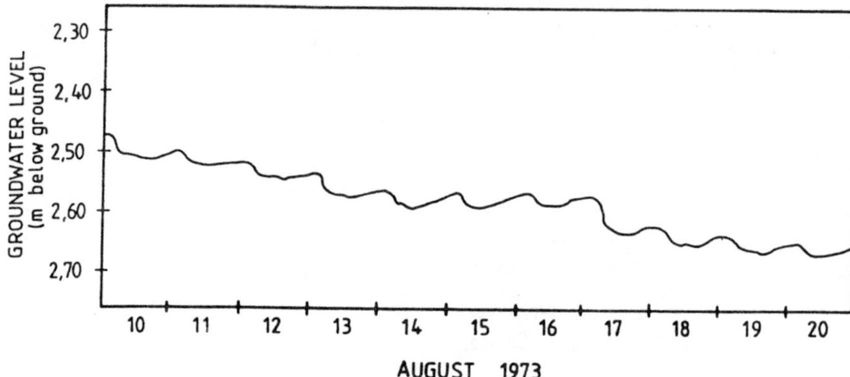


Figure 4. Diurnal fluctuations of the groundwater level in till, SE Sweden (from Johansson 1986).

### 3.7 Recharge and discharge areas

Groundwater recharge in humid climates takes place by percolation in the unsaturated zone more or less continuously (in tropical climate and in temperate oceanic climate) or during one or two seasons of the year, so called direct recharge. The terrain is divided into recharge and discharge areas (Figure 5). The extension of these areas may vary in time due to seasonality in precipitation and fluctuating groundwater levels. Discharge areas may also revert to recharge areas e.g. by lowering of the groundwater level by pumping. Groundwater flow systems are developed in local as well as regional scale under ideal conditions (homogeneous, isotropic, undisturbed) (Figure 5/Toth 1962, Gustafsson 1968/). This is contrary to the conditions in arid climates, where the input of water is intermittent and - as regards groundwater recharge - mainly localized to small, low-lying areas of the terrain (e.g. river valleys with permeable alluvial soils, wadis, alluvial fans) and to rock outcrops on the slopes. The transit time of water differs very much in the different flow systems - from days in the small local systems to thousands of years in the big regional systems. These differences must influence the choice of methods for estimation of recharge. The methods in which areal parameters (hydrometeorological and hydrogeological) are used, seem to be more suitable in humid climates than in arid climates. Environmental tracers are, however, of greatest inportance in such climates where an "event" (storm, snowmelt) can be followed by the tracers. By using $^{18}O$ it was found that groundwater is the dominating part of the stream discharge even during storms and snowmelt (Figure 6, Rhode 1987).

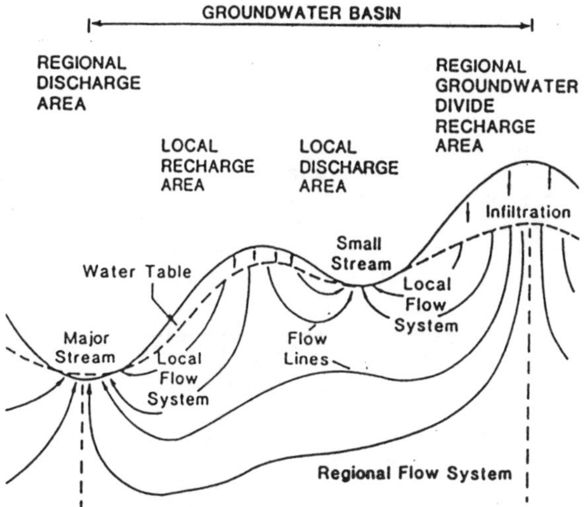


Figure 5. Recharge and discharge areas together with groundwater flow systems of different scales under homogeneous isotropic and undisturbed conditions (originally from van der Hejde 1986, here from Lönegren 1987).

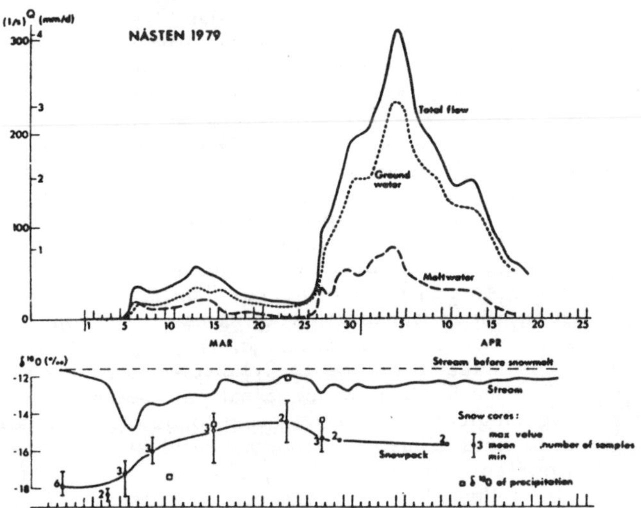


Figure 6. Oxygen-18 of streamwater and snowpack and total stream discharge with fractions of groundwater and meltwater estimated from oxygen-18 content (from Rodhe 1981).

### 3.8 Aquifers

The principal types of aquifers (pore, fracture, karst and combinations of these) exist in humid climates as well as in arid climates, but some types of aquifers are more frequent or more well-developed in one type of climate than the others. As karst formation depends on the occurrence of carbonate rocks, large quantities of water and time, karst aquifers seem to be most frequent in limestone areas in tropical wet climate (very stable) but they also occupy large areas in subtropical climate e.g. the Mediterranean (some fossil karst). Fracture aquifers in hard crystalline rocks may - on the other hand - "degenerate" in such types of climates, where the chemical weathering is strong and the fracture can be filled up with clay. This is contrary to arid climates, where mechanical weathering dominates. The geological and climatological history of an area also create special types of aquifers e.g. those of glacial origin. There is, however, a curious similarity between large areas of glaciated terrain in the cold temperate and boreal climates and large areas in the humid tropics: the dominating aquifers are fracture aquifers in hard rocks overlain by pore aquifers in low to medium permeable soils (till respectively laterite and regolith).

The conclusion is that, as regards the type of aquifers, the same techniques can be used for estimation of groundwater recharge in all types of climate. The crucial factors are how the topography, geology and aquifers are formed in detail in the region which is to be investigated. Two of the most important parameters seem to be the soil type and the thickness of soil cover (overburden) (see e.g. Sophocleous and Perry, 1985). A very favourable situation for groundwater recharge is if the infiltration capacity of the soil is higher than the intensity of precipitation (= no surface runoff) and if the thickness of permeable strata is great enough to prevent water loss by evapotranspiration from the soil cover. The underlaying aquifer must of course be able to store and transmit the recharged water. This is generally the case in glaciated terrain with till on hard, fractured rocks. The movement of water normally is also rather rapid in thin layers of soil. The most favourable situation is, if the ground surface consists of permeable rock, e.g. karstic rocks or volcanic rocks as lava, possibly with a shallow soil profile. In such areas all water infiltrates and no surface drainage occurs. The streams and rivers are fed by water from big springs with discharge of some tens of cubic meters. The spring discharge method seems to be the most suitable method for estimating groundwater recharge in such areas, if the catchment area can be determined correctly and the springs are permanent, that is in humid climates.

## 4. CONCLUDING REMARKS

The differences in sources and processes of groundwater recharge in humid climates compared with those in arid climates, mean that the applicability of the available recharge estimation techniques will be different. The need to proceed from a well-defined conceptualization of different recharge processes must be emphasized. The choice of technique must also be guided by the objective of the study, available data, possibility to get supplementary data and of course financial means. Since the results often are afflicted with substantial uncertainties, different techniques based on independent input data should be applied in the same area if possible.

## REFERENCES

Andersen, L.J. and Sevel, T., 1974. 'Environmental tritium in the unsaturated zone: Estimation of recharge to an unconfined aquifer.' In: Isotope techniques in groundwater hydrology, pp. 57-72, IAEA, Wien.

Balek, J., 1983. Hydrology and water resources in tropical regions. Development in water sciences. Elsevier.

Biswas, A.K., 1984. 'Climate and development'. In: Climate and development (Editor: A.K. Biswas). Tycooli International Publ. Ltd., vol 13.

Engelmark, H., 1984. 'Infiltration in unsaturated frozen soil'. Nordic Hydrol., 15:243-252.

Falkenmark, M., 1979. 'Some hydrological similarities and dissimilarities' between temperate and tropical countries'. Hydrology in developing countries. Nordic IHP report No.2.

Falkenmark, M., 1986. 'Comparative hydrology related to the main hydrological regimes of the world'. Introductory memorandum. Manuscript.

Grip, H. and Rodhe, A., 1985. Vattnets väg från regn till bäck. Forskningsrådens förlagstjänst, Stockholm (in Swedish), 156 pp.

Gustafsson, Y., 1968. 'The influence of topography on ground water formation'. In: Ground water problems. Proceedings of the international symposium held in Stockholm, October 1966, pp. 3-21.

Jacks, G., Knutsson, G., Maxe, L. and Fylkner, A., 1984. 'Effects of acid rain on soil and groundwater in Sweden'. In: Pollutants in porous media. Ecological Studies, 47, pp. 94-114. Springer Verlag.

Johansson, P-O., 1986. 'Diurnal groundwater level fluctuations in sandy till - a model analysis'. J. Hydrol., 87:125-134.

Kane, D.L. and Stein, J., 1983. 'Water movement into seasonally frozen soils'. Water Resour. Res., 19:1547-1557.

Knutsson, G., 1968. 'Tracers for groundwater investigations'. In: Ground water problems. Proceedings of the international symposium held in Stockholm. October 1966, pp. 123-152.Pergamon Press.

L'vovich, M.I., 1979. World water resources and their future. Translation by the American Geophysical Union. Litho Crafters. Inc., Chelsea, Michigan.

Lönegren, H., 1987. Control of Land Use and Groundwater Quality in Colorado and Sweden. Linköping Studies in Arts and Sciences. 11.

Rodhe, A., 1981. 'Spring flood - Meltwater of groundwater'. Nordic Hydrol., 12:21-30.

Rodhe, A., 1987. The origin of streamwater traced by oxygen-18. Uppsala University. Dept. of Physical Geography. Division of Hydrology. Report Series A No 41 + Appendix, 260 pp + 73 pp.

Sophocleous, M. and Perry, C.A., 1985. 'Experimental studies in natural groundwater-recharge dynamics: The analysis of observed recharge events'. J. Hydrol., 81:297-332.

Tóth, J., 1962. 'A theory of groundwater motion in small drainage basins in Central Alberta, Canada'. J. Geophys. Res., 67:4375-4387.

## List of Participants

Agacik, Mr.O., DSI Geotechnical and Groundwater Division, Yucetepe, Ankara, Turkey.

Allison, Dr.G.B., C.S.I.R.O., Private Bag 2, Glen Osmond, S.A. 5064, Australia.

Altug, Mrs.A., DSI Geotechnical and Groundwater Division, Yucetepe, Ankara, Turkey.

Atalay, Mr.E., DSI Geotechnical and Groundwater Division, Yucetepe, Ankara, Turkey.

Balek, Dr.J., Stavebni Geologie, Gorkeho Nam. 7, 11309 Prague 1, Czechoslavakia.

Balta, Mr.H., DSI Geotechnical and Groundwater Division, Yucetepe, Ankara, Turkey.

Bredenkamp, Dr.D.B., Directorate Geohydrology, Dept. of Water Affairs, Private Bag X313, Pretoria 0001, Rep. South Africa.

Bruin, Dr.H.A.R. de, Agricultural University of Wageningen, Dept. of Physics and Meteorology, Duivendaal 2, 6701 AP Wageningen, The Netherlands.

Cetincelik, Mr.M., DSI Geotechnical and Groundwater Division, Yucetepe, Ankara, Turkey.

Demirkol, Dr.C., Dept. of Geological Engineering, Cukurova University, Adana, Turkey.

Edmunds, Dr.W.M., British Geological Survey, Crowmarsh Gifford, Wallingford, OXON OX10 8BB, England.

Egboka, Assoc.Prof.Dr.B.C.E., Anambra State University of Technology, P.M.B. 01660, Dept of Geological Sciences, Enugu, Nigeria.

Erguvanli, Prof.Dr.K., Istanbul Technical University, Dept. of Applied Geology, Istanbul, Turkey.

Eroskay, Prof.Dr.O., Istanbul University, Faculty of Engineering, Vezneciler, Istanbul, Turkey.

Foster, Dr.S.S.B., British Geological Survey, Crowmarsh Gifford, Wallingford, OXON OX10 8BB, England.

Da Franca, Dr.N., UNESCO Division of Water Sciences, 7, Place de Fontenoy, B.P. 307 Paris, France.

Fry, Mr.R.G., Dept. of Water Affairs, Private Bag 13193, Windhoek, Namibia/SW Africa.

Ganoulis, Prof.J.G., Hydraulics Laboratory, School of Technology, Aristotle University of Thessalonika, 54006 Thessalonika, Greece.

Geirnaert, Dr.W., Institute of Earth Sciences, Free University, P.O. Box 7161, 1007 MC Amsterdam, The Netherlands.

Gieske, Drs. A., Faculty of Science, University of Botswana, Private bag 0022, Gaborone, Botswana.

Griend, Dr.A.A.van de, Institute of Earth Sciences, Free University, P.O. Box 7161, 1007 MC Amsterdam, The Netherlands.

Gulenbay, Mr.A., DSI Geotechnical and Groundwater Division, Yucetepe, Ankara, Turkey.

Gunay, Assoc.Prof.Dr.G., Hydrogeological Engineering Dept., Engineering Faculty, Hacettepe University, Beytepe 06532, Ankara, Turkey.

Johansson, Mr.P.O., Dept. of Land Improvement and Drainage, Royal Institute of Technology, S-10044 Stockholm, Sweden.

Kaptan, Mr.C., DSI Geotechnical and Groundwater Division, Yucetepe, Ankara, Turkey.

Kaya, Mr.A., DSI Geotechnical and Groundwater Division, Yucetepe, Ankara, Turkey.

Knutsson, Prof.Dr.G., Dept. of Land Improvement and Drainage, Royal Institute of Technology, S-10044 Stockholm, Sweden.

Kovacs, Drs.D.F., ITC, P.O.Box 6, 7500 AA Enschede, The Netherlands.

Kuran, Mr.I.H., DSI Geotechnical and Groundwater Division, Yucetepe, Ankara, Turkey.

Lerner, Prof.D.N., Dept. of Geological Sciences, University of Birmingham, P.O.Box 363, Birmingham B15 2TT, England.

Levin, Mr.M., Instituto de Geochronologia y Geologia Isotopica, Ciudad Universitaria, Pabellon Ingeis, 1428 Buenos Aires, Argentine.

Lobo-Ferreira, Eng.J.P., Departamento de Hidraulica, Laboratorio Nacional de Engenharia Civil, Av. do Brasil 101, 1799 Lisboa Codex, Portugal.

Nazik, Mr.M., DSI Geotechnical and Groundwater Division, Yucetepe, Ankara, Turkey.

Oncu, Mr.I., DSI Geotechnical and Groundwater Division, Yucetepe, Ankara, Turkey.

Rodrigues, Dr.J.D., Ministerio da Habitacao e Obras Publicas, Laboratorio Nacional de Engenharia Civil, Av. do Brasil 101, 1799 Lisboa Codex, Portugal.

Romijn, Drs.E., Prov. Waterboard of Gelderland, P.O.Box 9090, 6800 GX Arnhem, The Netherlands.

Rushton, Prof.Dr.K.R., Dept. of Civil Engineering, University of Birmingham, P.O.Box 363, Birmingham B15 2TT, England.

dos Santos, Mr.B., Av. Almirante Gago Coutinho 30, 1000 Lisboa, Portugal.

Selaolo, Mr.E., Botswana Geological Survey, Lobatse, Botswana.

Senarath, Dr.D.C.H., Dept of Civil Engineering, University of Moratuwa, Katubedda, Moratuwa, Sri Lanka.

Seyhan, Dr.Ir.E., Institute of Earth Sciences, Free University, P.O.Box 7161, 1007 MC Amsterdam, The Netherlands.

Sharma, Dr.M.L., C.S.I.R.O. Division of Water Resources Research, Private Bag, P.O. Wembley, Western Australia 6014.

Simm, Mr.R.I.C., Hydrotechnica Ltd., Pengwern Court, High Street, Shrewsbury, Shrop. SY1 1SR, England.

Simmers, Prof.Dr.I., Institute of Earth Sciences, Free University, P.O.Box 7161, 1007 MC Amsterdam, The Netherlands.

Sorman, Prof.Dr.A.U., King Abdulaziz University, Dept. of Hydrology and Water Resources Management, P.O.Box 9034, Jeddah 21413, Saudi Arabia.

Sozen, Mr.M., DSI Geotechnical and Groundwater Division, Yucetepe, Ankara, Turkey.

Thiery, Dr.D.P., B.R.G.M., B.P. 6009 - 45060 Orleans Cedex 2, France.

Tuzcu, Mr.G., DSI, Geotechnical and Groundwater Division, Yucetepe, Ankara, Turkey.

Tyano, Prof.Dr.B., Ouagadougou University, B.P. 2523, Ouagadougou, Burkina Faso.

Vries, Dr.J.J. de, Institute of Earth Sciences, Free University, P.O.Box 7161, 1007 MC Amsterdam, The Netherlands.

Yurtsever, Mr. Y., International Atomic Agency, Section of Isotope Hydrology, Wagramerstrasse 5, P.O.Box 100, A-1400 Vienna, Austria.

INDEX